Studies in Computational Intelligence

Volume 911

Series Editor

Janusz Kacprzyk, Polish Academy of Sciences, Warsaw, Poland

The series "Studies in Computational Intelligence" (SCI) publishes new developments and advances in the various areas of computational intelligence—quickly and with a high quality. The intent is to cover the theory, applications, and design methods of computational intelligence, as embedded in the fields of engineering, computer science, physics and life sciences, as well as the methodologies behind them. The series contains monographs, lecture notes and edited volumes in computational intelligence spanning the areas of neural networks, connectionist systems, genetic algorithms, evolutionary computation, artificial intelligence, cellular automata, self-organizing systems, soft computing, fuzzy systems, and hybrid intelligent systems. Of particular value to both the contributors and the readership are the short publication timeframe and the world-wide distribution, which enable both wide and rapid dissemination of research output.

The books of this series are submitted to indexing to Web of Science, EI-Compendex, DBLP, SCOPUS, Google Scholar and Springerlink.

More information about this series at http://www.springer.com/series/7092

Aboul-Ella Hassanien ·
Mohamed Hamed N. Taha ·
Nour Eldeen M. Khalifa
Editors

Enabling AI Applications in Data Science

 Springer

Editors
Aboul-Ella Hassanien
Faculty of Computers
and Artificial Intelligence
Cairo University
Giza, Egypt

Chair of the scientific Research
Group in Egypt
Cairo University
Giza, Egypt

Nour Eldeen M. Khalifa
Faculty of Computers
and Artificial Intelligence
Cairo University
Giza, Egypt

Mohamed Hamed N. Taha
Faculty of Computers
and Artificial Intelligence
Cairo University
Giza, Egypt

ISSN 1860-949X   ISSN 1860-9503 (electronic)
Studies in Computational Intelligence
ISBN 978-3-030-52069-4   ISBN 978-3-030-52067-0 (eBook)
https://doi.org/10.1007/978-3-030-52067-0

This Springer imprint is published by the registered company Springer Nature Switzerland AG
The registered company address is: Gewerbestrasse 11, 6330 Cham, Switzerland

Preface

Artificial Intelligence and Data Science are the most useful technologies that could powerfully improve the human life. This will be achieved by merging both sciences in order to solve real complex problems in various fields. This book provides a detailed overview on the latest advancements and applications in the field of Artificial Intelligence and Data Science. AI applications have achieved great accuracy and performance with the help of the advancements in data processing and storage. AI applications also gained power through the amount and quality of data which is the main core of data science. This book is aimed to introduce the state of the art in research on Artificial Intelligence with the Data Science. We accepted 28 chapters. The accepted chapters covered the following four parts:

- Part I—Artificial Intelligence and Optimization
- Part II—Big Data and Artificial Intelligence Applications
- Part III—IOT within Artificial Intelligence and Data Science
- Part IV—Artificial Intelligence and Security

We thank and acknowledge all persons who were involved in all the stages of publishing. That includes (authors, reviewers, and publishing team). We profoundly revalue their engagement and sustenance that was essential for the success of the "Enabling AI applications in Data Science" edited book. We hope the readers would equally love the chapters and their contents and appreciate the efforts that have gone into bringing it to reality.

Giza, Egypt

Aboul-Ella Hassanien
Mohamed Hamed N. Taha
Nour Eldeen M. Khalifa

Contents

Artificial Intelligence and Optimization

Stochastic SPG with Minibatches

Application to Large Scale Learning Models

Andrei Pătraşcu, Ciprian Păduraru, and Paul Irofti

1 Introduction

Statistical learning from data is strongly linked to optimization of stochastic representation models. The traditional approach consists of learning an optimal hypothesis from a reasonable amount of data and further aim to generalize its decision properties over the entire population. In general, this generalization is achieved by minimizing population risk which, unlike the empirical risk minimization, aims to compute the optimal predictor with the smallest generalization error. Thus, in this paper we consider the following stochastic composite optimization problem:

$$\min_{w \in \mathbb{R}^n} \ F(w) := \mathbb{E}_{\xi \in \Omega}[f(w; \xi)] + \mathbb{E}_{\xi \in \Omega}[h(w; \xi)], \tag{1}$$

where ξ is a random variable associated with probability space $(\mathbb{P}, \Omega)$, the function $f(w) := \mathbb{E}_{\xi \in \Omega}[f(w; \xi)]$ is smooth and $h(w) := \mathbb{E}_{\xi \in \Omega}[h(w; \xi)]$ convex and nonsmooth. Most learning models in population can be formulated over the structure of (1) using a proper decomposition dictated by the nature of prediction loss and regularization.

Let the component $f(\cdot; \xi) \equiv \ell(\cdot; \xi)$ be a convex smooth loss function, such as quadratic loss, and $h \equiv r$ the "simple" convex regularization, such as $\|w\|_1$, then the resulted model:

A. Pătraşcu (✉) · C. Păduraru · P. Irofti
Computer Science Department, University of Bucharest, Bucharest, Romania
e-mail: andrei.patrascu@fmi.unibuc.ro

C. Păduraru
e-mail: ciprian.paduraru@fmi.unibuc.ro

P. Irofti
e-mail: paul@irofti.net

A.-E. Hassanien et al. (eds.), *Enabling AI Applications in Data Science*,
Studies in Computational Intelligence 911,
https://doi.org/10.1007/978-3-030-52067-0_1

$$\min_{w \in \mathbb{R}^n} \mathbb{E}_{\xi \in \Omega}[\ell(w; \xi)] + r(w).$$

has been considered in several previous works [15, 23, 30, 34], which analyzed iteration complexity of stochastic proximal gradient algorithms. Here, the proximal map of r is typically assumed as being computable in closed-form or linear time, as appears in ℓ_1/ℓ_2 Support Vector Machines (SVMs). In order to be able to approach more complicated regularizers, expressed as sum of simple convex terms, as required by machine learning models such as: group lasso [11, 35, 43], CUR-like factorization [42], graph trend filtering [33, 38], dictionary learning [7], parametric sparse representation [36], one has to be able to handle stochastic optimization problems with stochastic nonsmooth regularizations $r(w) = \mathbb{E}[r(w; \xi)]$. For instance, the grouped lasso regularization $\sum_{j=1}^{m} \|D_j w\|_2$ might be expressed as expectation by considering $r(w; \xi) = \|D_\xi w\|_2$. In this chapter we analyze extensions of stochastic proximal gradient for this type of models.

Nonsmooth (convex) prediction losses (e.g. hinge loss, absolute value loss, ϵ–insensitive loss) are also coverable by (1) through taking $h(\cdot; \xi) = \ell(\cdot; \xi)$. We will use this approach with $f(w) = \frac{\lambda}{2}\|w\|_2^2$ for solving hinge-loss-ℓ_2-SVM model.

Contributions.(i) We derive novel convergence rates of SPG with minibatches for stochastic composite convex optimization, under strong convexity assumption.
(ii) Besides the sublinear rates, we provide computational complexity analysis, which takes into account the complexity of each minibatch iteration, for stochastic proximal gradient algorithm with minibatches. We obtained $\mathcal{O}\left(\frac{1}{N\epsilon}\right)$ complexity which highlights optimal dependency on the minibatch size N and accuracy ϵ.
(iii) We confirm empirically our theoretical findings through tests over ℓ_2–SVMs (with hinge loss) on real data, and parametric sparse representation models on random data.

As extension of our analysis, scaling the complexity per iteration with the number machines/processors, would guarantee direct improvements in the derived computational complexity. Although the superiority of distributed variants of SGD schemes for smooth optimization are clear (see [23]), our results set up the theoretical foundations for development of distributed proximal gradient algorithms in fully stochastic environments. Further, we briefly recall the milestone results from stochastic optimization literature with focus on the complexity of stochastic first-order methods.

1.1 Previous Work

The natural tendency of using minibatches in order to accelerate stochastic algorithms and to obtain better generalization bounds is not new [9, 32, 39, 40]. Empirical advantages have been observed in most convex and nonconvex models although clear theoretical complexity scaling with the minibatch size are still under development for structured nonsmooth models. Great attention has been given in the last decade to the behaviour of the stochastic gradient descent (SGD) with minibatches, see [9, 15–19, 22, 32, 32]. On short, SGD iteration computes the average of gradients on a small number of samples and takes a step in the negative direction. Although more samples in the minibatch imply smaller variance in the direction and, for moderate minibatches, brings a significant acceleration, recent evidence shows that by increasing minibatch size over certain threshold the acceleration vanishes or deteriorates the training performance [10, 12].

Since the analysis of SGD naturally requires various smoothness conditions, proper modifications are necessary to attack nonsmooth models. The stochastic proximal point (SPP) algorithm has been recently analyzed using various differentiability assumptions, see [1, 4, 13, 25, 27, 31, 37, 41], and has shown surprising analytical and empirical performances. The works of [39, 40] analyzed minibatch SPP schemes with variable stepsizes and obtained $\left(\frac{1}{kN}\right)$ convergence rates under proper assumptions. For strongly convex problems, notice that they require multiple assumptions that we avoid using in our analysis: strong convexity on each stochastic component, knowledge of strong convexity constant and Lipschitz continuity on the objective function. Our analysis is based on strong convexity of the smooth component f and only convexity on the nonsmooth component h.

A common generalization of SGD and SPP are the stochastic splitting methods. Splitting first-order schemes received significant attention due to their natural insight and simplicity in contexts where a sum of two components are minimized (see [3, 20]). Only recently the full stochastic composite models with stochastic regularizers have been properly tackled [33], where almost sure asymptotic convergence is established for a stochastic splitting scheme, where each iteration represents a proximal gradient update using stochastic samples of f and h. The stochastic splitting schemes are also related to the model-based methods developed in [6].

1.2 Preliminaries and Notations

For $w, v \in \mathbb{R}^n$ denote the scalar product $\langle w, v \rangle = w^T v$ and Euclidean norm by $\|w\| = \sqrt{w^T w}$. We use notations $\partial h(w; \xi)$ for the subdifferential set and $g_h(w; \xi)$ for a subgradient of $h(\cdot; \xi)$ at w. In the differentiable case we use the gradient notation $\nabla f(\cdot; \xi)$. We denote the set of optimal solutions with W^* and w^* for any optimal point of (1). Also, for any $\xi \in \Omega$, we use notation $F(\cdot; \xi) = f(\cdot; \xi) + h(\cdot; \xi)$ and $\partial F(\cdot; \xi) = \nabla f(\cdot; \xi) + \partial h(\cdot; \xi)$.

Assumption 1 The central problem (1) has nonempty optimal set W^* and satisfies:
(i) The function $f(\cdot; \xi)$ has L_f-Lipschitz gradient, i.e. there exists $L_f > 0$ such that:

$$\|\nabla f(w; \xi) - \nabla f(v; \xi)\| \leq L_f \|w - v\|, \qquad \forall w, v \in \mathbb{R}^n, \xi \in \Omega.$$

and f is σ_f–strongly convex, i.e. there exists $\sigma_f > 0$ satisfying:

$$f(w) \geq f(v) + \langle \nabla f(v), w - v \rangle + \frac{\sigma_f}{2} \|w - v\|^2 \qquad \forall w, v \in \mathbb{R}^n. \tag{2}$$

(ii) There exists subgradient mapping $g_h : \mathbb{R}^n \times \Omega \mapsto \mathbb{R}^n$ such that $g_h(w; \xi) \in \partial h(w; \xi)$ and $\mathbb{E}[g_h(w; \xi)] \in \partial h(w)$.

(iii) $F(\cdot; \xi)$ has bounded gradients on the optimal set: there exists $\mathcal{S}_F^* \geq 0$ such that $\mathbb{E}\left[\|g_F(w^*; \xi)\|^2\right] \leq \mathcal{S}_F^* < \infty$ for all $w^* \in W^*$;

(iv) For any $g_F(w^*) \in \partial F(w^*)$ there exists bounded subgradients $g_F(w^*; \xi) \in \partial F(w^*; \xi)$, such that $\mathbb{E}[g_F(w^*; \xi)] = g_F(w^*)$ and $\mathbb{E}[\|g_F(w^*; \xi)\|^2] < \mathcal{S}_F^*$. Moreover, for simplicity we assume throughout the paper $g_F(w^*) = \mathbb{E}[g_F(w^*; \xi)] = 0$
□

Condition (i) of the above assumption is typical for composite (stochastic) optimization [3, 20, 27]. Condition (ii) guarantees the existence of a subgradient mapping for functions $h(\cdot; \xi)$. Also condition (iii) of Assumption 1 is standard in the literature related to stochastic algorithms.

Remark 1 The assumption (iv) requires a more consistent discussion. Denoting $\mathbb{E}[\partial F(\cdot; \xi)] = \{\mathbb{E}[g_F(\cdot; \xi)] \mid g_F(\cdot; \xi) \in \partial F(\cdot; \xi)\}$, then for convex functions $\mathbb{E}[\partial F(x; \xi)] \subseteq \partial F(x)$ for all $x \in \text{dom}(F)$ (see [29]). However, (iv) is guaranteed by the stronger equality

$$\mathbb{E}[\partial h(x; \xi)] = \partial h(x). \tag{3}$$

Discrete case. Let us consider finite discrete domains $\Omega = \{1, \ldots, m\}$. Then [28, Theorem 23.8] guarantees that the finite sum objective of (1) satisfy (3) if $\bigcap_{\xi \in \Omega} \text{ri}(\text{dom}(h(\cdot; \xi))) \neq \emptyset$. The ri(dom($\cdot$)) can be relaxed to dom($\cdot$) for polyhedral components. In particular, let $X_1, \ldots, X_m$ be finitely many closed convex satisfying qualification condition: $\bigcap_{i=1}^m \text{ri}(X_i) \neq \emptyset$, then also (3) holds, i.e. $\mathcal{N}_X(x) = \sum_{i=1}^m \mathcal{N}_{X_i}(x)$ (see (by [28, Corollary 23.8.1])). Again, ri(X_i) can be relaxed to the set itself for polyhedral sets. As pointed by [2], the (bounded) linear regularity property of $\{X_i\}_{i=1}^m$ implies the intersection qualification condition.

Under support of these arguments, observe that (iv) can be easily checked for our finite-sum examples given below in the rest of our sections.

Continuous case. In the nondiscrete case, sufficient conditions for (3) are discussed in [29]. Based on the arguments from [29], an assumption equivalent to (iv) is considered in [33] under the name of 2–integrable representation of x^* (definition in [33, Section B]). On short, if $h(\cdot; \xi)$ is normal convex integrand with full domain then x^* admits an 2–integrable representation $\{g_h(x^*; \xi)\}_{\xi \in \Omega}$, and implicitly (iv) holds. Lastly, we mention that deriving a more complicated result similar to Theorem 1 we could avoid assumption (iv). However, since (iv) facilitates the simplicity and naturality of our results and while our target applications are not excluded, we assume throughout the paper that (iv) holds.

Given some smoothing parameter $\mu > 0$ and $I \subset [m]$, we define the prox operator:

$$\text{prox}_{h,\mu_k}(w; I) = \arg\min_{z \in \mathbb{R}^n} \frac{1}{|I|} \sum_{i \in I} h(z; i) + \frac{1}{2\mu} \|z - w\|^2$$

In particular, when $h(w; \xi) = \mathbb{I}_{X_\xi}(w)$ the prox operator becomes the projection operator $\text{prox}_{h,\mu}(w; \xi) = \pi_{X_\xi}(w)$. Further we denote $[m] = \{1, \ldots, m\}$. Given the constant $\mu_0 \geq 0$ then a useful inequality for the sequel is:

$$\sum_{i=1}^{T} \left(\frac{\mu_0}{i}\right)^\gamma \leq \mu_0 \left(1 + \frac{T^{1-\gamma}}{1-\gamma}\right). \tag{4}$$

2 Stochastic Proximal Gradient with Minibatches

In the following section we present the Stochastic Proximal Gradient with Minibatches (SPG-M) and analyze the complexity of a single iteration under Assumption 1. Let $w^0 \in \mathbb{R}^n$ be a starting point and $\{\mu_k\}_{k \geq 0}$ be a nonincreasing positive sequence of stepsizes.

Stochastic Proximal Gradient with Minibatches (SPG-M):
For $k \geq 0$ compute:
1. Choose randomly i.i.d. N–tuple $I^k \subset \Omega$ w.r.t. probability distribution $\mathbb{P}$
2. Update:

$$v^k = w^k - \frac{\mu_k}{N} \sum_{i \in I^k} \nabla f(w^k; i)$$

$$w^{k+1} = \arg\min_{z \in \mathbb{R}^n} \frac{1}{N} \sum_{i \in I^k} h(z; i) + \frac{1}{2\mu} \|z - v^k\|^2.$$

3. If the stopping criterion holds, then **STOP**, otherwise $k = k + 1$.

Computing v^k necessitates an effort equivalent with N vanilla SGD iterations. However, to obtain w^{k+1}, a strongly convex inner problem has to be solved and the linear scaling in N of the computational effort holds only in structured cases. In fact, we will adopt specific inner schemes to generate a sufficiently accurate suboptimal solution of the inner subproblem. The details are given in Sect. 2.1. In the particular scenario when $f(w) = \|w\|_2^2$ and h is a nonsmooth prediction loss, then SPG-M learns completely, at each iteration k, a predictor w^{k+1} for the minibatch of data samples I^k, while maintaining a small distance from the previous iterated predictor.
For $N = 1$, SPG-M iteration reduces to stochastic proximal gradient $w^{k+1} = \text{prox}_{h,\mu_k}$ $\left(w^k - \mu_k \nabla f(w^k; \xi_k); \xi_k\right)$ based on stochastic proximal maps [26, 33].

The asymptotic convergence of vanishing stepsize non-minibatch SPG (a single sample per iteration) has been analyzed in [33] with application to trend filtering. Moreover, sublinear $\mathcal{O}(1/k)$ convergence rate for non-minibatch SPG has been provided in [26]. However, deriving sample complexity for SPG-M with arbitrary minibatches is not trivial since it requires proper estimation of computational effort required by a single iteration. In the smooth case ($h = 0$), SPG-M reduces to vanilla minibatch SGD [15]:

$$w^{k+1} = w^k - \frac{\mu_k}{N} \sum_{i \in I^k} \nabla f(w^k; i).$$

On the other hand, for nonsmooth objective functions, when $f = 0$, SPG-M is equivalent with a minibatch variant of SPP analyzed [1, 27, 37, 39, 40]:

$$w^{k+1} = \text{prox}_{h,\mu_k}(w^k; I^k).$$

Next we analyze the computational (sample) complexity of SPG-M iteration and suggest concrete situations when minibatch size $N > 0$ is advantageous over single sample $N = 1$ scheme.

2.1 Complexity per Iteration

In this section we estimate bounds on the sample complexity $T_v(N)$ of computing v^k and $T_w(N)$ for computing w^{k+1}. Let $I \subset [m], |I| = N$, then it is obvious that sample complexity of computing v^k increase linearly with N, thus

$$T_v(N) = \mathcal{O}(N).$$

A more attentive analysis is needed for the proximal step:

$$w^{k+1} = \arg\min_{z \in \mathbb{R}^n} \frac{1}{N} \sum_{i \in I} h(z; \xi_i) + \frac{1}{2\mu} \|z - v^k\|^2. \tag{5}$$

Even for small $N > 0$ the solution of the above problem do not have a closed-form and certain auxiliary iterative algorithm must be used to obtain an approximation of the optimal solution. For the above primal form, the stochastic variance-reduction schemes are typically called to approach this finite-sum minimization, when h obey certain smoothness assumptions. However, up to our knowledge, applicability of variance-reduction methods leave out our general convex nonsmooth regularizers. SGD attains an δ–suboptimal point $\|\tilde{z} - \text{prox}_\mu(w; I)\|^2 \leq \delta$ at a sublinear rate, in $\mathcal{O}\left(\frac{\mu}{\delta}\right)$ iterations. This sample complexity is independent of N but to obtain high accuracy a large number of iterations have to be performed.

The dual form is more natural, as follows:

$$\min_{z \in \mathbb{R}^n} \frac{1}{N} \sum_{i=1}^{N} h(z; \xi_i) + \frac{1}{2\mu} \|z - w\|^2$$

$$= \min_{z \in \mathbb{R}^n} \frac{1}{N} \sum_{i=1}^{N} \max_{v_i} \langle v_i, z \rangle - h^*(v_i; \xi_i) + \frac{1}{2\mu} \|z - w\|^2$$

$$= \max_{v} \min_{z \in \mathbb{R}^n} \left\langle \frac{1}{N} \sum_{i=1}^{N} v_i, z \right\rangle - \sum_{i=1}^{N} h^*(v_i; \xi_i) + \frac{1}{2\mu} \|z - w\|^2$$

$$= \max_{v \in \mathbb{R}^{Nn}} -\frac{\mu}{2N^2} \|\sum_{j=1}^{N} v_j\|^2 + \left\langle \frac{1}{N} \sum_{j=1}^{N} v_i, w \right\rangle - \frac{1}{N} \sum_{j=1}^{N} h^*(v_j; \xi_j). \tag{6}$$

Note that in the interesting particular scenarios when regularizer $h(\cdot; \xi)$ results from the composition of a convex function with a linear operator $h(w; \xi) = l(a_\xi^T w)$, the dual variable reduces from Nn to N dimensions. In this case (6) reduces to

$$\max_{v \in \mathbb{R}^N} -\frac{\mu}{2N} \|\sum_{j=1}^{N} a_{\xi_j} v_j\|^2 + \left\langle \sum_{j=1}^{N} a_{\xi_i} v_i, w \right\rangle - \sum_{j=1}^{N} l^*(v_j). \tag{7}$$

Starting from the dual solution v^*, then the primal one is recovered by $z(w) = w - \frac{\mu}{N} \sum_{i=1}^{N} v_i^*$ for (6) or $z(w) = w - \frac{\mu}{N} \sum_{i=1}^{N} a_{\xi_i} v_i^*$ for (7). In the rest of this section we will analyze only the general subproblem (6), since the sample complexity estimates will be easily translated to particular instance (7). For a suboptimal $\tilde{v}$ satisfying $\|\tilde{v} - v^*\| \leq \delta$, primal suboptimality with δ accuracy is obtained by: let $\tilde{z}(w) = w - \frac{\mu}{N} \sum_{i=1}^{N} \tilde{v}_i$ then

$$\|\tilde{z}(w) - z(w)\| = \mu \left\| \frac{1}{N} \left(\sum_{i=1}^{N} \tilde{v}_i - \sum_{i=1}^{N} v_i^* \right) \right\|$$

$$\leq \frac{\mu}{N} \sum_{i=1}^{N} \|\tilde{v}_i - v_i^*\| \leq \frac{\mu}{\sqrt{N}} \|v^* - \tilde{v}\| \leq \frac{\mu\delta}{\sqrt{N}}. \tag{8}$$

Notice that the hessian of the smooth component $\frac{\mu}{2N} \|\sum_{j=1}^{N} v_j\|^2$ is upper bounded by $\mathcal{O}(\mu)$. Without any growth properties on $h^*(\cdot; \xi)$, one would be able to solve (6), using Dual Fast Gradient schemes with $\mathcal{O}(Nn)$ iteration cost, in $\mathcal{O}\left(\max \left\{ Nn, Nn\sqrt{\frac{\mu R_d^2}{\delta}} \right\} \right)$ sample evaluations to get a δ accurate dual solution [21]. This implies, by (8), that there are necessary,

$$T_w^{in}(N; \delta) = \mathcal{O}\left(\max\left\{Nn, N^{3/4}n\frac{\mu R_d}{\delta^{1/2}}\right\}\right)$$

sample evaluations to obtain primal δ–suboptimality. For polyhedral $h^*(\cdot; \xi)$, there are many first-order algorithms that attain linear convergence on the above composite quadratic problem [5]. For instance, the Dual Proximal Gradient algorithm have $\mathcal{O}(Nn)$ arithmetical complexity per iteration and attain a δ–suboptimal dual solution in $\mathcal{O}\left(\frac{L}{\hat{\sigma}(I)}\log\left(\frac{1}{\delta}\right)\right)$, where L is the Lipschitz gradient constant and $\hat{\sigma}(I)$ represents the quadratic growth constant of the dual objective function (6). Therefore there are necessary $\mathcal{O}\left(\frac{N\mu}{\hat{\sigma}(I)}\log\left(\frac{1}{\delta}\right)\right)$ sample evaluations for δ–dual suboptimality, and:

$$T_{w, poly}^{in}(N; \delta) = \mathcal{O}\left(\max\left\{N, \frac{N\mu}{\hat{\sigma}(I)}\log\left(\frac{\mu}{N^{1/2}\delta}\right)\right\}\right)$$

sample evaluations to attain primal δ–suboptimal solution.

3 Iteration Complexity in Expectation

In this section we derive estimates on the necessary number of SPG-M iterations to obtain an ϵ–suboptimal solution of (1). First we recall a widely known result stating that the variance of the subgradients at optimum scales linearly with the minibatch size.

Lemma 1 Let $w^* \in W^*$ and $I = \{\xi_1, \ldots, \xi_N\}$ be an i.i.d. random N–tuple of samples, then

$$\mathbb{E}\left[\|g_F(w^*; I)\|^2\right] = \frac{\mathbb{E}\left[\|g_F(w^*; \xi)\|^2\right]}{N}.$$

Proof By using the squared norm expansion results:

$$\mathbb{E}\left[\|g_F(w^*; I)\|^2\right]$$

$$= \mathbb{E}\left[\frac{1}{N^2}\sum_{i=1}^{N}\mathbb{E}[\|g_F(w^*; \xi)\|^2] + \frac{1}{N^2}\sum_{i=1}^{N}\sum_{j=1}^{N}\mathbb{E}[\langle g_F(w^*; \xi_i), g_F(w^*; \xi_j)\rangle]\right]$$

$$\overset{Assump.\,1(iv)}{=} \mathbb{E}\left[\frac{1}{N}\mathbb{E}[\|g_F(w^*; \xi)\|^2] + \frac{1}{N^2}\sum_{i=1}^{N}\sum_{j=1}^{N}\mathbb{E}[\|g_F(w^*)\|^2]\right]$$

$$= \frac{1}{N}\mathbb{E}[\|g_F(w^*; \xi)\|^2].$$

We will use the following elementary relations: for any $a, b \in \mathbb{R}^n$ and $\beta > 0$ we have

$$\langle a, b \rangle \leq \frac{1}{2\beta}\|a\|^2 + \frac{\beta}{2}\|b\|^2 \tag{9}$$

$$\|a + b\|^2 \leq \left(1 + \frac{1}{\beta}\right)\|a\|^2 + (1 + \beta)\|b\|^2. \tag{10}$$

The main recurrences which will finally generate our sublinear convergence rates are presented below.

Theorem 1 *Let Assumptions 1 hold and $\mu_k \leq \frac{1}{4L_f}$. Assume $\|w^{k+1} - \text{prox}_{h,\mu}$ $(v^k; I^k)\| \leq \delta_k$, for all $k \geq 0$, then the sequence $\{w^k\}_{k \geq 0}$ generated by SPG-M satisfies:*

$$\mathbb{E}[\|w^{k+1} - w^*\|^2] \leq \left(1 - \frac{\sigma_f \mu_k}{2}\right) \mathbb{E}[\|w^k - w^*\|^2]$$
$$+ \mu_k^2 \frac{\mathbb{E}\left[\|g_F(w^*; \xi)\|^2\right]}{N} + \left(3 + \frac{2}{\sigma_f \mu_k}\right) \delta_k^2.$$

Proof Denote $\bar{w}^{k+1} = \text{prox}_{h,\mu_k}(v^k; I^k)$ and recall that $\frac{1}{\mu_k}\left(v^k - \bar{w}^{k+1}\right) \in \partial h(\bar{w}^{k+1}; I^k)$, which implies that there exists a subgradient $g_h(\bar{w}^{k+1}; I^k)$ such that

$$g_h(\bar{w}^{k+1}; I^k) + \frac{1}{\mu_k}\left(\bar{w}^{k+1} - v^k\right) = 0. \tag{11}$$

Using these optimality conditions we have:

$$\|w^{k+1} - w^*\|^2 = \|w^k - w^*\|^2 + 2\langle w^{k+1} - w^k, w^k - w^* \rangle + \|w^{k+1} - w^k\|^2$$

$$\overset{(10)}{\leq} \|w^k - w^*\|^2 + 2\langle \bar{w}^{k+1} - w^k, w^k - w^* \rangle + 2\langle w^{k+1} - \bar{w}^{k+1}, w^k - w^* \rangle$$
$$+ \frac{3}{2}\|\bar{w}^{k+1} - w^k\|^2 + 3\|\bar{w}^{k+1} - w^{k+1}\|^2$$

$$= \|w^k - w^*\|^2 + 2\langle \bar{w}^{k+1} - w^k, \bar{w}^{k+1} - w^* \rangle + 2\langle w^{k+1} - \bar{w}^{k+1}, w^k - w^* \rangle$$
$$- \frac{1}{2}\|\bar{w}^{k+1} - w^k\|^2 + 3\|\bar{w}^{k+1} - w^{k+1}\|^2$$

$$\overset{(9)}{\leq} \left(1 + \frac{\sigma_f \mu_k}{2}\right)\|w^k - w^*\|^2 + 2\langle \mu_k \nabla f(w^k; I_k) + \mu_k g_h(\bar{w}^{k+1}; I_k), w^* - \bar{w}^{k+1} \rangle$$
$$- \frac{1}{2}\|\bar{w}^{k+1} - w^k\|^2 + \left(3 + \frac{2}{\sigma_f \mu_k}\right)\delta_k^2. \tag{12}$$

Now by using convexity of h and Lipschitz continuity of $\nabla f(\cdot; I_k)$, we can further derive:

$$2\mu_k \langle \nabla f(w^k; I_k) + g_h(\bar{w}^{k+1}; I_k), w^* - \bar{w}^{k+1} \rangle - \frac{1}{2}\|\bar{w}^{k+1} - w^k\|^2$$

$$\leq 2\mu_k \langle \nabla f(w^k; I_k), w^* - \bar{w}^{k+1} \rangle - \frac{1}{2}\|\bar{w}^{k+1} - w^k\|^2$$

$$+ 2\mu_k(h(w^*; I_k) - h(\bar{w}^{k+1}; I_k))$$

$$\leq -2\mu_k \left(\langle \nabla f(w^k; I_k), \bar{w}^{k+1} - w^k \rangle + \frac{1}{8\mu_k}\|\bar{w}^{k+1} - w^k\|^2 \right)$$

$$+ h(\bar{w}^{k+1}; I_k) \bigg) + 2\mu_k \langle \nabla f(w^k; I_k), w^* - w^k \rangle - \frac{1}{4}\|\bar{w}^{k+1} - w^k\|^2 + 2\mu_k h(w^*; I_k).$$

By taking expectation w.r.t. ξ_k in both sides, we obtain:

$$- 2\mu_k \mathbb{E}\left[\langle \nabla f(w^k; I_k), \bar{w}^{k+1} - w^k \rangle + \frac{1}{8\mu_k}\|\bar{w}^{k+1} - w^k\|^2 + h(\bar{w}^{k+1}; I_k) \right]$$

$$+ 2\mu_k \langle \nabla f(w^k), w^* - w^k \rangle - \frac{1}{4}\mathbb{E}\left[\|\bar{w}^{k+1} - w^k\|^2\right] + 2\mu_k h(w^*)$$

$$\overset{\mu_k \leq \frac{1}{4L_f}}{\leq} 2\mu_k \mathbb{E}\left[f(w^k; I_k) - F(\bar{w}^{k+1}; I_k) \right] + 2\mu_k \langle \nabla f(w^k), w^* - w^k \rangle$$

$$- \frac{1}{4}\mathbb{E}[\|\bar{w}^{k+1} - w^k\|^2] + 2\mu_k h(w^*)$$

$$\leq 2\mu_k \left(f(w^k) - \mathbb{E}\left[F(\bar{w}^{k+1}; I_k) \right] \right)$$

$$+ 2\mu_k \left(F(w^*) - f(w^k) - \frac{\sigma_f}{2}\|w^k - w^*\|^2 \right) - \frac{1}{4}\|\bar{w}^{k+1} - w^k\|^2$$

$$= -\sigma_f \mu_k \|w^k - w^*\|^2 + 2\mu_k \mathbb{E}\left[F(w^*) - F(\bar{w}^{k+1}; I_k) - \frac{1}{8\mu_k}\|\bar{w}^{k+1} - w^k\|^2 \right].$$

$$(13)$$

By combining (13) with (12) and by taking the expectation with the entire index history we obtain:

$$\mathbb{E}[\|w^{k+1} - w^*\|^2] \leq \left(1 - \frac{\sigma_f \mu_k}{2} \right) \mathbb{E}\left[\|w^k - w^*\|^2 \right]$$

$$+ 2\mu_k \mathbb{E}\left[F(w^*) - F(\bar{w}^{k+1}; I_k) - \frac{1}{8\mu_k}\|\bar{w}^{k+1} - w^k\|^2 \right]. \quad (14)$$

A last further upper bound on the second term in the right hand side: let $w^* \in W^*$

$$\mathbb{E}\left[F(\bar{w}^{k+1}; I_k) - F^* + \frac{1}{8\mu_k}\|\bar{w}^{k+1} - w^k\|^2\right]$$

$$\geq \mathbb{E}\left[\langle g_F(w^*; I_k), \bar{w}^{k+1} - w^*\rangle + \frac{1}{8\mu_k}\|\bar{w}^{k+1} - w^k\|^2\right]$$

$$\geq \mathbb{E}\left[\langle g_F(w^*; I_k), w^k - w^*\rangle + \langle g_F(w^*; I_k), \bar{w}^{k+1} - w^k\rangle + \frac{1}{8\mu_k}\|\bar{w}^{k+1} - w^k\|^2\right]$$

$$\geq \mathbb{E}\left[\langle g_F(w^*; I), w^k - w^*\rangle + \min_z \langle g_F(w^*; I), z - w^k\rangle + \frac{1}{8\mu_k}\|z - w^k\|^2\right]$$

$$\geq \langle g_F(w^*), w^k - w^*\rangle - 2\mu_k\mathbb{E}\left[\|g_F(w^*; I)\|^2\right]$$

$$\overset{Assump.\ 1.(iv)}{=} -2\mu_k\mathbb{E}\left[\|g_F(w^*; I)\|^2\right] \overset{Lemma\ 1}{=} -2\mu_k\frac{\mathbb{E}\left[\|g_F(w^*; I)\|^2\right]}{N}, \tag{15}$$

where we call that by Assumption 1.(iv) $g_F(w^*) = \mathbb{E}[g_F(w^*; \xi)] = 0$. Finally, from (14) and (15) we obtain our above result. □

Remark 2 Consider deterministic setting $F(\cdot; \xi) = F(\cdot)$ and $\mu_k = \frac{1}{2L_f}$, then SPG-M becomes the proximal gradient algorithm and Theorem 1(i) holds with $g_F(w^*; \xi) = g_F(w^*) = 0$, implying that $\Sigma = 0$. Thus the well-known iteration complexity estimate $\mathcal{O}\left(\frac{L_f}{\sigma_f} \log(1/\epsilon)\right)$ [3, 20] of proximal gradient algorithm is recovered up to a constant.

Theorem 2 *Let assumptions of Theorem 1 hold. Also let $\delta_k = \mu_k^{3/2}/N^{1/2}$. Then the sequence $\{w^k\}_{k\geq 0}$ generated by SPG-M attains $\mathbb{E}[\|w^T - w^*\|^2] \leq \epsilon$ within the following number of iterations:*

Constant stepsize: *Let $K > \mathcal{O}\left(\frac{\mu_0(\Sigma^2 + \mu_0 + 1/\sigma_f)}{\sigma_f^2 N}\right)$ and $\mu_k := \frac{2\mu_0}{K} \in \left(0, \frac{1}{4L_f}\right]$, then*

$$T \leq T_c^{out} := \mathcal{O}\left(\max\left\{1 + \frac{\max\{\Sigma^2, 2/\sigma_f\}\log(2r_0^2/\epsilon)}{\epsilon\sigma_f^2 N}, 1 + \sqrt{\frac{72\log^2(2r_0^2/\epsilon)}{\epsilon\sigma_f^3 N}}\right\}\right)$$

Nonincreasing stepsize: *For $\mu_k = \frac{2\mu_0}{k}$, then*

$$T \leq T_v^{out} = \mathcal{O}\left(\max\left\{\frac{\|w^0 - w^*\|^2}{\epsilon}, \frac{\Sigma^2 + \mu_0 + 1/\sigma_f}{N\epsilon}\right\}\right)$$

Mixed stepsize: *Let $\mu_k = \begin{cases} \frac{\mu_0}{4L_f} & k < T_1 \\ \frac{2\mu_0}{k}, & T_1 \leq k \leq T_2 \end{cases}$, then*

$$T \leq T_m^{out} := \underbrace{\frac{4L_f}{\mu_0\sigma_f} \log\left(\frac{2\|w^0 - w^*\|^2}{\epsilon}\right)}_{T_1} + \underbrace{\mathcal{O}\left(\frac{C}{\epsilon N}\right)}_{T_2},$$

where $C = \frac{\mu_0 \Sigma^2 L_f \sigma_f + \mu_0^2 \sigma_f + \mu_0 L_f}{L_f^2 \sigma_f^2} + \Sigma^2 + \mu_0 + 1/\sigma_f$.

Proof **Constant step-size.** Interesting convergence rates arise for proper constant stepsize μ_k. Let $\mu_k := \mu \in \left(0, \frac{1}{4L_f}\right]$ and $\delta_k = \mu^{3/2}/N^{1/2}$, then Theorem 1 states that

$$
\begin{aligned}
\mathbb{E}[\|w^k - w^*\|^2] &\le \left(1 - \frac{\sigma_f \mu}{2}\right)^k r_0^2 + \frac{1 - (1 - \sigma_f \mu/2)^k}{\sigma_f \mu/2} \frac{\mu^2}{N} \left(\Sigma^2 + 3\mu + \frac{2}{\sigma_f}\right) \\
&\le \left(1 - \frac{\sigma_f \mu}{2}\right)^k r_0^2 + \frac{2\mu}{\sigma_f N} \left(\Sigma^2 + 3\mu + \frac{2}{\sigma_f}\right),
\end{aligned}
\tag{16}
$$

which imply a linear decrease of initial residual and, at the same time, the linear convergence of w^k towards a optimum neighborhood of radius $\frac{2\mu}{\sigma_f N} \left(\Sigma^2 + 3\mu + \frac{2}{\sigma_f}\right)$. The radius decrease linearly with the minibatch size N. Given integer $K > 0$, let $\mu = \frac{2\mu_0}{K}$ then after

$$
T = \left\lceil \frac{2K}{\sigma_f \mu_0} \log \left(\frac{2r_0^2}{\epsilon}\right) \right\rceil
\tag{17}
$$

equation (16) leads to

$$
\begin{aligned}
\mathbb{E}[\|w^T - w^*\|^2] &\le \frac{2\mu_0}{K\sigma_f N} \left(\Sigma^2 + \frac{6\mu_0}{K} + \frac{2}{\sigma_f}\right) + \frac{\epsilon}{2} \\
&\le \frac{4 \log(2r_0^2/\epsilon)}{(T-1)\sigma_f^2 N} \left(\Sigma^2 + \frac{12 \log(2r_0^2/\epsilon)}{\sigma_f(T-1)} + \frac{2}{\sigma_f}\right) + \frac{\epsilon}{2}.
\end{aligned}
\tag{18}
$$

Overall, to obtain $\mathbb{E}[\|w^T - w^*\|^2] \le \epsilon$, SPG-M has to perform

$$
\mathcal{O}\left(\max\left\{1 + \frac{\max\{\Sigma^2, 2/\sigma_f\} \log(2r_0^2/\epsilon)}{\epsilon \sigma_f^2 N}, 1 + \sqrt{\frac{72 \log^2(2r_0^2/\epsilon)}{\epsilon \sigma_f^3 N}}\right\}\right)
$$

SPG-M iterations.

Variable stepsize. Now let $\mu_k = \frac{2\mu_0}{k}$, $\delta_k = \frac{\mu_k^{3/2}}{N^{1/2}}$, then Theorem 1 leads to:

$$
\mathbb{E}[\|w^k - w^*\|^2] \le \prod_{j=1}^{k} \left(1 - \frac{\sigma_f \mu_j}{2}\right) r_0^2 + \sum_{i=1}^{k} \frac{\mu_i^2}{N} \left(\Sigma^2 + 3\mu_i + \frac{2}{\sigma_f}\right) \prod_{j=i+1}^{k} \left(1 - \frac{\sigma_f \mu_j}{2}\right)
\tag{19}
$$

By using further the same (standard) analysis from [26, 27], we obtain:

$$
\mathbb{E}[\|w^k - w^*\|^2] \le \underbrace{\mathcal{O}\left(\frac{r_0^2}{k}\right)}_{\text{optimization error}} + \underbrace{\mathcal{O}\left(\frac{\Sigma^2 + \mu_0 + 1/\sigma_f}{Nk}\right)}_{\text{sample error}}
\tag{20}
$$

The rate (20) implies the following iteration complexity:

$$\mathcal{O}\left(\max\left\{\frac{r_0^2}{\epsilon}, \frac{\Sigma^2 + \mu_0 + 1/\sigma_f}{N\epsilon}\right\}\right).$$

Mixed stepsize. By combining constant and variable stepsize policies, we aim to get a better "optimization error" and overall, a better iteration complexity of SPG-M. Inspired by (17)–(18), we are able to use a constant stepsize policy to bring w^k in a small neighborhood of w^* whose radius is inversely proportional with N.

Let $\mu_k = \frac{\mu_0}{4L_f}$, using similar arguments as in (17)–(18), we have that after:

$$T_1 \geq \frac{4L_f}{\mu_0\sigma_f} \log\left(\frac{2r_0^2}{\epsilon}\right)$$

the expected residual is bounded by:

$$\mathbb{E}[\|w^{T_1} - w^*\|^2] \leq \frac{\mu_0}{4L_f\sigma_f N}\left(\Sigma^2 + \frac{3\mu_0}{4L_f} + \frac{2}{\sigma_f}\right) + \frac{\epsilon}{2}. \qquad (21)$$

Now restarting SPG-M from w^{T_1} we have from (20) that:

$$\mathbb{E}[\|w^k - w^*\|^2] \leq \mathcal{O}\left(\frac{\mathbb{E}[\|w^{T_1} - w^*\|^2]}{k}\right) + \mathcal{O}\left(\frac{\Sigma^2 + \mu_0 + 1/\sigma_f}{Nk}\right)$$

$$\leq \mathcal{O}\left(\frac{\mu_0\Sigma^2 L_f\sigma_f + \mu_0^2\sigma_f + \mu_0 L_f}{L_f^2\sigma_f^2 kN}\right) + \mathcal{O}\left(\frac{\epsilon}{2k}\right) + \mathcal{O}\left(\frac{\Sigma^2 + \mu_0 + 1/\sigma_f}{Nk}\right). \qquad (22)$$

iterations. Overall, SPG-M computes w^{T_2} such that $\mathbb{E}[\|w^{T_2} - w^*\|^2] \leq \epsilon$ within a number of iterations bounded by

$$T_1 + T_2 = \frac{4L_f}{\mu_0\sigma_f}\log\left(\frac{2r_0^2}{\epsilon}\right) + \mathcal{O}\left(\frac{C}{\epsilon N}\right),$$

where $C = \frac{\mu_0\Sigma^2 L_f\sigma_f + \mu_0^2\sigma_f + \mu_0 L_f}{L_f^2\sigma_f^2} + \Sigma^2 + \mu_0 + 1/\sigma_f$. $\square$

Remark 3 For variable stepsize $\mu_k = \frac{\mu_0}{k}$, the "optimization rate" $\mathcal{O}(r_0^2/k)$ of (20) is optimal (for strongly convex stochastic optimization) and it is not affected by the variation of minibatch size. Intuitively, the stochastic component within optimization model (1) is not eliminated by increasing N, only a variance reduction is attained. The authors of [42], under bounded gradients of the objective function, obtained

$\mathcal{O}\left(\frac{L^2}{\sigma_f Nk}\right)$ convergence rate for their Minibatch-Prox Algorithm in the average sequence, using classical arguments. However, their algorithm is based on knowledge of σ_f, which is used to compute stepsize sequence $\mu_k = \frac{2}{\sigma_f(k-1)}$. Moreover, in the first step the algorithm has to compute $\text{prox}_{h,+\infty}(\cdot; I^0) = \arg\min_z \frac{1}{N}\sum_{i \in I} h(z; i)$, which for small σ_f might be computationally expensive. Notice that, under knowledge of σ_f, we could obtain similar sublinear rate in $\mathbb{E}[\|w^k - w^*\|^2]$ using similar stepsizes.

Remark 4 We make a few observations about the T_m^{out}. For a small conditioning number $\frac{L_f}{\sigma_f}$ the constant stage performs few iterations and the total complexity is dominated by $\mathcal{O}\left(\frac{C}{\epsilon N}\right)$. This bound (of the same order as [42]) present some advantages: unknown σ_f, evaluation in the last iterate and no uniform bounded gradients assumptions. On the other hand, for a sufficiently large minibatch $N = \mathcal{O}(1/\epsilon)$ and a proper constant μ_0, one could perform a constant number of SPG-M iterations. In this case, the mixed stepsize convergence rate provides a link between population risk and empirical risk.

3.1 Total Complexity

In this section we couple the complexity per iteration estimates from Sect. 2.1 and Sect. 3 and provide upper bounds on the total complexity of SPG-M.

The sample complexity, often used for stochastic algorithms, refers to the entire number of data samples that are used during all iterations of a given algorithmic scheme. In our case, given the minibatch size N and the total outer SPG-M iterations T^{out}, the sample complexity is given by NT^{out}. In the best case NT^{out} is upper bounded by $\mathcal{O}(1/\epsilon)$. We consider the dependency on minibatch size N and accuracy ϵ of highly importance, thus we will present below simplified upper bounds of our estimates. In Sect. 2.1, we analyzed the complexity of a single SPG-M iteration for convex components $h(\cdot; \xi)$ denoted by $T_v^{in} + T_w^{in}$. Summing the inner effort $T_v^{in} + T_w^{in}$ over the outer number of iterations provided by Theorem 2 leads us to the total computational complexity of SPG-M. We further derive the total complexity for SPG-M with mixed stepsize policy and use the same notations as in Theorem 2:

$$T^{total} = \sum_{i=0}^{T_m^{out}} (T_v^{in}(N) + T_w^{in}(N; \delta))$$

$$\leq \sum_{i=0}^{T_m^{out}} \left(\mathcal{O}(Nn) + \mathcal{O}\left(\max\left\{ Nn, N^{3/4}n\frac{\mu_i R_d}{\delta_i^{1/2}} \right\} \right) \right)$$

$$\leq \sum_{i=0}^{T_m^{out}} \left[\mathcal{O}(Nn) + \mathcal{O}\left(Nn\mu_i^{1/4} R_d\right) \right]$$

$$\overset{(1.4)}{\leq} \mathcal{O}(T_m^{out} Nn) + \mathcal{O}\left(Nn(T_m^{out})^{3/4} R_d\right)$$

$$\leq \mathcal{O}\left(\left(\frac{L_f}{\sigma_f} \log (1/\epsilon) + \frac{C}{N\epsilon}\right) Nn\right) + \mathcal{O}\left(Nn \left(\frac{L_f}{\sigma_f} \log (1/\epsilon) + \frac{C}{N\epsilon}\right)^{3/4} R_d\right).$$

Extracting the dominant terms from the right hand side we finally obtain:

$$T^{total} \leq \mathcal{O}\left(\frac{L_f Nn}{\sigma_f} \log (1/\epsilon) + \frac{Cn}{\epsilon}\right) + \mathcal{O}\left(N^{1/4} n \left(\frac{C}{\epsilon}\right)^{3/4} R_d\right).$$

The first term $\mathcal{O}\left(\left(\frac{L_f}{\sigma_f} \log (1/\epsilon) + \frac{C}{N\epsilon}\right) Nn\right)$ is the total cost of the minibatch gradient step v^k and is highly dependent on the conditioning number $\frac{L_f}{\sigma_f}$. The second term is brought by solving the proximal subproblem by Dual Fast Gradient method depicting a weaker dependence on N and ϵ than the first. Although the complexity order is $\mathcal{O}\left(\frac{Cn}{\epsilon}\right)$, comparable with the optimal performance of single-sample stochastic schemes (with $N = 1$), the above estimate paves the way towards the acceleration techniques based on distributed stochastic iterations. Reduction of the complexity per iteration to $\frac{1}{\tau}(T_v^{in}(N) + T_w^{in}(N; \delta))$, using multiple machines/processors, would guarantee direct improvements in the optimal number of iterations $\mathcal{O}(\frac{Cn}{\tau\epsilon})$. The superiority of distributed variants of SGD schemes for smooth optimization are clear (see [23]), but our results set up the theoretical foundations for distributing the algorithms for the class of proximal gradient algorithms.

4 Numerical Simulations

4.1 Large Scale Support Vector Machine

To validate the theoretical implications of the previous sections, we first chose a well known convex optimization problem in the machine learning field: optimizing a binary Support Vector Machine (SVM) application. To test several metrics and methods, a spam-detection application is chosen using the dataset from [14]. The dataset contains about 32000 emails that are either classified as spam or non-spam. This was split in our evaluation in a classical 80% for training and 20% for testing. To build the feature space in a numerical understandable way, three preprocessing steps were made:

1. A dictionary of all possible words in the entire dataset and how many times each occurred is constructed.
2. The top 200($= n$; the number of features used in our classifier) most-used words indices from the dictionary are stored.
3. Each email entry i then counts how many of each of the n words are in the i-th email's text. Thus, if X_i is the numerical entry characteristic to email i, then X_{ij} contains how many words with index j in the top most used words are in the email's text.

The pseudocode for optimization process is shown below:

For $k \geq 0$ compute
1. Choose randomly i.i.d. N–tuple $I^k \subset \Omega$
2. Update:

$$v^k = (1 - \lambda \mu_k) w^k$$

$$u^k = \arg \max_{u \in [0,1]} \quad -\frac{\mu}{2N} \|\tilde{X}_{I_k} u\|_2^2 + u^T (e - \tilde{X}_{I_k}^T v^k)$$

$$w^{k+1} = v^k + \frac{\mu_k}{N} X_{I^k} u^k$$

3. If the stopping criterion holds, then **STOP**, otherwise $k = k + 1$.

To compute the optimal solution w^*, we let running the binary SVM state-of-the-art method for the dataset using SGD hinge-loss [24] for a long time, until we get the top accuracy of the model (93.2%). Considering this as a performance baseline, we compare the results of training process efficiency between the SPG-M model versus SGD with mini-batches. The comparison is made w.r.t. three metrics:

1. **Accuracy**: how well does the current set of trained weights performs at classification between spam versus non-spam.
2. **Loss**: the hinge-loss result on the entire set of data.
3. **Error** (or optimality measure): computes how far is the current trained set of weights (w^k at any step k in time). From the optimal ones, i.e. $\|w^k - w^*\|^2$.

The comparative results between the two methods and each of the metrics defined above are shown in Figs. 1, 2 and 3. These were obtained by averaging several executions on the same machine, each with a different starting point. Overall, the results show the advantage of SPG-M method over SGD: while both methods will converge to the same optimal results after some time, SPG-M is capable of obtaining better results all the three metrics in a shorter time, regardless of the batch size being used. One interesting observation can be seen for the SGD-Const method results, when the loss metric tends to perform better (2). This is because of a highly tuned constant learning rate to get the best possible result. However, this is not a robust way to use in practice.

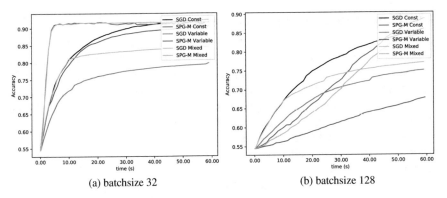

(a) batchsize 32   (b) batchsize 128

Fig. 1 Comparative results between SPG-M and SGD for the **Accuracy** metric, using different batchsizes and functions for chosing the stepsize

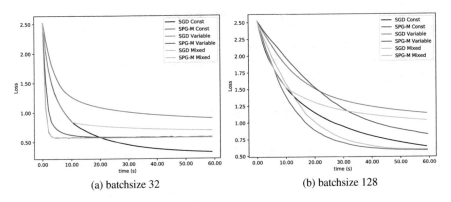

(a) batchsize 32   (b) batchsize 128

Fig. 2 Comparative results between SPG-M and SGD for the **Loss** metric, using different batchsizes and functions for chosing the stepsize

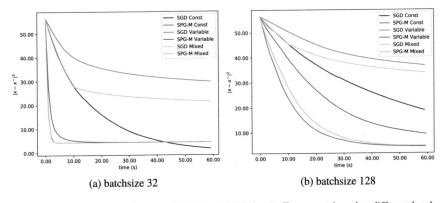

(a) batchsize 32   (b) batchsize 128

Fig. 3 Comparative results between SPG-M and SGD for the **Error** metric, using different batchsizes and functions for chosing the stepsize

4.2   Parametric Sparse Representation

Given signal $y \in \mathbb{R}^m$ and overcomplete dictionary $T \in \mathbb{R}^{m \times n}$ (whose columns are also called atoms), sparse representation [8] aims to find the sparse signal $x \in \mathbb{R}^n$ by projecting y to a much smaller subspace generated by a subset of the columns from T. Sparse representation is a key ingredient in dictionary learning techniques [7] and here we focus on the multi-parametric sparse representation model proposed in [36]. Note that this was analyzed in the past in non-minibatch SPG form [26] which we denote with SSPG in the following. The multi-parametric representation problem is given by:

$$\min_x \quad \|Tx - y\|_2^2$$
$$\text{s.t.} \quad \|\Delta x\|_1 \le \delta, \tag{23}$$

where T and x correspond to the dictionary and, respectively, the resulting sparse representation, with sparsity being imposed on a scaled subspace Δx with $\Delta \in \mathbb{R}^{p \times n}$. In pursuit of 1, we move to the exact penalty problem $\min_x \frac{1}{2m}\|Tx - y\|_2^2 + \lambda\|\Delta x\|_1$. In order to limit the solution norm we further regularize the unconstrained objective using an ℓ_2 term as follows:

$$\min_x \quad \frac{1}{2m}\|Tx - y\|_2^2 + \frac{\alpha}{2}\|x\|_2^2 + \frac{\lambda}{p}\|\Delta x\|_1.$$

The decomposition which puts the above formulation into model (1) consists of:

$$f(x; \xi) = \frac{1}{2}(T_\xi x - y_\xi)_2^2 + \frac{\alpha}{2}\|x\|_2^2 \tag{24}$$

where T_ξ represents line ξ of matrix T, and

$$h(x; \xi) = \lambda|\Delta_\xi x|. \tag{25}$$

To compute the SPG-M iteration for the sparse representation problem, we note that

$$\text{prox}_{h,\mu}(x; i) = \arg\min_z \frac{\lambda}{N}\|\Delta_I z\|_1 + \frac{1}{2\mu}\|z - x\|^2.$$

Equivalently, once we find dual vector

$$z^k = \arg\min_{-1 \le z \le 1} \frac{\mu\lambda^2}{2N^2}\|\Delta_I^T z\|^2 - \frac{\lambda}{N}z^T\Delta_I t$$

then we can easily compute $\text{prox}_{h,\mu}(x; I) = x - \frac{\mu\lambda}{N}\Delta_I^T z$. We are ready to formulate the resulting particular variant of SPG-M.

SPG-M—Sparse Representation (SPGM-SR):   For $k \geq 0$ compute
1. Choose randomly i.i.d. N–tuple $I^k \subset \Omega$
2. Update:

$$y^k = \left[I_n - \frac{\mu_k}{N} \left(T_{I_k}^T T_{I_k} + N\alpha I_n \right) \right] x^k + \frac{\mu_k}{N} T_{I_k}^T y_{I_k}$$

$$z^k = \arg \min_{-1 \leq z \leq 1} \frac{\mu \lambda^2}{2N^2} \| \Delta_I^T z \|^2 - \frac{\lambda}{N} z^T \Delta_I t$$

$$x^{k+1} = y^k - \frac{\mu_k \lambda}{N} \Delta_I^T z^k$$

3. If the stopping criterion holds, then **STOP**, otherwise $k = k + 1$.

We proceed with numerical experiments that depict SPG-M with various batch sizes N and compare them with SSPG [26], which is equivalent to SPG-M where $N = 1$. In our experiments we use batches of $N = \{1, 10, 50, 100\}$ samples from a population of $m = 400$ using dictionaries with $n = 200$ atoms. We fix $\lambda = 5 \dot{1}0^{-4}$ and stop the algorithms when the euclidean distance between current solution x and the optimum $x^\star$ is less than $\varepsilon = 10^{-3}$. CVX is used to determine $x^\star$ within a $\varepsilon = 10^{-6}$ margin.

Here we choose two scenarios: one where $\alpha = 0.2$ and all methods provide adequate performance, depicted in Figs. 4 and 5, and a second where $\alpha = 0.7$ is larger and stomps performance as can be seen in the first ten iterations of Figs. 6 and 7. In these figures we can also observe that the mixed stepsize indeed provides much better convergence rate and that the multibatch algorithm is always ahead of the single batch SSPG version.

We continue our investigation by adapting the minibatch stochastic gradient descent method to the parametric sparse representation problem which leads to the following algorithm:

Fig. 4 Variable stepsize ($\alpha = 0.2$)

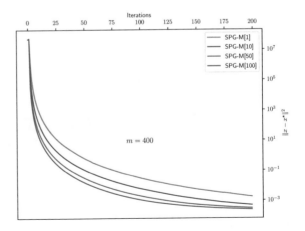

Fig. 5 Mixed stepsize ($\alpha = 0.2$)

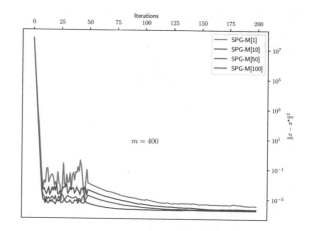

Fig. 6 Variable stepsize ($\alpha = 0.7$)

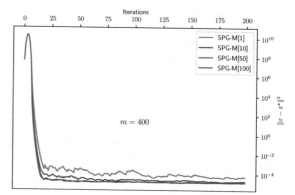

Fig. 7 Mixed stepsize ($\alpha = 0.7$)

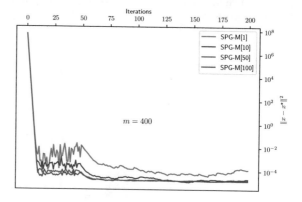

Fig. 8 SPG-M and SGD with Variable stepsize. The other batch sizes were identical with $N = 100$ for SGD ($\alpha = 0.7$)

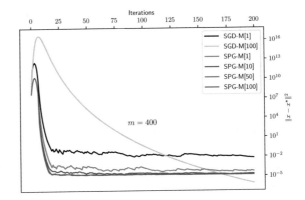

SGD-M—Sparse Representation (SGDM-SR): For $k \geq 0$ compute

1. Choose randomly i.i.d. N–tuple $I^k \subset \Omega$
2. Update:

$$x^{k+1} = x^k - \frac{\mu_k}{N} \left(T_{I_k}^T T_{I_k} + N\alpha I_n \right) x^k + \frac{\mu_k}{N} T_{I_k}^T y_{I_k} - \frac{\mu_k \lambda}{N} \sum_{i \in I^k} \operatorname{sgn}(\Delta_i x^k) \Delta_i^T,$$

where $\operatorname{sgn}(\Delta_i x) = \begin{cases} +1, & \Delta_i x > 0 \\ -1, & \Delta_i x < 0 \\ 0, & \Delta_i x = 0 \end{cases}$

3. If the stopping criterion holds, then **STOP**, otherwise $k = k + 1$.

When applying SGDM-SR on the same initial data as our experiment with $\alpha = 0.7$ with the same parametrization and batch sizes we obtain the results depicted in Fig. 8. Here the non-minibatch version of SGD is clearly less performant than SPG-M, but what is most interesting is that the minibatch version for all batch sizes behaves identically and takes 100 iterations to recover and reach the optimum around iteration 150.

5 Conclusion

In this chapter we presented preliminary guarantees for minibathc stochastic proximal gradient schemes, which extend some well-known schemes in the literature. For future work, would be interesting to analyze the behaviour of SPG-M scheme on nonconvex learning models.

We provided significant improvements in iteration complexity that future work can further reduce using distributed and parallelism techniques, as hinted by the distributed variants of SGD schemes [23].

24

A. Pătraşcu et al.

Acknowledgements C. Păduraru and P. Irofti are also with The Research Institute of the University of Bucharest (ICUB) and were supported by a grant of the Romanian Ministry of Research and Innovation, CCCDI-UEFISCDI, project number 17PCCDI/2018 within PNCDI III.

References

1. Asi, H., Duchi, J.C.: Stochastic (approximate) proximal point methods: convergence, optimality, and adaptivity. SIAM J. Optim. **29**(3), 2257–2290 (2019)
2. Bauschke, H.H., Borwein, J.M., Li, W.: Strong conical hull intersection property, bounded linear regularity, Jameson's property (g), and error bounds in convex optimization. Math. Program. **86**(1), 135–160 (1999)
3. Beck, A., Teboulle, M.: A fast iterative shrinkage-thresholding algorithm for linear inverse problems. SIAM J. Imaging Sci. **2**(1), 183–202 (2009)
4. Bianchi, P.: Ergodic convergence of a stochastic proximal point algorithm. SIAM J. Optim. **26**(4), 2235–2260 (2016)
5. Dhillon, I.S., Hsieh, C.-J., Si, S.: Communication-efficient distributed block minimization for nonlinear kernel machines. In: KDD '17: Proceedings of the 23rd ACM SIGKDD International Conference on Knowledge Discovery and Data Mining, pp. 245–254 (2017)
6. Davis, D., Drusvyatskiy, D.: Stochastic model-based minimization of weakly convex functions. SIAM J. Optim. **29**(1), 207–239 (2019)
7. Dumitrescu, B., Irofti, P.: Dictionary Learning Algorithms and Applications. Springer (2018)
8. Elad, M.: Sparse and Redundant Representations: From Theory to Applications in Signal and Image Processing. Springer Science & Business Media (2010)
9. Friedlander, M.P., Schmidt, M.: Hybrid deterministic-stochastic methods for data fitting. SIAM J. Sci. Comput. **34**(3), 1380–1405 (2012)
10. Goyal, P., DollÃar, P., Girshick, R., Noordhuis, P., Kyrola, A., Wesolowski, L., Tulloch, A., Jia, Y., He, K.: Accurate, large minibatch SGD: training ImageNet in 1 hour (2017). arXiv:1706.02677 [cs.CV]
11. Hallac, D., Leskovec, J., Boyd, S.: Network lasso: clustering and optimization in large graphs. In: Proceedings of the 21th ACM SIGKDD International Conference on Knowledge Discovery and Data Mining, pp. 387–396 (2015)
12. Hoffer, E., Hubara, I., Soudry, D.: Train longer, generalize better: closing the generalization gap in large batch training of neural networks (2017). arXiv:1705.08741 [stat.ML]
13. Koshal, J., Nedic, A., Shanbhag, U.V.: Regularized iterative stochastic approximation methods for stochastic variational inequality problems. IEEE Trans. Autom. Control **58**(3), 594–609 (2012)
14. Metsis, V.: Ion Androutsopoulos, and Georgios Paliouras. Spam filtering with Naive Bayes-which naive bayes? In: CEAS, Mountain View, CA, vol. 17, pp. 28–69 (2006)
15. Moulines, E., Bach, F.R.: Non-asymptotic analysis of stochastic approximation algorithms for machine learning. In Advances in Neural Information Processing Systems, pp. 451–459 (2011)
16. Nedic, A., Necoara, I.: Random minibatch subgradient algorithms for convex problems with functional constraints. Appl. Math. Optim. **80**, 801–833 (2019)
17. Nedić, A.: Random projection algorithms for convex set intersection problems. In: 49th IEEE Conference on Decision and Control (CDC), pp. 7655–7660. IEEE (2010)
18. Nedić, A.: Random algorithms for convex minimization problems. Math. Program. **129**(2), 225–253 (2011)
19. Nemirovski, A., Juditsky, A., Lan, G., Shapiro, A.: Robust stochastic approximation approach to stochastic programming. SIAM J. Optim. **19**(4), 1574–1609 (2009)
20. Nesterov, Y.: Gradient methods for minimizing composite functions. Math. Program. **140**(1), 125–161 (2013)

21. Nesterov, Y.: Introductory Lectures on Convex Optimization: A Basic Course, vol. 87. Springer Science & Business Media (2013)
22. Nguyen, L.M., Nguyen, P.H., van Dijk, M., Richtárik, P., Scheinberg, K., Takáč, M.: SGD and hogwild! convergence without the bounded gradients assumption (2018). arXiv:1802.03801
23. Niu, F., Recht, B.H., RÃ1, C., Wright, S.J.: Hogwild!: a lock-free approach to parallelizing stochastic gradient descent. In: NIPS'11: Proceedings of the 24th International Conference on Neural Information Processing Systems, pp. 693–701 (2011)
24. Patil, R.C., Patil, D.R.: Web spam detection using SVM classifier. In: 2015 IEEE 9th International Conference on Intelligent Systems and Control (ISCO), pp. 1–4. IEEE (2015)
25. Patrascu, A.: New nonasymptotic convergence rates of stochastic proximal point algorithm for stochastic convex optimization. To appear in Optimization (2020)
26. Patrascu, A., Irofti, P.: Stochastic proximal splitting algorithm for composite minimization, pp. 1–16 (2020). arXiv:1912.02039v2
27. Patrascu, A., Necoara, I.: Nonasymptotic convergence of stochastic proximal point methods for constrained convex optimization. J. Mach. Learn. Res. 18(1), 7204–7245 (2017)
28. Tyrrell Rockafellar, R.: Convex Analysis. Princeton University Press (1970)
29. Rockafellar, R.T., Wets, R.J.B.: On the interchange of subdifferentiation and conditional expectation for convex functionals. Stoch. Int. J. Probab. Stoch. Process. 7(3), 173–182 (1982)
30. Rosasco, L., Villa, S., Vũ, B.C.: Convergence of stochastic proximal gradient algorithm. Appl. Math. Optim., 1–27 (2019)
31. Ryu, E.K., Boyd, S.: Stochastic proximal iteration: a non-asymptotic improvement upon stochastic gradient descent. Author website, early draft (2016)
32. Zhang, H., Ghadimi, S., Lan, G.: Mini-batch stochastic approximation methods for nonconvex stochastic composite optimization. Math. Program. 155, 267–305 (2016)
33. Salim, A., Bianchi, P., Hachem, W.: Snake: a stochastic proximal gradient algorithm for regularized problems over large graphs. IEEE Trans. Autom. Control 64(5), 1832–1847 (2019)
34. Shalev-Shwartz, S., Singer, Y., Srebro, N., Cotter, A.: Pegasos: primal estimated sub-gradient solver for SVM. Math. Program. 127(1), 3–30 (2011)
35. Shi, W., Ling, Q., Gang, W., Yin, W.: A proximal gradient algorithm for decentralized composite optimization. IEEE Trans. Signal Process. 63(22), 6013–6023 (2015)
36. Stoican, F., Irofti, P.: Aiding dictionary learning through multi-parametric sparse representation. Algorithms 12(7), 131 (2019)
37. Toulis, P., Tran, D., Airoldi, E.: Towards stability and optimality in stochastic gradient descent. In: Artif. Intell. Stat., 1290–1298 (2016)
38. Varma, R., Lee, H., Kovacevic, J., Chi, Y.: Vector-valued graph trend filtering with non-convex penalties. IEEE Trans. Signal Inform. Process. Over Netw. (2019)
39. Wang, J., Srebro, N.: Stochastic nonconvex optimization with large minibatches. In: In International Conference On Learning Theory (COLT), pp. (98):1–26 (2019)
40. Wang, J., Wang, W., Srebro, N.: Memory and communication efficient distributed stochastic optimization with minibatch prox. In: In International Conference On Learning Theory (COLT), pp. 65:1–37 (2017)
41. Wang, M., Bertsekas, D.P.: Stochastic first-order methods with random constraint projection. SIAM J. Optim. 26(1), 681–717 (2016)
42. Wang, X., Wang, S., Zhang, H.: Inexact proximal stochastic gradient method for convex composite optimization. Comput. Optim. Appl. 68(3), 579–618 (2017)
43. Zhong, W., Kwok, J.: Accelerated stochastic gradient method for composite regularization. In: Artificial Intelligence and Statistics, pp. 1086–1094 (2014)

The Human Mental Search Algorithm for Solving Optimisation Problems

Seyed Jalaleddin Mousavirad, Gerald Schaefer,
and Hossein Ebrahimpour-Komleh

1 Introduction

Optimisation plays a crucial role in the performance of most data science algorithms so that in many cases there is a direct relationship between the efficacy of the optimisation algorithm and the efficacy of the data science algorithm. For example, in a multi-layer neural network, optimal network weights are sought during training, while optimisation is also important for finding suitable parameters in regression problems. Optimisation is thus a necessity to solve data-related problems effectively.

In a data science algorithm, in particular, a machine learning algorithm, representation, optimisation, and generalisation are often considered independently [29]. In other words, when studying representation or generalisation, we often do not consider whether optimisation algorithms can be effectively applied or not. On the other hand, when considering optimisation algorithms, we often do not explicitly discuss generalisation errors (sometimes representation error is assumed zero) [29].

Machine learning algorithms traditionally use conventional algorithms such as gradient descent for optimisation. These however have several drawbacks such as local optima stagnation and sensitivity to initialisation, leading to unsatisfactory ability in finding the global optimum.

Stochastic algorithms such as metaheuristics represent a problem-independent alternative to address the limitations of conventional algorithms. Metaheuristic algorithms benefit from stochastic operators applied in an iterative manner and have

S. J. Mousavirad (✉)
Faculty of Engineering, Sabzevar University of New Technology, Sabzevar, Iran

G. Schaefer
Department of Computer Science, Loughborough University, Loughborough, UK

H. Ebrahimpour-Komleh
Department of Electrical and Computer Engineering, University of Kashan, Kashan, Iran

A.-E. Hassanien et al. (eds.), *Enabling AI Applications in Data Science*,
Studies in Computational Intelligence 911,
https://doi.org/10.1007/978-3-030-52067-0_2

become popular since they are more robust and do not require the gradient of the cost function.

Generally speaking, metaheuristic algorithms can be divided into two categories: single-solution-based and population-based algorithms. Single-solution-based algorithms such as simulated annealing [3], variable neighbourhood search [18], or iterated local search [27], start with a random candidate solution and try to improve it during some iterations. In contrast, population-based algorithms begin with a set of random candidate solutions which interact with each other to share information. Consequently, population-based algorithms exhibit a higher likelihood of escaping from local optima compared to single-solution-based approaches. Population-based metaheuristics include particle swarm optimisation (PSO) [13], harmony search (HS) [9], artificial bee colony algorithm (ABC) [12], and biogeography-based optimisation (BBO) [26].

Population-based metaheuristic algorithms have been employed in various data science domains. For example, metaheuristics have shown superior ability in finding neural network weights [1, 20, 21] or architecture [4, 11], while feature selection is another domain, with the aim to find the optimal features for decision making [10, 19]. Other applications include clustering [5, 22], regression [30], and deep belief networks [7, 15].

In this chapter, we present the Human Mental Search (HMS) algorithm as a new class of population-based metaheuristic algorithms. We explain the idea behind HMS and how the algorithm can successfully tackle optimisation problems. HMS is inspired by exploration strategies in the bid space of online auctions, and comprises three main operators, mental search, grouping, and movement. Mental search explores the vicinity of candidate solutions based on a Levy flight distribution, whereas grouping partitions the current population using a clustering algorithm to finding a promising area in search space. During movement, candidate solutions move towards the promising area. We conduct an extensive set of experiments on both normal and large-scale problems and compare HMS to other state-of-the-art algorithms. The obtained results show HMS to yield very competitive performance to tackle optimisation problems.

The remainder of the chapter is organised as follows. Section 2 explains the inspiration, while Sect. 3 introduces the HMS algorithm in detail. Section 4 evaluates the HMS algorithm on both normal and large-scale problems. Finally, Sect. 5 concludes the chapter.

2 Inspiration

Radicchi et al. [23, 24] showed that humans employ a Levy flight strategy to explore the bid space of online auctions. In particular, they studied the behaviour of attendants in a new type of auction called Lowest Unique Bid (LUB). Some of their observations were:

- The period of an auction is determined a priori.
- A bid has a value between L and H.
- Each attendant has a specific strategy of α.
- Each attendant is allowed to participate in several bids.
- The next bid of every attendant follows from a Levy flight distribution.
- Failed attendants try to bring their strategy closer to the winner.

The HMS algorithm tries to use the above-mentioned observations to explore the search space in order to find the global optimum.

3   The Human Mental Search Algorithm

The workings of the HMS algorithm is summarised in Fig. 1 in form of a flowchart, while in the following we explain in more detail its main components. In HMS, each candidate solution is called a bid (offer), and each group corresponds to an attendant.

3.1   Initialisation

HMS, similar to other population-based metaheuristic algorithms, works on a set (population) of candidate solutions (bids). Associated with each candidate solution is an objective function value, which shows the quality of the candidate solution; e.g. for a minimisation problem, the lower the objective function value the better quality of the candidate solution.

3.2   Mental Search

Mental search explores the vicinity of each bid based on a Levy flight mechanism. Levy flight is a type of random walk with its step size determined by a Levy distribution so that there are many small steps and sometimes a long step. In comparison to Brownian motion, Levy flight is more effective since its long and small steps lead to an increase in both exploration and exploitation.

A new bid in mental search is created as

$$NS = bid + S, \tag{1}$$

with S is calculated as

$$S = (2 - iter(2/MaxIter))\alpha \oplus Levy, \tag{2}$$

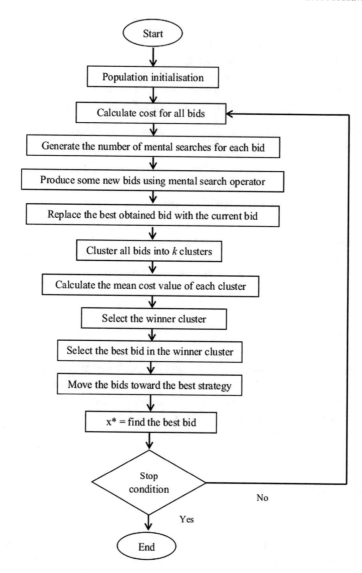

Fig. 1 Flowchart of HMS algorithm

where $MaxIter$ is the maximum number of iterations, $iter$ is the current iteration, α is a random number, and $\oplus$ denotes entry-wise multiplication. The $(2 - iter(2/\max iter))$ factor is an ascending factor, starting from 2 and ending in 0 to allow high exploration at the beginning and high exploitation capability towards the end.

Creating the step size S using a Levy distribution leads to

$$S = (2 - iter(2/MaxIter))0.01\frac{u}{v^{1/\beta}}(x^i - x^*), \qquad (3)$$

where x^* is the best solution found so far, and u and v are two random numbers drawn from a normal distribution as

$$u \sim N(0, \sigma_u^2), \quad v \sim N(0, \sigma_v^2), \qquad (4)$$

with

$$\sigma_u = \left\{ \frac{\Gamma(1 + \beta)\sin(\frac{\pi\beta}{2})}{\Gamma[(\frac{1+\beta}{2})]\beta 2^{(\beta-1)/2}} \right\}^{1/\beta}, \quad \sigma_v = 1, \qquad (5)$$

where Γ is a standard gamma function.

Figure 2 illustrates the mental search operator.

3.3 Grouping

The grouping operator is used to cluster the current population. For this, clustering, an unsupervised machine learning technique in which patterns close to each other organised into groups, and in particular, k-means is employed. After clustering, the mean objective function value for each cluster is calculated, and the cluster with the best objective function value selected as the winner cluster to represent a promising area in search space. Figure 3 illustrates the grouping operator for 12 candidate solutions. This mechanism is different from other algorithms that use the best candidate solutions to find a promising area.

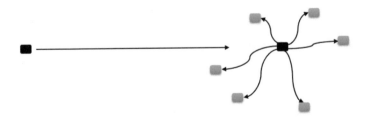

Fig. 2 The mental search operator generates several new bids in the neighbourhood of a current bid

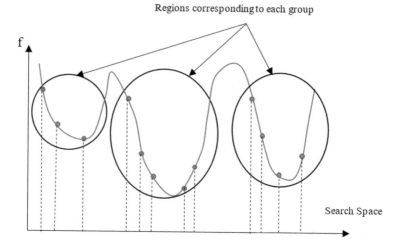

Fig. 3 The grouping operator partitions the bids based on a clustering algorithm. Each red point indicates a candidate solution. As can be seen, the candidate solutions are clustered into 3 groups so that each cluster includes candidate solutions close to each other

3.4 Movement

The movement strategy employs the best bid in the winner cluster as the destination bid. Then, other candidate solutions move toward the promising area as

$$bid_n^{t+1} = bid_n^t + C(r \times winner_n^t - bid_n^t), \tag{6}$$

where bid_n^{t+1} is the n-th bid element at iteration $t + 1$, $winner_n^t$ is the n-th element of the best bid in the winner cluster, t is the current iteration, C is a constant number, and r is a random number between 0 and 1 taken from the normal distribution.

4 Experimental Results

In our experiments, we assess the performance of the HMS algorithm compare it to some state-of-the-art algorithms and some recent metaheuristic algorithms. In particular, we benchmark HMS against particle swarm optimisation (PSO) [25], harmony search (HS) [9], shuffled frog-leaping algorithm (SFLA) [8], artificial bee colony algorithm (ABC) [12], imperialist competitive algorithm (ICA) [2], biogeography-based optimisation (BBO) [26], firefly algorithm (FA) [31], grey wolf optimiser (GWO) [17], and whale optimisation algorithm (WOA) [16]. In all experiments, the population size and the number of iterations are set to 30 and 500, respectively. For HMS, the number of clusters is set to 5, while C, M_L, and M_H are set to 2, 2, and 10,

Table 1 Parameter settings for the experiments

Algorithm	Parameter	Value
PSO	Cognitive constant (C_1)	2
	Social constant (C_2)	2
	Inertia constant (w)	1 to 0
HS	harmony memory considering rate	0.9
	Pitch adjusting rate	0.1
SFLA	Number of memeplexes	100
	Number of frogs	30
ABC	Limit	Number onlooker bees $\times N_{var}$
ICA	Number of empires	5
	Coefficient associated with average power	0.1
	Revolution rate	0.2
	Deviation assimilation parameter	$\pi/4$
	Direction assimilation parameter	0.5
BBO	Habitat modification probability	1
	Maximum immigration rate	1
	Maximum emigration rate	1
FA	Light absorption coefficient	1
	Attractiveness at $r = 0$	1
	Scaling factor	0.2
GWO	No parameter	
WOA	Constant defining shape of logarithmic spiral (b)	1

respectively. The parameters used for the other algorithms are given in Table 1. Each algorithm is run 50 times for each problem and we report the mean and standard deviation of the obtained results.

4.1 Unimodal Functions

Unimodal functions have only one global optimum without any local optima and are thus useful to consider the exploitation capability of algorithms. The employed benchmark functions are shown in Table 2, while Fig. 4 depicts their 2-D search spaces.

The results for all algorithms are given in Table 3, which also gives a ranking of the algorithms for each test function. As can be seen, HMS is top ranked for 5 of the 7 functions, indicating its power in solving unimodal functions. Also, from the last row of the table, HMS yielded the first overall rank to tackle unimodal functions, while the second and third ranks went to GWO and SFLA, respectively.

Table 2 Unimodal benchmark functions. D is the number of dimensions, range defines is the boundary of the search space, and f_{min} is the optimum value

Function		D	Range	f_{min}
F1	$\sum_{i=1}^{n} x_i^2$	30	[−100,100]	0
F2	$\sum_{i=1}^{n} |x_i| + \prod_{i=1}^{n} |x_i|$	30	[−10,10]	0
F3	$\sum_{i=1}^{n} (\sum_{j-1}^{i} (x_j^2))$	30	[−100,100]	0
F4	$max_i(|x_i|), 1 \leq i \leq n$	30	[−100,100]	0
F5	$\sum_{i=1}^{n-1} [100(x_{i+1} - x_i^2)^2 + (x_i - 1)^2]$	30	[−30,30]	0
F6	$\sum_{i=1}^{n} ([x_i + 0.5])^2$	30	[−100,100]	0
F7	$\sum_{i=1}^{n} i x_i^4 + random[0, 1]$	30	[−1.28,+1.280]	0

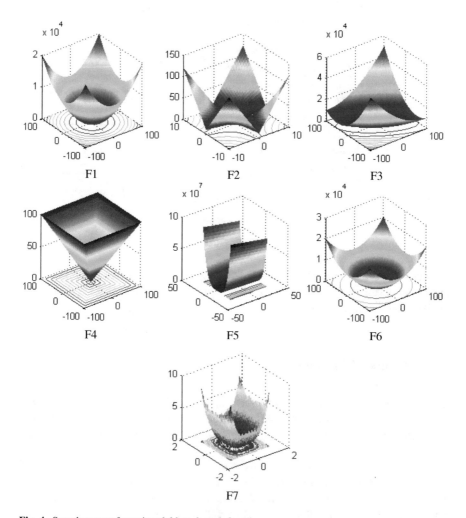

Fig. 4 Search spaces for unimodal benchmark functions

Table 3 Results for all algorithms on unimodal benchmark functions

Function		PSO	HS	SFLA	ABC	ICA	BBO	FA	GWO	WOA	HMS
F1	Avg.	321.8775	291.9645	6.7096E−19	2.7422E04	1.9004E−06	2.4929	0.0055	8.6478E−28	1.3396E−73	0
	Stddev.	96.3709	82.9488	1.9050E−18	3.6898E03	2.2412E−06	0.5285	0.0030	1.0918E−27	4.6770E−73	0
	Rank	9	8	4	10	5	7	6	3	2	1
F2	Avg.	7.8858	3.0447	2.3890E−11	84.2773	1.1933E−04	0.4938	0.5284	1.2648E−16	4.3201E−50	0
	Stddev.	3.2087	0.6654	2.1578E−11	6.7554	8.1609E−05	0.0672	0.2599	1.6947E−16	2.2718E−49	0
	Rank	9	8	4	10	5	6	7	3	2	1
F3	Avg.	2901.9230	21.3541	1.2236	2.7631E04	7.7461	486.7795	8195.2356	1.4881E−04	4.3242E04	0
	Stddev.	1080.425	2.3304	1.4039	5.2132E03	2.4242	486.7795	8195.2356	1.4881E−04	4.3242E04	0
	Rank	7	5	3	9	4	6	8	2	10	1
F4	Avg.	14.0870	21.7955	0.0291	60.5905	8.4321	1.5562	7.1223	7.6278E−07	43.9332	0
	Stddev.	3.2558	2.3199	0.0214	3.3712	3.2616	0.1586	2.5367	5.0380E−08	29.5684	0
	Rank	7	8	3	10	6	4	5	2	9	1
F5	Avg.	1.9504E04	2.4305E04	25.8845	3.3098E07	5.8725	245.1556	84.0091	27.1047	27.9634	28.2242
	Stddev.	2.4883E04	1.2178E04	18.5675	9.0472E06	2.4985	252.8022	32.6139	0.7762	0.4133	0.2476
	Rank	8	9	2	10	1	7	6	3	4	5
F6	Avg.	368.2913	253.8506	7.6109E−19	2.6756E04	7.8532	2.4665	8.0577	0.7865	0.4143	1.4690
	Stddev.	140.6384	87.7723	1.4446E−18	4.8617E03	2.6143	0.5996	0.1428	0.3173	0.1744	1.1588
	Rank	8	8	1	10	6	5	7	3	2	4
F7	Avg.	0.1284	0.2177	0.0097	19.3983	1.1611	0.0141	0.1232	0.0020	0.0033	1.6218E−05
	Stddev.	0.0577	0.0567	0.0054	4.4983	1.1418	0.0045	0.0461	9.7713E−04	0.0034	1.7430E−05
	Rank	7	8	4	10	9	5	6	2	3	1
Average rank		8	7.71	3	9.86	5.14	5.71	6.43	2.57	4.57	2
Overall rank		9	8	3	10	5	6	7	2	4	1

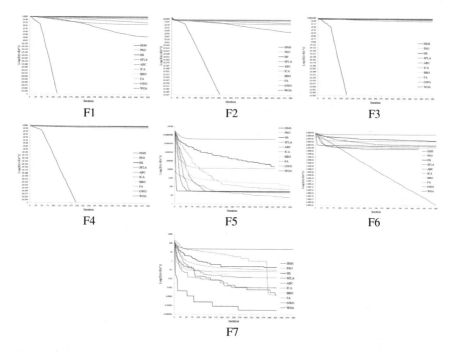

Fig. 5 Convergence curves for unimodal benchmark functions

The convergence curves for the algorithms can be seen in Fig. 5. As can be observed from there, HMS algorithm is converging fastest in most functions.

4.2 Multi-modal Functions

Multi-modal benchmark functions contain several local optima and consequently can be employed to assess the exploration ability of algorithms and how well they are able to escape from local optima. The employed multi-modal benchmark functions are listed in Table 4, while 2-D graphs of the search spaces are shown in Fig. 6.

Table 5 compares the HMS algorithms with others other algorithms on multi-modal benchmark functions. HMS yields the best results for 3 of the 6 functions and ranks second for 2 further functions, clearly demonstrating its high ability in escaping from local optima. Overall, HMS is ranked first overall rank, followed by WOA and FA. Comparing Tables 3 and 5, we can observe that while GWO and SFLA are ranked second and third for solving unimodal functions, they cannot preserve their performance for multi-modal functions.

The convergence curves of the algorithms in Fig. 7 verify the high convergence speed of HMS. For example, for functions such as F9 and F11, the convergence curve decreased steeply for HMS.

Table 4 Multi-modal benchmark functions

Function		D	Range	f_{min}
F8	$\sum_{i=1}^{n} -x_i \sin(\sqrt{\lvert x_i \rvert})$	30	[−500, 500]	−418.9829 × D
F9	$\sum_{i=1}^{n} [x_i^2 - 10\cos(2\pi x_i) + 10]$	30	[−5.12, 5.12]	0
F10	$-20 \exp(-0.2\sqrt{\frac{1}{n}\sum_{i=1}^{n} x_i^2}) - \exp(\frac{1}{n}\sum_{i=1}^{n} \cos(2\pi x_i)) + 20 + e$	30	[−32, 32]	0
F11	$\frac{1}{4000}\sum_{i=1}^{n} x_i - \prod_{1}^{n} \cos(\frac{x_i}{\sqrt{i}}) + 1$	30	[−600, 600]	−4.687
F12	$\frac{\pi}{n}(10\sin(\pi y_1) + \sum_{i=1}^{n-1}(y_i - 1)^2[1 + 10\sin^2(\pi y_i + 1)] + (y_n - 1)) + \sum_{i=1}^{n} u(x_i, 10, 100, 4),$ with $y_i = 1 + \frac{x_i+1}{4}$ and $u(x_i, a, k, m) = \begin{cases} k(x_i - a)^m, & x_i > 0 \\ 0, & -a < x_i < a \\ k(-x_i - a)^m, & x_i < -a \end{cases}$	30	[−50, 50]	−1
F13	$0.1\sin^2(3\pi x_1) + \sum_{i=1}^{n}(x_i - 1)^2[1 + \sin^2(3\pi x_i + 1)] + (x_n - 1)^2[1 + \sin^2(2\pi x_n)] + \sum_{i=1}^{n} u(x_i, 10, 100, 4)$	30	[−50, 50]	−1

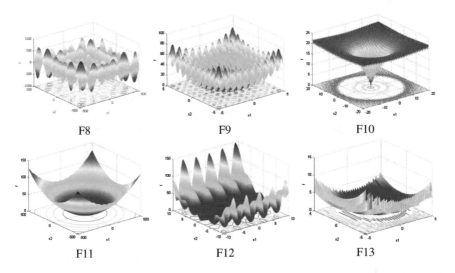

F8   F9   F10

F11   F12   F13

Fig. 6 Search spaces of multi-modal benchmark functions

4.3 Large-Scale Benchmark Functions

Large-scale problems occur in many data science domains and large-scale global optimisation (LSGO), which is dealing with optimisation problems with more than 100 decision variables [14] is a challenging task since as the number of decision variables increases, the search space grows exponentially, making it difficult for optimisation algorithms to work effectively. Algorithms that work well for both normal and LSGO problems are hence highly sought after.

To evaluates HMS for LSGO problems, we increase the number of dimensions from 30 to 200, while employing the same benchmark functions are in Sects. 4.1

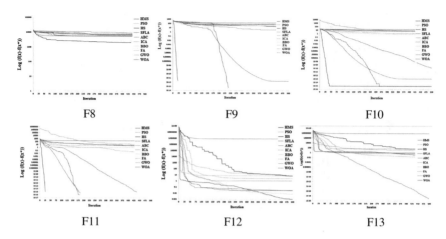

Fig. 7 Convergence curves for multi-modal benchmark functions

and 4.2. Table 6 shows the results for the unimodal function and $D = 200$. As can be observed, HMS yields the best performance for most functions by a high margin. Comparing with Table 3, we can conclude that HMS is even better suited for large-scale unimodal problems in comparison with the other algorithms.

Table 7 compares the HMS algorithm with others on large-scale multi-modal functions. Here, HMS performs best for 3 of the 6 functions and is overall ranked second. While WOA performs best overall here, it did significantly worse on the unimodal functions.

4.4 Rotated and Shifted Benchmark Functions

For a more comprehensive comparison, we further evaluate HMS on a set of 15 CEC-2005 benchmark functions [28] which includes shifted and rotated functions. A summary of the functions is given in Table 8, while 2-D graphs of the functions are shown in Fig. 8.

Table 9 gives the obtained results for all algorithms. HMS performs best for 7 of the 15 functions and second best for a further 5 functions, which further shows the ability of HMS in solving complex optimisation problems.

4.5 Statistical Tests

Non-parametric statistical approaches are divided into two groups, namely, pairwise comparisons and multiple comparisons. Pairwise comparisons evaluate only two

Table 5 Results for all algorithms on multi-modal benchmark functions

Function		PSO	HS	SFLA	ABC	ICA	BBO	FA	GWO	WOA	HMS
F8	Avg.	−5839.7546	−12405.1614	−7119.1248	−9113.9842	−8512.5212	−8099.3154	−8969.5712	−6266.9171	−9045.2153	−9308.3218
	Stddev.	802.4674	64.5581	1475.2159	459.6007	632.2129	576.9913	678.2812	608.7592	1581.2441	600.9386
	Rank	10	1	8	3	9	7	5	9	4	2
F9	Avg.	87.9261	19.3135	68.7188	151.9296	3.2897	50.2997	75.5336	4.2873	0	0
	Stddev.	16.7137	3.1289	19.7114	13.0229	1.8242	14.7106	13.2396	7.8023	0	0
	Rank	9	5	7	10	3	6	8	4	1.5	1.5
F10	Avg.	6.6234	4.9583	0.3039	18.4379	6.0145	0.6089	0.2699	1.0415E-13	3.8488E-15	8.8816E-16
	Stddev.	0.9991	0.5705	0.6440	0.2734	2.6799	0.0909	0.3801	1.4897E-14	2.8119E-15	0
	Rank	9	7	5	10	8	6	4	3	2	1
F11	Avg.	4.1369	3.5463	0.0102	240.8548	11.5789	1.0024	0.0084	0.0028	0.0114	0
	Stddev.	1.2931	0.8018	0.0140	39.8329	3.8413	0.0267	0.0047	0.0074	0.0436	0
	Rank	8	7	4	10	9	6	3	2	5	1
F12	Avg.	9.1955	6.4039	0.3153	184237.4110	6.7856	0.0080	3.2959	0.0480	0.0219	0.0178
	Stddev.	4.3929	1.6343	0.3836	104481.1280	2.8135	0.0037	0.7816	0.0197	0.0135	0.0281
	Rank	9	7	5	10	8	1	6	4	3	2
F13	Avg.	143.5001	220.7775	0.0095	920691.2156	6.4195	0.1236	0.0019	0.6069	0.5574	0.2194
	Stddev.	341.9978	250.9874	0.0201	31057.1275	2.3270	0.0315	7.5304E-04	0.2392	0.1735	0.2599
	Rank	8	9	2	10	7	3	1	6	5	4
Average rank		8.83	6.00	5.17	8.83	6.83	4.83	4.50	4.67	3.42	2.25
Overall rank		9.5	7	6	9.5	8	5	3	4	2	1

Table 6 Results for all algorithms on unimodal benchmark functions for $D = 200$

Function		PSO	HS	SFLA	ABC	ICA	BBO	FA	GWO	WOA	HMS
F1	Avg.	1.5726E04	1.3116E05	4.4589E05	5.0335E05	9.8617E03	1.1125E03	2.3430E04	9.2015E−08	5.3182E−71	0
	Stddev.	1.8113E03	6.5347E03	1.1290E04	9.8706E03	3.2471E03	84.7199	4.2076E03	6.1483E−08	2.5841E−10	0
	Rank	6	8	9	10	5	4	7	3	2	1
F2	Avg.	156.8753	296.3713	5.5107E84	6.2941E25	228.348	39.231	227.3975	3.3703E−05	7.8590E−49	0
	Stddev.	12.3644	9.6628	1.6213E85	3.01999E26	55.1634	2.9843	14.3488	9.0712E−06	2.8757E−48	0
	Rank	5	8	10	9	7	4	6	3	2	1
F3	Avg.	1.9921E05	1.7694E06	1.6958E06	1.7433E06	5.9663E05	2.3123E05	2.8345E08	1.8616E04	4.6119E06	0
	Stddev.	4.9198E04	2.3577E05	1.3892E05	2.9197E05	7.8567E04	3.6379E04	3.7962E07	8.4802E03	1.2566E06	0
	Rank	3	8	6	7	5	4	10	2	9	1
F4	Avg.	33.1202	82.0031	67.7457	98.8834	94.2710	36.5689	72.0320	24.2905	81.7410	0
	Stddev.	2.7025	0.9841	1.0111	0.8226	1.3616	2.0162	5.3872	6.8181	20.9828	0
	Rank	3	8	5	10	9	4	6	2	7	1
F5	Avg.	4.1310E06	3.2875E08	1.5132E09	1.9762E09	8.5196E06	4.4911E04	8.0270E06	198.0339	197.7540	197.0797
	Stddev.	1.1278E06	2.6852E07	6.2308E07	9.5497E07	6.3057E06	6.4151E03	2.7874E06	0.4679	0.1895	0.2143
	Rank	5	8	9	10	7	4	6	3	2	1
F6	Avg.	1.5875E04	1.3089E05	4.4758E05	5.0564E05	9.4243E03	1.1154E03	2.4963E04	29.2361	11.4810	45.7007
	Stddev.	2.1562E03	8.8120E03	8.4940E03	9.4677E03	3.6849E03	78.9891	4.3514E03	1.5876	4.0354	0.4879
	Rank	6	8	9	10	5	4	7	2	1	3
F7	Avg.	12.8865	931.9836	4.7249E03	6.0920E03	46.7428	0.6002	139.1196	0.0184	0.0031	1.3296E−05
	Stddev.	2.6560	82.1684	196.0919	409.1337	39.9407	0.0962	25.6236	0.0073	0.0029	1.3816E−05
	Rank	5	8	9	10	6	4	7	3	2	1
Average rank		4.71	8.00	8.14	9.43	6.29	4.00	7.00	2.57	3.57	1.29
Overall rank		5	8	9	10	6	4	7	2	3	1

Table 7 Results for all algorithms on multi-modal benchmark functions for $D = 200$

Function		PSO	HS	SFLA	ABC	ICA	BBO	FA	GWO	WOA	HMS
F8	Avg.	−2.2858E04	−5.3458E04	−1.1670E04	−1.8499E04	−1.25E04	−4.1833E04	−5.3993E04	−2.9357E04	−7.0312E04	−2.952E04
	Stddev.	2.4079E03	1.2999E03	918.5048	795.8096	2.2361E03	1.8554E03	1.7927E03	2.4428E03	1.1908E04	2.883E03
	Rank	7	3	10	8	9	4	2	6	1	5
F9	Avg.	1.4365E03	1.2649E03	3.0613E03	2.6686E03	629.8474	871.7112	1.2319E03	22.6712	7.5791E−15	0
	Stddev.	70.7638	47.0148	36.9157	70.9750	48.5355	51.0479	82.1383	14.6199	4.1513E−14	0
	Rank	8	7	9	10	4	5	6	3	2	1
F10	Avg.	11.3312	17.7619	20.7198	20.1281	17.8343	4.5262	14.5738	2.2673E−05	4.3225E−15	8.8818E−16
	Stddev.	0.5800	0.1685	0.0362	0.0464	0.9127	0.1128	0.6897	6.2867E−06	2.3756E−15	0
	Rank	5	7	10	9	8	4	6	3	2	1
F11	Avg.	148.8701	1.1755E03	4.0241E03	4.5481E03	94.7528	11.1871	206.8672	0.0028	0	0
	Stddev.	20.2986	49.5853	73.8511	112.0118	32.2501	0.8330	30.1673	0.0110	0	0
	Rank	6	8	10	9	5	4	7	3	2	1
F12	Avg.	4.5756E04	5.0453E08	2.8806E09	4.4458E09	1.6951E07	19.8444	1.4453E05	0.5522	0.0628	1.0309
	Stddev.	5.1036E04	5.8615E07	1.3324E08	2.6288E08	3.7016E07	3.7585	1.8727E05	0.0494	0.0317	0.0285
	Rank	5	8	9	10	7	4	6	2	1	3
F13	Avg.	2.3014E06	1.1824E09	6.1629E09	8.5301E09	2.6582E07	335.7491	7.7316E06	16.9810	6.8141	19.4815
	Stddev.	1.0705E06	1.1889E08	2.8682E08	4.9642E08	2.8040E07	110.9103	4.3347E06	0.6140	2.4383	0.1009
	Rank	5	8	9	10	7	4	6	2	1	3
Average rank		6.00	6.83	9.33	9.50	6.67	4.17	5.50	3.17	1.42	2.42
Overall rank		6	8	9	10	7	4	5	3	1	2

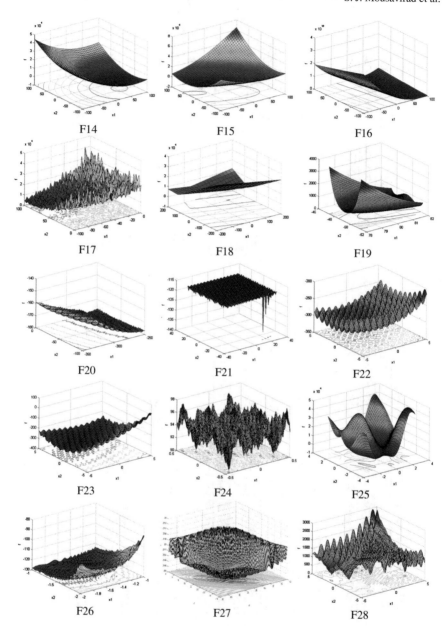

Fig. 8 Search space of CEC-2005 benchmark functions

Table 8 CEC-2005 benchmark functions

Function		D	Range	f_{min}
F14	Shifted Sphere Function	30	[−100,100]	0
F15	Shifted Schwefel's problem 1.2	30	[−10,10]	0
F16	Shifted Rotated High Conditioned Elliptic Function	30	[−100,100]	0
F17	Shifted Schwefel's Problem 1.2 with Noise in Fitness	30	[−100,100]	0
F18	Schwefel's problem 2.6 with Global Optimum on Bounds Basic Functions	30	[−30,30]	0
F19	Shifted Rosenbrock's Function	30	[−100,100]	0
F20	Shifted Rotated Griewank's Function without Bounds	30	[−1.28,+1.280]	0
F21	Shifted Rotated Ackley's with Global Optimum on Bounds	30	[−100,100]	0
F22	Shifted Rastrigin's Function	30	[−100,100]	0
F23	Shifted Rotated Rastrigin's Function	30	[−100,100]	0
F24	Shifted Rotated Weierstrass's Function	30	[−100,100]	0
F25	Schwefel's Problem 2.13	30	[−100,100]	0
F26	Expanded Extended Griewank's + Rosenbrock's (F8F2)	30	[−100,100]	0
F27	Expanded Rotated Extended Scaffe's F6	30	[−100,100]	0
F28	Hybrid Composition Function 1	30	[−100,100]	0

algorithms, whereas multiple comparisons consider more than two algorithms. In this paper, we conduct two well-known statistical tests [6], the Wilcoxon signed rank test (pairwise comparison) and the Friedman test (multiple comparisons), to evaluate the algorithms statistically.

Table 10 gives the results of the Wilcoxon signed rank test. From there, it is clear that HMS significantly outperforms all other algorithms as the obtained p-values are below 0.05 in all cases.

Table 9 Results for all algorithms on on CEC-2005 benchmark functions

Function		PSO	HS	SFLA	ABC	ICA	BBO	FA	GWO	WOA	HMS
F14	Avg.	903.4478	−235.6340	−447.6251	5.2336E04	−450.0000	−447.2722	4.4705E04	1.3410E04	3.1509E03	−450.0000
	Std.	1.2332E03	54.7480	3.9495E−14	7.8579E03	9.4105E−06	0.6148	148.5581	3.5560E03	1.066E03	2.1354E−12
	Rank	6	5	4	10	1.5	3	9	8	7	1.5
F15	Avg.	4.1802E03	2.1442E04	−434.6152	4.9181E04	6.6837E03	463.1837	1.3242E06	2.3535E04	1.0420E05	4.1116E03
	Std.	3.1802E03	6.1949E03	2.6125E−11	8.5048E03	2.9684E03	335.6409	7.8393E04	6.5639E03	2.2595E04	2.2208E03
	Rank	4	6	1	8	5	2	10	7	9	3
F16	Avg.	2.5736E07	1.0655E08	2.6154E07	3.4381E08	2.6214E08	8.8601E06	4.1796E08	5.4708E07	1.5142E08	2.3768E07
	Std.	1.1615E07	4.0750E07	1.2541E07	1.5891E08	9.5621E−06	3.3548E06	2.8676E07	3.6531E07	5.7674E07	1.3493E07
	Rank	3	6	4	9	8	2	10	5	7	1
F17	Avg.	1.1601E04	2.8580E04	2.3145E05	8.0669E04	6.6214E05	1.9975E04	3.4759E06	3.1940E04	2.0044E05	1.2486E04
	Std.	6.0649E03	7.9527E03	6.3256E04	1.0211E04	1.6254E−06	9.5093E03	5.9106E05	1.0160E04	6.7777E04	5.0319E03
	Rank	1	4	8	6	9	3	10	5	7	2
F18	Avg.	7.9969E03	5.6031E03	4.1965E04	2.7923E04	1.2651E04	5.5950E03	2.2592E04	1.1515E04	2.3648E04	5.2124E03
	Std.	2.0394E03	950.3237	2.3146E04	2.1070E03	2.6541E−07	1.3572E03	530.9176	3.2941E03	4.1496E03	2.3239E03
	Rank	4	3	10	9	6	2	7	5	8	1
F19	Avg.	5.1803E07	1.5974E06	6.6519E06	1.6459E10	5.3242E07	2.2963E03	2.2766E10	6.4994E08	2.3799E08	3.5238E06
	Std.	8.4480E07	5.8541E05	2.3654E07	5.3045E09	4.3265E−06	1.8645E03	1.8100E08	9.4421E08	2.0841E08	1.0954E07
	Rank	5	2	4	9	6	1	10	8	7	3
F20	Avg.	2.7967E03	2.5444E03	2.9562E03	3.5756E03	6.6521E04	2.5334E03	4.6183E03	392.4048	2.5398E03	2.5334E03
	Std.	180.6535	5.6160	6.6521	75.7184	2.3254E−06	0.1533	12.9747	209.6337	12.4783	0.0186
	Rank	7	5	8	6	10	2.5	9	1	4	2.5
F21	Avg.	−118.9290	−118.9007	−118.2153	−119.1328	−118.6232	−119.0103	−119.3484	−118.9376	−119.0289	−119.1862
	Std.	0.0754	0.0467	0.0952	0.0697	0.0712	0.0667	0.1018	0.0675	0.0862	0.1737
	Rank	7	8	10	3	9	5	1	6	4	2

(continued)

Table 9 (continued)

Function		PSO	HS	SFLA	ABC	ICA	BBO	FA	GWO	WOA	HMS
F22	Avg.	-165.1436	-294.6861	-112.6214	-108.3558	2.5760E04	-256.1489	-92.4980	-198.3721	-46.5267	-256.9424
	Std.	29.8408	6.8411	25.5241	36.2118	2.6164E-12	22.1705	15.4849	27.3246	56.0462	18.6105
	Rank	5	1	6	7	10	3	8	4	9	2
F23	Avg.	-53.7168	-94.4106	7.7495E04	177.9152	4.6281E04	-202.6800	102.0053	-113.5040	188.4374	-183.7616
	Std.	43.6019	17.4468	271.9707	66.4876	3.2658E-10	28.6949	18.8055	59.8083	72.5813	31.1374
	Rank	5	4	10	7	9	1	6	3	8	2
F24	Avg.	118.6942	132.6930	8.8141E04	124.2766	8.7940E04	118.7584	115.6625	116.2943	129.8792	114.0726
	Std.	3.9264	1.2428	33.4959	1.9378	2.9601E-11	2.6734	2.3523	6.9588	2.2229	4.0225
	Rank	4	8	10	6	9	5	2	3	7	1
F25	Avg.	9.5614E05	4.0175E05	7.4508E05	1.4557E04	3.6401E04	3.4787E04	2.9238E04	1.0303E06	7.4994E05	3.3920E04
	Std.	2.3017E05	8.5910E04	5.6959E05	6.1302E03	1.8570E04	1.8215E04	1.9832E04	4.0719E05	2.2676E05	1.0668E04
	Rank	9	6	7	1	3	5	2	10	8	4
F26	Avg.	-111.3386	-119.1150	-123.3829	-41.1588	-123.9024	-122.8261	-13.2207	-95.7207	-99.9633	-124.6913
	Std.	3.3220	2.0039	1.3633	36.3903	0.6872	1.7160	8.8421	7.5058	6.2787	1.6775
	Rank	6	6	7	9	3	5	2	10	8	4
F27	Avg.	-286.7639	-286.1862	-286.5418	-286.4730	-286.3202	-285.5187	-286.5392	-286.8498	-286.3400	-286.9038
	Std.	0.3497	0.1630	0.3045	0.2351	0.3085	0.0653	0.3948	0.4003	0.3474	0.3453
	Rank	3	9	4	6	8	10	5	2	7	1
F28	Avg.	788.0670	516.8825	531.6254	930.9383	447.9970	552.0418	843.3938	654.4284	909.9902	423.4599
	Std.	101.0248	34.1908	73.26	137.3253	134.6112	73.0665	166.0494	77.3970	196.0838	80.6935
	Rank	7	3	4	10	2	5	8	6	9	1
Average rank		5.1	5	6.2	7.1	6.5	3.57	7.13	5.4	7.13	1.87
Overall rank		4	3	6	8	7	2	9.5	5	9.5	1

Table 10 Results of Wilcoxon signed rank test

HMS vs.	PSO	HS	SFLA	ABC	ICA	BBO	FA	GWO	WOA
p-value	1.4833e–07	3.4701e–05	1.3861e–06	1.2000e–07	1.1845e–07	0.0014	2.4728e–06	1.1566e–04	1.9455e–04

Table 11 Results of Friedman test

Algorithm	PSO	HS	SFLA	ABC	ICA	BBO	FA	GWO	WOA	HMS	p-value
Rank	6.17	6.38	6.27	8.59	6.34	4.27	6.37	4.00	4.74	1.88	9.0799e–25

The results of the Friedman test are shown in Table 11 which gives the average Friedman rank for each algorithm as well as the yielded p-value. HMS obtained the best rank, while the p-value is negligible, indicating a significant difference among the algorithms.

5 Conclusions

Designing an efficient optimisation algorithm for a data science task is a crucial and challenging task. In this chapter, we have presented the Human Mental Search (HMS) algorithm as a new stochastic optimisation algorithm. HMS works in three main stages, mental search, which explores the vicinity of candidate solutions, grouping, which clusters the current population, and movement, which moves candidate solutions towards a promising area in search space identified in grouping. Since many problems in data science are large-scale problem, we have benchmarked the algorithm on both normal and large-scale problems. Our experiments confirm very good optimisation performance in comparison to other metaheuristic algorithms.

References

1. Amirsadri, S., Mousavirad, S.J., Ebrahimpour-Komleh, H.: A levy flight-based grey wolf optimizer combined with back-propagation algorithm for neural network training. Neural Comput. Appl. **30**(12), 3707–3720 (2018)
2. Atashpaz-Gargari, E., Lucas, C.: Imperialist competitive algorithm: an algorithm for optimization inspired by imperialistic competition. In: IEEE Congress on Evolutionary Computation, pp. 4661–4667 (2007)
3. Brooks, S.P., Morgan, B.J.: Optimization using simulated annealing. J. R. Stat. Soc. Ser. D (The Statistician) **44**(2), 241–257 (1995)
4. Carvalho, A.R., Ramos, F.M., Chaves, A.A.: Metaheuristics for the feedforward artificial neural network (ANN) architecture optimization problem. Neural Comput. Appl. **20**(8), 1273–1284 (2011)
5. Das, S., Abraham, A., Konar, A.: Automatic clustering using an improved differential evolution algorithm. IEEE Trans. Syst. Man Cybern. Part A Syst. Hum. **38**(1), 218–237 (2008)

6. Derrac, J., García, S., Molina, D., Herrera, F.: A practical tutorial on the use of nonpara-metric statistical tests as a methodology for comparing evolutionary and swarm intelligence algorithms. Swarm Evolut. Comput. **1**(1), 3–18 (2011)
7. Espinoza-Pérez, S., Rojas-Domınguez, A., Valdez-Pena, S.I., Mancilla-Espinoza, L.E.: Evolutionary training of deep belief networks for handwritten digit recognition. Res. Comput. Sci. **148**, 115–131 (2019)
8. Eusuff, M., Lansey, K., Pasha, F.: Shuffled frog-leaping algorithm: a memetic meta-heuristic for discrete optimization. Eng. Optim. **38**(2), 129–154 (2006)
9. Geem, Z.W., Kim, J.H., Loganathan, G.V.: A new heuristic optimization algorithm: harmony search. Simulation **76**(2), 60–68 (2001)
10. Hancer, E., Xue, B., Zhang, M.: Differential evolution for filter feature selection based on information theory and feature ranking. Knowl.-Based Syst. **140**, 103–119 (2018)
11. Kapanova, K., Dimov, I., Sellier, J.: A genetic approach to automatic neural network architecture optimization. Neural Comput. Appl. **29**(5), 1481–1492 (2018)
12. Karaboga, D., Basturk, B.: A powerful and efficient algorithm for numerical function optimization: artificial bee colony (ABC) algorithm. J. Glob. Optim. **39**(3), 459–471 (2007)
13. Kennedy, J., Eberhart, R.: Particle swarm optimization (PSO). In: IEEE International Conference on Neural Networks, pp. 1942–1948 (1995)
14. Mahdavi, S., Shiri, M.E., Rahnamayan, S.: Metaheuristics in large-scale global continues optimization: a survey. Inform. Sci. **295**, 407–428 (2015)
15. Minija, S.J., Emmanuel, W.S.: Imperialist competitive algorithm-based deep belief network for food recognition and calorie estimation. Evolut. Intell., 1–16 (2019)
16. Mirjalili, S., Lewis, A.: The whale optimization algorithm. Adv. Eng. Softw. **95**, 51–67 (2016)
17. Mirjalili, S., Mirjalili, S.M., Lewis, A.: Grey wolf optimizer. Adv. Eng. Softw. **69**, 46–61 (2014)
18. Mladenović, N., Hansen, P.: Variable neighborhood search. Comput. Oper. Res. **24**(11), 1097–1100 (1997)
19. Mousavirad, S., Ebrahimpour-Komleh, H.: Feature selection using modified imperialist competitive algorithm. In: ICCKE 2013, pp. 400–405. IEEE (2013)
20. Mousavirad, S.J., Bidgoli, A.A., Ebrahimpour-Komleh, H., Schaefer, G.: A memetic imperialist competitive algorithm with chaotic maps for multi-layer neural network training. Int. J. Bio-Inspir. Comput. **14**(4), 227–236 (2019)
21. Mousavirad, S.J., Bidgoli, A.A., Ebrahimpour-Komleh, H., Schaefer, G., Korovin, I.: An effective hybrid approach for optimising the learning process of multi-layer neural networks. In: International Symposium on Neural Networks, pp. 309–317 (2019)
22. Mousavirad, S.J., Ebrahimpour-Komleh, H., Schaefer, G.: Effective image clustering based on human mental search. Appl. Soft Comput. **78**, 209–220 (2019)
23. Radicchi, F., Baronchelli, A.: Evolution of optimal lévy-flight strategies in human mental searches. Phys. Rev. E **85**(6), 061121 (2012)
24. Radicchi, F., Baronchelli, A., Amaral, L.A.: Rationality, irrationality and escalating behavior in lowest unique bid auctions. PloS One **7**(1) (2012)
25. Shi, Y., Eberhart, R.: A modified particle swarm optimizer. In: IEEE International Conference on Evolutionary Computation, pp. 69–73 (1998)
26. Simon, D.: Biogeography-based optimization. IEEE Trans. Evolut. Comput. **12**(6), 702–713 (2008)
27. Stützle, T.: Local search algorithms for combinatorial problems. Darmstadt University of Technology PhD Thesis, vol. 20 (1998)
28. Suganthan, P.N., Hansen, N., Liang, J.J., Deb, K., Chen, Y.P., Auger, A., Tiwari, S.: Problem definitions and evaluation criteria for the CEC 2005 special session on real-parameter optimization. Nanyang Technological University Singapore, Technical report (2005)
29. Sun, R.: Optimization for deep learning: theory and algorithms (2019). arXiv:1912.08957
30. Tran, T.H., Nguyen, H., Nhat-Duc, H., et al.: A success history-based adaptive differential evolution optimized support vector regression for estimating plastic viscosity of fresh concrete. Eng. Comput., 1–14 (2019)
31. Yang, X.S.: Firefly algorithm, stochastic test functions and design optimization (2010). arXiv:1003.1409

Reducing Redundant Association Rules Using Type-2 Fuzzy Logic

Eman Abd El Reheem, Magda M. Madbouly, and Shawkat K. Guirguis

1 Introduction

Besides improving database techniques, a massive count of databases has been created in several fields, led to the search for beneficial information from this mountain of raw data. So, the researchers existed a solution inside the area of DM [1]. DM intends to evoke and explore helpful hidden information in such data. DM is the phase in the knowledge discovery process that attempts to specify new and major patterns in data [2]. It includes several techniques like clustering, classification, ARM, etc. where clustering collects data objects into many clusters so that objects within a group (cluster) have elevated similarity, but dissimilar to objects in other clusters like K-medoids, and K-means. Classification classifies new data based on the training set and the class labels in a classifying attribute. Classification techniques include decision trees, neural networks, and rule-based systems, and ARM is meant to figure out frequent patterns or associations from data sets located in databases like market basket data analysis [3, 4]. In this research, we adopt ARM for finding associations among variables in the database.

The ARs represent an indispensable task in DM to find interesting associations and implicit information amongst a substantial set of data items. The goal of ARs is parsing sales operands and mining patterns in addition to telling the customer's purchase behavior. For illustration, from the transaction data of a supermarket, the rule {cheese, milk} => {eggs} is generated, which represents those customers who buy cheese and milk also buy eggs [5, 6] The apriori algorithm is a famous and

E. A. E. Reheem (✉) · M. M. Madbouly · S. K. Guirguis
Department of Information Technology, Alexandria University, Alexandria, Egypt
e-mail: eman.abdelreheem@alexu.edu.eg

M. M. Madbouly
e-mail: mmadbouly@alexu.edu.eg

© The Editor(s) (if applicable) and The Author(s), under exclusive license   49
to Springer Nature Switzerland AG 2021
A.-E. Hassanien et al. (eds.), *Enabling AI Applications in Data Science*,
Studies in Computational Intelligence 911,
https://doi.org/10.1007/978-3-030-52067-0_3

popular algorithm for ARs. It is implemented for derivation all frequent itemsets from the database and producing the ARs for discovering the knowledge using the pre-defined threshold measures (minsup and minconf) [7].

In usual, the ARM encompasses of two principal parts: First, minsup is calculated to find out whole frequent itemsets in a database by implementing apriori algorithm where takes the itemsets and begets candidate K-itemsets and next check it with the minsup, if candidate K-itemset is fewer than minsup then pruning it, but if it is higher than minsup then it is frequent itemset. Second, from those frequent itemsets, the ARs are produced which meet both the minsup and minconf thresholds [8].

In the literature, frequent itemsets are interesting if its sup value is greater than or equal to the minsup and an ARs are robust if its con value is bigger than or equal to minconf. In the apriori-based mining algorithm, users can recognize the minsup and minconf for their databases. Thus the performance of this algorithm is heavy since it relies on certain user-specified thresholds, For instance, if the user identifies the minsup is very high, nothing might be located in a database and rules are misleading, whereas if the minsup is exceedingly low; might perform to poor mining performance and generating many uninteresting ARs. Therefore, users are unreasonably demanded to know details of the database to be mined, to specify a suitable threshold. Wherefore, gaining every frequent itemsets in a database is challenging since it involves searching all possible itemsets (item combinations). So, the problem can be described as how to gain every AR that meets user-specified minsup and minconf values [7].

A fundamental problem in ARM is presence a tremendous quantity of obtained rules; some of which might be redundant and present no new information so severely hampers the efficient usage of the explored knowledge. Moreover, several extracted rules provide no benefit to the user or can change by other rules, thus deemed redundant rules [9]. To handle problems in ARM, some researchers presented the fuzzy set theory to ARM (called FARs).

Fuzzy system can aid in minimizing the key disadvantage current ARM suffers from depending on fuzzy set concept. FARs is a stretch of the classical ARM by defining sup and conf of the fuzzy rule. It utilized for transforming quantitative data into fuzzy data [9, 10]. So, the FARs has a suitable property in terms of quantization of numerical attributes in database compared with Boolean quantized generalized ARM. Eventually, the mining results of FARs produce linguistic terms rather than intervals, which are easy to understand and nearer to user's mind [11].

The contribution of this article is to prune redundant rules resulting from the apriori algorithm for an actual DM application and to find frequent itemsets by generating minsup and minconf suitable to a database to be mined. The proposed system relies on the using of T2FARM to reduce the redundant rules extracted for an actual application in addition to find out all ARs that satisfy minsup and minconf values. A type-2 fuzzy logic system (T2FLS) is beneficial in cases where it is incapable of determining an exact certainty and measurement uncertainties. It recognized that T2 fuzzy sets permit to model and to reduce uncertainties in rule-based FLS. The results conducted to displays the improvement of redundant rules pruning compared to traditional FARs.

This paper is ordered as follows: Sect. 2 briefly discusses the research related to ARs. The proposed model is introduced in Sect. 3. The experimental result and evaluation of the proposed system are displayed in Sect. 4. Finally, the conclusion and future work are given in Sect. 5.

2 Literature Review

DM became a requisite research area so, there are too many works done in academia and industry on developing new approaches to DM over the last decade. The research in DM still stays high due it portrays problem- rich research area, the richness of practical applications to derive and obtain the useful hidden information from the data, and finding associations among lots of domains in immense relational databases. For instance, in [12], DM techniques can find valuable information to educational systems to improve learning that adopted in the formative evaluation to assistance educators create an educational foundation for decisions when adjusting an environment or teaching approach.

The authors in [13] compared the performing of numerous DM techniques in e-learning systems to foretell the marks that university students will obtain in the final exam of a course. Many classification methods have been used, like decision trees, fuzzy rules, and neural networks. There are three steps intended for the mining process: First, pre-processed to transform the valuable data into KEEL data files. Then, DM algorithms executed to find out hidden knowledge inside the data of concern for the instructor. Finally, post-processing models obtained saved into result files that necessity be understood by the teacher to achieve decisions regarding the students. Their tests show not gets better classification accuracy, and there has not one single algorithm that obtains the best classification accuracy with every datasets.

ARM is a common researched technique for DM. It assists in detecting associations between items in addition to identifying strong rules discovered from databases. This technique benefits users by allowing buying products relying on their preferences. The academics in [14] applied a generalized ARM that employs pruning methods for generalizing rules. This algorithm produces fewer candidate itemsets, and a vast count of rules pruned by minimal confidence. This study suffers from utilizing original frequent and ARs as opposed to rescanning the database. However, it shows better than another algorithm, where it can prune a tremendous count of rules.

Another relates work in [15] DM algorithm presented to find out positive and negative ARs. It studies the methods of sup and conf level, the explanation of the positive and negative ARs, and explains the conflicted rules problems which are in the positive and negative ARM and the solutions of these conflicting rules problems. This study can apply in various applications to find robust patterns and generates all varieties of limited rules.

Some works that employed FLS for processing the uncertain information in ARs. For illustration, in [16] a fuzzy rule algorithm depend on fuzzy decision tree is provided for DM, which integrates the comprehensibility of rules produced relies on decision tree and fuzzy sets. First, they implement histogram analysis to specify membership functions for each attribute of data. Then they build a fuzzy decision tree. The authors also implement the genetic algorithm to improve the initial membership functions. Their article is effective in the performance and comprehensibility of rules compared with other methods.

In [3] submitted the fuzzy transaction as a fuzzy subset of items to find out FARs in relational databases that contain quantitative data. This model operates in distinct mining varieties of patterns, from regular ARs to fuzzy and approximates functional dependencies and gradual rules. The academic in [11] presented FARs to derive rules from the database in addition to prune the redundant rules extracted. They define redundancy of FARs and show theorems concerning the redundancy of FARs. But their algorithm shows limitation in terms of computational time, and non-redundant rules are unexpectedly deleted. So, the authors see that this algorithm must enhance by implementing the other method.

The academics in [17] generated rare ARs from an educational database by applying the fuzzy-based apriori algorithm. This method is utilized to less frequent itemsets by implementing measure 'maximum sup' for providing rare items and to prune rare items and measure 'Rank' is utilized to prune the particular outliers from the rare items created. The ARs can produce after the rare items produced.

Aim of the Work Due to the significant challenge for the ARM algorithm, which is the enormous number of the extracted rules, could be redundant; the traditional fuzzy algorithm can meet some difficulties. From this point, we intend to design efficient ARs to prune redundant rules extracted from the apriori algorithm for DM application and detect frequent itemsets by applying the T2FLS. Results displayed that such a techniques could be worked effectively in DM, minimizing the impacts of typical drawbacks of the FLS that is less able to pruning redundant rules.

3 Proposed System

The proposed model mainly focuses on every previously mentioned problem. This model purposes to reduce the number of redundant rules to maximize the accuracy of the results that happened by combines an apriori algorithm with T2FLS to prune redundant rules for a real DM application. The schema of the proposed system is presented in Fig. 1.

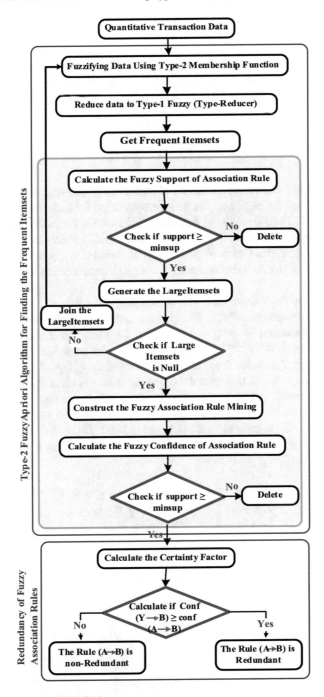

Fig. 1 The proposed of T2FARM

3.1 Quantitative Transaction Data

We apply the dataset, popularly known as "Adult" data from the UCI machine learning repository. The Adult data set contains 32,659 records. 60% of the data are used for training, and 40% of data are used for testing.

3.2 Fuzzifying Association Rules Mining

T2FLS model and minimizes the effects of uncertainties. A fuzzy membership function characterizes T2 FLS, i.e., the membership value for each element of this set is FS in [0, 1]. The membership functions (MF) of T2FLS is three dimensional: upper membership functions (UMF), lower membership functions (LMF) and the area between these two functions is footprint of uncertainty (FOU), that give additional degrees of freedom that make it feasible to handle uncertainties (see Fig. 2) [18, 19].

Type-2 Fuzzy set is concerned with quantifying using natural language in which words can have vague meanings. We operated to transforms the quantitative values v_{ij} of each transaction $T_i = (i = 1, 2, \ldots, n)$ for each item I_j into fuzzy values UMF f_{ijl}^{upper} and LMF f_{ijl}^{lower} by using type-2 MF for each R_{jl}, where R_{jl} is the l-th fuzzy region of item I_j. These items are fuzzified using type-2 Triangular Fuzzy Number (TFN), provided to decide the degree of membership of items in the apriori algorithm as exposed in Fig. 3. For example, if attribute age in database takes values from [10, 70], then it could be partitioned into four new attributes such as young [10, 30], youth [20, 40], middle age [30, 50], and old [50, 70]. In this work, we utilize type-2 Triangular MF to describe all variables (age, income, and education degree) that is defined as [20]:

$$LMF = \begin{cases} h(x+a)/a, if -a \leq x \leq 0 \\ h(a-x)/a, if\ 0 \leq x \leq a \\ 0, otherwise \end{cases} \quad (1)$$

Fig. 2 UMF and LMF representing the FOU

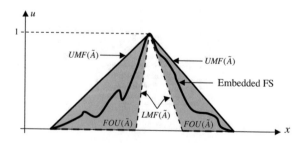

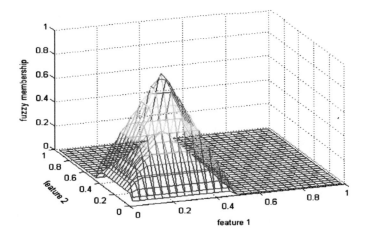

Fig. 3 Triangular T2FMF

Table 1 Age linguistic set and TFN

No.	Age linguistic term	TFN
1	Young	(10, 20, 30)
2	Youth	(20, 30, 40)
3	middle age	(30, 40, 50)
4	Old	(50, 60, 70)

$$UMF = \begin{cases} h(x+b)/b, if -b \leq x \leq 0 \\ h(b-x)/b, if 0 \leq x \leq b \\ 0, otherwise \end{cases} \tag{2}$$

where, $0 \leq a \leq b$ and $0 \leq h \leq 1$. All of the MF' parameters are numerically identified rely on the experiences. In our case, four linguistic sets used for variable "age", four linguistic sets for variable "income", and four linguistic sets for variable "education degree" are provided in Table 1, Table 2, and Table 3 respectively.

Table 2 Income linguistic set and TFN

No.	Income linguistic term	TFN
1	Low	(0, 10, 30)
2	Medium	(10, 30, 50)
3	High	(30, 50, 70)
4	Very high	(50, 70, 90)

Table 3 Education degree linguistic set and TFN

No.	Education_degree linguistic term	TFN
1	High school	(0, 5, 10)
2	Bachelor	(5, 10, 15)
3	Master	(10, 15, 20)
4	PhD	(15, 20, 25)

3.3   Type-Reducer

The type-reducer produces a type-1 fuzzy set output by center of sets (COS) type reduction. To get the type-reducer set, it is enough to reduce its UMF f_{ijl}^{upper} and LMF f_{ijl}^{lower} to fuzzy value f_{ijl}.

3.4   Apriori Algorithm for Finding the Frequent Itemsets

In this stage, after reduce type-2 fuzzy values to type-1 fuzzy values by the center of sets type-reduction method, the apriori DM is related to finding frequent itemsets and discovers interesting ARs. The apriori algorithm requires two thresholds: minsup and minconf. These two thresholds identify the association that must hold before the rule will be mined. The steps of the apriori algorithm are given as follows:

3.4.1   Get Frequent Itemsets

The objective of this stage is to find all the itemsets that meet the minsup threshold (these itemsets are called frequent itemsets). An itemset with sup higher than or equal the minsup is a frequent itemset.

To produce the frequent itemsets is as follows

1. Compute the *supp* of ARs for each R_{jl} in the transactions. Thus $supp_{jl}$ is the sum of f_{ijl} in transactions.
2. Check if $supp_{jl} \geq$ minsup, If the value of $supp_{jl}$ is equal or larger than the minsup value, then this $supp_{jl}$ is a frequent and put it in the large 1-itemsets (L_1). If L_1 is not null, then do the next step..
3. Set $r = 1$, where r is used to represent the number of items in the current itemsets to be processed.
4. Join the large r-itemsets Lr to generate the candidate $(r + 1)$ itemsets
5. Do the following substeps for each newly formed $(r + 1)$ itemset

   a. Compute f_{ijl}^{upper} and f_{ijl}^{lower} by using type-2 MF.
   b. Reduce each f_{ijl}^{upper} and f_{ijl}^{lower} of each item I_j to fuzzy value.

c. Compute fuzzy value of each item by using the minimum operator for the intersection.

d. Compute the *supp* of ARs and then Check if $supp_{jl} \geq$ minsup, If the value of $supp_{jl}$ is equal or larger than the minsup value, then put it in $Lr + 1$, if $Lr + 1$ is null then do the next step; otherwise, set $r = r + 1$ and repeat STEPs 4–5.

3.4.2 Construct the FARMs

After getting on fuzzy frequent itemsets, we capable of generates FARs. For instance, some FARs is expressed as the next form:

- IF age is young, THEN income is high.
- IF age is youth, THEN income is low.
- IF income is high, THEN age is young.
- IF age is youth, and income is low THEN education degree is high school.

Where *age, income,* and *education degree* are sets of attribute and *young, youth, low, high school, and high* are linguistic labels.

3.4.3 Calculate the Fuzzy Confidence of ARs

By getting to ARs, the Conf of a rule A→B is defined as:

$$Conf_{(A \to B)} = \frac{Supp_{fuzzy}(A \cup B)}{Supp_{fuzzy}(A)} \qquad (3)$$

where $\left(Conf_{(A \to B)}\right)$ the percentage of the fuzzy number of transactions that contain $A \cup B$ to the total fuzzy number of transactions that contain A.

Then, we compare the conf of a rule with the minconf. The rule is satisfied when its conf is higher than or equals the minconf.

3.5 Redundancy of FARs

ARM finds an association among items based on the sup and the conf of the rule. However, these measures have many limitations as the database grows larger and larger. Thus the mined rules increase faster, and many redundant rules are extracted, and it becomes impossible for humans to find interesting rules. For this reason, in this paper, we use the certainty factor (CF) to measure the redundancy of extracted FARs. CF was developed to measure the uncertainty in the ARs. The value of the certainty factor is a number from -1 to 1. The CF of the rule $X \to Y$ is described as follows:

$$CF(X \to Y) = \frac{Conf(X \to Y) - supp(Y)}{1 - supp(Y)}, if Conf(X \to Y) > supp(Y) \quad (4)$$

$$CF(X \to Y) = \frac{Conf(X \to Y) - supp(Y)}{supp(Y)}, if Conf(X \to Y) \leq supp(Y) \quad (5)$$

$$CF(X \to Y) = 1, if supp(Y) = 1 \quad (6)$$

$$CF(X \to Y) = -1, if supp(Y) = 0 \quad (7)$$

Then now, we can remove the redundant FARs by assuming some theorems that consider the consequent part of rules is identical for redundancy theorem. The theorems are represented as follows:

- **Theorem 1**: A → B and A′ → B be two FARs:
  IF $Conf(A \to B) \geq Conf(A' \to B)$, **THEN** $Cf(A \to B) \geq Cf(A' \to B)$

  - Ex: **IF** *Conf(age is young → income is low) ≥ Conf(education-degree is high school → income is low)*
  - **THEN** *Cf (age is young → income is low) ≥ Cf (education-degree is high school → income is low)*

- **Theorem 2**: Combine FARs: we consider A → C, B → C and A, B → C be two FARs, where *A, B,* and *C* are fuzzy itemsets.
  IF max $(Conf(A \to C), Conf(B \to C) \geq Conf(A, B \to C))$
  THEN max $(Cf(A \to C), Cf(B \to C)) \geq Cf(A, B \to C)$

  - Ex: **IF max** *(Conf (age is young → income is low), Conf (education degree is high school → income is low)) ≥ Conf (age is young & education degree is high school → income is low)*
  - **THEN max** *(Cf (age is young → income is low), Cf (education degree is high school → income is low)) ≥ Cf (age is young & education degree is high school → income is low)*

- **Theorem 3**: We consider A → B FAR, where *A* and *B* are fuzzy itemsets. Let fuzzy itemset family *Q,*
  IF $\max_{Y \in Q}(Conf(Y \to B)) \geq Conf(A \to B)$,
  THEN $\max_{Y \in Q}(Cf(Y \to B)) \geq Cf(A \to B)$

  - Ex: **IF max** *(Conf (age is young & education degree is high school), **or** (age is youth & education degree is master) → income is low) ≥ Conf (age is young → income is low)*
  - **THEN max** *(Cf (age is young & education degree is high school), **or** (age is youth & education degree is master) → income is low) ≥ Cf (age is young → income is low)*

From these theorems, we now able to extract redundant rule and non-redundant of FARs in terms of a joining of FARs to prune redundant rules by considering $A \to B$ FAR, where A and B are fuzzy itemsets based on $Q = 2^x - X - \emptyset$

- **IF** $\max_{Y \in Q}(Conf(Y \to B)) \geq Conf(A \to B)$,
  THEN the rule $A \to B$ is a redundant rule.
- **IF** $\max_{Y \in Q}(Conf(Y \to B)) < Conf(A \to B)$,
  THEN the rule $A \to B$ is non-redundant rule.

Finally, The CF value of the non-redundant rules is larger than the CF of the corresponding redundant rule. So, the association between the antecedent and consequent of the non-redundant rule is stronger than any corresponding redundant rule.

4 Experimental Results

In this section, we do some experiments that have been carried out to check the performance of the proposed system and assert improvements over the traditional approach. A program is implemented using MATLAB to assess the performance of the proposed system. We used "Adult" database from the UCI machine learning.

In the first experiment, we display the capacity of the proposed technique in eliminating redundant rules compared to traditional FARs by set minsup as 0.5 and set minconf (0.2, 0.7, and 0.9). The number of the extracted rules dependent on the minconf where the number of association rules increased when reduces the minconf (see Table 4). The results presented that the proposed technique is able to delete redundant rules better than the traditional FARs.

Figure 4 and Table 5 present the rule all extracted from proposed system and number of deleted redundant rules by the proposed system. Table 6 shows the performing the proposed system with traditional fuzzy in terms of pruning redundant ARs. Results demonstrated that the proposed T2FARM has good performance and more accurate to prune redundant rule than a traditional FARs (see Fig. 5).

Table 4 Capacity of the proposed technique in eliminating redundant rules compared to traditional FARs

	Minconf	No. of extracted rules	No. of non-redundant rules
Fuzzy-approach	0.2	248	240
	0.7	238	225
	0.9	230	190
Proposed system	0.2	250	180
	0.7	220	130
	0.9	200	100

Fig. 4 The extracted rules from the proposed system

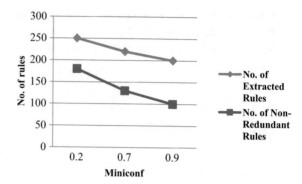

Table 5 No. of extracted rules from T2FARM

Minconf	No. of extracted rules	No. of non-redundant rules
0.2	250	180
0.7	220	130
0.9	200	100

Table 6 No. of non-redundant rules

Minconf	Fuzzy-approach	Proposed system
0.2	240	180
0.7	225	130
0.9	190	100

Fig. 5 Number of non-redundant ARs

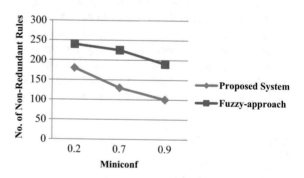

In the second experimental, the execution time is used to compute time for the proposed system and traditional FAR.as displayed in Table 7. The proposed system is corresponds to highest time. This is because the computations of T2FLS are highly complicated computations.

Table 7 Computation time

Association rule system approaches	Time in Sec.	Time in Sec.
Fuzzy-approach	0.099	0.070
Proposed system	1.194	1.024

5 Conclusion

In this paper, we suggested a novel approach for extracting hidden information from items. The submitted approach is based on the hybridization between ARs, and T2FLS to recognize frequent itemsets that fulfill minsup and minconf values in addition to minimize the redundant rules mined from the apriori algorithm.

The experimental results demonstrate that a proposed approach gives better performance during compared with traditional FAR in all the sensitive parameters. Our results on a real-world data set confirm that the proposed approach has excellent potential in terms of accuracy. Our future work includes integrating genetic algorithm with T2FLS to develop the prune redundant rule.

References

1. Shu, J., Tsang, E.: Mining fuzzy association rules with weighted items. In: 2000 IEEE International Conference on Systems, Man, and Cybernetics, pp. 1906–1911 (2000)
2. Deepashri, K.S., Kamath, A.: Survey on techniques of data mining and its applications. Int. J. Emerg. Res. Manag. Technol. 6(2), 198–201 (2017)
3. Delgado, M., Marín, N., Sánchez, D., Vila, M.: Fuzzy association rules: general model and applications. IEEE Trans. Fuzzy Syst. 11–2 (2003)
4. Zhao, Y., Zhang, T.: Discovery of temporal association rules in multivariate time series. In: International Conference on Mathematics, Modeling and Simulation Technologies and Applications (MMSTA 2017), pp. 294–300 (2017)
5. Helm, B.L.: Fuzzy Association Rules an Implementation in R, Master's thesis, Vienna University of Economics and Business Administration (2007)
6. Darwish, S.M., Amer, A.A., Taktak, S.G.: A novel approach for discovery quantitative fuzzy multi-level association rules mining using genetic algorithm. Int. J. Adv. Res. Artif. Intell. (IJARAI) 5(6), 35–44 (2016)
7. Sowan, B., Dahal, K., Hossain, M.A., Zhang, L., Spencer, L.: Fuzzy association rule mining approaches for enhancing prediction performance. Expert Syst. Appl. 40(17), 6928–6937 (2013)
8. Suganya, G., Paulraj Ananth, K.J.: Analysis of association rule mining algorithms to generate frequent itemset. Int. J. Innov. Res. Sci. Eng. Technol. 6(8), 15571–15577 (2017)
9. Xu, Y., Li, Y., Shaw, G.: Concise representations for approximate association rules. In: IEEE International Conference on Systems, Man and Cybernetics, Singapore, pp. 94–101 (2008)
10. Watanabe, T.: Mining fuzzy association rules of specified output field. In: IEEE International Conference on Systems, Man and Cybemetics, pp. 5754–5759 (2004)
11. Watanabe, T.: Fuzzy association rules mining algorithm based on output specification and redundancy of rules. In: IEEE International Conference on Systems, Man and Cybemetics, pp. 283–289 (2011)

12. Romero, C., Ventura, S.: Educational data mining: a survey from 1995 to 2005. Expert Syst. Appl. **33**(1), 135–146 (2007)
13. Romero, C., Espejo, P.G., Zafra, A., Romero, J., Ventura, S.: Web usage mining for predicting final marks of students that use moodle courses. J. Comput. Appl. Eng. Educ. **27**(3), 135–146 (2013)
14. Huang, Y., Wu, C.H.: Mining generalized association rules using pruning techniques. In: IEEE International Conference on Data Mining, pp. 227–234 (2002)
15. Antonie, L., Zaïane, O.R.: Mining positive and negative association rules: an approach for confined rules. In: 8th European Conference on Principles and Practice of Knowledge Discovery in Databases Cite this publication, pp. 1–13 (2004)
16. Kim, M.W., Lee, J.G., Min, C.H.: Efficient fuzzy rule generation based on fuzzy decision tree for data mining. In: IEEE Conference on International Fuzzy Systems, South Korea (1999)
17. Rajeswari, A.M., Sridevi, M. Chelliah, C.D.: Outliers detection on educational data using fuzzy association rule mining. In: Proceedings of International Conference on Advanced in Computer Communication and Information Science (ACCIS-14), pp. 1–9 (2014)
18. Chen, Y., Yao, L.: Robust type-2 fuzzy control of an automatic guided vehicle for wall-following. In: International Conference of Soft Computing and Pattern Recognition, pp. 172–177 (2009)
19. Darwish, S.M., Mabrouk, T.F., Mokhtar, Y.F.: Enriching vague queries by type-2 fuzzy orderings. Lecture Notes on Information Theory, vol. 2(2), pp. 177–185 (2014)
20. Jerry, M.: Type-2 fuzzy sets and systems: an overview. IEEE Comput. Intell. Mag. **2**(1), 20–29 (2007)

Identifiability of Discrete Concentration Graphical Models with a Latent Variable

Mohamed Sharaf

1 Introduction

This work is inspired by the work of Stanghellini and Vantaggi's criteria for identifiability in undirected graphical models presented in [20]. However our paper presents a detailed algorithmic prespective along with an implementation of the algorithm in MATLAB, see [19] for more details. Also, our paper is a self-contained presentation of the algorithm upon which we built iugm. Latent models [5] have been used in several fields, e.g. econometrics, statistics, epidemiology, psychology, sociology, etc. In these models, there exist hidden variables which are not observed and could have an effect (sometimes causal) on the observed variables. We consider models in which there is a single hidden variable with no parents, the hidden variable is the parent of all the other variables, and all variables are binary, i.e. have exactly two states or levels. While the probability distribution of the observed variables is known, it is also important to assess (learn) the probability distribution of all variables including the hidden one. Within the scope of this paper, we call this problem the identifiability problem. (Note that this is different from the problem of identifiability of causal effects, which is the focus of, for example, [9, 17]. Identifiability comes in different forms: global, local, and generic. Quoting from [4].

Definition 1 A parameterization is called *identifiable* if it is one-to-one. That is, let ξ_1 and ξ_2 be two parameter values with their corresponding distributions P_{ξ_1} and P_{ξ_2}, then $\xi_1 \neq \xi_1$ implies $P_{\xi_1} \neq P_{\xi_2}$

In a globally identifiable model [7], for every pair of instantiations of the parameter vector θ, θ_a and θ_b, the distributions of the observable variables, $P(\theta_a)$ and $P(\theta_b)$

M. Sharaf (✉)
Computer and Information Sciences, Jouf University, Sakaka, Saudi Arabia

Computer and Systems Dept, Al-Azhar University, Cairo, Egypt
e-mail: msharaf@azhar.edu.eg

© The Editor(s) (if applicable) and The Author(s), under exclusive license
to Springer Nature Switzerland AG 2021
A.-E. Hassanien et al. (eds.), *Enabling AI Applications in Data Science*,
Studies in Computational Intelligence 911,
https://doi.org/10.1007/978-3-030-52067-0_4

are the same iff $\theta_a = \theta_b$ (e.g., no two distinct sets of parameter values yield the same distribution of the data). This implies that, once the parameter space is defined, we cannot find two different vectors of parameter values that lead to the same moments of the distribution of the observed variables, i.e., the parameter values are uniquely identified by the distribution of the observed variables. This one-to-one mapping holds in the whole domain.

In contrast, *local identification* is a weaker form of identification than *global identification*. In order to define local identifiability, the concept of neighborhood of parameter vectors is introduced. Conceptually, in locally identifable models the uniqueness property is preserved only in a neighborhood of parameter vectors. Given the parameter vector θ, the model is locally identifiable if for a vector of values θ_a, there is no vector θ_b in the neighborhood of θ_a, such that $P(\theta_a) = P(\theta_b)$ unless $\theta_a = \theta_b$. Therefore, global identifiability implies local identifiability but not vice versa. *Generic identifiability* is weaker than local identifiabilty, because there may exist a set of null measure of non-identifiable parameters. Generically identifiable models are the weakest form of identifiable models and are introduced in [2].

2 Undirected Graph and Markov Properties

A graph $G = (V, E)$ is defined by a set of vertices V (*nodes*) or variables and a set of edges $E \subseteq V \times V$. Since we are dealing with undirected graphical models, we have only one type of edges $(-)$. The complement of a graph $G = (V, E)$ is the graph $\hat{G} = (V, \hat{E})$, which consists of the same vertex set V of G and the set $\hat{E}$ defined as: $\forall \{u, v\} \in \hat{E} \iff \{u, v\} \notin E$, where u and v are distinct. Given a graph $G = (V, E)$ and $U \subset V$. Let $E_U = \{(v_1 v_2) \in E; v_1, v_2 \in U\}$. The graph $G_U = (U, E_U)$ is called subgraph of G induced by U. A clique (complete set) C is a subset of the vertex set $C \subseteq V$, s.t. if $a, b \in C \Rightarrow \{a, b\} \in E$. C is a maximal clique (maximal complete set) if $(\forall C, \exists C'$ and $C \subset C' \subset V) \Longrightarrow (C = C')$.

A cycle of length n, in a graph G, has starting point v_1 and endpoint v_2, s.t. $v_1 = v_2$ and number of edges equal n. A graph $G = (V, E)$ without cycle of length ≥ 3 is called a forest. If the forest is connected, we call it a tree. The tree is rooted (V, E, w) if $w \in V$ and w is the root. A binary rooted tree is a tree where each vertex has either 0 or two children.

Let $G = (V, E)$ be a graph and let A, B, C be disjoint subsets of V. One can say that B separates A and C or B is a separator of (A, C) if for any path $P \subset V$ with end points $x \in A$ and $y \in C$ there exists a point z, s.t. $z \in B \cap P$.

A *clique tree* for a graph $G = (V, E)$ is a tree $T = (V_T, E_T)$ where V_T is a set of cliques of G that contains all maximal cliques of G. A *junction tree* for a graph G is a clique tree for G that satisfies the following condition. For any cliques C_1 and C_2 in the tree, every clique on the path connecting C_1 and C_2 contains $C_1 \cap C_2$.

A graph is *chordal* (also called *triangulated*) if it contains no chordless cycles of length greater than 3. A graph is *complete* if E contains all pairs of distinct elements of V. A vertex is *simplicial* in a graph if its neighbors form a complete subgraph.

A graph is *recursively simplicial* if it contains a simplicial vertex v and when v is removed the subgraph that remains is recursively simplicial.

Let N and U stand for the sets of observable (observed) and unobservable (hidden) variables in graph G, i.e., N and U partition V. An undirected graph G in the iugm package may be represented either in adjacency matrix of dimension ($|V| \times |V|$), sparse matrix, or incidence matrix.

Undirected graphical models are also called Markov random fields (sometimes called Markov networks), where each node represents a random variable, and the graph factorizes according to its maximal cliques. In the next section we discuss the Markov properties.

2.1 Markov Properties

Conditional independence relations are encoded in a graph by Markov properties [12, 16]. Therefore, it is the conditional independence relations that give graphical models their modular structure, or in other words, graphical models are representations of the conditional independence relations. Let A be a subset of V and let $\chi_A = \times_{\alpha \in A} \chi_\alpha$ and further $\chi = \chi_V$. Typical elements of χ_A are denoted as $x_A = (x_\alpha)_{\alpha \in A}$. Similarly, $X_A = (X_\alpha)_{\alpha \in A}$.

Properties of conditional independence
Conditional independence $X \perp Y | Z$ has the following properties.

- if $X \perp Y | Z$ then $Y \perp X | Z$ (*symmetry*),
- if $X \perp Y | Z$ and U=g(x) then $U \perp Y | Z$ (*decomposition*),
- if $X \perp Y | Z$ and U=g(x) then $X \perp Y | (Z, U)$ (*weak union*),
- if $X \perp Y | Z$ and $X \perp W | (Y, Z)$ then $X \perp (Y, W) | Z$ (*contraction*),
- Under some regularity conditions if $X \perp Y | (Z, W)$ and $X \perp Z | (Y, W)$ then $X \perp (Y, Z) | W$ (*intersection*).

Pairwise Markov property
Non-adjacent vertices A and B (indicated by $A \nsim B$) in an undirected graph G are conditionally independent given the set $V \setminus \{A, B\}$, i.e.,

$$A \nsim B \implies A \perp B | V \setminus \{A, B\} \tag{1}$$

Global Markov property
For disjoint vertex subsets A, B, and C in the graph G such that C separates A and B, A is conditionally independent on B given C

$$A \perp B \mid C, \tag{2}$$

where the conditional dependence of A and B give C is defined as:

$$X_A \perp X_B \mid X_C \tag{3}$$

Local Markov property
Every node i is independent on all remaining nodes given i' s immediate neighbors, N_i, i.e.,

$$i \perp V \backslash \{i, N_i\} \mid N_i \tag{4}$$

It follows form the definition that pairwise Markov property (PM) is weaker than local Markov property (LM) which is weaker than the global Markov property (GM). Syntactically, $GM \implies LM \implies PM$.

2.2 Graph Decomposition

Decomposable graph
A graph $G = (V, E)$ is decomposable if either:

1. G is complete, or
2. V can be expressed as v = $A \cup B \cup C$ where

   a. A, B and C are disjoint,
   b. A and C are non-empty,
   c. B is complete,
   d. B separates A and C in G, and
   e. $A \cup B$ and $B \cup C$ are decomposable.

So, graph G is said to be decomposable if it is complete or admits a proper decomposition into its decomposable subgraphs. Any decomposable graph can be recursively decomposed into its maximal prime subgraphs. In other words, a graph is said to be *prime* if no proper decomposition exists. Subgraphs that are not decomposable anymore are called *maximal prime subgraphs* [12].

Lauritzen [12] shows that a graph $G = (V, E)$ is decomposable if there exists a binary rooted tree, called a decomposition tree, whose vertices are labelled by some non empty subsets of V with the following properties:

1. The root is labeled by the set of vertices $\in V$,
2. Whenever U labels a node of the tree, then U is complete if and only if the node has no children,
3. If U labels a node of the tree with two children labelled U_1 and U_2 then

   a. There exists B, which is equal to $U_1 \cap U_2$ and B is complete,
   b. There exist A and C, s.t. $A = U_1 \backslash U_2$, and $C = U_2 \backslash U_1$ are not empty,
   c. B separates A and C.

Theorem 1 *According to Lauritzen and Jensen [10, 12], the following properties of G are equivalent.*

1. *G is chordal.*
2. *G is decomposable.*
3. *G is recursively simplicial.*
4. *G has a junction tree.*
5. $A \cup B$ *and* $B \cup C$ *are decomposable.*

Factorization and decomposable graph
The unique prime components resulting from recursive decomposition of a graph G yield a probability distribution that factors with respect to G. This probability distribution is written as the product of factors, one for each prime component, i.e., maximal clique $C \in G$:

$$P_X = P(X_1, X_2, ..., X_N) = K \prod_{C \in G} \phi_C(X_C), \tag{5}$$

where K is a normalization constant.

Note that we do not distinguish nodes in G from variables in P_X. Let I_P be the conditional independence relation extracted from the probability distribution, P_X, and I_G be the conditional independence extracted from G using the global Markov property. If $I_G \subseteq I_p$ then we say that G is an independence map of (I-map) of P_X. P_X is said to be faithful to G if $I_G = I_P$; G is then also said to be a perfect map of P_X, see [13, 15] for more details.

Theorem 2 *According to Jordan [11], for any undirected graph* $G = (V, E)$*, let* P_X *be the distribution that factors over that graph as in Eq. 4. Then,* P_X *satisfies the global Markov property. In other words G is an I-Map for* P_X*. Let F denote the property that* P_X *factorizes w.r.t. G and let GM, LM, and PM denote global, local and pairwise Markov properties (w.r.t. the graph G) respectively.*

Therefore, it holds that: $F \implies GM \implies LM \implies PM$. Moreover, Theorem 3 indicates that any distribution that respects the PM property factorizes according to Eq. 4.

Theorem 3 *According to Besag [3]; attributed to Hammersley and Clifford) If* $P_X >$ 0 *is a positive probability distribution, and the pairwise Markov property holds for* $G = (X, E)$ *and* P_X*, then* P_X *factorizes over G.*

Therefore, we can conclude that in the case of positive distributions all Markov properties coincide: $F \iff GM \iff LM \iff PM$.

Theorem 4 *According to Meek [13], the set of multinomial and multivariate normal distributions which are not faithful to their I-map (graph G) has measure zero, i.e., the set of parameters of such distributions has Lebesgue measure zero.*

The above theorem can be generalized to different graphical models. For example, all positive discrete probability distributions that factorize according to a chain graph are faithful, as shown in [18].

2.3 Log-Linear Models

The log-linear model is a special case of the generalized linear models (GLM). For variables that have Poisson distribution, the analysis of contingency tables [1] is one of the well-known applications of log-linear models. The conditional relationship between two or more discrete categorical variables is analyzed by taking the natural logarithm of the cell frequencies within a contingency table. For a two-variable model, this is shown in the following equation:

$$log(F_{jk}) = \mu + \lambda_j^{X_1} + \lambda_k^{X_2} + \lambda_{jk}^{X_1 X_2} \tag{6}$$

The symbols of Eq. 6 are defined as follows:
$log(F_{jk})$: the *log* of the expected cell frequency, i.e., cell (j, k) in Eq. 6,
μ: the overall mean of the natural log of the expected frequencies,
λ: the term that represents the effect of a specific variable(s) on the given cell frequency,
X_1, X_2: variables,
$\lambda_j^{X_1}$, $\lambda_k^{X_2}$: the main effect of variable X_1, X_2 respectively,
$\lambda_{jk}^{X_1 X_2}$: The interaction factor between X_1, X_2.

In Eq. 6, the main effects and the two-way effect are included; models in which all possible effects are included are called *saturated models*. If there is no effect (interaction/association) between X_1, X_2, we set the term $\lambda_{jk}^{X_1 X_2}$ to zero.

The term *hierarchical* indicates that if a term is set to zero, its associated higher orders are also set to zero. The hierarchical log-linear model M-for the set of random variables represented by the set of vertices V-can be specified by the generators that are associated with the cliques of the graphical model G. The generators of the models are the maximal interaction terms. In particular, a zero two-factor interaction term indicates the absence of an edge in G. As an additional example in a three-dimensional contingency table with generic cell $t = (i, j, k)$ for the variables $V = (X_1, X_2, X_3)$ the model is saturated if all terms are non-zero:

$$log(F_{ijk}) = \mu + \lambda_i^{X_1} + \lambda_j^{X_2} + \lambda_k^{X_3} + \lambda_{ij}^{X_1 X_2} + \lambda_{ik}^{X_1 X_3} + \lambda_{jk}^{X_2 X_3} + \lambda_{ijk}^{X_1 X_2 X_3} \tag{7}$$

If only the terms $\lambda_i^{X_1} + \lambda_j^{X_2} + \lambda_k^{X_3}$ (called the main effects of the model or zero order terms) are nonzero, this model represents complete independence among X_1, X_2 and X_3.

In this section, we described the relation between the factorization of the undirected graph according to its prime, undecomposable components and the distribution. This relationship is important since we will search for a generalized identifying sequence and calculate the rank deficiency of the Jacobian matrix $D(\beta)$ if the model is unidentifiable. In the next section, we will describe model formulation of undirected graphs with the purpose of studying their identifiability property.

3 Undirected Graph Model Formulation

In this section, we give the model on which the iugm is built. The model is presented concisely in [20]. We assume that the undirected graphical model satisfies the global Markov property and the joint probability distribution is factorized according to the undirected graph G.

Assume that we have a sample of v observations $x_1, x_2, ..., x_v$ which can be perceived as independent Poisson random variables, with interactions described by the undirected graph G. For example, the saturated model in Eq. 7 would be described by a clique on X_1, X_2, X_3 in a graphical model, G.

Since we require positive means, the general form of such models (called log-linear models) is:

$$\log(\mu_X) = Z\beta \tag{8}$$

In Eq. 8, β is a p-dimensional vector of the unknown parameters, and Z is the $2l \times p$ design matrix guaranteeing that the joint distribution factors according to the given graphical model.

For each variable v with l_v levels, $\forall v \in V$, let $l = \prod_{i=1}^{v} l_i$. Let Y be the marginal table resulting from the full cross-classification according to the observed variables only, and let $Y = LX$, where $L = (1, 1) \otimes I$, and I is the identity matrix. In other words, Y is the concatenation, side by side, of two identity matrices, each of dimension l; $L = (I_{l \times l} | I_{l \times l})$. The mean of Y is calculated as follows:

$$\mu_Y = L\mu_X \tag{9}$$

Substitute from Eq. 8 in Eq. 9:

$$\mu_Y = L \exp^{Z\beta} \tag{10}$$

Taking the partial derivative of μ_Y with respect to β yields the following:

$$D(\beta)^T = \frac{\partial \mu_Y^T}{\partial \beta} = \frac{\partial (L \exp^{Z\beta})^T}{\partial \beta} = (LRZ)^T \tag{11}$$

In Eq. 11, $R = diag(\mu_X)$, L is a $l \times 2l$ matrix, and Z is the design matrix.

A distribution P is completely characterized by its moments. Therefore, a graphical model is identifiable if its parameters can be expressed uniquely in terms of the moments of the distribution of its observable variables. In addition to showing that a model is identifiable, we must prove the mapping between the moments, μ_Y and the new parameters, β. For local identifiability this mapping must be one-to-one in a small neighborhood of any point of the parameter space Ω. Because of the inverse function theorem [21], this property holds if the matrix of all partial derivatives (the Jacobian matrix) of these moments has a nonzero determinant (e.g., the Jacobian matrix is full everywhere in the parameter space).

3.1   Identifiability Conditions

An undirected graphical model G is locally identifiable if $D(\beta)$ is full everywhere. As mentioned in the Introduction, [20] proved graphical necessary and sufficient conditions for an undirected graphical model to be identifiable. The iugm package implements the Stanghellini-Vantaggi graphical conditions, which are listed below:

1. $\hat{G} = (V, \hat{E})$ (the complement of the graph $G = (V, E)$) has a clique at least of size 3.
2. A clique C_0 of graph G has a *generalized identifying sequence* (GIS) $\{S_s\}_{s=0}^q$ of complete subgraphs, s.t.:

   a. $\forall i \in S_s, \exists j \in S_{s+1}, s.t.(i, j) \in \hat{E}$,
   b. $|S_{s+1}| \leq |S_s|$ for $s \in \{0, 1, ..., q - 1\}$, $C_0 = S_0$, and $|S_q| = 1$.

The iugm package also computes the rank deficiency (which indicates the absence of local identifiability) when the conditions are not met.

4   Algorithm

In this section, we introduce the algorithm on which iugm is based. The algorithm decides whether an undirected graph G is identifiable or not. Moreover, if the model is unidentifiable, the algorithm decides where the identifiability breaks and computes the rank deficiency of the matrix $D(\beta)$.

The input to the main program consists of three arguments: A which is the representation of the undirected graph, $type$ which is the type of representation the aforementioned graph, i.e., 'incidence' for incidence matrix, 'sparse' for sparse representation, and the empty string for the default, which is adjacency matrix, and v, verbose mode, which is true or false. For graph visualization, the $drawGraph$ function, iugm uses the Fruchterman-Rheingold force-directed layout algorithm [8]: each iteration computes $O(|E|)$ attractive forces and $O(|V|^2)$ repulsive forces.

If the observed graph is composed of two complete components, this graphical model is unidentifiable, because clumping each component will result in a graph consisting of the hidden variable and two observable nodes, with no edge between the observable variables. It is well known and easy to show, by a parameter counting argument, that a graphical model with two observable nodes and one hidden node is not identifiable from the observed distribution alone. Hence, the resulting reduced graph is not identifiable. The function $maximalCliques$ finds maximal cliques for the complement graph, $compGraph$, using the Bron-Kerbosch algorithm [6], the fastest known, with worst-case running time $O(3^{n/3})$.

Even if condition 1 is not satisfied, i.e., there is no clique of size 3 or more in the complement graph, we calculate the rank of the Jacobian matrix $D(\beta)$. Hence, we loop on the cliques of the graph, and calculate the boundary nodes in the complement graph $\hat{G}$ for that clique. Then we check for condition 2, which states that $\forall i \in S_s, \exists j \in$

S_{s+1}, $s.t.(i, j) \in E$. In Example 1 in Sect. 5, the boundary is a complete graph in $\hat{G}$, and therefore we need to build the constraints that specify where identifiability breaks down, which are stored in iugm in cell array. The constraints specify the interactions on the parameters that must sum to zero. We pass these constrainst to the *Jacob* function along with the decomposable components, i.e., cliques of the graph G, and total number of observed noded in G. If the model is identifiable and we can find a GIS, we proceed to the recursive function *buildTree*. This function recursively build the identifying sequence using tree data structure. The root of the tree is C_0 and the leaf is S_q, such that $|S_q| = 1$.

Algorithm 1 Deciding identifiability of undirected graph G, $mainProg(A, type, v)$

Require: The complementary graph $\hat{G}$ has one m_clique C and $m \geq 3$ and number of observed variables, $O \geq 3$
Ensure: G is not composed of two connected components   ((see explanation))
1: $compGraph \leftarrow complement(G)$
2: $drawGraph(A)$
3: $drawGraph(compGraph)$
4: $cliqueComp \leftarrow maximalCliques(compGraph)$
5: $m \leftarrow max(cliqueComp)$
6: **if** $m \geq 3$ **then**
7:     **print** Condition 1 is not satisfied!
8:     $clique \leftarrow maximalCliques(A)$
9:     $numCliques \leftarrow getNumberCliques(clique)$
10:    **for** $i = 1$ to $numCliques$ **do**
11:        $bd \leftarrow boundaries(compGraph, clique[i])$
12:        $constraints \leftarrow constructConstraints()$
13:    **end for**
14:    $N \leftarrow 1$
15:    **while** $N < numCliques$ **do**
16:        $buildTree()$
17:    **end while**
18:    $GIS \leftarrow checkGIS()$
19:    **print** GIS
20: **end if**
21: **print** Computing the Jacobian matrix $D(\beta)$
22: $D(\beta) = Jacob(cliques, nodes, constraints)$
23: **print** Checking the rank deficiency of the Jacobian matrix $D(\beta)$
24: **return**

---

5   Experimental Results

In this section, we give two graphical models. We represent the graph as an adjacency matrix (A). This matrix is passed as an argument to the main program, *mainProg*. For the cases of unidentifiability, the main program returns the constraints and prints

the deficiency in the Jacobian matrix. For the identifiable models, the main program
returns the generalized identifying sequence (GIS).

5.1 Example 1

In Fig. 1a, we show the adjacency matrix representation for the graphical model. The
program draws the graphical model G as shown in Fig. 1b and its complement $\hat{G}$ as
shown in Fig. 1c (cf. lines 2 and 3 in Algorithm 1).

The main program follows the execution path decribed in Algorithm 1. Since
the model is unidentifiable, we look for rank deficiency and where identifiability
breaks down. The program computes the Jacobian matrix and the constraints as
shown in Fig. 2. For the graphical model in Fig. 1b, there is no GIS, because the
boundary for $C_0 = \{1, 4, 5\}$ in the complementary graph of the observed variables
is $bd_{\hat{G}_0}(C_0) = \{2, 3, 6\}$, which is complete in $\hat{G}^O$ as shown in Fig. 1c. The relevant
output of iugm on this graphical model is shown in Fig. 2. The set of constraints
is stored in the MATLAB cell array variable $constraints$. The constraints can be
represented in terms of the interaction coefficients in β; in this case, the constraints
state that identifiability is lost when the interaction terms satisfy the following system
of linear equations:

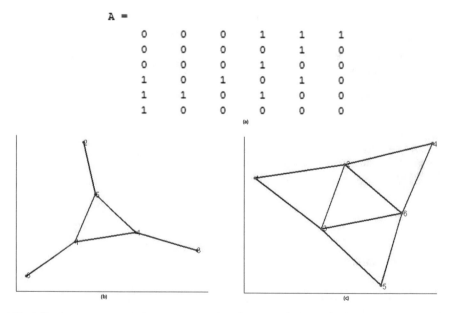

Fig. 1 A simple graphical model as a case study: **a** adjacency matrix A, **b** graph G and **c** complement
graph $\hat{G}$

Fig. 2 Rank deficiency and constraints for the graph in Fig. 1a

```
>> example01

**The constraints of this graphical model:
    [0  6      ][0  1  6    ]
    [0  3      ][0  3  4    ]
    [0  2      ][0  2  5    ]

**rank deficiency:
        1      28      27
```

$$\begin{cases} \beta_{\{0,6\}} + \beta_{\{0,1,6\}} = 0 \\ \beta_{\{0,3\}} + \beta_{\{0,3,4\}} = 0 \\ \beta_{\{0,2\}} + \beta_{\{0,2,5\}} = 0 \end{cases}.$$

These constraints result in a rank deficiency of 1, as computed by iugm and shown in Fig. 2.

6 Complexity Analysis

Let N be the number of cliques in the graphical model. The dimension of the Z matrix is $2^{l+1} \times K$, where l is the number of the observed random variables and K is computed by the following Eq. (12):

$$K = 2 * \sum_{j=1}^{N} (2^{S_j} - 1 - 2^{P_j}) + 1, \tag{12}$$

where S_j is the size (number of nodes) of clique j and P_j is the number of nodes in clique j that are also in cliques numbered less than j.

The hierachical log-linear model is used to build the design matrix. We provide a formula that helps in listing all the interactions.

$$Interactions = \cup_{i=1}^{N}[(\cup_{j=1}^{Cl_i} C_j^{Cl_i}) \cup (\cup_{j=1}^{Cl_i} [C_j^{Cl_i} \cup \{H\}])], \tag{13}$$

where N is the number of cliques, Cl_i is the set of vertices in clique i, $C_j^{Cl_i}$ is the choice of j vertices out of the set of vertices in Cl_i and H is the hidden variable.

Proposition 1 *Let S_i be a set of cardinality n in a GIS. Then S_{i+1} is of cardinality less than n, unless $(\cap_{t=1}^{n} bd(v_t) = \phi), \forall t, 1 \leq t \leq n, v_t \in S_i$, in which case the cardinality of S_{i+1} is exactly n.*

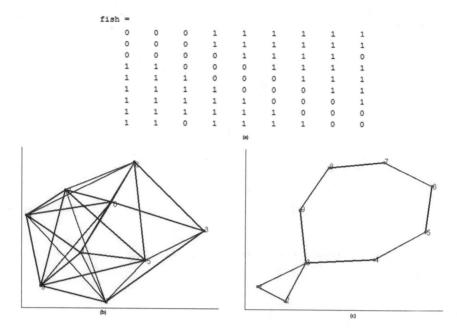

fish =

$$
\begin{array}{ccccccccc}
0 & 0 & 0 & 1 & 1 & 1 & 1 & 1 & 1 \\
0 & 0 & 0 & 1 & 1 & 1 & 1 & 1 & 1 \\
0 & 0 & 0 & 0 & 1 & 1 & 1 & 1 & 0 \\
1 & 1 & 0 & 0 & 0 & 1 & 1 & 1 & 1 \\
1 & 1 & 1 & 0 & 0 & 0 & 1 & 1 & 1 \\
1 & 1 & 1 & 1 & 0 & 0 & 0 & 1 & 1 \\
1 & 1 & 1 & 1 & 1 & 0 & 0 & 0 & 1 \\
1 & 1 & 1 & 1 & 1 & 1 & 0 & 0 & 0 \\
1 & 1 & 0 & 1 & 1 & 1 & 1 & 0 & 0
\end{array}
$$

(a)

(b)

(c)

Fig. 3 A dense graphical model tested for identifiability: **a** adjacency matrix A, **b** graph G and **c** complement graph $\hat{G}$

6.1 Example 2

In the graphical model shown in Fig. 3, both condition 1 and condition 2 are satisfied. For the GIS $C_0 = \{1, 4, 6, 8\}$, $C_1 = \{3, 7\}$, $C_2 = \{1, 6\}$ or $\{1, 8\}$, etc, the cardinality of C_2 is equal to that of C_1 GIS. As claimed in Proposition 1, $bd(3) \cap bd(7) = \phi$. The generically identifiable sequences are computed by iugm as shown in Fig. 4. In conformance with the existence of GIS, the rank of the Jacobian matrix is full everywhere, and iugm verifies that.

6.2 Running-Time Analysis

The iugm packages uses the Fruchterman-Rheingold force-directed layout algorithm, where each iteration computes $O(|E|)$ attractive forces and $O(|V|^2)$. In turn, iugm computes the cliques of the given graphical model. The iugm package employs the Bron-Kerbosch algorithm whose worst-case (Moon-Moser graph) running time (with a pivot strategy that minimizes the number of recursive calls made at each step) is $O(3^{n/3})$ for any n-vertex graph [14]. Note that $3^{n/3}$ is the upper bound on the number of maximal cliques of an n-vertex graph. The worst-case running time of the recursive function $buildTree$ depends on the height of the tree of sets that the

```
>> example04

**A Generalized Identifying Sequence is:
  [1 4 6 8]--->[3 7]--->[1 6]--->[2 5]--->[1 4]--->[3]

**A Generalized Identifying Sequence is:
  [1 4 6 8]--->[3 7]--->[1 6]--->[3 5]--->[4]

**A Generalized Identifying Sequence is:
  [1 4 6 8]--->[3 7]--->[1 8]--->[2 7]--->[1 6]--->[2 5]--->[1 4]--->[3]

**A Generalized Identifying Sequence is:
  [1 4 6 8]--->[3 7]--->[1 8]--->[2 9]--->[3]

**A Generalized Identifying Sequence is:
  [1 4 6 8]--->[3 7]--->[2 6]--->[1 5]--->[2 4]--->[3]

**A Generalized Identifying Sequence is:
  [1 4 6 8]--->[3 7]--->[2 6]--->[3 5]--->[4]

**A Generalized Identifying Sequence is:
  [1 4 6 8]--->[3 7]--->[2 8]--->[1 7]--->[2 6]--->[1 5]--->[2 4]--->[3]

**A Generalized Identifying Sequence is:
  [1 4 6 8]--->[3 7]--->[2 8]--->[1 9]--->[3]

**A Generalized Identifying Sequence is:
  [1 4 6 8]--->[3 7]--->[4 6]--->[5]

**A Generalized Identifying Sequence is:
  [1 4 6 8]--->[3 7]--->[6 9]--->[3 5]--->[4]

**rank deficiency:
     0    142    142
```

Fig. 4 The GIS and the rank deficiency for graph G at Fig. 3b

program explores when looking for a GIS. The GIS for any graphical model starts with a S_0 of size n elements, and continues looking for an identifiability sequence if it exists.

7 Conclusion

We present an algorithm which is based on the theoretical work by [20] and accompany the algorithm with a MATLAB library to supplement the state of the art algorithm. We claim that the implementation, iugm package, provides a sound and fair judgement on the algorithm that is designated to decide the identifiability of undirected graphical models with binary variables and exactly one hidden variable connected to all other (observed) variables.

References

1. Agresti, A.: An Introduction to Categorical Data Analysis, 2nd edn. Wiley-Interscience (2007)
2. Allman, E.S., Matias, C., Rhodes, J.: Identifiability of parameters in latent structure models with many observed variables. Ann. Stat. **37**(6A), 3099–3132 (2009)
3. Besag, J.: Spatial interaction and the statistical analysis of lattice systems. J. R. Stat. Soc. Ser. B (Methodological) **36**(2), 192–236 (1974)
4. Bickel, P.J., Doksum, K.A.: Mathematical Statistics: Basic Ideas and Selected Topics, vol. 1, 2 edn. Prentice-Hall (2001)
5. Bollen, K.A.: Structural Equations with Latent Variables, p. 0471011711. Wiley-Interscience, ISBN (1989)
6. Bron, C., Kerbosch, J.: Algorithm 457: finding all cliques of an undirected graph. Commun. ACM **16**, 575–577 (1973)
7. Drton, M., Foygel, R., Sullivant, S.: Global identifiability of linear structural equation models. The Annals of Statistics, 1003.1146 (2011). http://www.imstat.org/aos/future_papers.html
8. Fruchterman, T., Reingold, E.: Graph drawing by force-directed placement. Softw.-Pract. Exp. **21**(11), 1129–1164 (1991)
9. Huang, Y., Valtorta, M.: Pearl's Calculus of intervention is complete. In: Proceedings of the Twenty-second Conference on Uncertainty in Artificial Intelligence (UAI-06), pp. 217–224 (2006)
10. Jensen, F.V., Nielsen, T.D.: Bayesian networks and decision graphs, 2nd edn. Springer Publishing Company, Incorporated (2007). ISBN 9780387682815
11. Jordan, M., Bishop, C.: An Introduction to Graphical Models. MIT Press (2002)
12. Lauritzen, S.L.: Graphical Models. Oxford University Press (1996). ISBN 0-19-852219-3
13. Meek, C.: Strong completeness and faithfulness in Bayesian networks. In: Proceedings of Eleventh Conference on Uncertainty in Artificial Intelligence, pp. 411–418. Morgan Kaufmann (1995)
14. Moon, J., Moser, L.: On cliques in graphs. Israel J. Math. **3**(1), 23–28 (1965). https://doi.org/10.1007/BF02760024
15. Pearl, J., Paz, A.: Graphoids: a graph-based logic for reasoning about relevance relations. Technical Report (R–53–L), Los Angeles (1985)
16. Pearl, J.: Probabilistic Reasoning in Intelligent Systems: Networks of Plausible Inference. Morgan Kaufmann Publishers Inc., San Francisco, CA, USA (1988). ISBN 0-934613-73-7
17. Pearl, J.: Causality: Models, Reasoning, and Inference, 2 edn. Cambridge University Press (2009)
18. Peña, J.M.: Faithfulness in Chain graphs: the discrete case. Int. J. Approx. Reason. **50**(8), 1306–1313 (2009)
19. Sharaf, M.: Identifiability of discrete concentration graphical models with a latent variable. Ph.D. thesis, University of South Carolina, South Carolina (2014)
20. Stanghellini, E., Vantaggi, B.: Identification of Discrete Concentration Graph Models with One Hidden Binary Variable. Bernoulli (to appear) (2013)
21. Stromberg, K.R.: An Introduction to Classical Real Analysis. Wadsworth International, Belmont, CA (1981)

An Automatic Classification of Genetic Mutations by Exploring Different Classifiers

Badal Soni, K. Suganya Devi, and Angshuman Bora

1 Introduction

Before even humans learned about the existence of diseases in their body, they prevailed. Man became smarter and civilized soon. It was then that they knew the cause of most of their deaths. Ever since then, they took precautions in whatever way they could. With the advent of time, this became a profession and researchers kept searching for the treatment of diseases. Cancer is one such disease that dominates a vast number of other diseases.

Scientists had been researching on cancer treatment but there is not hundred percent success yet. Being in the 21st Century and not able to solve a problem, increases the gravity of the problem. A complete check up of the body is often done nowadays by most individuals as a precaution for life threatening diseases. In case of cancer, scientists find it easier to treat the disease in early stage. With the advent of technology, mere observations have changed to a large expansion of every problem. Technology has opened new doors in every field and when it comes to cancer treatment, it has helped in a mesmerizing way by decreasing the amount of deaths.

Accurate prediction of the type of cancer from the early symptoms had proved to be the most efficient and is hence focused upon, so that the treatment could be done in the early stage itself. This introduces the concept of training Machine Learning

B. Soni (✉) · K. Suganya Devi · A. Bora
National Institute of Technology, Silchar, Assam, India
e-mail: soni.badal88@gmail.com

K. Suganya Devi
e-mail: suganyanits@gmail.com

A. Bora
e-mail: angshu.btf@gmail.com

models in this field. Machine Learning is nothing but an application of Artificial Intelligence which makes the machine learn itself with the intervention of humans. Different methods for training the models and their accuracy are discussed later in this paper.

In a nutshell, we are trying to identify the type of cancer on a test sample by making use of the concept of Machine Learning. It is to be noted that using Machine Learning models have improved the prediction of cancer by 15–20%. The detection and diagnosis of cancer are given more importance by most Machine Learning models in the domain of medical science, rather than error prediction and treatment of cancer. Prescription of precise medicine and proper genetic testing will alter the way of treating life threatening diseases like cancer. However, the level of precision has not reached yet due to the myriad amount of manual work involved. Because of this, Memorial Sloan Kettering Cancer Center (MKSCC) had launched a competition, accepted by the NIPS 2017 Competition Track in order to take personalised medicine to a whole new level. We made use of the dataset provided in the competition to contribute in whatever small way we can.

There can be thousands of genetic mutations on a cancer tumour. But all of them are not cancer causing. Hence, our prime focus is on distinguishing the driver mutations (those that lead to growth of the tumour) from passenger mutations (neutral mutations). In the current scenario, this task of distinguishing is manually done. It is highly labour intensive and time consuming as the pathologists have to classify each test mutation as per evidence from the clinical literature. So, we were motivated to automate this classification process.

The interpretation of genetic mutations by genomic researchers is traditionally labour intensive, because it requires a great amount of work that is to be done manually. To optimize this *labour intensive* task, we have blended together technology with their medical experience. We used different algorithms to train a classification model that automatically classifies genetic variations of interest into *nine* classes. These classes help a pathologist to distinguish among the mutations that contribute to tumour growth (driver mutations) and those that do not (passenger mutations). The model we developed by implementing various Machine learning concepts, provides probabilistic output. Hence, the model is *interpretable*. In the current scenario, the distinction between driver and passenger mutations are being manually done. Manually going through the clinical literature for every test mutation is highly labour intensive, tiring and time consuming. Moreover, there is the possibility of error as well as life risks in case of wrong analysis. Hence, our objective is to automate this process of classifying genetic mutations by implementing various ML algorithms, using clinical evidence as the baseline of knowledge.

To develop a classification model that uses clinical evidence as the baseline of knowledge.

To reduce computational time by automatically classifying genetic variations and making output interpretable.

2 Literature Survey

In this section we discuss many related papers with their advantages and limitations.

Asri et al. [1], assesses the correctness in classifying data with respect to efficiency and effectiveness of each algorithm in terms of accuracy, precision, sensitivity and specificity. Experimental results show that SVM gives the highest accuracy (97.13%) with lowest error rate. Kuorou et al. [2], used classification algorithms and techniques of feature selection, outlining three integral case studies which includes prediction of susceptibility to cancer, recurrence of cancer and its survival using apt Machine Learning concepts.

Sheryl et al. [3], developed a comprehensive pipeline that performs the classification of genetic mutations based on the clinical literature. The text features in their solution was mined from three views, namely entity name, entity text and original document view. Different ML algorithms had been used to get text features from the perspectives of document level, knowledge, sentence and word level. The results obtained convey that multi view ensemble classification framework gives good accuracy.

Bruno et al. [4], presents the possibility of using KNN algorithm with TF-IDF method and framework for text classification. The results of testing showed the good and bad features of algorithm, providing guidance for the further development of similar frameworks. Jiang et al. [5], succeeded to reduce computation on text similarity and improve KNN, thereby outperforming popular classifiers like Naive Bayes, SVM and KNN. His work was based on clustering algorithms.

Wan et al. [6], uses SVM classifier at the training stage of KNN. It has a low influence on implementing K parameter, so the accuracy of the KKN classification retained. However, it needs higher time complexity.

Zhang et al. [7], separated a class of document vectors and its compliment with the help of hyperplanes. Naive Bayes seemed to be worse. Performance of LLSF was as close as state-of-art. Performance LR and SVM was quite equivalent.

Leopold et al. [8], TF-IDF weighting scheme had greater impact on the performance of SVM than Kernel Functions alone. For SVM Classification, selection of features and preprocessing is not required. Aurangzeb et al. [9], comparing with ANN, SVM capture the inherent characteristics of the data better. Bekir et al. [10], presents detection of drug therapy of leukemia cancer by using Naive Bayes Classifier. Proposed study supports use of personalized drug therapy in clinical practices. This tool can be used for treatment of variety diseases with similar characteristics. Watters et al. [11], after performing their own experiments, arrived to a conclusion that the performance of SVM Classifier was way more good than Artificial Neural Network for IQ57 as well as IQ87. Not only the performance, but also SVM is less expensive in terms of computation. So, for datasets that has less categories with documents, these authors suggested the use of SVM rather than ANN. Mertsalov et al. [12], says that the automatic classification of documents is much handy for larger enterprises. The man-made machines for document classification can now perform even better than man himself and hence its use in broader domains will keep enhancing with time. Image and Speech processing based applications have been discussed in [13–15].

Li et al. [16], arrived to a conclusion that all of the four classifiers, namely Decision Tree, Naive Bayes (NB), Subspace method for classification and the nearest neighbour classifier had performed quite good on the data-set of Yahoo. Here, NB gave the maximum accuracy. Moreover, they observed that combining various classifiers does not lead to a significant improvement than using a single classifier. Karel et al. [17], derived that reducing the number of features proved better than not reducing them. Moreover, this algorithms of feature extraction lead to better performance than other feature selection methods. Quan et al. [18], showed that methods of smoothing enhanced the accuracy of NB for the classification of shorter texts. Two stage (TS) as well as Absolute Discounting (AD) proved to be the best smoothing methods. The method of N-gram frequency by William et al. [19], gives a less expensive and more effective way for classifying the documents. It is achieved by the use of sample of required categories instead of choosing costly and complicated methods like assembling detailed lexicons or parsing of natural language.

Liu et al. [20], proposed SVM-based classification model. Different classification techniques like KNN, decision tree, NB, and SVM have been presented. In SVM by using non linearity mapping the author has changed inseparable sample of low-dimensional sample space to high-dimensional feature space to make it linearly separable. Thorsten et al. [21], the author of this paper highlighted the merits of Support Vector Machine (SVM) classifier for categorization of text, On the Reuters dataset, KNN had proved to be the best among all conventional methods. SVM, on the other hand, gave the best results of classification, thereby leaving behind all of the conventional methods by good margins.

3 Datasets Description

There are nine different classes a genetic mutation can be classified on. This is not a trivial task since interpreting clinical evidence is very challenging even for human specialists. Therefore, modeling the clinical evidence (text) will be critical for the success of your approach. The nine mutation classes are as follows:

- Gain of function
- Likely gain of function
- Loss of function
- Likely loss of Function
- Neutral
- Likely_Neutral
- Switch of Function
- Likely Switch of Function
- Inconclusive.

We have provided the dataset for training and testing in two separate files. The first file (training/test_variants) provides information regarding the genetic mutations while the second file (training/test_text) provides clinical literature (text) that pathologists use for the classification of mutations.

3.1 File Descriptions

The data we used was collected from a competition in Kaggle, titled **Personalised medicine Re-defining cancer treatment** [22]. This data was prepared by Memorial Sloan Kettering Cancer Center (MSKCC). The size of the merged dataset is 528 MB.

Training_variants: It is a CSV file that contains genetic mutation descriptions that are used in training. The fields are ID (this id of each row links a mutation to clinical literature), Gene (where a mutation is present), Variation (change in the amino acid sequence for the mutation), Class (genetic mutations are classified into 9 different classes) (Fig. 1 and Table 1).

Training_text: This is one double pipe (‖) delimited file containing the clinical literature (in text format) which classifies the genetic mutations. The fields include ID (this id of each row links a mutation to clinical literature), and Text (this contains clinical evidence which contributes to the classification of the genetic mutations) (Fig. 2 and Tables 2, 3).

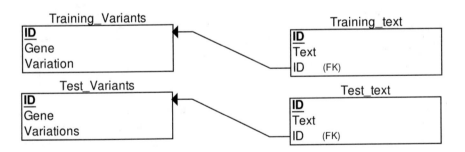

Fig. 1 Relational Schema Diagram

Table 1 A sample view of Training_variants dataset

ID	Gene	Variation	Class
25	CBL	H398Y	4
26	SHOC2	S2G	4
27	TERT	Y846C	4
28	TERT	C228T	7
29	TERT	H412Y	4

Fig. 2 Distribution of datapoints in different classes

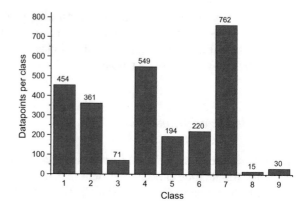

Table 2 A sample view of Training_text data set

ID	Text
25	Recent evidence demonstrated acquired uniparen...
26	Abstract n-myristoylation common form co-trans...
27	Heterozygous mutations telomerase components t...
28	Sequencing studies identified many recurrent c...
29	Heterozygous mutations telomerase components t...

Table 3 General stat of data set

Stat	Training	Testing
Number of samples	3321	368
Number of different genes	264	140
Number of different mutations	2996	328

Test_variants: The Test_variants is a unlabelled data file where the file contains all the fields as mentioned in the Training_variants except the Class. The fields are ID (the id of the row used to link the mutation to the clinical evidence), Gene (the gene where this genetic mutation is located), Variation (the aminoacid change for this mutations)

Test_text: The Test_text data file is similar to our Training_text. The fields include ID (this id of each row links a mutation to clinical literature), and Text (this contains clinical evidence which contributes to the classification of the genetic mutations).

4 Proposed Methodology

4.1 Data Preprocessing

Without preprocessing the raw data, we cannot imagine to implement Machine Learning algorithms unless we do not care for the model accuracy, which definitely is our prime concern. Working on raw data is many a time as good as not working. Hence, preprocessing becomes one of the most integral initial steps to support different algorithms. In our model, the accuracy would have been significantly less had we not done any preprocessing. An example of preprocessing includes reducing the dimension of the dataset, i.e. the selection of those features that lead to better accuracy. Preprocessing of raw data is characterized by a few steps which are explored as follows.

4.1.1 Data Cleaning

Data Cleaning forms the most initial step of Preprocessing. Sometimes, the dataset may not contain some data in between, while at sometimes we have noise in the data. Moreover, there are outliers in most classes which are preferred to be detected and removed. We would also want to minimize the duplication of the same data. To tackle all these problems, we have used Data Cleaning.

4.1.2 Data Integration

Sometimes, people collect data from various sources since the entire dataset is not always available in one place. Different data lie scattered in different places and hence, after collection of all these data, they have to be made consistent, so that after cleaning the data, it can be useful for analysis.

4.1.3 Data Reduction

After the integration, we perform the following:

- Stopwords Removal: This step includes the removal of words like is, are, the, of, etc. as they are the most frequently appearing words that in no way help our classification
- Special Character Removal: This step includes the removal of special symbols like , . / ; [ ]) (= + as they do not discriminate the classes and hence are not useful for classification
- Conversion of the textual data to lowercase (Table 4).

Table 4 A sample view of the preprocessed dataset

ID	Gene	Variation	Class	Text
1510	alk	Npm-alk-fusion	7	Nucleophosmin-anaplastic lymphoma kinase npm-a...
383	tp53	h115n	6	Sequence-specific dna binding exonuclease acti...
19	cbl	y371s	4	Acquired uniparental disomy aupd common feat...
2907	nf2	k79e	1	Despite intense study neurofibromatosis type 2 ...
3063	med12	l1224f	2	Human thyroid hormone receptor-associated...

4.2 Encoding and Embedding

The machine does not understand words or strings. Hence, in order to implement various ML algorithms on our dataset, we first need to convert every word to numerical form so that it is understood by the machine. This conversion of text to numerical form is done by encoding and embedding. The different encoding and embedding techniques we have used are discussed below.

4.2.1 Bag of Word Model

Bag of Words is an encoding technique that represents a document as frequencies of every word the document contains. However, a corpus often contains less significant words (words that carry no special information about a particular document) whose frequency can be very high. These words do not discriminate among the classes and hence are least important for the classification process. Thus, one demerit of Bag of Words is that it cares only for the frequency of words rather than their semantics. Second, it does not carry about the words order either. There may be two sentences that convey the same meaning but the sequence of words is different. In both cases, sets of words are extracted from the text and are used to train a simple classifier, as it could be xgboost which it is very popular in Kaggle competitions. There are variants of the previous algorithms, for example the term frequency inverse document frequency, also known as TFIDF, tries to discover which words are more important per each type of document. For example, if you have 3 documents.

- D1—most cancers are characterized by increased STAT3 activation
- D2—phosphorylate STAT3 levels are associated with reduced survival
- D3—increased phosphorylate of STAT3 in tumors.

First, it creates a vocabulary using unique words from all the documents then, for each word the frequency of the word in the corresponding document is inserted. Bag of Words representation does not consider the semantic relation between words. Generally, the neighbor words in a sentence should be useful for predicting your target word.

4.3 The TF-IDF Model

Term Frequency (TF): This gives more significance to the most frequently appearing word in the document. For example, if a word appears once in a document and total number of words in that document is five, then TF value of that word is $1/5$.

$$TF(Wi, Dj) = (Number\ of\ times\ Wi\ occurs\ in\ Dj)/(Word\ Count\ in\ Dj) \quad (1)$$

Here W_i is the ith word and D is document.

Inverse Document Frequency (IDF): This gives least significance to the most frequently appearing word in the document corpus i.e, there are four documents and a particular word appears in three of those documents. Then IDF value of that word is $\log(4/3)$.

$$IDF(Wi, Dj) = log(Number\ of\ document/Number\ of\ document\ having\ Wi) \quad (2)$$

Therefore, finally the TF-IDF value of that word is $TF \times IDF$, i.e. $(1/5) \times log(4/3)$ or 0.0249.

Advantages of TF-IDF are as follows: easy to compute, we have basic metric to extract the most descriptive terms in a document, we can easily compute the similarity between 2 documents using it. Disadvantages of TF-IDF are as follows: it is based on Bag of Words model. Hence, it does not care about the semantics, position of a word in a document, and even co-occurrences in separate documents. It is good as a lexical level feature. Unable to capture the semantic information (unlike embedding techniques).

4.4 Word2Vector Embedding

Using a neural network language model, Word2Vec computes the vector representation for every word present in a document. Every word is trained in the model to maximize log probability with the words in its neighbour. So, from the words that is embedded, we get the vector representations which help us to find the similarities among them.

Word2Vec is not an algorithm for text classification but an algorithm to compute vector representations of words from very large datasets. A characteristic of Word2Vec is that those words with similar context have their vectors in the same space. For example, countries would be close to each other in the vector space. Another property of this algorithm is that some concepts are encoded as vectors. Using word representations provided by Word2Vec we can apply math operations to words and so can use algorithms like Support Vector Machines (SVM) or the deep learning algorithms (Tables 5, 6, 7, 8, 9 and Fig. 3).

Table 5 Row wise representation of dataset

ID	Text
R1	Cyclin-dependent kinases cdks regulate variety fundamental cellular process...
R2	Abstract background non-small cell lung cancer nsclc is a heterogeneous group...
R3	Recent evidence demons acquired uniparental disomy aupd novel mecha...

4.5 Doc2Vec Embedding

Doc2Vector or Paragraph2Vector is a variation of Word2Vec that can be used for text classification. This algorithm tries to fix the weakness of traditional algorithms that do not consider the order of the words and also their semantics. The aim of Doc2Vec is creating numerical representations of the documents, irrespective of the length. However, the documents have no logical structure unlike words. So, some other method ought to be found.

This algorithm is similar to Word2Vec, it also learns the vector representations of the words at the same time it learns the vector representation of the document. It considers the document as part of the context for the words. Once we train the algorithm we can get the vector of new documents doing the same training in these new documents but with the word encoding fixed, so it only learns the vector of the documents. Then we can apply a clustering algorithm or find the closest document in the training set in order to make a prediction. It behaves as a memory which can remember the missing part from the current context (Fig. 4).

5 Classification Models

5.1 Logistic Regression (LR)

LR is a popular classification method which predicts binary outcomes (True/False, Yes/No, 1/0) for a given set of independent variables. It can be considered as some especial case of Linear Regression when outcome variable is categorical, where log of odds is used as a dependent variable. In a nutshell, it predicts probability of an events occurrence by fitting the data to logistic function.

Now, Logistic regression gives an output which is the probability that the given input points belong to a certain class. The output is most often used with a threshold value that decides which class a probability should be assigned to. This threshold value is usually 0.5, but can be set to other values in some applications. Logistic regression is also considered to be a linear classifier. Let us say we have two classes, 1, and 0, by providing a set of data points x, logistic regression can predict which of the classes 1, or 0 is the most probable one, or with other words, can perform classification with a probabilistic outcome. Finding the parameters that estimates the

Table 6 Bag of word calculations of the data used

ID	CSDK	Disorder	Aups	Patient	Funda.	Background	Heterogeneous	Acquired	Cell	Process	Abstract	NSCLC
V0	1	0	0	0	1	0	0	0	1	1	0	0
V1	0	1	0	0	0	1	1	0	0	0	1	1
V2	0	0	1	1	0	0	0	1	0	0	0	0

Table 7 Bag of word encoding

ID	Cancers	Characterized	Increased	STAT3	Activation	Phosphorylate	lvl	Associated	Reduced	Survival	Tumors
D1	1	1	1	1	1	0	0	0	0	0	0
D2	0	0	0	1	0	1	1	1	1	1	0
D3	0	0	1	1	0	1	0	0	0	0	1

Table 8 TF-IDF words representation in dataset

ID	Text
S1	Cell lung cancer heterogeneous mutations.......
S2	Novel mechanism pathogenic mutations cancer.......
S3	Recurrent coding mutations human cancer genes......

probability of the classes can be done in several different ways. When training the algorithms, one can use different optimization methods to fit the data to a model.

5.2 Naive Bayes (NB)

In the Naive Bayes Classifier, the probability classifier is simplified by the assumption of class conditional independence among features and words of a given class. The classifier works according to Bayes Theorem where the predictors are assumed to be independent. In a nutshell, a given feature is considered independent of other features. Let us take the following example: an animal can be considered as carnivorous if it eats meat.

The Naive Bayes model is quite easier to implement. It is very useful for datasets that are large enough. Besides its simplicity, it is renowned for outperforming newer and sophisticated classifiers.

In Naive Bayes, the discriminant function is given by its posterior probability, which is obtained from the Bayes theorem. The posterior probability is given by (Fig 5).

$$P = \frac{e^{a+bX}}{1 + e^{a+bX}}$$

$$P(c/x) = \frac{P(x/c)P(c)}{P(x)}$$

Here,

- P(c/x) denotes posterior probability
- P(x/c) denotes Class Conditional Probability
- P(c) denotes Class Prior Probability
- P(x) denotes predictor prior probability.

Table 9 TF-IDF scores for words in dataset

ID	Cancer	Cell	Coding	Gene	Eterogeneous	Human	Lung	Mechanism	Mutations	Pathogenic	Recrrent
S1	0.31	0.52	0.0	0.0	0.52	0.0	0.52	0.0	0.31	0.0	0.0
S2	0.31	0.0	0.0	0.0	0.0	0.0	0.0	0.52	0.31	0.52	0.0
S3	0.27	0.0	0.46	0.46	0.0	0.46	0.0	0.0	0.27	0.0	0.46

Fig. 3 Word Embedding with Word2Vec

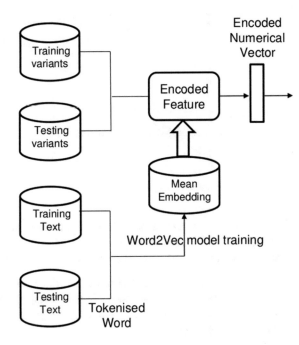

Fig. 4 Doc Embedding with Doc2Vec

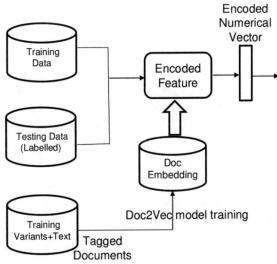

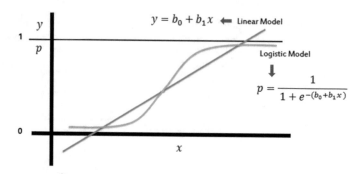

Fig. 5 Logistic Regression(LR) Curve

The Bayesian classification helps in the evaluation and understanding of different learning methods. Besides, it computes explicit probability for hypothesis. It is also robust to the presence of noise in the input data. It has its own limitations like

- It works well on small datasets. For most of the practical applications it hardly fits.
- NB has high bias and low variance. Hence it makes its application limited. Having said this there are no regularization or hyper-parameters tuning involved here to adjust the bias thing.

5.3 Support Vector Machine (SVM)

This is the most used classification algorithm. The working principal Support vector machine is mainly based on marginal calculations. By a hyper plane it divides the data points into different subclass. In case of linear SVM, the mathematical expression of the hyperplane is written below (Fig. 6):

$$\mathbf{x}_i \mathbf{w} + b = 0. \tag{3}$$

For minimal classification error the distance of the hyper plane and classes should be maximum. The hyperplane to have the most extreme edge, which can amplify the separation of the hyper-plane and closest focuses from the two classes.

The mechanism of optimal hyper plane for division of training data without error can be examined using soft margin which will permit an interpretive technique of learning with errors. The optimal hyper plane can be describe as below, if there are n number of training patterns say

$$(x_1, y_1), (x_2, y_2), â Ăeâ Ăe, (x_n, y_n), x_i \in [-1, 1] \tag{4}$$

can be linearly separable if there exist a scalar b and vector w than,

$$w \times y_i + bâL'ĕ1, if\ x_i = 1\ and\ w \times y_i + bâL'd' - 1, if\ x_i = -1. \tag{5}$$

Fig. 6 Example of SVM classification

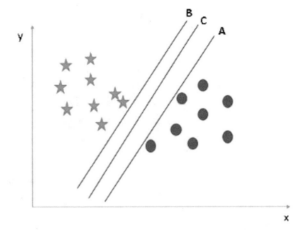

Than the optimal hyper plane Eq. 4 should divide the training data with ultimate margins.

$$w_0 \times x + b_0 = 0. \qquad (6)$$

Development of the mechanism of Support vector machines were done for training data separation with minimal or no error. Later the support vector approach is extended to overlay the idea of division without error on the training vectors is not achievable. As robust and ground breaking as neural networks the support vectors can be considered as a new learning machine with this new extension.

5.4 Random Forest (RF)

Random forest algorithm can use both for classification and the regression kind of problems. RF is basically a **supervised learning algorithm** which generates a forest for the evaluation of the results. It constructs many decision trees by choosing K data points from the data set. Then, they are all merged together for obtaining more stable as well as accurate predictions. For each K data points decision tree we have many predictions and then we take the average of all the predictions.

RF is one Ensemble learning algorithm, because many models are merged together to predict a single result. RF generally use ID3 algorithm for decision tree training. There are three principle decisions to be made while constructing a random tree. These are the strategy for part the leafs, the kind of indicator to use in each leaf, and the technique for infusing haphazardness into the trees. And Gini measure is used for calculating the utilization of split feature. The mathematical explanation of Gini can be given here (Fig. 7),

$$Gini(P_m) = \sum_n \mathbf{Q_{m_n}}(1 - Q_{m_n}). \qquad (7)$$

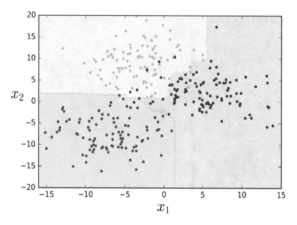

Fig. 7 Random Forest Classifier

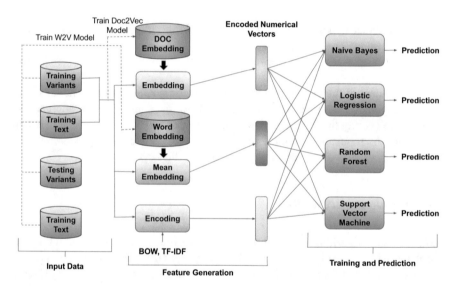

Fig. 8 Proposed working Model

5.5 Proposed Architecture

Our proposed architecture consists of three phases. They are: Input Data, Feature Generation, Training and Prediction. Each of them are explained as follows (Fig. 8).

Input Data: The dataset that we have collected from the Kaggle competition was splitted into Training and Testing Data, which served as input for encoding and embedding. The input data was a set of three features: Gene, Variation and Text. Each of these features is of the form of strings. Hence, they ought to be converted to numerical form for the implementation of classification algorithms.

Feature Generation: In this phase, encoding and embedding is done. The encoders we have used are Bag of Words and Term Frequency - Inverse Document Frequency. The embedding techniques used are Word2Vec Embedding and Doc2Vec Embedding. Both the techniques of encoding and embedding are used to convert text data to numerical form after normalization so that classification algorithms can be implemented.

Training and Prediction: This is the third phase of the proposed architecture which involves the use of four different classifiers. They are: Naive Bayes, Logistic Regression, Random Forest and Support Vector Machine. These classifiers classify the encoded and embedded data into nine different classes. They give probabilistic output, and hence our model is interpretable.

6 Experimental Results and Discussions

Our work involving the implementation of various Machine Learning algorithms, explores combination of different classifiers, encoding and embedding techniques, to bring out the best possible accuracy in the process of classifying the genetic mutations.

- Using good encoding can increase the model accuracy of Naive Bayes. But it is hard to control or deal with, hence bad estimators of class.
- SVM perform classification by finding the hyper plane that differentiate the two classes. In text classification.
- Random Forest: A feature is selected for every node with the use of random sub-sample of the features. It constructs many decision trees by choosing K data points from the data set after which it takes average of the class prediction probabilities for every combination of feature-value to get the overall prediction of the model.
- All models have less accuracy between Classes (1–4) & (2–7).
- Every ML model requires specific encoding of feature like: SGD classifier—can handle high Dimensional feature data, RF—can not handle high Dimensional feature data.
- Bag of Word encoding technique produces very high dimensional sparse matrix data. Using countVectorizer function with unigram, produces 125000+ feature, which is sparse in nature.
- Using bi-gram or tri-gram encoding result 250000+ feature, and produce same result as unigram encoding.
- BOW is not able to find semantic and syntactic information text (Tables 10, 11, 12, 13, 14 and Figs. 9, 10).

Our proposed model did not attain better accuracy in Word2Vec and Doc2Vec Embedding because the text entry of different sample data points of the Text feature was completely similar. For example, for ID 0, there were more than 65 data points where the text entry was completely same. Because of this, the corpus of words is not large, leading to similar vectors of a document even though they belong to different

Table 10 Model classification for single Data point

Class type	Prediction probability	Ground truth
1	0.802710	1
2	0.015388	0
3	0.000548	0
4	0.176723	0
5	0.000539	0
6	0.000621	0
7	0.000254	0
8	0.000522	0
9	0.000399	0

Table 11 Accuracy representation of various classifier using Bag of Word model

Classifier model	Accuracy(%)
Naive Bayes	59.69
Random forest	56.54
Logistic regression	62.25
Support vector machine	64.21

Table 12 Accuracy representation of various classifier using TF-IDF model

Classifier model	Accuracy(%)
Naive Bayes	64.36
Random forest	64.81
Logistic regression	67.27
Support vector machine	66.46

Table 13 Accuracy representation of various classifier using Word2Vec model

Classifier model	Accuracy(%)
Naive Bayes	49.022
Random Forest	65.864
Logistic Regression	53.23
Support Vector Machine	52.48

Table 14 Accuracy representation of various classifier using Doc2Vec model

Classifier model	Accuracy(%)
Naive Bayes	48.12
Random forest	54.43
Logistic regression	49.17
Support vector machine	49.77

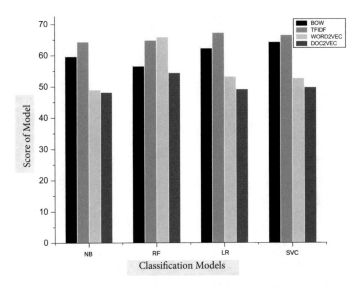

Fig. 9 Comparison of various classifier in different embeddings techniques

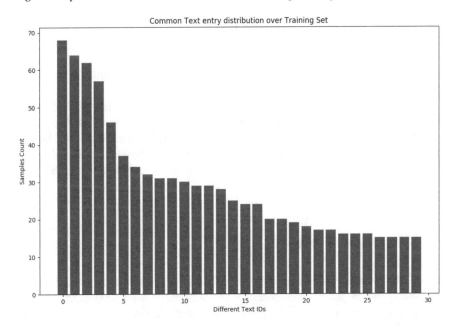

Fig. 10 Distribution of count of same Text attribute

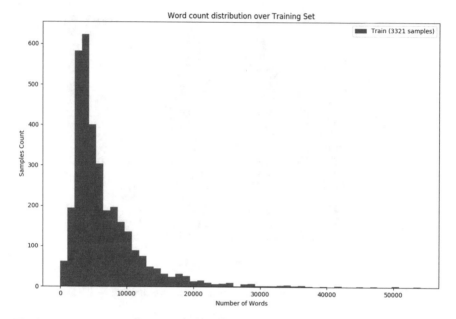

Fig. 11 Distribution of word count of Text attribute

classes. Hence, Word2Vec and Doc2Vec Embedding performed poorer in our dataset (Fig. 11).

The next issue with our dataset is that it contains myriad words that are totally insignificant for classification. For example the text entry of the Text feature contains lot of bibliography content. Thus, the dataset has noisy information leading to useless vectors of a document. Hence, as the vectors start to change from what it should be, the probability of misclassification increases. So, Word2Vec and Doc2Vec Embedding did not work well on our dataset.

6.1 Confusion Matrix

A confusion matrix can be used as a way to visualize the results of a classification algorithm. For the binary case where 1 and 0 is the two possible outcomes, the algorithm can be used to predict whether a test sample is either 0, or 1. As a way to measure how well the algorithm performs, we can count four different metrics, here 1 defined as positive and 0 defined as negative:

- True positive (TP), the algorithm classifies 1 where the correct class is 1.
- False positive (FP), the algorithm classifies 1 where the correct class is 0.
- True negative (TN), the algorithm classifies 0 where the correct class is 0.

Fig. 12 BOW encoding with SVM

PREDICTED CLASSES

ACTUAL CLASS	1	2	3	4	5	6	7	8	9
1	59	2	0	34	6	2	11	0	0
2	5	45	0	1	0	1	39	0	0
3	0	0	1	9	0	0	8	0	0
4	11	1	1	111	3	1	9	0	0
5	12	1	1	9	14	2	8	0	1
6	4	5	0	4	1	31	10	0	0
7	1	18	0	4	4	1	161	0	2
8	0	1	0	1	0	0	2	0	0
9	0	0	0	0	0	0	2	0	5

Fig. 13 TF-IDF encoding with LR

PREDICTED CLASSES

ACTUAL CLASS	1	2	3	4	5	6	7	8	9
1	74	1	0	20	8	2	9	0	0
2	3	43	0	2	1	1	41	0	0
3	1	0	5	4	0	0	8	0	0
4	10	2	1	98	11	1	14	0	0
5	9	0	0	8	17	5	9	0	0
6	9	1	1	2	2	33	7	0	0
7	1	9	4	0	3	0	174	0	0
8	0	1	0	0	0	0	3	0	0
9	1	0	0	2	0	0	2	1	3

- False negative (FN), the algorithm classifies 0 where the correct class is 1 (Figs. 12, 13, 14 and 15).

7 Conclusions and Future Work

Our work has contributed in developing a comprehensive pipeline for performing the classification of genetic mutations based on clinical evidence. Different ML algorithms are explored to get features of the text. Different combinations of classifiers,

Fig. 14 WORD2VEC embedding with RF

PREDICTED CLASSES

ACTUAL CLASS	1	2	3	4	5	6	7	8	9
1	65	1	1	29	4	1	13	0	0
2	3	46	0	2	1	1	38	0	0
3	1	0	8	1	0	0	8	0	0
4	14	4	2	108	0	0	9	0	0
5	14	1	1	8	14	1	9	0	0
6	8	2	0	2	1	29	13	0	0
7	2	19	1	4	1	0	164	0	0
8	1	0	0	1	0	0	1	0	0
9	0	0	0	0	0	0	1	0	4

Fig. 15 DOC2VEC embedding with RF

PREDICTED CLASSES

ACTUAL CLASS	1	2	3	4	5	6	7	8	9
1	54	3	0	21	0	1	24	0	0
2	3	22	0	6	0	0	61	0	0
3	4	0	5	6	1	0	8	0	0
4	24	1	1	95	1	1	20	0	1
5	16	0	1	10	12	0	19	0	2
6	8	1	3	5	2	22	10	0	0
7	7	11	1	12	2	0	150	0	0
8	0	1	0	1	0	0	0	0	0
9	1	0	0	0	0	0	3	0	2

encoding and embedding techniques have been tried to give us the best possible score in classifying a test mutation automatically as a driver or a passenger.

Logistic Regression with TF-IDF encoding gave the best score, despite of using Word2Vec and Doc2Vec embedding to capture semantics. The issues with our dataset as discussed could be removed using a better dataset. So, our model should give much more accuracy when the potential of Word2Vec and Doc2Vec are used to their full potential. Future work of this work is discuss below.

- Million Gigabytes of data are produced every day. In our case of detection and diagnosis of cancer also, generation of new data and better observations are going on. So, the clinical literature would eventually be improved for proper analysis and classification of the test mutations. As soon as we get another dataset, we can make use of all our encoding and embedding techniques to its full potential, thereby giving a hike to the accuracy.

- Use of XgBoost and Ensemble methods like multiple Gradient Boost Decision Trees classifier.
- Entity text view for genes and variation can be utilized to distinguish the different process as there is the presence of similar text entry in different data points. This could improve Word2Vec and Doc2Vec embedding technique.

Acknowledgments This work was supported by Multimedia and Image Processing Laboratory, Department of Computer Science and Engineering National Institute of Technology Silchar, India.

References

1. Asri, H., Mousannif, H., Al Moatassime H., Noel, T.: Using machine learning algorithms for breast cancer risk prediction and diagnosis. Procedia Computer Science, vol. 83, pp. 1064–1069 (2016)
2. Kourou, K., Exarchos, T.P., Exarchos, K.P., Karamouzis, M.V., Fotiadis, D.I.: Machine learning applications in cancer prognosis and prediction. Comput. Struct. Biotechnol. J. **13**, 8–17 (2015)
3. Zhanga, X.S., Chena, D., Zhua, Y., Chea, C., Suc, C., Zhaod, S., Mina, X., Wanga, F.: A multi-view ensemble classification model for clinically actionable genetic mutations (2018). arXiv:1806.09737
4. Trstenjak, B., Mikac, S., Donko, D.: Knn with tf-idf based framework for text categorization. Procedia Eng. **69**, 1356–1364 (2014)
5. Jiang, S., Pang, G., Wu, M., Kuang, L.: An improved k-nearest-neighbor algorithm for text categorization. Expert Syst. Appl. **39**(1), 1503–1509 (2012)
6. Wan, C.H., Lee, L.H., Rajkumar, R., Isa, D.: A hybrid text classification approach with low dependency on parameter by integrating k-nearest neighbor and support vector machine. Expert Syst. Appl. **39**(15), 11880–11888 (2012)
7. Zhang, T., Oles, F.J.: Text categorization based on regularized linear classification methods. Inf. Retr. **4**(1), 5–31 (2001)
8. Leopold, E., Kindermann, J.: Text categorization with support vector machines. how to represent texts in input space? Mach. Learn. **46**(1–3), 423–444 (2002)
9. Khan, A., Baharudin, B., Lee, L.H., Khan, K.: A review of machine learning algorithms for text-documents classification. J. Adv. Inf. Technol. **1**(1), 4–20 (2010)
10. Karlik, B., Öztoprak, E.: Personalized cancer treatment by using naive bayes classifier. Int. J. Mach. Learn. Comput. **2**(3), 339 (2012)
11. Basu, A., Walters, C., Shepherd, M.: Support vector machines for text categorization. In: Proceedings of the 36th Annual Hawaii International Conference on System Sciences, pp. 7– pp. IEEE (2003)
12. Mertsalov, K., McCreary, M.: Document classification with support vector machines. ACM Computing Surveys (CSUR), pp. 1–47 (2009)
13. Soni, B., Das, P.K., Thounaojam, D.M.: Keypoints based enhanced multiple copy-move forgeries detection system using density-based spatial clustering of application with noise clustering algorithm. IET Image Process. **12**(11), 2092–2099 (2018)
14. Soni, B., Das, P.K., Thounaojam, D.M.: Improved block-based technique using surf and fast keypoints matching for copy-move attack detection. In: 2018 5th International Conference on Signal Processing and Integrated Networks (SPIN), pp. 197–202. IEEE (2018)

15. Soni, B., Debnath, S., Das, P.K.: Text-dependent speaker verification using classical lbg, adaptive lbg and fcm vector quantization. Int. J. Speech Technol. **19**(3), 525–536 (2016)
16. Li, Y.H., Jain, A.K.: Classification of text documents. Comput. J. **41**(8), 537–546 (1998)
17. Fuka, K., Hanka, R.: Feature set reduction for document classification problems. In: IJCAI-01 Workshop: Text Learning: Beyond Supervision (2001)
18. Yuan, Q., Cong, G., Thalmann, N.M.: Enhancing Naive Bayes with various smoothing methods for short text classification. In: Proceedings of the 21st International Conference on World Wide Web, pp. 645–646 ACM (2012)
19. Cavnar, W.B., Trenkle, J.M. et al.: N-gram-based text categorization. In: Proceedings of SDAIR-94, 3rd Annual Symposium on Document Analysis and Information Retrieval, vol. 161175. Citeseer (1994)
20. Liu, Z., Lv, X., Liu, K., Shi, S.: Study on SVM compared with the other text classification methods. In: 2010 Second International Workshop on Education Technology and Computer Science, vol. 1, pp. 219–222. IEEE (2010)
21. Joachims, T.: Text categorization with support vector machines: learning with many relevant features. In: European Conference on Machine Learning, pp. 137–142. Springer (1998)
22. https://www.kaggle.com/c/msk-redefining-cancer-treatment/data (2017)

Towards Artificial Intelligence: Concepts, Applications, and Innovations

Djamel Saba, Youcef Sahli, Rachid Maouedj, Abdelkader Hadidi, and Miloud Ben Medjahed

Abbreviations

AADRP	Agency for Advanced Defense Research Projects
AAOD	American Air Operations Division
AI	Artificial Intelligence
AIVA	Artificial Intelligence Virtual Artist
ATC	Air-Traffic-Control
DAAI	Design Assisted by Artificial Intelligence
HVAC	Heating, Ventilation and Air Conditioning
ITS	Intelligent Tutoring Systems
KBS	Knowledge Based Systems
KE	Knowledge Engineering
MEC	Mobile Edge Computing
ML	Machine Learning
MOOC	Massive Open Online Courses
ZAML	Zest Automated Machine Learning

D. Saba (✉) · Y. Sahli · R. Maouedj · A. Hadidi · M. B. Medjahed
Unité de Recherche en Energies Renouvelables en Milieu Saharien, Centre de Développement des Energies Renouvelables, URER-MS, CDER, 01000 Adrar, Algeria
e-mail: saba_djamel@yahoo.fr

Y. Sahli
e-mail: sahli.sofc@gmail.com

R. Maouedj
e-mail: ra_maouedj@yahoo.fr

A. Hadidi
e-mail: hadidiabdelkader@gmail.com

M. B. Medjahed
e-mail: benmedjahed_78@yahoo.fr

© The Editor(s) (if applicable) and The Author(s), under exclusive license
to Springer Nature Switzerland AG 2021
A.-E. Hassanien et al. (eds.), *Enabling AI Applications in Data Science*,
Studies in Computational Intelligence 911,
https://doi.org/10.1007/978-3-030-52067-0_6

1 Introduction

AI was appeared right after the first uses of the computers, more precisely at the Dartmouth congress in 1956, following the contribution of Alan Mathison Turing [1]. The four participants in this congress concluded that AI is: "the possibility of designing an intelligent machine". However, the science of "making machines work intelligently" is generally called AI. Indeed, AI does not has a common definition. Marvin Lee Minsky is the first definer for IA, it presents the IA as "the construction of computer programs which engage in tasks which are for the moment, accomplished in a way which can equalize human intelligence" [2]. Then, Entwistle defined AI as the intelligence presented by machines [3]. Furthermore, one of the first textbooks defined it as "the study of ideas that allow computers to be intelligent". This definition was updated in the following years, and the AI was then seen as an attempt to "make the computer do things for which people are better at the moment" [4]. It is also important to note that the term AI remains controversial, and left this question unanswered: "if a machine can ever be intelligent"? [5].

The first use of this innovative technology (AI) was back to John McCarthy [2] for effecting rather complex calculations. Subsequently, AI has been used to develop and advance many areas, including finance, health, education, transportation…etc. Among the most important applications of AI are the automaton and the robot. These two applications have the same function, but their designs are different. The automaton represents a mechanical concept, while the robot is much more electronic, perfected in particular with the aid of computers [6]. Some robots are automata and others are not. An automatic robot is a system that executes the tasks to be performed automatically. Even after finishing his tasks, he/she will continue them without stopping until the intervention of the human being. Some machines that could be considered as intelligent are not in reality. For example, the calculators are not AI concepts, because they perform requested calculations without being aware of them. They are equipped with programs that allow them to quickly solve the requested calculations.

At the beginning of the 21st century, the concept of AI has better developed and it is finding more applicability in the military and civil fields. Afterwards, it is considered to be a simulation of the processes of human intelligence by machines, in particular, computer systems. These processes include learning (acquiring information and rules for using information), reasoning (using rules to reach approximate or definitive conclusions) and self-correction. Specific AI applications include expert systems, voice recognition, and machine vision [1]. Hence, AI refers to any computer system that uses a logical process to learn and improve taking into account the system environment and previous errors. Therefore, one could argue that AI is a generic term encompassing intelligent robotics, ambient intelligence, machine automation, autonomous agents, reactive and hybrid behavioral systems, and big and small data [7, 8]. It can be said that developments in robotic systems have occurred in parallel with developments in the field of AI. During the initial stages of AI-based technological developments, robotics was more seen as a technology involving automatic systems where the machine would perform the preprogrammed activity. Current robotics is known for its decision-making capabilities. For example, speech recognition systems first interact with the user to gather information about the characteristics of their voice.

For the design and implementation of projects based on AI technology, several approaches are necessary in the realization, such as the domain ontology which is used as a representative form of knowledge about a world or a certain part of this world [9, 10]. They also describe [10]: individuals (basic objects), classes (sets, collections, or types of objects), attributes (properties, functionalities, characteristics or parameters that objects can possess and share), relationships (the links that objects can have them), events (changes to attributes or relationships), meta-class (collections of classes that share certain characteristics) [11]. The use of ontologies for knowledge management has proven to be advantageous in the field of research in AI where knowledge management is based on the representation of knowledge to simulate human reasoning to model knowledge in a usable way by machine [12]. They allow the representation of knowledge and the modeling of reasoning which are fundamental characteristics of KBS (Knowledge-Based Systems) [12]. The idea of distributed problem solving dates back to the mid-1970s with, the languages of the actors and the architecture model or blackboard, initially proposed for automatic understanding [13]. Therefore, a distributed AI system includes a set of autonomous entities, called agents that interact with them to complete a task that helps solve a complex problem [14]. The concept of agent has been studied not only in AI but also in philosophy or psychology [15]. Then, with the development of the Internet, there are several names appearing: resource agent, broker agent, personal assistant, mobile agent, web agent, interface agent, avatar, etc. In general, an agent is an informatics entity, locates in an environment, and acts autonomously to achieve the objectives for that it was designed. Its agents can also be physical entities (machines, manipulating robots, etc.): the domain is then that of multi-robot systems.

The remainder of this paper is organized as follows. Section 2 presents Intelligence and AI. Section 3 explains the evolution of Big Data and Data Mining. Section 4 explains the research paradigms of AI. Section 5 details each application's examples of AI. Section 6 clarifies the advantages and disadvantages of AI. Section 7 provides a discussion about our research. Finally, Sect. 8 concludes the paper.

2 Intelligence and AI

The word "intelligence" is derived from the Latin intellegere or intelligere ("to choose between") [16]. It should be noted that the Internet, by amplifying the mechanisms for reinforcing information between human beings, has greatly contributed to installing the concept of collective intelligence. Intelligence is the dynamic ability to be able to make inferences between stimuli, to deduce abstractions, to create a language that allows naming, exchange and make connections between these abstractions. Intelligence has made it possible to define the concept of context, thus being able to explain that the links are not necessarily repetitive. It is in all these capacities that humans are distinguished from other mammals. Not only can a dog not say "tomorrow", but it is likely that the concept of "tomorrow" is not developed in its cognitive capacities. Now have tools to understand the brain and know that the development of the cerebral cortex has allowed humans to define abstractions. Language functions, integral to intelligence, have been identified in certain areas of the brain, the Broca and Wernicke areas. Some animals have these areas, but they are highly developed in humans. It should be noted that this development of the brain explains an interesting phenomenon: most mammals are born almost adults, except humans. A foal can walk after a few hours, for example. On the other hand, it is because the development of the brain cannot take place in utero (because it would be too large to cross the cervix) that humans are born "too early". In doing so, they are quickly subjected to all stimuli from a more varied environment than the mother's womb. The impact on intelligence development may be better understood with the discovery of the fundamental role of epigenetics, a science studying the relationships between genes and the environment.

It is therefore not easy to define intelligence with language as the only tool. A dictionary has the particularity of being self-referencing since words are defined using words. On the other hand, you have to understand artificial intelligence as an attempt to understand intelligence using computer programs.

2.1 AI in History

The birth certificate of AI corresponds to the meeting program organized at Dartmouth College (Hanover, New Hampshire, USA) in 1956 [1]. On this occasion the term "artificial intelligence" was used to designate the new field of research; however, some researchers consider AI only complex processing of information.

Dartmouth's meetings were the result of a comparison of the functionality of the human brain with the first computers that just appeared (which mainly turned towards numerical computation) [1]. Thus the birth of AI was influenced in one way or another directly by various actions: notably, the works of Warren McCulloch and Walter Bates who suggested, inspired by neurophysiology, early models of artificial neural networks, and Norbert Wiener models on cybernetics (science-focused science) in Study of the communication and control mechanisms of machines and living organisms), the mechanisms of Claude Shannon in information theory, those of John von Neumann in computer engineering, and the mechanisms of Alan Turing in machine-calculated functions [17–20].

It was also in 1956 that Newell and Simon (in collaboration with J. Cliff Shaw), proposed a first computer program capable of demonstrating theorems in logic, before soon presenting a "general problem solver" ('General Problem Solver'), based on the assessment of the difference between the situation the solver has arrived at and the goal it has to achieve [21].

From the start, AI was involved in developing software, including chess. Other programs, notably Arthur Samuel and Alex Bernstein, appeared in the early 1960s, and over the decades they managed to conquer players from the higher and higher levels, such as MacHack's Richard Greenblatt [1]. The 1970s research in this field was distinguished by the idea of providing the machine with the ability to implement advanced strategies that develop dynamically through play (as in the works of Hans Berliner). However, it is primarily a computer computing power capable of exploring the giant harmonic spaces that conquer the discipline world champion (Deep Blue computer victory over Gary Kasparov, in 1997) [1]. AI's 1970s were also the first experiments to be carried out on portable robots (such as the Shakey Robot from SRI at Menlo Park in California), which jointly formed computer vision problems, knowledge representation, and planning [22]. A decade or so later, Rodney Brooks of the Massachusetts Institute of Technology interested in robot companies that respond to their environment [23].

As this brief historical outline illustrates, AI was first developed on a large scale in the United States, before the mid-1970s, and then researchers from Europe and Asia came after them.

2.2　AI Development Schools

Many schools have developed curricula and applications of artificial intelligence. Some of them rely on human intelligence to produce similar or better intelligence. Including those who tried to model intelligence in the form of rules and link them together to obtain better intelligence. As for the last school, it paid attention to modeling the human mind.

- **Human approach**: the first school tried to imitate the reasoning of a human being. A new profession was born: the cognitician, whose role consisted of discussing with an expert his problem-solving methods, then in projecting his expertise in a logical language understandable by a computer, for example, by making assistance tools decision-making in the form of graphs, "decision trees". This method encountered several problems [24]. First, the experts could feel dispossessed if they revealed their expertise; the cognitician then had to do all the work of a psychological approach so as not to offend their sensitivity. Second, traditional computer languages were procedural, and thus ill-suited to a logic of rules. Then, invented languages such as LISP or Prolog, whose syntax is that of a rules engine. These languages have now fallen into disuse. In addition, a rule engine does not appear to be logically rigorous, depending on the precedence of one rule over another.
- **Factor and data analysis**: the second school used statistical tools and, above all, rejected all that was modeling of the human, in other words, all that was the engine of rules [24]. The range of tools was wide, ranging from traditional classification methods, such as factor analysis, to sophisticated modelings, such as the Markov chains that have long been the basis of automatic dictation software. It was the occasion of an additional conflict between "soft sciences", those which privileged

a rather inductive approach based on human models, and "hard sciences", which, conversely, started from the data and refused all anthropomorphism.

- **Imitation of nature**: the third school wanted to model nature and the human brain. It was the start of neural networks, a category of algorithms that use the brain paradigm to solve problems. Neural networks are currently dominant in all applications related to pattern recognition: language comprehension, machine translation, image analysis, etc. They are, for example, the basis of autonomous cars.

3 From Big Data to Data Mining

Big data is the big break of our time, which has allowed artificial intelligence to make huge leaps with a world of mass data production. Some are public, others are private. In February 20, 2020, could observe in a single second: 8,856 messages sent on Twitter, download of 972 photos on Instagram, 1,658 messages sent on Tumblr, 4,437 calls on Skype, 87,429 GB of Internet traffic, 80,772 searches on Google, 82,767 videos viewed on YouTube, 2,884,128 Emails sent [25, 26] (Table 1).

The possession of this data gives enormous power to GAFA [27] and their Chinese equivalents. Nevertheless, Google offers us programming interfaces allowing us to query its own. Google Maps, for example, allows access to maps of the entire planet. Google offers a site where can compare trends on several keywords, with the possibility of zooming in on geographic areas, on time scales or even filtering on categories. As an example, can thus compare the evolution of interest in "Macron" versus "Trump" research between February 2019 and February 2020 (Fig. 1).

Google, Apple, Facebook, and Amazon couldn't help but invest heavily in AI when they have huge amounts of data. Here are a few examples.

- **Google**: this is very involved in AI, usually proceeds by the takeover. In 2014, Google bought the British company DeepMind, which had developed neural networks for playing video games [28]. DeepMind's stated goal today is to "understand what intelligence is." DeepMind is famous for its AlphaGo program, which beat the world champion of go. In October 2017, the program took an additional step: by playing against itself, not only was its learning shorter but above all, it became stronger than the previous version [29]. The first example of unsupervised learning, facilitated by the fact that the context, namely the rules of the game of go, is perfectly mathematizable. Google also has its own recommendation engine called Google Home, a speakerphone and a voice assistant available in three different versions.
- **Amazon**: uses AI in its recommendation engine, named Echo, and in its assistants based on its voice recognition system, Alexa, available in seven different versions [30]. Through its cloud service offering, Amazon also offers services based on AI, such as speech recognition or chatbots, the famous chatbot.

Table 1 Rate of use of informatics tools linked with big data and data mining

Date	Tool	Task	Utilization rate (in seconds)	Utilization rate (since the opening web page)
January 20, 2020	Twitter	Send messages	8856	1,132,273 Tweets since opening this page 0:02:08 s ago [25, 26]
	Instagram	Download photos	972	436,478 New photos since opening this page 0:05:30 s ago [25, 26]
	Tumblr	Send messages	1658	1,797,367 Tumblr posts since opening this page 0:08:44 s ago [25, 26]
	Skype	Calls (voice communications)	4437	3,112,981 Calls since opening this page 0:11:42 s ago [25, 26]
	Internet	Internet traffic	87429	76,145,595 Gigabytes since opening this page 0:14:31 s ago [25, 26]
	Google	Google searches	80772	101,686,086 Searches since opening this page 0:20:59 s ago [25, 26]
	YouTube	The vision of the videos	82767	92,237,136 Videos since opening this page 0:18:34 s ago [25, 26]
	Internet	Send Mails	2884128	4,209,690,894 Emails since opening this page 0:23:24 s ago. $\approx 67\%$ of email is spam [25, 26]

- **Facebook**: is a huge user of AI. It chooses the messages it displays using a recommendation engine type engine. Facebook recently implemented an artificial intelligence engine to detect suicidal tendencies. As Joaquin Candela, director of the applied AI department says, "Facebook would not exist without AI". According to Terena Bell [31] there are several ways in which Facebook uses AI among

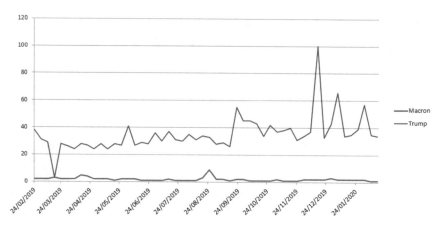

Fig. 1 Evolution in all countries of interest in "Macron" versus "Trump" research between February 2019 and February 2020

them, either to connect to brain waves, or to decide which TV shows to watch, or to solve Facebook Live problems, or to ensure employee safety.

- **Apple**: it invests heavily in AI and even offers a blog where its research is explained. Siri is the most obvious expression: this high-performance speech recognition system uses the latest generation of neural networks which enable commands to be understood even in very noisy environments [32]. Apple will also launch its own recommendation engine, called HomePod, a speaker connected with Apple Music [33].
- **IBM**: Watson analyzed 200 million pages to finally beat the former champions and win the first prize. Now Watson is offered by IBM in other fields, such as medicine and law. IBM also has a blog on the subject of AI.

The major question is the degree of openness of these systems. IBM offers programming interfaces to Watson that are open, just like Amazon offers interfaces on its deep learning system. So anyone can inexpensively connect to and benefit from these systems. Facebook, on the other hand, does not share most of its research, does not yet offer programming interfaces to use it, but has recently made its learning environment. This apparent simplicity of connection is not sufficient, it must be emphasized that it is not the algorithms that hold the main value. These AI systems are all the more efficient because they are contextualized and therefore need targeted data to benefit from effective learning.

In 1988, Indian computer researcher Raj Reddy thus defined the main challenges of AI: the translating telephone (under development), the chess champion (completed), the resolution of mathematical theorems (some theorems are already), autonomous cars (realized), self-organizing systems (in progress) and self-replicating systems (in progress) [34].

4   AI Research Paradigms

Research directions for AI include representation and data acquisition, problem-solving algorithms, collective AI, formalization and reasoning mechanisms, evaluation, decision and planning, reasoning on the change of time and place, the interrelation between language and AI, indexing and multimedia, and virtual reality.

4.1   Representation and Data Acquisition

With the diagnostic system of MYCIN in 1975, the structure of problem-solving systems met the requirements of establishing AI, by separating the (motor) program on the one hand with general thinking capabilities (reasoning capabilities), and on the other hand the application-specific knowledge (knowledge base), often they are represented in the form of phrases and rules "if the conditions are then conclusions [35]." Representations, based on logic, or structured as "frames" (groups of context-specific objects, which Minsky suggested to represent visual concepts at first), scripts (describing the sequence of events), and other representations of this type, such as semantic networks (also entered Networks of associations associated with associations in AI that were later developed in classic computer applications such as "object programming," which address the concept of framework by representing entities to which behaviors are linked, or like databases. These systems have won with their modular architecture, making the simple development of knowledge-based applications can be envisioned in various fields such as mechanics, medicine, chemistry, etc., and for various tasks: fault diagnosis and repair, planning, etc. The only difficulty was to inform the base using the concepts and rules of the domain in question, taking into account its use in thinking. However, this difficulty has become responsible for not using many systems, however technically valid they are: they do not contain the correct knowledge! So the key issue of knowledge acquisition, which has become a real area of research, is to help define user support systems that are used because it correctly deals with the right issues. This field of computing uses psychology (to study and collect the patterns of thinking of individuals), as well as the work environment and social sciences (to determine the knowledge that adapts to the needs of users in the workplace), or even linguistics when one wants to extract from knowledge. The original problem now includes designing "smart" user-assisted applications that seek to define principles, methods, and tools to guide the process of collecting, analyzing, and structuring the knowledge necessary for the final application. The mainstream of research is based on the main idea of the conceptual model. The model is seen as a simplified representation, justified by application needs. It is understandable because it can be interpreted by the individuals involved in the project and can be translated into an operational language, which is interpreted by the device. The system should include rational (goal-driven), efficient, relevant, and possibly adaptable, behavior towards users and experts in the field. Important applications were presented in various fields, such as:

- Computerization of medical records in hospitals;
- Diagnose malfunctions in the operation of smelting furnaces;
- Intelligent search for information in heterogeneous data sources on the internet (with ontology) [36].

The exchange of knowledge, the sharing but especially the localization of knowledge or the identification of skills is presented as being within the reach of everyone and any company. Far from yielding to the charms of attractive speeches linked to knowledge management in companies, knowledge engineering studies closely, in close collaboration with managers, specialists in information systems or human sciences, scientific issues and theoretical of this deceptive facility. It is a question of supplying technological resources in a relevant way, of facilitating a finalized use which helps to filter, find, and organize the knowledge which each one needs, in a more or less operational form. Thus, by turning to corporate information systems and information technology, knowledge engineering continues to address, within new application frameworks, the recurring problem of access to meaning and knowledge.

4.2 Problem Solving Algorithms

To attack a problem, it can solve it into sub-problems and then analyze it, etc. A set of possible decompositions can be represented by a 'sub-problem graph'. Replacing a problem with an equivalent problem can be seen as a particular form of decomposition. In this case, looking for a sequence of equivalence operations which leads to an already resolved problem. By representing the problems by vertices and the equivalence operations by edges, then, come back to the search for a path, possibly optimal or close to the optimum, in a "state graph". In general, there are so many possible routes that one cannot examine all the alternatives. On the other hand, it sometimes happens that research can be guided by knowledge of more or less analytical or experimental or empirical origin that is called "heuristics": this is called "Heuristically Ordered Resolution". So to build an itinerary, trying to reduce the distance as the crow flies between the objectives is a familiar and often effective heuristic. The heuristics and the various algorithms which are capable of exploiting them are interesting if they favor the discovery of satisfactory, even optimal, solutions in relation to the difficulty of the problem and the resources mobilized.

After the success of programs playing backgammon, the victory of the Deep Blue chess player system in 1997 against reigning world champion Gary Kasparov is a spectacular testament to the progress made over the past half-century to explore vast areas sub problem graphs [37]. Heuristically Ordered Resolution and Resolution by Satisfaction of Constraints can be used, possibly together, in various contexts: identification, diagnosis, decision, planning, demonstration, understanding, learning… These approaches are today able to tackle problems of greater size or more complex, possibly in dynamic environments with incomplete or uncertain information. Their development goes hand in hand with advances in the power of computer systems.

4.3 Collective AI

Since its inception, AI has mainly focused on the theories and techniques allowing the realization of individual intelligence. There is another form of intelligence— collective that one - like simple multi-cellular beings, colonies of social insects, human societies. These sources of inspiration show that a higher form of intelligence can result from the correlated activity of simpler entities. In artificial systems, this field is called Distributed AI or Multi-Agent Systems, which include in the term of Collective Artificial Intelligence (CAI) [15]. Then, in the mid-1970s, part of AI explored the potentials of knowledge distribution leading to distributed problem-solving. The distribution of control is the main characteristic of the second generation of systems in which a single entity does not have control of the others and does not have a global view of the system [38]. But therefore, it is not enough to have a bunch of agents to obtain a system of the same name; just like a pile of bricks is not a house [39]. Thus, the field of study currently relates to the theories and techniques allowing the realization of a coherent collective activity, for agents who are by nature autonomous and who pursue individual objectives in an environment of which they have only one perception partial [7].

These agents have the ability to interact with other agents and their environment. They are social in the sense that they can communicate with others [11]. When they react to perceived events, they are called reactive. They are proactive if they have the capacity to define objectives, to take initiatives... Today; the concepts of autonomy, interaction, dynamics, and emergence are increasingly taken into account for the specification of these systems, leading to specific design methodologies. In addition, any agent that requires 'intelligence' can benefit from the contribution of the other AI techniques. However, the specificity of the field relates on the one hand to organizational theories allowing the realization of a 'collective intelligence' under the previous constraints, and on the other hand to the techniques of interactions between agents to manage all the unforeseen (conflict, concur open system or in a dynamic environment.

The CAI has developed in particular to facilitate the simulation of applications distributed geographically, logically, semantically or temporally [12]. Multi-agent platforms are privileged tools for simulating companies of individuals immersed in a common environment, such as a team of robot footballers. This is particularly the case with environmental simulations where wish to observe the influence of individual behavior on global phenomena (pollution, forecasting of floods, etc.). In addition, the CAI is aimed at applications for which it is difficult to design and validate large-scale software with conventional development techniques. Thus, the CAI is also concerned with the design of software components which will be intended to be reused. Software "Agentification" should make it possible—in the long term—to assemble them from a library without knowing their specification finely: they are responsible for adjusting their individual behaviors in the prior framework of the task defined by the application. Agent and multi-agent technology aims to become a new programming paradigm in the years to come. This field is also motivated by the evolution of informatics tools: cost of machines, machine networks, Internet... Indeed, the Internet is today another privileged field of investigation of the CAI where agents collaborate in a common activity (electronic commerce, information retrieval, etc.) without a global control algorithm. These are often applications called open systems because the number of agents in the system changes over time [40]. This problem is also found in the field of collective robotics.

The previous fields of application have in common that they place the individual agents but also the collective in dynamic environments where it will be necessary to face the unexpected and therefore where it will be necessary to adapt. Thus, the agents must have the possibility of exchanging on their respective problems which they will necessarily meet in their activity, especially since they have only a partial knowledge of the world in which the system evolves. If agents never have to face the unexpected, then multi-agent technique is not required. The work on agents focusing mainly on interactions, they were inspired by research on human dialogue, in particular the work of D. Saba [15, 38]. It follows attempts to formalize the acts of communication and standardization (FIPA). Another important element concerns the organizational model for carrying out the collective task. It has long been proven that there is no ideal organization and that organizational change is a way to transform the function of the system. The question is thus approached in three directions according to the presuppositions adopted:

- Either the organization is a controllable element of the system or then the multi-agent system brings concepts and techniques from organizational theory, leading to specifications of organizational models.
- Either the organization is an unknown element or it is necessary to give the agents strategies to participate in a coherent collective activity. The "Beliefs Desires Intentions" models or the dependency networks between agents have been defined to allow the implementation of such organizations.
- When the work simultaneously considers that the organization is not knowable and the crisis exit protocols are not either, then it is in the field of the emergence of coherent collective functions from entities having elementary behaviors, such as work in "emergent computation".

4.4   Formalize and Mechanized Reasoning

The common sense reasoning allows him to reconcile situations and transpose solutions and to take advantage of both general facts and examples, or even to capture quantitative information [41]. Such reasoning processes exceed the capacities of representation and inference of classical logic (specially developed in the 21st century in relation to the question of the foundations of mathematics), and of the theory of probabilities with regard to certain aspects of uncertainty. This does not mean, however, that the reasoning that man makes outside of mathematical deduction is completely devoid of rigor, or of practical value. Thus AI is interested in formalizing different forms of reasoning studied for a long time by philosophers: deductive reasoning (used for example in syllogisms), inductive reasoning (or generalization, used in learning), or adductive reasoning (who seeks to explain, in terms of causes, and observed situation and who is at work to establish diagnoses), or even reasoning by analogy. The latter makes it possible, from a repertoire of observed cases, to extrapolate a plausible conclusion for a situation encountered presenting similarities with known cases. The AI has also formalized various extensions or weakening of deductive reasoning, whether to propagate uncertainty (probabilistic or not) associated with information or to be able to produce plausible conclusions, in the event of incomplete information. In fact, the "if... then..." rules, which constitute a convenient format for expressing knowledge (even specialist knowledge), do not always explicitly specify all the exceptions to the rules. Otherwise, they would be too cumbersome to describe and would often require more information to be applied than is available. Another transcendence of the classical deduction is based on the idea of proximity (a conclusion, strictly speaking false, but close to another conclusion, proven as for it, will be considered almost true), and then makes it possible to obtain more conclusions by interpolation between proven conclusions. Then, deductive reasoning requires observing contradictory information [42]. In fact, the information often comes from multiple sources, from unequal unreliability, and the aggregation often leads to a recommended conflict, either in the processes of "merging" the information or taking it into consideration if the person seeks to develop systems capable of proposing arguments in favor or against the same statement. This can be especially faced in situations of dialogue, where each of them does not have the same information and seeks to clarify his information, and perhaps also to understand the intentions of the other. Finally, theoretical research for many "expert systems" (i.e. exploiting knowledge provided by experts), increasingly powerful and sophisticated, capable of offering opinions, solutions, diagnoses, forecasts, are developed in practice. Such systems, of course, pose human interface problems (user, expert, designer) in terms of the representation and acquisition of information. They must also be given the ability to provide explanations so that the conclusions reached by the calculation can be clarified. Finally, the way in which uncertainty is understood and returned to the user must also be studied from the point of view of cognitive psychology.

4.5 Evaluate, Decide, and Plan Situations

Decision theory (multi-criteria decision, uncertain decision, and group decision) has been studied for over fifty years by economists and researchers in Operational Research [43]. For the former, it is a question of modeling decision-making among "rational" decision-makers, that is to say, economic agents whose behavior is permanently in agreement with postulates, while the latter considers the decision more in view of optimization problems where, for example, several criteria are involved. The purpose of the decision models thus developed is to classify possible decisions, often using numerical evaluations (the use of which is justified by postulates). These methods do not allow us to easily go back to the reasons that make one decision better than another. In addition, these models often assume the knowledge of numerical functions to evaluate the choices, and of probability distributions to describe the uncertainty on the outcome of the actions. However, the preferences of the agents are not always known in a complete (and coherent!) manner in the form of evaluation functions.

The problem of decision-making was not one of the central concerns of AI until the early 90s: AI, being turned towards symbolic and logical modeling of reasoning, was indeed quite far from the numerical operations of compromises used in the decision [44]. However, it has appeared for a few years that AI can bring tools to the decision-making problem allowing a more flexible and more qualitative representation of information, preferences, and goals, and proposes formulations which then lend themselves more easily explanatory possibilities of the proposed decisions. Such approaches are useful in guiding a user in his choices, by offering him solutions that meet his preferences, which he can change if necessary. A decision support system can also be based on the known results of decisions made previously in situations similar to that where a decision must be proposed. Finally, planning (which includes monitoring the execution of tasks) is a decision problem studied for thirty years in AI. It is a question of determining a chain of actions (the simplest possible) which makes it possible to achieve a goal from a given situation. The planning of elementary operations for a robot (or more generally a set of robots) in order to achieve a more global task is the classic example in AI of this problem, which can also be encountered more generally in the organization of complex activities. One of the difficulties lies in the fact that a series of actions leading to the partial satisfaction of the desired goal may not constitute a piece of a solution making it possible to achieve the goal completely. Since the early 1990s, the use of probabilities or other models of representation have made it possible to extend this problem to uncertain environments (where, for example, an action can fail), taking into account cost and the possibility of including informative actions (which provide information on the state of the world) at each decisional stage.

The previous situations concerned only an isolated agent (faced with a possibly uncertain environment). For the past ten years, AI researchers have also been concerned with decision-making concerning a group of individuals. Note that this is a centralized decision (as opposed to the "Collective Artificial Intelligence"' section where it is a distributed decision): a group of agents, each with their own preferences, must reach a joint decision [45]. The quality criterion of this joint decision depends on the type of application considered: thus, for voting problems or more generally the search for compromise, it is important to obtain a fair solution; whereas for so-called combinatorial auction problems (increasingly used in electronic commerce), it is a question of maximizing the total gain of a set of sellers, by determining an optimal allocation of goods to agents (buyers), knowing that each agent has previously expressed the amount of money he is willing to pay for each possible combination of goods.

4.6 Reasoning on Change, Time and Place

An intelligent system has two basic capabilities. On the one hand, it is the possibility of perceiving the outside world, and on the other hand the ability to act on it. The perception is done through sensors, while the action is accomplished through motor control (the effectors of a robot) [46]. In computer language and control theory, it is these sensors that provide the inputs, while the effectors provide the outputs. An instance of this perception/action scheme is for example communication between agents. From this analysis, obtain the basic architecture of an intelligent system if add as a third component of the internal state in which the agent is at a given instant. The actions taken by the agent are then a function of this internal state, which is modified according to perceptions. To be considered intelligent, the internal state of this system must allow it to reason. This is why in the standard approach, this internal state is based on a symbolic representation of the external world, consisting of "facts", that is to say, assertions corresponding to what the agent believe to be true. In the world these internal facts are often called the agent's beliefs [10]. Of course, these beliefs are generally imperfect (incomplete, mistaken), and new perceptions frequently lead to a change, or update, of beliefs. At an elementary level, an agent performs an action, most often within a sequence of actions, also called a plan. At a higher level, the agent who performs an action seeks to achieve a goal, which he believes to achieve by executing a previously generated plan. It is, therefore, his beliefs that come into play when he plans, that is to say, he reasons to generate a plan of elementary actions. Note that a plan can also call on perceptual capacities through communicative actions or test actions. In addition, the formalisms which have been proposed in the literature to take into account the evolution of knowledge bases (the formalisms for updating and reasoning about actions) come up against several difficulties. The best known is the famous problem of the persistence of information during a change (the problem of "decor"): how to avoid describing the non-effects of a change? As surprising as it may seem, many formalisms force us to point out that turning on my computer will not change my neighbor's state. Today generally consider that the decor problem has been solved in its simple version, namely that the actions have no "side effects": turning on the central unit of my PC will not turn on its screen. It is, therefore, the interaction of this last problem called "branching" and the problem of decor in all its generality which constitutes a formidable challenge not yet resolved to date.

The evolution of an agent is punctuated by the perception of objects, events, and facts, and by the execution of actions. On the other hand, generating a plan means determining the actions to be carried out as well as the sequence they constitute, that is to say, their temporal scheduling. Time, a fundamental parameter in the reasoning of agents, must, therefore, be represented. As with so many fields, the ontological nature of time is not unquestionably imposed on us, and different approaches to the representation of time have been proposed in the literature [13]. One factor that has motivated many theoretical developments concerns the notion of the duration of temporal entities. Either assume that the various events (actions and perceptions) are instantaneous (or that their temporal component can be expressed in terms of instants), or consider their temporal extent as being non-zero and irreducible. This second approach is the only valid one in general if the agent does not have an internal clock determining an "absolute" time, and can only situate various events in relation to each other, such as the opening of the tap, flowing water, filling the glass and drowning the fly. Studies have shown that temporal reasoning systems known as "interval calculation" (although the term "interval" in mathematics denotes convex sets of instants, these are extended primitive temporal entities) are unfortunately more complex than those based on instants. All scheduling problems expressed in terms of intervals and which are "calculable" have not yet been identified and classified.

A distinction is made between epistemic (effect on knowledge) and ontic (effect on the physical world) actions, considering that the former have effect only on internal states, while the latter have effects on the state of the world and therefore fit into the space surrounding the system [47]. Being able to reason about these last actions, therefore, requires space modeling. As with time, they are then faced with choices of representation. Although it is possible to resort to classical mathematical analysis (Euclidean space, Cartesian coordinates), many approaches have preferred modeling which is not only closer to human cognition but also more robust in the presence of incomplete information or imprecise. This is why the spatial reasoning called "qualitative" is most often based on a space-based on objects, bodies or volumes having a perceptible spatial extension, rather than on the primitive abstract entities that are the points and the figures of the classic geometry. If the works have advanced a lot with regard to the modeling of merotopological concepts (inclusion, contact) on objects or bodies, at present, reason qualitatively on geometric concepts (distance, alignment, shape…) with extended entities remains a difficult problem [48].

4.7 Summarize, Learn, Discover

When an interest in simulating a person's intelligent behavior is concerned with a machine, a distinction can be made between the ability to understand and the ability to learn, assuming this is possible. The field of learning in artificial intelligence is just an attempt to simulate with a machine this remarkable ability to know the man who illustrates. Where do robots recognize that, after several unsuccessful attempts to track a path, that the door is closed and that it is necessary to find another solution, from where does the program that stimulates, starting with the history of the individual's Internet connections, start with the same profile, and therefore programs that use the huge database of the human genome correspond to different observations and ultimately "discover" that a particular gene is responsible for a specific disease. At the heart of the learning problem, it remains that seeing the precise meaning give to the word "learn" and this is the first challenge for the computer world in this field. Hence the importance of foundational theoretical work that attempts to give a formal framework for machine learning. Then distinguish about concepts that can be learned from others that cannot. Somewhat similar to the way Alan Turing's work distinguishes what can be calculated without what can be calculated! Then realize that for the concept to be "learnable", it must be able to summarize it (the computer scientist says "compressed") but this is clearly not enough [49]. "Learn" not just "summarize", it's more than that! Indeed, what have learned should also allow us to provide answers to questions other than those used in the learning process: in other words, "learning" is also "generalization".

Therefore, it is possible to search for algorithms and therefore to build programs that "learn"... Their execution on a machine then gives the idea that the machine is at school. A new question then arises: learn, yes, but from what? What information does have? In general, the learning process involves at least two agents: an apprentice and a teacher. Again, are inspired by the human model. And a question and answer process begins. By simplifying a bit, learn from examples and counterexamples. A wide range of techniques was then developed to learn from examples. Statistical methods, historically the oldest, are very effective [50]. But techniques specific to Computing were developed from the 80 s: neural networks which simulate the architecture of the human brain, genetic algorithms which simulate the natural selection process of individuals, inductive logic programming which makes " walk backward "the usual process of deduction, Bayesian networks which are based on probability theory to choose, among several hypotheses, the most satisfactory. There is no connection between designing machines that learn and using computers to teach (this is part of Computer-Aided Instruction or CAI) [51]. Today, machines are starting to learn, but the results, although often remarkable, are still far from the performance of an elementary school child who learns a lot without understanding how. This means that there is still work to clone the man by a machine.

4.8 Interrelations Between Language and AI

Human beings are able to understand and speak one or more languages. This language ability is inseparable from human intelligence [52]. It is therefore legitimate to ask the question of the relationships which unite AI and language. The texts emphasize that a central task of AI is to formalize human knowledge and reasoning. Language is the main means at our disposal for expressing our reasoning and conveying our knowledge. Messages or exchanges in natural language, therefore, represent, for some AI researchers, a privileged field of observation. Thus, for example, the study of the expression of space (through prepositions, verbs, localization names,...) in language allows it to advance studies on spatial reasoning or does the systematic study of a corpus of specialized texts make it possible to acquire knowledge specific to a domain through the study of the organization of the specialized lexicon of this domain. On the other hand, if one seeks to endow the computer with capacities resembling those of humans, it is natural to consider making the computer capable of understanding, or in any case of automatically processing information in natural language, and make it able to generate natural language messages wisely. This is where AI and linguistic computing come together. To illustrate the mutual fertilization of AI and linguistic computing, let's take the fields of semantics and pragmatics. It is about building a representation of messages in natural language that can be manipulated by the machine. This implies the modeling of a very large number of semantic phenomena, such as reference mechanisms (pronominal, spatial, temporal: from there, I saw you with his binoculars), quantification, complex operators (negation, modalities, epistemic operators, argumentative because, because of), the generative dynamics of language (production of metaphors, metonymies, shifts or fluctuations in established meanings: rusty ideas, selling a Picasso, etc.). In addition to the phenomena related to the meaning and the relationships that the statements establish with the world, it is necessary to be interested in phenomena concerning the contextualization of the statement in a situation of communication. Thus, for human-machine dialogue, seek to formalize the acts of language, to take into account the model of the user, speaker or listener, to represent the evolution of knowledge and beliefs during the dialogue, and to be able to generate responses, model how to present, argue and be cooperative with a user [53].

A part of linguistic computing is becoming a technological discipline, with major challenges in the industry and in society. It imposes requirements of efficiency, simplicity, and expressiveness in the mechanisms and in the representation of considerable volumes of data. From this point of view, the techniques of automatic knowledge acquisition and modeling are precious. While the applications that revolve around automatic indexing, information retrieval, and even automatic translation, still make moderate use of AI techniques, other fields such as dialog, automatic summary or Cooperative human-machine interfaces must and call heavily on the work of AI in formalizing reasoning [54].

From a fundamental point of view, language analysis and linguistic computing have repositioned themselves [55]. Starting from idealized representations and reasoning, in the tradition of Cartesian thought or of the mind - pure logic, numerous works, inscribed in a double cognitive and anthropological perspective, have shown that our concepts, even the most abstract, are largely metaphorical, that is to say, constructed by abstraction from concrete representations. Other work, in particular generativist, has shown that language appears more like an instinctual component than an acquired aptitude. The relationships between AI and linguistic computing are therefore evolving. An axis which includes the paradigms proposed, among others, by Darwin, Turing and Chomsky is constructed and is singularly highlighted: evolution and creative dynamic of language, calculation, and modeling by principles and parameters postulated to be instinctual, implemented by constraints at the expense of heavy systems of rules, postulated acquired [56].

4.9   Multimedia Indexing and AI

Multimedia resources available today on the computer are numerous; they often remain difficult to access due to the absence of a systematic index listing relevant information on their content. It is not always possible to manually associate a text description or a few keywords with multimedia content to make further research on this information [57]. In this case, automatic indexing must take over. The researchers involved in the fieldwork on all the points of the chain which links the production of the document to the consultation of this one while passing by its encoding and its storage. Whether it is for extracting knowledge about the content of documents, managing and structuring the data produced and making them available to users, AI is increasingly involved in the multimedia automatic indexing chain. One of the most important steps for already existing documents to be indexed automatically is to extract descriptive data called metadata. These metadata are, for example, the time segments or the regions corresponding to an important or at least structuring object of the document. This automatic extraction of metadata requires the implementation of pattern recognition tools or the learning of models of the documents to be analyzed [58]. Can thus use so-called "stochastic" models that characterize the evolution over time or in space of the content: in speech, a locution is expressed by a succession of realizations of phonemes; in a television program, one sequence always takes place before another. These models, learned a priori according to the objects to be recognized, thus allowing the machine to locate such or such event. Other techniques consist of exploiting expert knowledge transcribed through a database of heuristics. Can thus signal to the machine that the presence of an object in an image is characterized by the presence of a region of such shapes, such as color, and such texture. Sometimes this combination of several low-level descriptors cannot be formulated in a precise and categorical manner. Then uses the tools of fuzzy logic to apprehend specifications referring to non-clear-cut properties. For example, to detect an advertising page broadcast on television, can characterize it as being a segment having a fairly rapid assembly, significant movements of objects or cameras, and visually presenting numerous bright and contrasting colors. Multimedia indexing also poses the problems of representing, structuring and using metadata in the same system, or between different systems. Judicious choices and specific technologies are essential in order to manage a knowledge not always reliable and precise, to build indexes and to define distances to optimize the time and the quality of the search for information, to translate these indexes by widening their semantic content, establishing links between them, studying the nature of structures and links to improve the exploitation of data.

The "intelligent" exploitation of the data goes hand in hand with the development of "learned" consultation tools capable of adapting to different modes of access and to different types of use [59]. This type of information, coupled with a symbolic representation of content, can for example dynamically build optimal paths to the data sought by taking into account ergonomic and cognitive considerations. Finally, the field of application of multimedia indexing is vast: it concerns production aid as well as analysis aid, the design of archiving tools as well as encoding tools. Even if solutions had to be found in each area to meet obvious needs, no homogeneous approach had been proposed until then. It is, therefore, a question here of convergence of technologies among which AI has an important role to play.

4.10   Virtual Reality and AI

Virtual Reality offers new forms of interaction between humans and systems. The advent of networked workstations with very strong 3D graphics capabilities coupled with new visualization and interaction devices whose use is intuitive (headphones, glove…) makes it possible to provide several users with sensory information necessary to convince them of their presence in a synthetic world [60]. In addition, the possibility of manipulating certain aspects of these virtual worlds almost as in real life offers stakeholders the possibility of using their experience and natural capacities to work in a cooperative manner. The synthetic world created and managed by a Distributed Virtual Reality system can, therefore, be virtually populated by numerous users who interact, through specialized peripherals, with these virtual worlds and in particular with autonomous entities, animated and endowed with complex behaviors: collaborative behaviors, but also adaptive and not only reactive, that is to say with the ability to reason and learn, or even seeking to anticipate the dynamics of the world in which they find themselves, or even emerging behavior [60]. Work in virtual reality is thus at the border of many fields: distributed systems (and applications), networks, geometric modeling and image synthesis, human-system interaction, for the systems themselves, but also, techniques from classical artificial intelligence or artificial life, for the management of autonomous or semi-autonomous entities (agents) endowed with behaviors.

An important aspect of work in image synthesis, in CAD and therefore in virtual reality, concerns the modeling of the virtual universe. The goal of this first step, crucial for the creation of complex 3D worlds, is to provide a 3D geometric description of the universe [61]. To facilitate and automate this complex task, but also to control it, it is necessary to manage constraints or check properties on objects or on the Stage. Thus, declarative modeling and constraint modeling are interested in issues of Artificial Intelligence, such as natural language and the semantics of spatial properties and in their translation into a large set of complex constraints supplying resolvers that can use an exploration of the space of solutions either exhaustive or stochastic.

One of the most promising motivations for the modeling of these synthetic actors is to artificially reproduce living properties linked to adaptation. The first form of adaptation considers a point of view rather centered on the individual and on the way in which he can learn during his life, according to his different environmental conditions. The second is more focused on the species and its genetic characteristics and focuses on modeling theories of evolution and how they can improve the performance of individuals by making them evolve. Finally, the last one emphasizes the collective and social phenomena that can appear in groups of individuals, such as the processes of communication or cooperation. Some of these concerns are also at the heart of current research conducted in Artificial Intelligence. Finally, studies and developments around Virtual Reality therefore often make significant and original contributions to the various fields mentioned, through all these many interactions.

5 AI Applications

Today, AI occupies a large place. Moreover, it is likely to grow in scale in the years to come. So it has evolved in many fields (health, informatics, industry…). As a result, the one today is capable of performing many functions such as treating patients, assembling parts to achieve automobile construction (Fig. 2). This intelligence, therefore, makes it possible to improve the living standards and work of people.

a. **Smart home**

The concept of a smart home is not new and has existed since long before the birth of IoT [62]. The idea of the smart home is to monitor, manage, optimize, remotely access, and fully automate the home environment while minimizing human effort. The IoT provide the underlying ecosystem that supports, and the easy realization of a smart home application.

The IoT-based smart home uses both local storage and processing units (such as a gateway or hub) as well as cloud infrastructure [63]. With the increase in the size of the environment of study, informatics performance should be improved significantly.

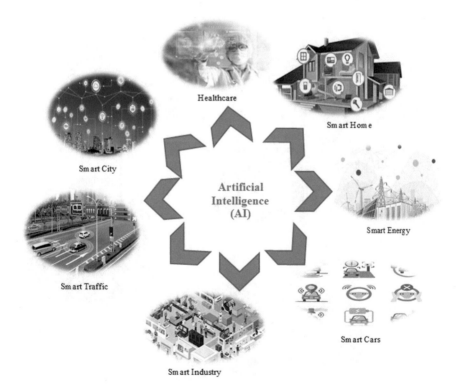

Fig. 2 Applications of artificial intelligence

The smart home is sometimes seen as an extension of the concept of the smart grid [64] From this point of view, the main objective of the smart home is to optimize energy consumption by taking into account various inputs such as the mode of use and the real-time presence of residents, the external environment (for weather conditions) [9].

The main elements of the smart home application are the surveillance systems, the intelligent HVAC (heating, ventilation and air conditioning) system, the intelligent personalization of the environment according to the user profile, the intelligent management of the environment and object (Fig. 3). However, the key factors, or relevant challenges for the smart home, are high security, high reliability, high interoperability, strong adaptation to a wireless environment with multiple risks, etc.

b. **Aviation**

Aircraft simulators use AI to process data from flights. In the field of wars using aircraft, computers can find the best success scenarios [65]. Computers can also create strategies based on the location, size, speed, and strength of forces and counter-forces. Pilots can receive help in the air during combat by computer. Artificial intelligent programs can sort information and provide the pilot with the best possible maneuvers, without forgetting to get rid of certain maneuvers that would be impossible for a

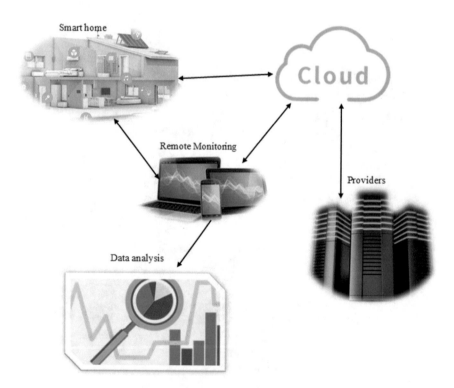

Fig. 3 Essential elements of the smart home

human being to perform. Several aircraft are necessary to obtain good approximations of certain calculations so that the computer-simulated pilots are used to collect data [66]. Computer simulated pilots are also used to train future air traffic controllers [67].

c. **Smart city**

Many countries have already started their smart city project plans, including Germany, the United States, Belgium, Brazil, Italy, Saudi Arabia, Spain, Serbia, the United Kingdom, Finland, Sweden, China, and India [68]. The main elements of a smart city based on IoT are smart hygiene, traffic, management, smart governance, smart switching, smart monitoring, and smart management of public services,... (Figure 4). In addition, the main elements in the development of the existing Smart City based on the Internet of Things are high scalability, high reliability, high availability and high security. Also, a range of technologies such as RFID, ZigBee, Bluetooth, Wi-Fi and 5G can be used in the future.

d. **Computer science**

AI researchers have created many tools to solve the most difficult computer problems. Then, Many of their inventions have been adopted by mainstream computing

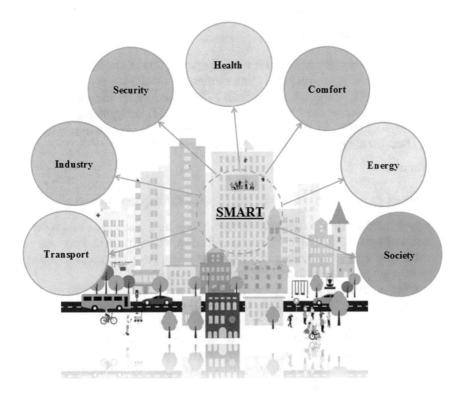

Fig. 4 IoT-based smart cities

and are no longer considered part of AI. According to Ligeza and Antoni [69], all of the following were developed: rapid development environments, symbolic programming, time-sharing, interactive interpreters, lists of data structures, functional programming, object-oriented programming.

AI can be used to create another AI. For example, in November 2017, Google's AutoML project is develop new neural network topologies, and a system optimized for ImageNet [70]. According to Google, NASNet performance has exceeded all previously published ImageNet performance.

e. Education

There are several companies that create robots to teach children, although these tools are not yet widely used. There has also been an increase in intelligent tutoring systems, or STIs, in higher education. For example, an STI called SHERLOCK teaches air force technicians to diagnose electrical system problems in aircraft [71]. Another example is DARPA, the Defense Advanced Research Projects Agency, which used AI to develop a digital tutor to train its navy recruits in technical skills in a shorter time.

Data sets collected from e-learning systems have enabled analysis, which will be used to improve the quality of large-scale learning. Examples of learning analytics can be used to improve the quality of learning [72].

f. Smart energy management

The tasks involved in the energy sector are the production of electricity, the transmission of electricity, distribution to end consumers, monitoring and maintenance including detection and rectification of faults [73]. However, the early use of IoT is fairly visible, since counters intelligent multi-purpose and intelligent thermostats are already deployed (Fig. 5). Continuing its momentum, the IoT has an important role to play both for the public service provider and the consumer.

g. Finance

AI applications in finance can be restricted to the following sub-branches:

- **Algorithmic trading**: algorithmic trading involves the use of complex AI systems to make business decisions at speeds faster than any human often making millions of transactions per day without any human intervention. So, Automated trading systems are generally used by large institutional investors [74].
- **Market analysis and data mining**: several large financial institutions have invested in AI engines to help them with their investment practices. Black Rock's AI engines, Aladdin, are used both within the company and with clients to help them make investment decisions [75]. Its wide range of features includes the use of natural language processing to read text such as news, broker reports, and social media feeds. It then measures sentiment on the companies mentioned and assigns a score. Banks like UBS and Deutsche Bank use an AI engine called Sqreem (quantum sequential reduction and extraction model) which can extract data to

Fig. 5 Components of smart energy management system

develop consumer profiles and associate them with the wealth management products they would like [76]. Goldman Sachs uses Kensho, a market analysis platform that combines statistical computing with the processing of big data and natural language. Its machine learning systems extract data accumulated on the Internet and assess the correlations between world events and their impact on asset prices [77]. Information retrieval, which is part of AI, is used to extract information from the live news feed and to facilitate investment decisions.

- **Personal finance**: several emerging products are using AI to help people with their personal finances. For example, DIGIT is an application based in AI, which automatically helps consumers to optimize their spending and their savings according to their own habits and personal goals [78]. The application can analyze factors such as monthly income, current balance, and spending habits, then make its own decisions and transfer money to the savings account.
- **Portfolio management**: Robo-advisers provide financial advice and portfolio management with minimal intervention from people [79]. This category of financial adviser works on the basis of algorithms designed to automatically develop a financial portfolio according to the investment objectives and the risk tolerance of the clients. It can adapt to real-time market changes and calibrate the portfolio accordingly.
- **Subscription**: an online lender, Upstart, analyzes vast amounts of consumer data and uses machine learning algorithms to develop credit risk models that predict the likelihood of consumer default. So that technology is presented to banks so that they can take advantage of their underwriting processes [80]. In addition, the ZestFinance application it is the ZAML platform (Zest Automated Machine Learning) specifically intended for credit underwriting [81]. This platform uses to

analyze tens of thousands of traditional and non-traditional variables. Finally, this platform is particularly useful for assigning credit scores to those with a limited credit history.

h. Heavy industry

Robots have become common in many industries. Robots are effective in very repetitive tasks that can lead to errors or accidents due to poor concentration and other jobs. In 2014, China, Japan, the United States, the Republic of Korea and Germany together accounted for 70% of the total volume of robot sales [82]. In the automotive industry, a particularly automated sector, Japan has the highest density of industrial robots in the world: 1,414 per 10,000 employees [82].

i. Hospitals and medicine

Healthcare applications driven by the IoT are expected to be massively deployed and it is planned to capture most of the future IoT market by 2025 [83]. The IoT can properly manage all kinds of medical care [84]. However, the healthcare sector demands an IoT paradigm to detect symptoms and thus diagnose, suggest preventive measures and gradually adapt treatment based on AI. In addition, the IoT can guide pharmaceutical companies to develop and design new drugs based on the analysis of generated IoT data.

Various entities that can benefit directly or indirectly from the application of IoT in healthcare are doctors, support staff, patients, hospitals, medical insurance companies, and the pharmaceutical industries (Fig. 6). IoT-based medical devices

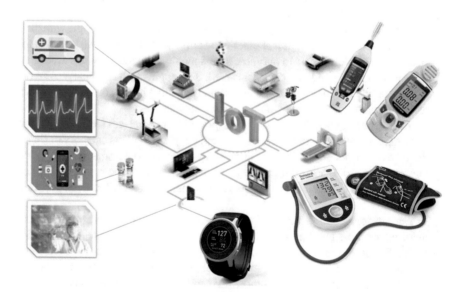

Fig. 6 Healthcare based on IoT

such as Smartwatches, shoes, clothing, bracelets, etc. can remotely detect the vital signs of a patient, etc.

Artificial neural networks and clinical decision support systems are used for medical diagnosis, as is the concept processing technology in EMR software [85]. Other tasks in medicine that can potentially be performed by AI include:

- Computer-aided interpretation of medical images. Such systems help to digitize digital images.
- Analysis of the sound of the heart.
- Robot for the care of the elderly.
- Extraction of medical records to provide more useful information.
- Design treatment plans.
- Help with repetitive tasks, including medication management.
- Provide consultations.
- Creation of medicines
- Use avatars instead of patients for clinical training
- Predict the probability of death from surgical procedures.
- Predict the progression of HIV (Human immunodeficiency virus).

j **Police and justice**

Some policies are currently developing AI-based solutions launched in recent years in the field of crime prevention [86]. AI is considered as a public safety resource in many ways. A specific AI-based application can be found to identify people. Intelligence analysts, for example, rely on portraits of the face to help determine the identity and whereabouts of an individual. In addition, examining the possibly relevant images and videos in timely manner is a time-consuming, painstaking task, with the potential for human error due to fatigue and other factors. Unlike humans, machines do not tire. Through initiatives such as the Intelligence Advanced Research Projects Activity's Janus computer-vision project, analysts are performing trials on the use of algorithms that can learn how to distinguish one person from another using facial feature in the same manner as a human analyst [87]. However, video analytics can prevent the detection of individuals in multiple locations across multiple cameras, the detection of crimes through motion and pattern analysis, the identification of crimes in progress, and assistance to investigators in identifying suspects. With technology like cameras, video and social media generating massive amounts of data, Amnesty International can reveal crimes that cannot be discovered by traditional methods that will help ensure greater public safety and criminal justice. Amnesty International also has the ability to assist crime labs in areas such as complex DNA analysis [88].

k. **IoT Automotive**

By 2040, it is estimated that 90% of global vehicle sales will be either highly auto-mated vehicles or fully automated vehicles [89]. Many companies, for example, Tesla, Google, BMW, Ford, Uber, etc. work in this direction [90, 91]. The IoT in collaboration with cloud computing will play an important role in the creation of remote and autonomous connected vehicles (also known as autonomous or driverless vehicles or robots) a significant development in the near future. Today, there are more than 200 sensors integrated into a single vehicle [92]. This trend corresponds to the fundamental prerequisite of IoT compatible vehicles. Various underlying technologies for vehicle communications are Bluetooth, ZigBee, dedicated short range communication (DSRC), Wi-Fi and 4G cellular technology [92].

l. Law

The issue of "the search for the law" has both conceptual and practical importance. Theoretically, what distinguishes legal thinking from other forms of thinking (such as practical or moral thinking) is precisely its dependence on legal materials [93]. Although AI research tools may not be quite suitable for answering basic questions in jurisprudence, they can be used in various ways. In practice, the ability to recognize and search for relevant legal authorities is one of the most fundamental skills for legal professionals [94]. If this skill is easily accomplished by a machine, this may significantly reduce the cost of legal services and democratize access to the law.

The application of AI to the legal field promises considerable efficiency gains in the legal and judicial spheres. Two technological approaches contribute to the computerization of law: it is first possible to model legal knowledge using rules-based systems, or expert systems, within the framework of deterministic algorithmic, to render account for the logical articulation of certain rules of law [95]. In addition, other technologies, which rely on statistical processing tools for natural language, make it possible to explore large quantities of documents such as digital court decisions, to more or less automatically identify rules of law, or the answers generally were given.

m. Human resources and recruitment

AI is used in three ways by human resources (HR) and recruiting professionals. The AI is used to filter CVs and classify candidates according to their level of qualification [96]. AI is also used to predict the success of candidates in given roles through job search platforms, and now the AI is deploying discussion robots that can automate repetitive communication tasks [97]. Typically, CV selection involves a recruiter or other HR professional who analyzes a CV database. Now, startups are creating machine learning algorithms to automate CV selection processes.

In the area of recruitment and talent development through assessment, AI has five key advantages [98, 99]:

- **Accuracy**: the power of today's computers makes it possible to precisely evaluate very large quantities of candidate data to make better choices during selection.
- **Efficiency**: AI for recruitment allows human resources teams to assess job-specific data in a consistent, objective manner and this very early in the process.
- **Fight against bias**: prejudices are often the source of poor selection choices. In theory, AI helps to eliminate the conscious and unconscious biases of recruiters.

You have to be very careful when programming the system because the quality of the algorithm is based on the data that feeds it. To guarantee a non-discriminatory process, several evaluators must be included when defining the model and not just one, for example.

- **Transparency**: Allowing candidate assessment models to learn for themselves by following the best practices of human recruiters remains the best guarantee for a fair and objective process.
- **Commitment to the process**: the AI allows recruiters to offer immediate help, for example via interactive chatbots which can answer candidates' questions about the selection process or certain tests. AI also makes it possible to make faster decisions and therefore to be more reactive as well as limiting biases.

Beyond technological considerations, four key points must be taken into account [98, 99]:

- **Legitimacy**: standardized plug-and-play AI models will not allow you to stand out as an employer. If your competitors use the same solutions, you will all target the same talents. This promising feature is problematic to justify the selection or rejection of a candidate. Only custom AI models allow you to make transparent and defensible choices.
- **Deadline for implementation**: custom AI systems replicate the behavior and best practices of your recruiters. It is therefore necessary to pre-populate the tool with relevant data.
- **Ethics**: defining the scope of actions of an AI model raises ethical questions.
- **Data processing**: If AI excels at analyzing large volumes of data, the results can be misinterpreted or misused. Good data processing practices are essential for privacy issues, but also for preserving your company's reputation.

n. **Media**

Certain AI applications are oriented towards the analysis of audiovisual multimedia content such as films, television programs, advertising videos or user-generated content [100]. However, typical use case scenarios include image analysis using object recognition or facial recognition techniques, or video analysis for the scene, object or scene recognition relevant faces. The motivation for the use of media analysis based on AI can be, the facilitation of multimedia research, the creation of a set of descriptive keywords for a multimedia element, the verification of the media content for a particular content (such as checking the relevance of the content for a particular time of television viewing), automatic speech recognition for archiving or other purposes, and detection of logos, products or celebrity faces for the placement of relevant advertisements.

o. **Smart Agriculture**

Various tasks involved in smart agriculture based on IoT are monitoring and acquiring ambient data which are of great variety and enormous volume, followed by their aggregation and exchange on the network, taking short and long term decisions based on data analysis and AI and, sometimes, remote actuation of decisions using field robots. It can also help predict yield to ensure the economic value to be gained, as well as the early detection of diseases spread in crops, to enable timely preventive measures [101]. Various technologies used are Bluetooth, RFID, Zigbee, GPS as well as other technologies which are gaining popularity in this application, namely SigFox, LoRa, NB-IoT, edge computing and cloud computing [102].

Some of the main components of IoT-based smart agriculture are smart plowing, smart irrigation, smart fertilization, smart harvesting, smart stock maintenance, smart livestock management that deals with smart tracking of animals, intelligent health monitoring, intelligent feeding and fodder management, etc. An IoT solution for smart farming has different requirements and therefore takes into account the following key factors; low cost, low energy consumption, highly reusable, interoperative, highly efficient, resource-efficient, scalable and gradually scalable solutions.

p. **Music**

The evolution of music has always been affected by technology, AI has made it possible through scientific progress, to imitate, to a certain extent, the human composition [103]. Among the first remarkable efforts, David Cope created an AI called Emily Howell which has managed to make itself known in the field of algorithmic music [104]. Other projects, such as AIVA (Artificial Intelligence Virtual Artist), focus on the composition of symphonic music, mainly classical music for film music [105]. He achieved a world first by becoming the first virtual composer to be recognized by a professional musical association. With the efforts of Smaill A, they used AI to produce computer-generated music for different purposes [106]. Also, initiatives such as Google Magenta, led by the Google Brain team, want to know if AI can be able to create an irresistible art [107].

q. **News, publications and writing**

AI contributes to computer-generated reporting, including statistic-based event summaries [108]. Likewise, Insights creates summaries, overviews, and profiles [109]. Where this company completed one billion stories in 2014, compared to 350 million in 2013 [110]. Echobox is a software company that helps article publishers on social media platforms like Facebook and Twitter [111]. By analyzing large amounts of data, he learns that a specific audience responds to different articles at different times. Then he chooses the best stories and times to publish them. Another company, Yseop, uses AI to convert structured data into smart comments in the natural language. Yseop can write reports quickly as it is possible to create thousands of pages per second and multiple languages [112]. Boomtrain's is another example of AI designed to learn how to best engage each reader with the exact articles [113].

r. Online and telephone customer service

AI is used in automated online assistants that can be seen as avatars on web pages [114]. It can benefit companies by reducing their operating and training costs. A major underlying technology in such systems is natural language processing. Pypestream uses automated customer service for its mobile application designed to streamline communication with customers. In addition, large companies invest in AI to manage customers, have worked on different aspects to improve different services. Among the companies operating in this field, Digital Genius, an AI start-up, searches the information database more efficiently (based on past conversations and frequently asked questions) and provides instructions to agents to help them resolve queries more efficiently [115]. IPSoft creates technology with emotional intelligence to adapt the interaction of the client which is characterized by adaptation with different languages [116]. Inbenta focuses on the development of natural language. In other words, to understand the meaning behind what someone is asking and not just look at the words used, using context and natural language processing [117]. One element of customer service that Ibenta has already achieved is its ability to respond en masse to e-mail queries.

s. Telecommunications maintenance

The name "smart grid" was formulated in the 1980s, with real premonition [118]. Today, this concept encompasses technological trends such as AI (with machine learning) or predictability solutions. Who says "intelligence" in communication and information systems, says expert systems and now AI with its machine learning approaches ("automatic learning") or its deep learning subset ("neural networks"))...

According to the Gartner study "The road to enterprise AI, 2017", a proportion of 10% of emergency interventions in the field will be triggered and programmed by AI by 2020 [119].

According to the Kambatla, K. et al. combined with analytics or big data, the research firm evokes a "considerable impact of these new technologies" on whole sections of the industry: monitoring of production machines, prevention of equipment failure and maintenance predictive [120].

t. Games

In the game field, AI is used to provide a quality gaming experience, for example by giving non-player characters behavior similar to that of humans, or by limiting the skills of the program to give the player a sense of fairness.

At the beginning of the field, there has long been a gap between classical AI research and the AI implemented by developers in video games [121]. Indeed, there were between the two major differences in knowledge, the problems encountered but also the ways of solving these problems. However, classical AI research generally aspires to improve or create new algorithms to advance the state of the art. On the other hand, the development of an AI in a video game aims to create a coherent system which integrates as best as possible in the design of the game to be fun for the player. Thus, a high-performance AI that is not well integrated into the gameplay can serve the game more than it will improve it.

Developing AI for a video game therefore often requires finding engineering solutions to problems little or not at all addressed by conventional AI research [122]. For example, an AI algorithm in a game is very tight in terms of computing power, memory and execution time. Indeed, the game must run through a console or an "ordinary" computer and it must not be slowed down by AI. This is why some state-of-the-art AI-intensive solutions could not be implemented in video games until several years after their use in classic AI.

American researchers have just developed an AI capable of solving the famous game of Rubik's Cube in just 1.2 s, a time inaccessible to humans. Just recently, Facebook and Carnegie Mellon University announced that a joint team of researchers had successfully developed AI software capable of defeating some of the best poker professionals. The progress of AI continues to increase and beat, one after the other, all the games that humans have sometimes taken years to master. However, the link between AI and gaming is not new. Already in 1979, the robot "Gammanoid" beat Luigi Villa, a backgammon champion, in Monte-Carlo [123]. Today, it is mainly thanks to the impressive progress of deep learning that many records are broken, as recently in poker.

u. **Transports**

Fuzzy logic controllers have been developed for automatic transmissions in automobiles. For example, the 2006 Audi TT, VW Touareg, and VW Caravell have the DSP transmission which uses Fuzzy Logic. Several variants of Škoda (Škoda Fabia) also include a Fuzzy Logic controller [124]. However, today's cars now have AI-based driver assistance functions such as automatic parking and advanced cruise controls. AI has been used to optimize traffic management applications, reducing wait times, energy consumption and emissions by up to 25% [125]. In the future, fully autonomous cars will be developed. Transport AI should provide safe, efficient and reliable transport while minimizing the impact on the environment and communities. The main challenge to develop this AI is the fact that transport systems are intrinsically complex systems involving a very large number of different components and parts, each with different and often contradictory objectives.

6 Advantages and Disadvantages of AI

Will see in this part "Advantages and disadvantages of AI". What are the positive and negative points of AI in the "Important areas of applications" seen previously as well as in other areas.

6.1 Advantages of AI

- **Computational**: the main feature of AI is to reduce calculation errors thanks to all algorithms and to use a computer to solve calculations in a faster and more efficient way. However, this field of intelligence cannot be qualified, as it is unable to correct errors that humans may have made.
- **Job**: another feature is the advantage of chain machines, for example, to replace humans on arduous or dangerous missions. These machines still allow once again to improve the work that men could have done. In addition, they have another great advantage, they have no physical limitations. They do not need to eat, they do not need rest and cannot be ill. Some scientists believe that one day they will be able to replace firefighters who risk their lives with machines, but for the time being, this is more of a risk than an advantage because the robot has no intuition from a man who saves lives. Additionally, knowing how to respond to a particular situation, maybe the opposite of what is expected.
- **Daily**: everyday robots are developing rapidly to enable them to do housework, shopping, kitchen, and other functions. Another important thing is about elderly characters who need help. Today, many robots can mow the lawn or sweep the house, and they adapt to the environment in which they mainly operate with the help of sensors. In the future, it will be possible to communicate with them and share their feelings and thoughts, as well as create a real connection between humans and the robot.
- **Discovery**: in space, robots and AI are widely used because they allow partially independent robots to be sent in place of men, as men cannot go due to strong or very weak pressures or temperatures. The robots explore what they see and the humans on Earth are informed. As a real example, AI has enabled robots sent to Mars to understand that in the past this planet was hot and humid and therefore suitable for life. By continuing to research, researchers can realize with certainty that life is possible anywhere on Earth. Therefore, the contribution of AI is vital because without these robots it would be impossible for humans to study what is happening in space. Finally, the goal in the future is to create a fairly smart robot that can handle any situation and is completely independent.
- **Study**: previously seen in the field of medicine, AI has enabled the development of patient robots that are of benefit to medical students. Thanks to these robots, the student will be able to train without risk. This is a real advantage because with robots mistakes can be made but they are less dangerous than if they happen to a person.
- **Medicine**: the development of prostheses will allow people who have lost the use of a limb to return to normal life as they had before. Indeed, artificial limbs have intelligence because they will have to adapt to the person on whom they were transplanted. For example, when the person wants to grab a glass, it will be necessary for the strength of the arm to be adapted so as not to break or drop the glass.

- **Save (or kill)**: this segment was seen in the military field as a robot with the advantage of being able to save people at risk without putting themselves at risk or to kill them if necessary during the war.
- **Games**: AI in video games allows players to face strong opponents increasingly. The game looks more realistic and the player really feels playing against human. It also allows you to play against the computer when there is no opponent to play with it.
- **In a company**: in a business that works with AI-based systems, there are several advantages:
  - Better use of human resources (employees become less and less productive in the face of routine tasks, with machines equipped with AI, work is carried out more suited to the skills of the employees)
  - Reduction of employees (less staff for routine tasks like making photocopies)
  - Reduction of personnel costs (with several machines to replace several employees, the company makes a big saving, it has more than a qualified maintenance worker to employ)
  - Learning (the AI is able to learn automatically and therefore progress quickly, so it can do the work of several people and have contact with the client).
- **New humanoid form**: robots are now represented as a human called a humanoid. The advantage facing this new form is the fact that the person in front does not have the impression of having to do with a machine and that gives a more pleasant side especially when this machine is permanently in your daily life.

6.2 Disadvantages and Limitations

All the advantages mentioned above also have drawbacks or limits as to their operation. The most conceivable drawback is the presence of an error in the programming of a robot whatever it is which would be fatal for the good functioning of this robot. This drawback is present in all areas without exception. It has been seen previously that the computer (or any robot) did not know how to detect an error that would have been made in its program. The consequences of such an error could be catastrophic on a large scale. In addition, what limits AI research is the very high price of this research. To make robots capable of being autonomous in space, an astronomical sum would have to be spent, which limits the progress of research for the moment. In medicine, patient robots are an advantage for students, but it is very expensive to manufacture, which is why not all schools can equip them. This is also the case for all the prostheses which are expensive to develop (they also have the other disadvantage, seen previously, that is to say, that a person wearing a prosthesis should not be charged disturbances and if it breaks down, the patient would suddenly find himself unable to move his hand in which he was holding something).

In companies, in particular, AI and new mechanized robots are causing job cuts. Human beings are gradually being replaced by robots, thereby increasing the already high unemployment rate.

Can go to war with robots? What if the robot suddenly turns against its camp? Is it human to send machines to kill men? Is it okay to use this science for this purpose? Since it is the military field that funds research in AI the most, it has the "power" to develop what it wants. However, imagine if the robots took control of the world. This is the idea that many filmmakers have taken up. In " I Robot " or " Terminator " human-made robots take over and destroy humanity. All of these fictional assumptions have been the subject of science fiction novels like those of Isaac Asimov for example. This one having written 3 laws in 1947 on the behavior which must have the robot:

- A robot cannot harm a human being, not remain passive or allow a human to be exposed to danger.
- A robot must obey the orders given to it by the human being unless such orders conflict with the first law.
- A robot must protect its existence as long as this protection does not conflict with the first or second law.

7 Discussion

With what was previously shown, the benefits of AI are great. We, therefore, find it interesting to use this form of intelligence as long as it is beneficial to humans. Indeed, the robot can replace the Man in difficult tasks (because it has no physical constraints), in annoying tasks or even in places where the Man cannot go. The robot will then be faster, more precise and more efficient. However, it should not be misused because a robot cannot replace a human. On the other hand, the use of this technology by companies increases more, which creates an increase in the unemployment rate. In addition, AI must be used for justified and morally correct purposes (for example: developing a robot that can help people with disabilities in everyday life and help students pursue their lessons and perform their scientific research) and in return not to develop robots that cause harm to individuals. Finally, it is not possible for anyone to succeed in developing a human person with a brain that evolves as a human being, because scientists have been trying for decades to understand the workings of the human brain that can reproduce and implant it into a robot but without result.

At the same time, it should not be thought that a robot is equal to humans; AI is only the mechanical reproduction of the human brain. In fact, what is missing in AI and robots, androids is awareness, intuition and expression of feelings, so how is it presented in a robot?! Especially since there are many forms of intelligence and our minds use different areas to solve all problems. Hence, AI can replace humans to the point where it will be necessary to create very accurate thinking. This is also what Albert Einstein defends in his expression, "Machines will one day be able to solve all problems, but none of them will ever be able to solve them!" Robots exist only to improve humanitarian action and to solve problems. Finally, AI will never replace that of humans. Indeed, the intelligence of the person, his brain is much more prominent than that of an intelligent device. The speed of reflection of human intelligence is more important than that of a machine. However, the human brain has many more neurons than an artificial machine. In addition, it should not be forgotten that man is distinguished from machines, by the fact that man reflects by himself as long as the machine will use the data they have been transmitted to him by human. However, AI will not be able to replace humans in the future. And therefore, Technology is progressing more and more and will occupy a place almost everywhere. Unlike humans, AI does not need to eat, sleep or rest because it is a machine. This may be very beneficial for men, but at the same time it may also mean an obstacle. Take an example of factory robots: the number of robots to collect objects is increasing. It is also very useful because it is a great job with routine and repetitive work.

8 Conclusion

The multiplied power of computers, the development of new hardware architectures, for example, make it possible to deal with and solve practical problems that are always more difficult. The emergence of new fields for computer science and information processing science opens up new challenges for AI research. Can citing, for example, the Internet sphere, the web, electronic commerce, etc. Who has used a search engine realizes the need for a "smart search" for information. How to organize and process these gigantic masses of information? How to extract the relevant knowledge for the problems posed? This is clearly an area where AI research finds an important field of application. It should also be noted that through their spin-offs in fields as diverse as aeronautics and space, agriculture, industrial production, banking, and finance..., IT has succeeded in considerably expanding its impact in transformation in depth of human activities.

So far, this intelligence has not reached perfection, that is, she is unable to think about herself, manage herself and experience feelings as a human will. However, it is still able to replace it in several fields previously seen such as medicine, the military, video games... as well as to simplify its daily life. The many developments present real moral problems, especially in the military sphere because the robots that have been created are certainly semi-autonomous (managed by one person) but they are nevertheless made up of weapons of war capable of killing people. However, it can be concluded that AI also has many positive points as well as many advantages that can improve a person's life to make it easier. Little by little, man replaces this type of intelligence, but using it eliminated a large number of jobs.

By what was previously shown, a person wants to make a smart machine like him. Tomorrow man has no time. He works tirelessly and cannot even take care of his home because of the limited time. Smart robots will be able to do housework, prepare a meal... this might be an excellent idea for the elderly, with reduced mobility. These artificial humans can not only be at our service, but they can also speak to us and have a real conversation with them. These robots in the future may be able to feel and love. In the age of robots, prosthetics become robots by themselves and combine AI to aid their owners' movements. Indeed, it perfectly adapts to the human condition: whether the latter is rising or ascending to the basement. The prostheses contain sensors, motors and an integrated calculator.

Can say that AI, by modeling and simulating cognitive functions part-tasks deemed intelligent to, in a generally both more reliable and faster machines than would make the man, by exploiting generally large quantities information. This gives rise to achievements which can be spectacular, but which can also disturb the idea that have of our intelligence.

Great progress still remains to be made in AI to create systems with wider ranges of cognitive capacities and also more user-friendly for users. Without a doubt, it is perfectly conceivable that the AI realizes systems capable of a certain introspection (i.e. to observe itself in tasks of reasoning and thus to acquire meta-knowledge), to analyze or to simulate emotions, or even to write poems or create graphic works obeying certain constraints or principles. But all of this will remain far enough from an autonomous though, conscious of itself, capable of juggling with its representations of the world, of behaving in a playful way, not purely reactive, of creating in a way directed, to dream.

Appendix

This section is dedicated to presenting the terms which are mainly related to the technology of AI (see Table 2).

Table 2 List of terms that are related with AI

Term	Description
A* (A Star)	A *, pronounced A star, is an algorithm for finding the path in a graph between an initial node and a final node, both given as in a labyrinth
Alan Turing (1912–1954)	Alan Mathison Turing is a British mathematician. He will invent the Turing test
Turing test	Turing explores the problem of AI and offers an experiment (The Turing Test) in an attempt to define a standard allowing to qualify a machine as "conscious"
Algorithm	An algorithm is a finite and unambiguous sequence of operations or instructions for solving a problem
Recursive algorithm	An algorithm is said to be recursive if it calls itself, it is said to self-call. That is to say that the algorithm at the heart of its calculation can have recourse to itself, it starts again

(continued)

Table 2 (continued)

Term	Description
Evolutionary algorithm	Evolutionary algorithms or evolutionary algorithms are algorithms inspired by the theory of evolution to solve various problems. That is to say that they record the results previously found and thus constitute a database. They thus develop a set of solutions to a given problem, with a view to finding the best results
Catheter	A catheter is a medical device consisting of a tube, of variable width and flexibility, and manufactured in different materials according to the models or the uses for which they are intended. The catheter is intended to be inserted into the lumen of a body cavity or blood vessel and allows drainage or infusion of fluids, or even access for other medical devices
Deep Blue	Deep Blue is a computer specializing in the game of chess by adding specific circuits, developed by IBM in the early 1990s and which lost a match (2-4) against the world chess champion Garry Kasparov in 1996, then defeated the world champion (3.5–2.5) in the rematch in 1997
Dofus and Sims	Video games of character simulations
Drone	The Drone is an unmanned human aircraft on board, which carries a payload, intended for surveillance, intelligence or combat type missions.
ELIZA	ELIZA is a computer program written by Joseph Weizenbaum in 1966, which simulated a Rogerian psychotherapist by rephrasing most of the "patient's" statements in questions, and asking them
Ethics	Ethics is the science of morals and mores. It is a philosophical discipline which reflects on the aims, on the values of existence, on the conditions of a happy life, on the concept of "good" or on questions of morals or morals
G.P.S	The Global Positioning System (GPS) is a geolocation system operating on a global level
IBM	International Business Machines, exactly International Business Machines Corporation, known by the abbreviation IBM, is an American multinational company active in the fields of computer hardware, software and IT services
John McCarthy (1927–2011)	John McCarthy is the main pioneer of AI with Marvin Minsky; he embodies the current emphasizing symbolic logic
LISP (List Procesor)	Functional programming language designed for list processing, used in the fields of AI and expert systems
Louise Bérubé	Quebec psychologist, specialized in behavioral neurology
Marvin Minsky	Marvin Lee Minsky is an American scientist. He works in the field of cognitive sciences and AI. He is also a co-founder, with computer scientist John McCarthy of the AI Group of the Massachusetts Institute of Technology (MIT) and author of numerous publications in both AI and philosophy such as, for example, The Spirit Society (1986)
Microswitch	Electric component allowing to cut the current or to reverse it

(continued)

Table 2 (continued)

Term	Description
MIT	The Massachusetts Institute of Technology, MIT, is a famous American university in the state of Massachusetts, in the immediate vicinity of Boston. Founded in the 19th century to meet a growing demand for engineers, the institute will become multidisciplinary but will keep a teaching in science giving a large place to experience as well as to technological and industrial applications
Nanorobot	Microscopic robot
Pathology	Diseases, sets of symptoms and their origins
Rotor	Movable part of an engine
Haptic sensations	Adjective which indicates interfaces which give feelings by the touch (pressure, movements…)
Stator	Fixed part of an engine
Helical	Adjective designating the form of helix

References

1. Flasiński, M., Flasiński, M.: History of artificial intelligence. In: Introduction to Artificial Intelligence (2016)
2. O'Regan, G., O'Regan, G.: Marvin Minsky. In: Giants of Computing (2013)
3. Entwistle, A.: What is artificial intelligence? Eng. Mater. Des. (1988). https://doi.org/10.1007/978-1-4842-3799-1_1
4. Copeland, B.J.: Artificial intelligence | Definition, Examples, and Applications | Britannica (2020). https://www.britannica.com/technology/artificial-intelligence. Accessed 26 Apr 2020
5. Murphy, R.R.: Introduction to AI robotics. BJU Int. (2000). https://doi.org/10.1111/j.1464-410X.2011.10513.x
6. McConaghy, E.: Automaton. West. Hum., Rev (2012)
7. Saba, D., Berbaoui, B., Degha, H.E., Laallam, F.Z.: A generic optimization solution for hybrid energy systems based on agent coordination. In: Hassanien, A.E., Shaalan, K., Gaber, T., Tolba, M.F. (eds.) Advances in Intelligent Systems and Computing, pp. 527–536. Springer, Cham, Cairo—Egypte (2018)
8. Saba, D., Degha, H.E., Berbaoui, B., et al.: Contribution to the modeling and simulation of multi-agent systems for energy saving in the habitat. In: Djarfour, N. (ed.) International Conference on Mathematics and Information Technology, p. 1. IEEE, Adrar-Algeria (2017)
9. Saba, D., Sahli, Y., Abanda, F.H., et al.: Development of new ontological solution for an energy intelligent management in Adrar city. Sustain. Comput. Inform. Syst. **21**, 189–203 (2019). https://doi.org/10.1016/J.SUSCOM.2019.01.009
10. Saba, D., Laallam, F.Z., Degha, H.E., et al.: Design and development of an intelligent ontology-based solution for energy management in the home. In: Hassanien, A.E. (ed.) Studies in Computational Intelligence, 801st edn, pp. 135–167. Springer, Cham, Switzerland (2019)
11. Saba, D., Maouedj, R., Berbaoui, B.: Contribution to the development of an energy management solution in a green smart home (EMSGSH). In: Proceedings of the 7th International Conference on Software Engineering and New Technologies—ICSENT 2018, pp. 1–7. ACM Press, New York, NY, USA (2018)
12. Saba, D., Zohra Laallam, F., Belmili, H. et al.: Development of an ontology-based generic optimisation tool for the design of hybrid energy systems. Int. J. Comput. Appl. Technol. **55**, 232–243 (2017). https://doi.org/10.1504/IJCAT.2017.084773

13. Degha, H.E., Laallam, F.Z., Said, B., Saba, D.: Onto-SB: Human profile ontology for energy efficiency in smart building. In: Larbi Tebessi university A (eds.) 2018 3rd International Conference on Pattern Analysis and Intelligent Systems (PAIS). IEEE, Tebessa, Algeria (2018)

14. Saba, D., Laallam, F.Z., Berbaoui, B., Fonbeyin, H.A.: (2016) An energy management approach in hybrid energy system based on agent's coordination. In: The 2nd international conference on advanced intelligent systems and informatics (AISI' 16). Advances in Intelligent Systems and Computing, Cairo, Egypt

15. Saba, D., Laallam, F.Z., Hadidi, A.E., Berbaoui, B.: Contribution to the management of energy in the systems multi renewable sources with energy by the application of the multi agents systems "MAS". Energy Procedia **74**, 616–623 (2015). https://doi.org/10.1016/J.EGYPRO. 2015.07.792

16. Cockcroft, K.: Book review: international handbook of intelligence. South African J. Psychol. (2005). https://doi.org/10.1177/008124630503500111

17. Mcculloch, W.S., Pitts, W.: A logical calculus nervous activity. Bull. Math. Biol. (1990). https://doi.org/10.1007/BF02478259

18. Wiener, N.: Norbert Wiener, 1894–1964. IEEE Trans. Inf. Theory (1974). https://doi.org/10. 1109/TIT.1974.1055201

19. Chiu, E., Lin, J., Mcferron, B., et al.: Mathematical Theory of Claude Shannon. Work Pap (2001)

20. Gass, S.I.: John von Neumann. In: International Series in Operations Research and Management Science (2011)

21. Newell, A., Shaw, J.C., Simon, H.A.: Elements of a theory of human problem solving. Psychol. Rev. (1958). https://doi.org/10.1037/h0048495

22. Nilsson, N.J.: Shakey The Robot (1984)

23. Brooks, R.A.: New approaches to robotics. Science (80) (1991). https://doi.org/10.1126/sci ence.253.5025.1227

24. Li, B.H., Hou, B.C., Yu, W.T., et al.: Applications of artificial intelligence in intelligent manufacturing: a review. Front. Inf. Technol. Electron. Eng. (2017)

25. Internetlivestats: Internet Live Stats—Internet Usage & Social Media Statistics (2020). https:// www.internetlivestats.com/. Accessed 20 Feb 2020

26. Internetlivestats: 1 Second—Internet Live Stats (2020). https://www.internetlivestats.com/ one-second/#tweets-band. Accessed 20 Feb 2020

27. Trends.google.com: Macron, Trump—Découvrir - Google Trends (2020). https://trends.goo gle.com/trends/explore?q=Macron,Trump. Accessed 20 Feb 2020

28. Powles, J., Hodson, H.: Google DeepMind and healthcare in an age of algorithms. Health Technol (Berl) (2017). https://doi.org/10.1007/s12553-017-0179-1

29. DeepMind: AlphaStar: mastering the real-time strategy game StarCraft II. DeepMind (2019)

30. Lau, J., Zimmerman, B., Schaub, F.: Alexa, are you listening? Proc. ACM Hum.-Comput. Interact (2018). https://doi.org/10.1145/3274371

31. Bell, T.: 6 ways Facebook uses AI | CIO (2018). https://www.cio.com/article/3280266/6-ways-facebook-uses-artificial-intelligence.html. Accessed 26 Apr 2020

32. Hoy, M.B.: Alexa, Siri, Cortana, and more: an introduction to voice assistants. Med. Ref. Serv. Q. (2018). https://doi.org/10.1080/02763869.2018.1404391

33. Apple: Optimizing Siri on HomePod in Far-Field Settings—Apple, vol. 1, Issue 12

34. Reddy, R.: Foundations and grand challenges of artificial intelligence. AI Mag. (1988)

35. Van Remoortere, P.: Computer-based medical consultations: MYCIN. Math Comput. Simul. (1979). https://doi.org/10.1016/0378-4754(79)90016-8

36. Saba, D., Laallam, F.Z., Hadidi, A.E., Berbaoui, B.: Optimization of a multi-source system with renewable energy based on ontology. Energy Procedia **74**, 608–615 (2015). https://doi. org/10.1016/J.EGYPRO.2015.07.787

37. Campbell, M., Hoane, A.J., Hsu, F.H.: Deep blue. Artif. Intell. (2002). https://doi.org/10. 1016/S0004-3702(01)00129-1

38. Saba, D., Laallam, F.Z., Berbaoui, B., Abanda, F.H.: An energy management approach in hybrid energy system based on agent's coordination. In: Hassanien, A., Shaalan, K., Gaber, T., Azar, A.T.M. (eds.) Advances in Intelligent Systems and Computing, 533rd edn, pp. 299–309. Springer, Cham, Cairo, Egypte (2017)
39. Saba, D., Degha, H.E., Berbaoui, B., et al.: Contribution to the modeling and simulation of multiagent systems for energy saving in the habitat. International Conference on Mathematics and Information Technology (ICMIT 2017), pp. 204–208. IEEE, Adrar, Algeria (2018)
40. Saba, D., Degha, H.E., Berbaoui, B., Maouedj, R.: Development of an ontology based solution for energy saving through a smart home in the city of Adrar in Algeria, pp. 531–541. Springer, Cham (2018)
41. Kerber, M., Lange, C., Rowat, C.: An introduction to mechanized reasoning. J. Math. Econ. **66**, 26–39 (2016). https://doi.org/10.1016/J.JMATECO.2016.06.005
42. Siekmann J (2014) Computational Logic, pp. 15–30
43. Peng, H.G., Wang, J.Q.: Hesitant uncertain linguistic Z-Numbers and their application in multi-criteria group decision-making problems. Int. J. Fuzzy Syst. (2017). https://doi.org/10.1007/s40815-016-0257-y
44. Huitt, W.G.: Problem solving and decision making: consideration of individual differences using the myers-briggs type indicator. J. Psychol. Type (1992). https://doi.org/10.1017/CBO9781107415324.004
45. Wilson, D.R.: Hand book of collective intelligence. Soc. Sci. J. (2017). https://doi.org/10.1016/j.soscij.2017.10.004
46. Upadhyay, S.K., Chavda, V.N.: Intelligent system based on speech recognition with capability of self-learning. Int. J. Technol. Res. Eng. ISSN (2014)
47. Herzig, A., Lang, J., Marquis, P.: Action representation and partially observable planning using epistemic logic. In: IJCAI International Joint Conference on Artificial Intelligence (2003)
48. Mezzadra, S., Neilson, B.: Between inclusion and exclusion: on the topology of global space and borders. Theory Cult. Soc. (2012). https://doi.org/10.1177/0263276412443569
49. Copeland, B.J., Proudfoot, D.: Alan Turing's forgotten ideas in computer science. Sci. Am. (1999). https://doi.org/10.1038/scientificamerican0499-98
50. Berbaoui, B., Saba, D., Dehini, R., et al.: Optimal control of shunt active filter based on Permanent Magnet Synchronous Generator (PMSG) using ant colony optimization algorithm. In: Proceedings of the 7th International Conference on Software Engineering and New Technologies—ICSENT 2018. ACM Press, New York, NY, USA, pp. 1–8 (2018)
51. Barrow, L., Markman, L., Rouse, C.E.: Technology's edge: the educational benefits of computer-aided instruction. Am. Econ. J. Econ. Policy (2009). https://doi.org/10.1257/pol.1.1.52
52. Gibson, K.R.: Evolution of human intelligence: the roles of brain size and mental construction. In: Brain, Behavior and Evolution (2002)
53. Minker, W., Bennacef, S.: Speech and human—machine dialog. Comput. Linguist (2005). https://doi.org/10.1162/0891201053630309
54. Bengler, K., Zimmermann, M., Bortot, D., et al.: Interaction principles for cooperative human-machine systems. It—Inf. Technol. https://doi.org/10.1524/itit.2012.0680
55. Rodríguez, R.M., Martínez, L.: An analysis of symbolic linguistic computing models in decision making. Int. J. General Syst. (2013)
56. Chomsky, N.: Language and Mind, 3rd edn.
57. Mantiri, F.: Multimedia and technology in learning. Univers. J. Educ. Res. (2014). https://doi.org/10.13189/ujer.2014.020901
58. Miranda, S., Ritrovato, P.: Automatic extraction of metadata from learning objects. In: Proceedings—2014 International Conference on Intelligent Networking and Collaborative Systems, IEEE INCoS 2014 (2014)
59. O'Leary, D.E.: Artificial intelligence and big data. IEEE Intell. Syst. (2013). https://doi.org/10.1109/MIS.2013.39
60. Gutiérrez-Maldonado, J., Alsina-Jurnet, I., Rangel-Gómez, M.V., et al.: Virtual intelligent agents to train abilities of diagnosis in psychology and psychiatry. Stud. Comput. Intell. (2008). https://doi.org/10.1007/978-3-540-68127-4_51

61. Appan, K.P., Sivaswamy, J.: Retinal image synthesis for CAD development. In: Lecture Notes in Computer Science (including subseries Lecture Notes in Artificial Intelligence and Lecture Notes in Bioinformatics) (2018)

62. Saba, D., Sahli, Y., Berbaoui, B., Maouedj, R.: Towards smart cities: challenges, components, and architectures. In: HassanienRoheet, A.E., BhatnagarNour E.M., KhalifaMohamed H.N.T. (eds.), Studies in Computational Intelligence: Toward Social Internet of Things (SIoT): Enabling Technologies, Architectures and Applications, pp. 249–286. Springer, Cham (2020)

63. Cyril Jose, A., Malekian, R.: Smart home automation security: a literature review. Smart Comput. Rev. (2015). https://doi.org/10.6029/smartcr.2015.04.004

64. Alam, M.R., Reaz, M.B.I., Ali, M.A.M.: A review of smart homes—past, present, and future. IEEE Trans. Syst. Man Cybern. Part C Appl. Rev. (2012). https://doi.org/10.1109/TSMCC.2012.2189204

65. Jones, R.M., Laird, J.E., Nielsen, P.E., et al.: Pilots for Combat Flight Simulation. AI Mag (1999). https://doi.org/10.1609/aimag.v20i1.1438

66. Gallagher, S.: AI bests Air Force combat tactics experts in simulated dogfights | Ars Technica (2016). https://arstechnica.com/information-technology/2016/06/ai-bests-air-force-combat-tactics-experts-in-simulated-dogfights/. Accessed 13 Jan 2020

67. Jones, R.M., Laird, J.E., Nielsen, P.E., et al.: Automated intelligent pilots for combat flight simulation. AI Mag. (1999)

68. Adapa, S.: Indian smart cities and cleaner production initiatives—integrated framework and recommendations. J. Clean. Prod. (2018). https://doi.org/10.1016/j.jclepro.2017.11.250

69. Ligeza, A.: Artificial intelligence: a modern approach. Neurocomputing (1995). https://doi.org/10.1016/0925-2312(95)90020-9

70. Zoph, B., Vasudevan, V., Shlens, J., Le, Q.V.: NASNet. Proc. IEEE Comput. Soc. Conf. Comput. Vis. Pattern Recogn. (2018). https://doi.org/10.1109/CVPR.2018.00907

71. Farr, M.J., Psotka, J.: Intelligent Instruction by Computer : Theory and Practice

72. Horvitz, E.: One Hundred Year Study on Artificial Intelligence. Stanford University (2016)

73. Treleaven, P., Galas, M., Lalchand, V.: Algorithmic trading review. Commun. ACM (2013)

74. Greenwood, J.: Why BlackRock is investing in digital—the platform. Corp Advis (Online Ed) (2016)

75. Crosman, P.: Beyond robo-advisers: how AI could rewire wealth management | American Banker. In: American Banker (2017). https://www.americanbanker.com/news/beyond-robo-advisers-how-ai-could-rewire-wealth-management. Accessed 14 Jan 2020

76. Antoine, G.: Kensho's AI for investors just got valued at over $500 million in funding round from wall street. In: Forbes.com (2017). https://www.forbes.com/sites/antoinegara/2017/02/28/kensho-sp-500-million-valuation-jpmorgan-morgan-stanley/#2598a9305cbf. Accessed 14 Jan 2020

77. ERIC, R.: The 8 best AI Chatbot apps of 2020. In: Thebalancesmb (2019). https://www.thebalancesmb.com/best-ai-chatbot-apps-4583959. Accessed 14 Jan 2020

78. Gofer, E.: Machine Learning Algorithms with Applications in Finance. Thesis (2014)

79. Obermeyer, Z., Emanuel, E.J.: Predicting the future-big data, machine learning, and clinical medicine. New Engl. J. Med. (2016)

80. AM, E.: ZestFinance introduces machine learning platform to underwrite millennials and other consumers with limited credit history | Business wire. In: Business wire (2017). https://www.businesswire.com/news/home/20170214005357/en/ZestFinance-Introduces-Machine-Learning-Platform-Underwrite-Millennials. Accessed 14 Jan 2020

81. World Robotics Organization: Executive Summary—World Robotics (Industrial {&} Service Robots) 2014. World Robot Rep (2014)

82. Adhikary, T., Jana, A.D., Chakrabarty, A., Jana, S.K.: The Internet of Things (IoT) Augmentation in healthcare: an application analytics. In: ICICCT 2019—System Reliability, Quality Control, Safety, Maintenance and Management (2020)

83. Yin, Y., Zeng, Y., Chen, X., Fan, Y.: The internet of things in healthcare: an overview. J. Ind. Inf., Integr (2016)

84. Kiah, M.L.M., Haiqi, A., Zaidan, B.B., Zaidan, A.A.: Open source EMR software: profiling, insights and hands-on analysis. Comput. Methods Programs Biomed. (2014). https://doi.org/10.1016/j.cmpb.2014.07.002
85. Sukhodolov, A.P., Bychkova, A.M.: Artificial intelligence in crime counteraction, prediction, prevention and evolution. Russ. J. Criminol. (2018). https://doi.org/10.17150/2500-4255.2018.12(6).753-766
86. Rigano, C.: Using Artificial Intelligence to Address Criminal Justice Needs (NIJ Journal 280) (2019)
87. Škrlec, B.: Eurojust and External Dimension of EU Judicial Cooperation. Eucrim—Eur Crim Law Assoc Forum (2019). https://doi.org/10.30709/eucrim-2019-018
88. Milakis, D., Snelder, M., Van Arem, B., et al.: Development and transport implications of automated vehicles in the Netherlands: scenarios for 2030 and 2050. Eur. J. Transp. Infrastruct. Res. (2017). https://doi.org/10.18757/ejtir.2017.17.1.3180
89. Andrea, M.: Some of the companies that are working on driverless car technology—ABC News (2018). https://abcnews.go.com/US/companies-working-driverless-car-technology/story?id=53872985
90. Richtel, M., Dougherty, C.: Google's Driverless Cars Run Into Problem: Cars With Drivers—The New York Times. New York Times (2015)
91. Guerrero-Ibáñez, J., Zeadally, S., Contreras-Castillo, J.: Sensor technologies for intelligent transportation systems. Sensors (Basel) 18 (2018). https://doi.org/10.3390/s18041212
92. Dadgosari, F., Guim, M., Beling, P.A., et al.: (2020) Modeling law search as prediction. Artif. Intell. Law 1–32. https://doi.org/10.1007/s10506-020-09261-5
93. Walker-Osborn, C.: Artificial intelligence automation and the law. ITNOW (2018). https://doi.org/10.1093/itnow/bwy020
94. Alarie, B., Niblett, A., Yoon, A.H.: How artificial intelligence will affect the practice of law. Univ. Tor. Law J. (2018)
95. Tambe, P., Cappelli, P., Yakubovich, V.: Artificial intelligence in human resources management: challenges and a path forward. Calif. Manag. Rev. (2019). https://doi.org/10.1177/0008125619867910
96. Radevski, V., Trichet, F.: Ontology-based systems dedicated to human resources management: an application in e-recruitment. In: Lecture Notes in Computer Science (including subseries Lecture Notes in Artificial Intelligence and Lecture Notes in Bioinformatics) (2006)
97. Upadhyay, A.K., Khandelwal, K.: Applying artificial intelligence: implications for recruitment. Strateg. HR Rev. (2018). https://doi.org/10.1108/shr-07-2018-0051
98. Raviprolu A (2017) Role of Artificial Intelligence in Recruitment. Int J Eng Technol
99. Sophie, C.: Intelligence artificielle (IA) dans les médias: beaucoup de fantasmes (2019). https://www.samsa.fr/2019/12/02/intelligence-artificielle-ia-dans-les-medias-beaucoup-de-fantasmes-quelques-realites-et-pas-mal-de-questions/. Accessed 7 Feb 2020
100. Muangprathub, J., Boonnam, N., Kajornkasirat, S., et al.: IoT and agriculture data analysis for smart farm. Comput. Electron. Agric. 156, 467–474 (2019). https://doi.org/10.1016/J.COMPAG.2018.12.011
101. FT: Smart agriculture based on cloud computing and IOT. J. Converg. Inf. Technol. (2013). https://doi.org/10.4156/jcit.vol8.issue2.26
102. Lopez-Rincon, O., Starostenko, O., Martin, G.A.S.: Algoritmic music composition based on artificial intelligence: A survey. In: 2018 28th International Conference on Electronics, Communications and Computers, CONIELECOMP 2018 (2018)
103. Cope, D.: Algorithmic music composition. In: Patterns of Intuition: Musical Creativity in the Light of Algorithmic Composition (2015)
104. Norton, D., Heath, D., Ventura, D.: Finding creativity in an artificial artist. J. Creat. Behav. (2013). https://doi.org/10.1002/jocb.27
105. Smaill, A.: Music and Artificial Intelligence (2002)
106. Kamhi, G., Novakovsky, A., Tiemeyer, A., Wolffberg, A.: Magenta (2009)
107. Brian, S.: Narrative science, the automated journalism startup—technology and operations management. In: HBS Digital Initiaitve (2018). https://digital.hbs.edu/platform-rctom/submission/narrative-science-the-automated-journalism-startup/. Accessed 23 Jan 2020

108. Brian, S.: Automated Insights: Natural Language Generation (2020). https://automatedins ights.com/. Accessed 23 Jan 2020
109. Spreitzer, G.M., Garrett, L.E., Bacevice, P.: Should your company embrace coworking? MIT Sloan Manag. Rev. (2015)
110. Echobox: Echobox—Social Media for Publishers (2020). www.echobox.com. https://www. echobox.com/. Accessed 23 Jan 2020
111. Yseop: Advanced Natural Language Generation (NLG) AI automation I Yseop (2020). www. yseop.com. https://www.yseop.com/. Accessed 23 Jan 2020
112. Boomtrain Software: Boomtrain Software—2020 reviews, pricing & demo. In: Boomtrain Software (2020). https://www.softwareadvice.com/marketing/boomtrain-profile/. Accessed 23 Jan 2020
113. D'Alfonso, S., Santesteban-Echarri, O., Rice, S., et al.: Artificial intelligence-assisted online social therapy for youth mental health. Front Psychol. (2017). https://doi.org/10.3389/fpsyg. 2017.00796
114. Digitalgenius: DigitalGenius I Customer Service Automation Platform (2020). www.digita lgenius.com, https://www.digitalgenius.com/. Accessed 23 Jan 2020
115. Ipsoft: IPsoft Inc., Global Leader in AI and Cognitive Tech Systems (2020). https://www.ips oft.com/. https://www.ipsoft.com/. Accessed 23 Jan 2020
116. Bloomberg: Inbenta Technologies Inc.: Private Company Information—Bloomberg. In: Bloomberg (2019)
117. Raza, M.Q., Khosravi, A.: A review on artificial intelligence based load demand forecasting techniques for smart grid and buildings. Renew. Sustain., Energy Rev (2015)
118. Gartner: The Road to Enterprise AI (2017)
119. Kambatla, K., Kollias, G., Kumar, V., Grama, A.: Trends in big data analytics. J. Parallel Distrib. Comput. (2014). https://doi.org/10.1016/j.jpdc.2014.01.003
120. Safadi, F., Fonteneau, R., Ernst, D.: Artificial intelligence in video games: towards a unified framework. Int. J. Comput. Games Technol. (2015). https://doi.org/10.1155/2015/271296
121. Frutos-Pascual, M., Zapirain, B.G.: Review of the use of AI techniques in serious games: decision making and machine learning. IEEE Trans. Comput. Intell. AI Games (2017)
122. Frutos-Pascual, M.: Les robots deviennent-ils plus intelligents que les humains ?—Maddy-ness—Le Magazine des Startups Françaises (2019). https://www.maddyness.com/2019/10/ 18/maddyfeed-robots-plus-intelligents-humains/. Accessed 7 Feb 2020
123. Anderson, J.R., Law, E.H.: Fuzzy logic approach to vehicle stability control of oversteer. SAE Int. J. Passeng. Cars—Mech. Syst. (2011). https://doi.org/10.4271/2011-01-0268
124. Abduljabbar, R., Dia, H., Liyanage, S., Bagloee, S.A.: Applications of artificial intelligence in transport: an overview. Sustain (2019)
125. Saba, D., Laallam, F.Z., Belmili, H., Berbaoui, B.: Contribution of renewable energy hybrid system control based of multi agent system coordination. In: Souk Ahres University (ed.) Symposium on Complex Systems and Intelligent Computing (CompSIC). Souk Ahres University, Souk Ahres (2015)

Big Data and Artificial Intelligence
Applications

In Depth Analysis, Applications and Future Issues of Artificial Neural Network

Badal Soni, Prachi Mathur, and Angshuman Bora

1 Introduction

The brain is a highly complex, non-linear, distributed and parallel processing system. It is known that the conventional digital computers existing today are far from achieving the capability of motor control, pattern recognition, perception and language processing of the human brain. This can be due to several reasons. Firstly, despite having a memory unit and several processing units which helps to perform the complex operations rapidly, it lacks the capability to adapt. Secondly, it lacks he capability to learn. Thirdly, we do not understand how to simulate the number of neurons and their interconnections as it exists in biological systems. Comparison of brain and computer indicate that the brain has 10^{11} neurons with switching time of 10^{-3} s while computer has 10^9 transistors with switching time 10^{-9} only. Though it seems like a flawed comparison since response time and quantity do not clearly measure the performance of the system but it indicates the advantage of parallelism in terms of processing time. The largest part of the brain is continuously working while the largest part of the computer is only a passive data storage. Thus the brain is perform parallel computations up to its theoretical maximum while the computer is magnitudes away. Moreover, computer is static and lacks the capability to adapt. On the other hand, the brain possesses biological neural network that can reorganize and

B. Soni (✉) · P. Mathur · A. Bora
National Institute of Technology, Silchar, Assam, India
e-mail: soni.badal88@gmail.com

P. Mathur
e-mail: mathur.prachi72@gmail.com

A. Bora
e-mail: angshu.btf@gmail.com

A.-E. Hassanien et al. (eds.), *Enabling AI Applications in Data Science*, Studies in Computational Intelligence 911, https://doi.org/10.1007/978-3-030-52067-0_7

recognize itself to learn from the environment. Also, we are posed with the challenge of performing the operations in the natural asynchronous mode.

The study of the ANN is motivated by their similarity to successfully work the biological systems. The main characteristics that the computer systems try to adapt are the learning capability which can be supervised or unsupervised, generalization to find results for similar problems, and fault tolerance.

Artificial Neural Network (ANN) is an information-processing parallel processor developed as a generalization to the mathematical model of neuro-biology. ANN is made up of following units:

1. Information processing units called nodes
2. Interconnection links between nodes
3. Each connection link having strengths called weights which are manipulated using learning from training patterns.
4. Each neuron having an activation function to get an output signal

Mathematically, a neural network is a sorted triple represented as (N, V, w) where set N represents a set of neurons or nodes and V is a set given as

$$\{(i, j)|i, j \epsilon N\} \tag{1}$$

which represent interconnection between node i and node j. The function

$$w : V \to R \tag{2}$$

defines the synapse weights, where w_{ij} represents the weight between node i and node j.

The paper, begins with a brief history and time-line of the development and research of ANN. Then it gives the various levels of structural description which is followed by the overview of the classification of ANN in terms of ANN models. They are based on connectivity among the neurons and their corresponding architectures. Various neuron models like Perceptron, Multilayer Perceptron, Hopfield Networks, Adaptive Resonance Theory (ART) Networks, Kohonen Networks, Radial Belief Function (RBF) Networks, Self-Organizing Maps and Boltzmann Machine are discussed. Since ANN is a major constituent of Artificial Intelligence and has its roots in learning and acquiring knowledge from the environment to mimic the human brain, training algorithms for various neuron models have been presented. The later part of the paper covers several real time applications ANN finds its extensive use in such as pattern recognition, clustering, fuzzy logic, soft computing, forecasting and neuro-sciences. This paper also discusses the various issues that need to be considered before ANN can be used for any application. Towards the end, a brief analysis of ANN as a development project is presented which talks about the steps of the ANN development cycle beginning from problem definition, design, realization, verification, implementation and maintenance. It also briefly touches the issues in the ANN development with the solutions on how it can improve or minimized. To conclude, this paper discusses the future scope of development in the field of ANN and its prominence as a technology. Thus, and its prime importance in this era of technology advancement and why it is of a considerable significance.

2 Related Work

This section is a literature survey of the development of ANN from the year 1943 to 2013. It discusses the ANN timeline from the time the first step to ANN was taken to the recent domains where ANN is being currently used.

2.1 The Onset

- **1943–1947**: The first step towards development of ANN began with the article written by Warren McCulloch and Walter Pitts which introduced neural network modeled using electronic circuits which could compute arithmetic and logical operations [31]. Later, they also presented the spacial pattern recognition by ANN. During the same time, Norbert Wiener and John Von Neumann also suggested that design of brain-inspired computers might be fascinating.
- **1949**: A book titled *The Organization of Behavior* [16] was written by Hebb, which emphasized on learning by the neuron synapses. Hebb went about proposing a learning law which said that neural connection is strengthened each time they are simultaneously activated. The change in strength is proportional to the product of the activities of the neurons.

2.2 The Golden Era

- **1951**: Snark is Neuro-computer developed by Marvin Minsky having the capability to adjust its weights without human intervention. The Snark operated successfully from a technical stand point but it never actually carried out any particularly interesting information processing functions. Although there were many other people examining the issues surrounding the neuro-computing in the 1940s and early 1950s, their work had more the effect of setting the stage for later developments than of actually causing those developments.
- **1956**: Dartmouth Summer Research Project was conducted where discussions happened to simulate the functioning of brain. It provided a surge to both artificial intelligence(AI) and ANN. Both aspects of simulation-research by means of software for AI and by imitating the smallest element-the neurons at a much lower level, for ANN were talked about.

- **1957–1958**: John von Neumann facilitated neuron simulation using telegraph relays or vacuum tubes. The first hardware neuro-computer was the *Mark I* which was developed by Frank Rosenblatt and his colleagues. The consequences of the result was the introduction of Perceptron to recognize numbers which is still in use today.
- **1959**: In the following year, Frank Rosenblatt went on to describe various kinds of the perceptron. He also coined and proved the perceptron convergence theorem. He described that neurons learned by adjusting the weights connected to the synapses [38].
- **1960**: A different kind of ANN element called ADALINE (Adaptive Linear Neuron) [52] was formulated by Bernard Widrow and Marcian E. Hoff. It possessed a accurate and fast adaptive learning law which could be trained by means of the Widrow-Hoff rule or delta rule. If wrongly applied, it could lead to infinitesimal small steps close to the target.
- **1961**: Karl Steinbuch [46], introduced the Lern Matrix which a associative memory like structure. It successfully summarizes the *conditional reflexes* of human brain and how the learning operations can be realized. He also described concepts for neural techniques and analyzed their possibilities and limits.
- **1965**: Nils Nilsson in his book *Learning Machines* presented the progress and work done in neural network in this period and assumed that the self learning had been discovered and thus "intelligent" systems had already been in place [33].
- **1969**: Paper [32] was written by Minsky and Seymour Papert's in 1969 where they mentioned that the perceptron is limited by the disadvantage of linear separability and hence not capable of dealing with the problems of forecast and other real world problems. The research also said that it will be the dead end of AI.

2.3 The Halt and Slow Revival

Due to absence of funds and lack of conferences and events there were not enough publications. But, the research continued by independent researchers which led to the development of various paradigms though it didn't achieve any recognition.

- **1972**: Teuvo Kohonen laid the foundation of a model of the linear associator that is a model of an associative memory [25]. It follows rules of linearity: (i) if the net gives a pattern X of outputs for a pattern P of inputs, it will give for kP for kX and (ii) Suppose that for pattern Q of inputs, the net gives pattern X of outputs; and that for pattern R, it gives pattern Y. Then for inputs $P + Q$, we get outputs $X + Y$. In the same year, a similar model was presented independently by James A. Anderson.
- **1973–1974**: A non-linear model of a neuron was proposed by Christoph von der Malsburg. Paul Werbos developed the back-propagation method.
- **1976–1980 and thereafter**: Stephen Grossberg developed various neuron models commonly known as *Additive and Shunting Models*. He discovered equations

for Short Term Memory, Medium Term Memory and Long Term Memory. He partnered with Gail Carpenter for the formulation of models of adaptive resonance theory (ART). These derivation showed, for the first time, that brain mechanisms could be derived by analyzing how behavior adapts autonomously in real time to a changing world [13].

- **1982**: The self-organizing feature maps (SOM) also known as Kohonen networks were developed by Teuvo Kohonen in paper [26]. It is a non-linear map based on competitive learning and ties to mimic the self organization mechanisms in the brain. During the same year, the Hopfield network came into picture which was developed by John Hopfield. It is a recurrent network which has loops. It brings together the ideas of neurobiology and psychology and present a model of human memory, known as an associative memory.
- **1983**: Fukushima, Miyake and Ito gave a Neuro-cognitron neural model which is an extension of cognitron developed in 1975 [11] that recognized handwritten characters.

2.4 Rejuvenation

In the early 1980s many researchers working on neuro-computing began submitting proposals on the neuro-computers development and on neural network applications.

- **1985**: An article was published by John Hopfield using the Hopfield nets, described acceptable solutions for the Travelling Salesman problem [18].
- **1986**: The delta rule generalization i.e back-propagation of error learning procedure was formulated and published by the Parallel Distributed Processing Group (PDP).
- **1987**: The first open conference on ANN was conducted and International Neural Network Society (INNS) was formed. This led to an exponential surge from this time, in the development in the domain of ANN [27, 57].
- **1988–1990**: Nobel laureate Leon Cooper, of Brown University introduced the first multilayer nets, the *Reduced Coulomb Energy Network*. The pioneer of first convolution neural network(CNN) was Yann LeCun. He named it LeNet5. It is three layered and used multilayer neural network as the final classifier.
- **1992**: Max-pooling was introduced to help with least shift invariance and tolerance to deformation to aid in 3D object recognition. Boser et al. [7], suggested a way to create nonlinear classifiers by applying the kernel trick to maximum-margin hyperplanes.
- **1995**: The concept of intelligent agents was introduced which included the mathematical formulation for reasoning about their properties [56].
- **2001–2004**: Reservoir computing, an extension of ANN was pioneered by Schrauwen, Benjamin, David Verstraeten, and Jan Van Campenhout in *An overview of reservoir computing: theory, applications, and implementations* as cited in [39]. It is a computational framework where an input signal is fed to a dynamical sys-

tem called a reservoir and the dynamics of the reservoir map the input to a higher dimension. Then a simple readout mechanism is trained to read the state of the reservoir and map it to the desired output.

- **2010–12**: The extension of ANN-Deep Learning became prominent. Dan Claudiu Ciresan and Jurgen Schmidhuber pioneered the very fist implementation of GPU ANNs. Both forward and backward ANN was implemented on NVIDIA GTX 280 processor with nine layers. Alex Krizhevsky extended the GPU ANNs which were an extension of the LeNet (1983).

3   Levels of Structural Description

The paper [30], described ANN in terms of structure or architecture at three different levels which can be given as-

- *Micro-Structure*: It deals with node-characteristics of ANN. It is the lowest level.
- *Meso-Structure*: It deals with network organization in the network. It is a form related to the function.
- *Macro-Structure*: Deals with interconnection between networks to accomplish complex tasks.

3.1   *Micro-Structure*

Neuron accepts a set of input signals from other neurons, summates the received signal, adds a threshold value and passes the summated value to the transfer function. The value obtained are interdependent on the previous value. Neurons differs from each other mainly on two parameters, changing the transfer function and adding new parameters such as bias, gain or threshold.

3.1.1   Neuron Signal Function

Transfer Function is the most common factor to distinguish between neurons. It specifies how the neurons scales the incoming signal and neuron activation. The following are the commonly used transfer functions (Table 1).

3.1.2   Bias, Gain and Threshold Function

A threshold function is used by ANN which is summed with the summated input signal. This is known as the *bias* and is like a constant output independent of the other input signals. Bias reduces the burden off the training of weights from lower layer to upper layer to avoid providing an appropriate scaling value. A gain term is

Table 1 Description of different transfer function

Name	Function	Characterstics
Binary threshold	$f(x_j)=$ $\begin{cases} 1 \text{ if } x_j >= 0 \\ 0 \text{ if } x_j < 0 \end{cases}$	Non Diffrentiable, step like $\epsilon\ \{0, 1\}$
Bipolar threshold	$f(x_j)=$ $\begin{cases} 1 \text{ if } x_j >= 0 \\ -1 \text{ if } x_j < 0 \end{cases}$	Non Diffrentiable, step like s_j $\epsilon\ \{-1, 1\}$
Linear	$f(x_j) = \alpha_j x_j$	Diffrentiable, Unounded s_j $\epsilon\ \{\infty, \infty\}$
Bipolar threshold	$f(x_j)=$ $\begin{cases} 0 \text{ if } x_j <= 0 \\ \alpha_j x_j \text{ if } 0 < x_j < x_m \\ 1 \text{ if } x_j >= x_m \end{cases}$	Differentiable, Piecewise Linear $s_j\ \epsilon\ [0, 1]$
Sigmoid	$f(x_j) = \frac{1}{1+e^{-\lambda_j x_j}}$ (3)	Differentiable, Monotonic, Smooth $s_j\ \epsilon\ (0, 1)$
Hyperbolic tangent	$f(x_j) = tanh(\lambda_i x_j)$ (4)	Differentiable, Monotonic, Smooth $s_j\ \epsilon\ (-1, 1)$
Gaussian	$f(x_j) = e^{-(x_j - c_j)^2/2\sigma^2}$ (5)	Differentiable, Non-Monotonic, Smooth $s_j\ \epsilon\ (0, 1)$
Stochastic	1 with probability $P(x_j)$ -1 with probability $1 - P(x_j)$	Non-Deterministic, step like, s_j $\epsilon\ \{0, 1\}\ or\ \{-1, 1\}\}$

multiplied by the summated output in some ANNs. Gain can be fixed or adaptive. If the gain is adaptive then neuron does not produce a constant value but may produce some signal with stronger strength than others.

3.2 Meso-Structure

Meso-Structure focuses on the physical organization and arrangement of the neurons in the network. These considerations help us distinguish between different types and classes of networks. The ANNs are characterized by-

- Number of layers- single-layered, bi-layered, multi-layered
- Number of neurons per layer
- Type of connections- Forward, backward, Lateral
- Degree of Connectivity

Based on the above properties Fig. 1 mentions the common topologies used in various neuron models-

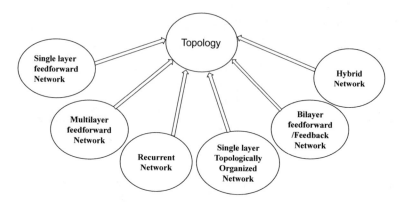

Fig. 1 Commonly used topologies

3.2.1 Topologies of Neural Network

Single-Layer Feed-Forward Network

One input layer projects to one output layer of neurons.

Multi-Layered Feed-Forward Network

Multilayer Network consists of one input layer, one or more hidden layers (not visible) and one output layer. Nodes of one layer has directed connections to the nodes of the next layer only (Figs. 2 and 3).

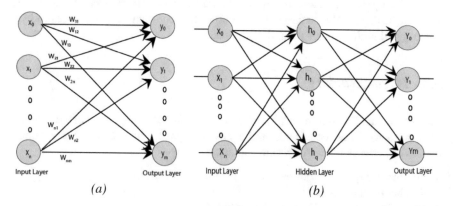

Fig. 2 a Single layer feed forward network. **b** Multilayer feed forward network

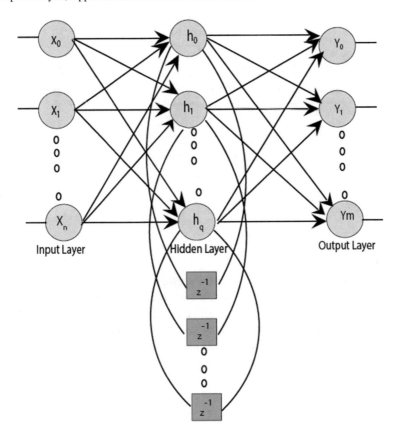

Fig. 3 Recurrent network

Recurrent Network

The feature of recurrent network is that it consists of at least one loop. It may or may not be a self-feedback loop. It impacts the learning capability and the performance of the network significantly and is responsible for the nonlinear dynamic behavior of neural network. Assuming that the neural network consists of the nonlinear units.

Single Layer, Topologically Organized Network

It is based on topologically-ordered vectors that are single layered from another class of networks. During learning, the distance between vectors is responsible for adjusting the relative positions in vector space Maren et al. [30].

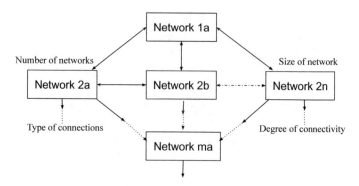

Fig. 4 Issues of building Macro-structures

Bilayer Feed-Forward/ Feedback Network

It is a two layered network which passes information using both feed-forward connection from first to second layer and feedback connection from second to the first layer. This type of a network is used for pattern hereto-association where pattern in one layer is associated with a pattern in another layer. It is commonly used in Bidirectional Associative Memory (BAM) and Adaptive Resonance Theory (ART) (Fig. 4).

3.3 Macro-Structure

Often a single type of neural network may not be adequate to solve complex tasks. Thus, there is a need to develop interacting ANNs. Two or more interacting networks are usually present in true macro-structures. Macro-structures are aimed to handle such complex tasks in modular way so that the interactions are isolated. Macro-structures are the two main types: strongly coupled and loosely coupled networks.

- Strongly coupled networks is a fusion of two or more networks into a single structure. It tries to eliminate the weaknesses of the interacting systems.
- Loosely coupled networks retain their structural properties and distinctness.

4   Training of Neural Network

ANN learns when its components or free parameters adapt to the changes in the environment. It can undergo changes if

- developing new interconnections
- deleting already established interconnections
- modifying the weights of the interconnections
- Threshold values of nodes are altered
- adding one or more hidden layers
- Adding one or more nodes at a particular layer or adding one or more hidden layers.

Altering the synaptic weights is the most common procedure to train the network. Furthermore, deletion and development of interconnections can be realized by ensuring that the connection weight is no longer trained when set to zero and setting a non-existent connection weight to other than zero respectively. Threshold values can be modified by adjusting synaptic weights. Changing the network architecture is difficult to perform. ANN learns by modifying weights according to algorithm rules The training can be done in the following ways.

4.1   Supervised Training/Associative Learning

In Supervised learning the ANN is trained by providing it with input as well as the desired or output patterns for the corresponding inputs so that the network can receive the error vector precisely. It can be self supervised or supervised by a teacher IT is the easiest and most practical form of training the ANN. The algorithm is shown in Fig. 8.

4.2   Unsupervised Learning/Self-organization Learning

Unsupervised learning consists only input patterns. There is no external teacher. In this training paradigm, the network is expected to detect statistical similarities or salient features to classify the inputs. Unlike the supervised learning, there is no a priori set of categories for classification Jha [22].

4.3   Reinforcement Learning

The training set consists of input patterns only. The learning machine performs some action and gets the result/feedback of the performance from the environment. The value returned indicates how good or bad is the result was right or output. Based on this the network parameters are adjusted Jha [22].

5 Neuron Models

The most commonly used Neuron models to solve real-life problems are-

5.1 Perceptron

Perceptron was introduce in late 1950s by Rosenblatt. It is the simplest model of ANN which acts as linear classifier as it can only separate of linearly separable patterns. It comprises of a single feed-forward layer with adjustable weights and bias. It forms the hyperplane between the two patterns if the they are linearly separable. Mathematically,

$$v_k = \sum_{j=1}^{m} w_{kj} x_j \tag{6}$$

and

$$y_k = \phi (u_k + b_k) \tag{7}$$

where $w_{k1}, w_{k2}, \dots, w_{km}$ are synaptic weights of neuron k, $x_1, x_2, \dots, x_m$ are input, u_k is the linear combiner output due to inputs, b_k is bias, ϕ (.) is activation function and y_k is output signal of neuron. Perceptron has a disadvantage that it does not converge if the patterns are linearly separable [15, 53] (Fig. 5).

5.2 Adaptive Linear Element (Adaline)

One of the first generation ANN model was proposed by Widrow and Hoff in 1959. Adaline was mainly developed to recognize binary patterns. Adaline is a single layer neuron which uses bipolar inputs and outputs. The training of Adaline happens based on Delta Rule or least mean square (LMS) [53] (Fig. 6).

Adaline has the output as a linear combination of inputs i.e the weighted sum of inputs. The weights are essentially continuous and may have negative or positive values. During the training process, patterns corresponding to the desired output are presented. An adaptive algorithm automatically adjusts the weights to minimize the squared error. The flowchart of the training process is shown in Fig. 7.

Perceptron and Adaline have several similarities which makes adaline an extension of Perceptron and differences which distinguishes it from perecetron. The similarity and differences are stated as follows:

Similarity

- Architecture is the same
- Both are binary pattern classifier

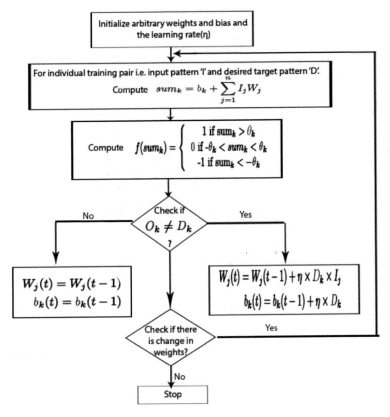

Fig. 5 Flowchart of perceptron algorithm

Fig. 6 Adaptive linear element (Adaline)

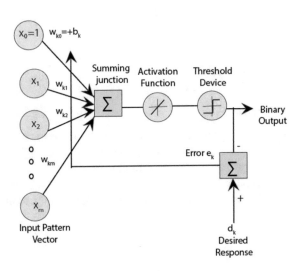

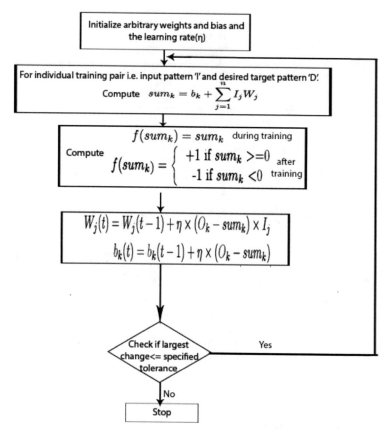

Fig. 7 Flowchart of adaline algorithm

- Both have linear decision boundary
- Both used threshold function
- Learn iteratively i.e. sample by sample.

Differences

- Transfer function for Adaline is taken to be the identity function during training and bipolar Heaviside step function after training while in Perceptron its identity function during and after training.
- Perceptron takes binary response and computes an error used to update the weights while Adaline uses a continuous response value to update the weights.

5.3   Support Vector Machines (SVM)

The SVM is a non-probabilistic linear machine which was pioneered by Vapnik in the 1990s. Figure 8, indicates that SVM is a generalization of a simple and intuitive classifier called the support vector classifier which in return, is an extension of maximal margin classifier SVM can accommodate non-linear class boundaries. It is used to construct a hyperplane as a decision surface in a way such that the margin of separation between positive and negative examples are maximized [21].

Consider the training samples (x_i, y_i) where $i = 1, 2...N$, x_i is the training pattern for the ith sample and y_i is the desired response. The separating hyperplanes have the property-

$$w_0 + w_1 x_{i1} + w_2 x_{i2} + \cdots + w_n x_{in} > 0 \, where \, y_i = +1 \qquad (8)$$

and

$$w_0 + w_1 x_{i1} + w_2 x_{i2} + \cdots + w_n x_{in} < 0 \, where \, y_i = -1 \qquad (9)$$

Equivalently, a separating hyperplane has the property that

$$y_i (w_0 + w_1 x_{i1} + w_2 x_{i2} + \cdots + w_n x_{in}) > 0 \, where \, y_i = -1 \qquad (10)$$

where **x** in input vector and **w** is an adjustable weight vector. The separation between the hyperplane and the closest data point i.e. d1 for H1 and d2 for H2 is called the *Margin of Separation*. The goal of the SVM is to find the hyperplane for which margin of separation is maximized i.e. $d1 = d2$ shown in Fig. 8. Such a hyperplane is called *Optimal Hyperplane*. The main aim of SVM is to maximize the distance given by

$$\frac{w^T x + b}{||w||} \qquad (11)$$

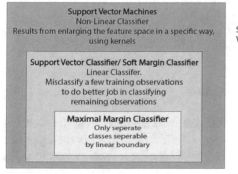
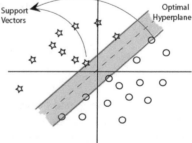

Fig. 8   SVM and optimal hyperplane of SVM

Since the input vector Support Vectors are those data points that lie closest to the decision boundary and most difficult to classify.

5.4 Radial Basis Functions (RBF)

RBF was first introduced by Powell in 1985 as a solution of real multivariate interpolation problem and later it was also worked upon by Broomhead and Lowe in 1988. RBF finds its utility in classification of complex patterns [15]. RBF is based on the idea of Covers Theorem which states that a complex pattern classification problem can cast in high-dimensional space nonlinearly is more likely to be linearly separable than in low-dimensional space. RBF network has 3 layers with different functions-

- *Input Layer*—It is mad up of sensory nodes which connects the network to external environment by taking the input.
- *Hidden Layer*—They are also called RBF neurons. They allow non linear transformations from the input space to output space. Generally, Hidden Layer is of high dimension.
- *Output Layer*—Output neurons are linear and only contain the response of activation function and one weighted sum as propagation function.

5.5 Hopfield Network

Hopfield network was first introduced in the year 1982. It is a form of Associative Memory Neural Network with the only change of being a recurrent network. The network is fully connected with symmetric synaptic weights and no self feedback. The feedback enabled the network to hold memories. In the Hopfield network only one of the activation unit is updated at a time when signal is received from the other unit. A neuron in the Hopfield network is governed by

$$\frac{du_i}{dt} = \sum_{j \neq i} T_{i,j} V_j + I_i \qquad (12)$$

and

$$V_i = g(u_i) \qquad (13)$$

where $T_{i,j}$ is the weighted connection between two neurons i and j which is a symmetric matrix with zero diagonal since no self feedback, I_j is the input of a single neuron i and g is a monotonically non-decreasing activation function. The neurons of Hopfield network tend towards a collective state which minimizes the energy function

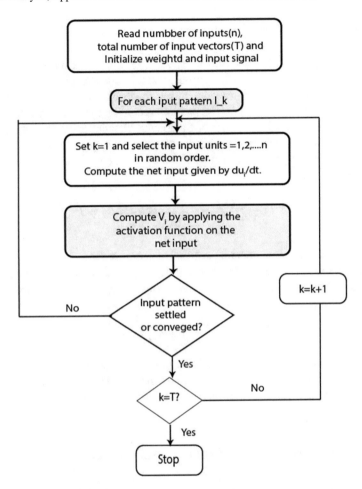

Fig. 9 Flowchart of Hopfield algorithm

$$E = -\sum_{i=1}^{n} I_i V_i - \sum_{i=1}^{n}\sum_{j>i} T_{i,j} V_i V_j \tag{14}$$

The Hopfield network is used for optimization by constructing the $T_{i,j}$ and I_i connections such that the minimum points of the energy function correspond to the optimal solutions of problem. The training algorithm is been given in Fig. 9.

The network is guaranteed to converge at the local minimum but to a specific pattern is not guaranteed due to oscillations [34].

5.6   Boltzmann Machine

The Boltzmann machine is a symmetrically connected, stochastic machine whose composition consists of stochastic neurons. It can be thought of as an extension of Hopfield Network due to it symmetric, no self feedback and random asynchronous update of unit activation. It can model patterns according to the Boltzmann distribution The two distinguishing features of Boltzmann Machine are each unit can reside in two possible states only in probabilistic manner and symmetric synaptic weights between its neurons. It can be characterized by an energy function which can be given as follows:

$$E = \frac{1}{2} \sum_j \sum_k w_{kj} x_k x_j \; such \; that \; j \neq k \tag{15}$$

where w_{kj} is the weights connecting neuron j to neuron k and x_j is the state of neuron j. It does not contain any self loops. At some step and temperature T, a random neuron is selected during the learning process and its state is flipped with probability:

$$P(x_k \rightarrow -x_k) = \frac{1}{1 + exp(-\Delta E/T)} \tag{16}$$

Initially, the temperature is set to relatively high value and the over time, it is gradually decreased according to some annealing schedule. The annealing schedule and determining the activation of units is central to performance [15].

5.7   Adaptive Resonance Theory Neural Network

ART is an unsupervised learning model was developed to overcome the problem of stability in learning due to network plasticity which causes the prior learning by the new learning. The key difference of ART is the "$Vigilance Parameters$" which is a user specified value to decide the similarity of the input pattern to assign it to a cluster. The term "$resonance$" corresponds to the state where the input matches prototype vector significantly to permit learning.

The basic architecture of ART consists of two layers, the input layer which is partitioned into input processing layer (LI1) and input interface layer (LI2), Output layer (L2) and Reset layer (R). The u_{ij} represents the weights from LI2 and L2 and d_{ij} is the weights from L2 to LI2. The competitive output layer L2 is the layer where the units check the proximity with the input patterns. The reset layer R compares input with vigilance parameters. If the vigilance parameters are met then only the weights are updated [34] (Fig. 10).

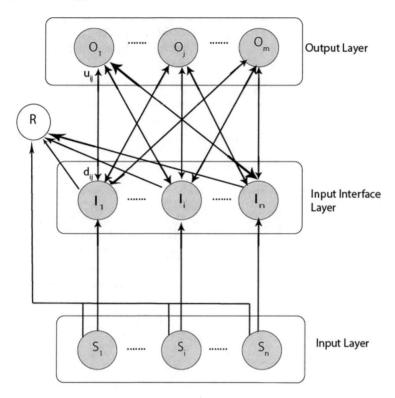

Fig. 10 Architecture of adaptive resonance theory

5.8 Kohonen Self Organizing Map

In 1989, Finnish professor Teuvo Kohonen had developed a topological structure analogous to a typical NN with competitive units or cluster units in network layers. The cluster is granted as a winner if the Euclidean Distance between the weight vector within the cluster and the input vector is minimum as compared to other clusters in the neighborhood. The weights are updated according to

$$w_{kj}(t+1) = w_{kj}(t) - \eta \times (x_i - w_{kj}(t)) \tag{17}$$

where w_{kj} is the weight of neuron k, η is the learning parameter and x_k is the input. SOM uses unsupervised learning procedure to reduce the dimensionality of the input pattern. KNN performs adaptive transformation produce a two-dimensional discretized representation of the input space in topological manner shown in Fig. 11.

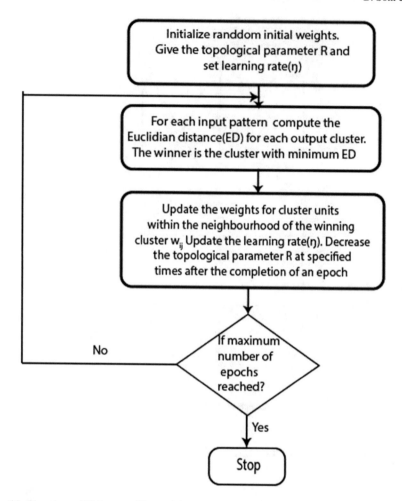

Fig. 11 Flowchart of Kohonen self organizing map

Therefore, this network is called *Self-organizing Map* or simply Kohonen Neural Network (KNN). Since there is no priori information about the desired output, thus a numerical measurement of the mapping error magnitude cannot be predicted and used [37] (Table 2).

Table 2 Summary of different neuron models

Model	Architecture	Characteristic	Learning algorithm	Application
Perceptron	Single Layer, Feed-forward	Binary, Threshold	Supervised, error correction	Pattern, Classification
Adaline	Single Layer, Feed-forward	Linear	Supervised, Gradient Descent	Regression
Multilayer Perceptron	Multi layered Feed-forward	Nonlinear sigmoid	Supervised, Gradient Descent	Function Approximation
Support Vector Machines	Multi layered	Binary Threshold	Supervised, reward punishment	Classification, Regression
Radial Basis Function Network	Multi layered distance based	Linear	Supervised, quadratic optimization	Classification, Regression, Interpolation
Hopfield Network	Single layered, feedback	Binary Threshold/Sigmoid	Outer product correlation	CAM, Optimization
Boltzmann Machine	Two layered, feedback	Binary Threshold	Stochastic, Gradient Descent	Optimization
Adaptive Resonance Theory (ART)	Two layered	Binary, faster than linear	Unsupervised Competitive	Clustering, Classification
Vector Quantization	Single layered, feedback	Faster than linear	Supervised and Unsupervised Competitive	Clustering, Quantization
Kohonen SOM	Single layered	Linear Threshold	Unsupervised Soft Competitive	Topological Mapping
Pulsed Neuron Models	Single layered/Multi layered	Linear Threshold	None	Coincidence Detection, Temporal Processing

6 Modern-Day Applications of ANN

- **Pattern Recognition, Feature Extraction and Classification**
  Significant attention is paid to the use of ANN in solving real world problems related to Pattern Recognition. The advantage of ANN is the varied number of non linear learning algorithms available based on the application. The two most commonly used models in pattern recognition are the Multilayer Perceptron and RBFs. Feature extraction and classification algorithms are also mapped on ANN for hardware implementation. Several applications such as- Interactive Voice Response (IVR), Stock price pattern recognition, Face recognition, entry of data is automated in large administrative systems, Electrocardiogram (ECG) pattern recognition using SOM, banking, Optical character recognition (OCR) to recognize simple pattern recognition system using feed-forward ANN, postal code recognition and

automatic cartography are some of the applications [6]. Adaline is used for Determination of the type of classroom namely-regular or special based on the score of reading tests and written test which calculated into a final score. The calculation of the final score and the class determination is still done manually which can causes errors, so adaline is used to predict the correct classroom for each student based on the final score [14]. SVM is often used to measure Predictive accuracy which is used as an evaluation criterion for the predictive performance of classification or data mining algorithms. The recently developed SVM performs better than traditional learning algorithms in accuracy [20]. Li et al. [29], have applied SVMs by taking DWFT as input for classifying texture, using translation-invariant texture features. They used a fusion scheme based on simple voting among multiple SVMs, each with a different setting of the kernel parameter, to alleviate the problem of selecting a proper value for the kernel parameter in SVM training and performed the experiments on a subset of natural textures from the Brodatz album. They claim that, as compared to the traditional Bayes classier and LVQ, SVMs, in general, produced more accurate classification results. Bio-informatics uses SVM approach to represents data-driven method for solving classification tasks. It has been shown to produce lower prediction error compared to classifiers based on other methods. Small organic molecules that modulate the function of protein coupled receptors are identified using the SVM. The SVM classifies approximately 90% of the compounds. This classifier can be used for fast filtering of compound libraries in virtual screening applications. Some of the applications are given in [23, 45].

- **Biological Modeling**
ANN models are inspired paradigm that emulate the brain which is based on neurons. Recent studies have shown that astrocytes (the most common glial cell) contributes significantly to brain information processing. Astrocytes can signal to other astrocytes and can communicate reciprocally with neurons, which suggests a more active role of astrocytes in the nervous system physiology. The circuit is designed based on fully connected, feed forward multilayer network and without back-propagation or lateral connections. First phase deals with training which uses non-supervised learning. This is followed by the training each pattern with the supervised learning. The second phase uses the evolutionary technique of GA to perform cross-overs and mutation. Multiple iterations of the above process is performed to minimize the error. Thus, ANN is used to mimic the neurobiology of the brain to understand the response mechanism [36].

Perceptron algorithm is used to find weighting function which distinguishes E. coli translational initiation sites from all other sites in a library of over 78,000 nucleotides of mRNA sequence. Weighting function can be used to find translational initiation sites within sequences that were not included in the training set [10].

- **Medicine**
A medical decision making system based on Least Square Support Vector Machine (LSSVM) has been developed in paper [35]. Breast Cancer diagnosis task using

the fully automatic LSSVM was done. WBCD dataset was used which strongly suggest that LSSVM can aid in the diagnosis of breast cancer.

- **Time Series Forecasting**
ANN is commonly used for Time Series Prediction. Based on the past values, with some error, it can predict the future values. The non-linear learning and noise tolerance of the real time series cannot be sufficiently represented by other techniques such as Box and Jenkins. ANN is significantly used and several works have been done in the domain. However, the search for the ideal network structure still remains a complex and crucial task. In year 2006, a hybrid technique constituting autoregressive model and ANN model with a single hidden layer was developed which allows you to specify parsimoniously models at a low computational cost. Hassan, Nath and Kerley proposed the hybrid model of HMM and GA to predict the financial market behavior. Several other models such as AR* model and generalized regression model, hybrid of fuzzy and ANN model was also proposed. In year 2013, a hybrid using ARIMA model was proposed in order to integrate the advantages of both models. In 2017, a hybrid model based on Elman recurrent neural networks (ERNN) with stochastic time effective function (STNN) was produced which displayed the best results as compared to linear regression, complexity invariant distance (CID), and multi-scale CID (MCID) analysis methods. Thus, ANN plays a significant role in Time series Forecasting [48].

- **Data Mining**
Data mining is aimed at gaining insight into large collections of data. The data mining based on neural network is composed by data preparation, rules extracting and rules assessment three phases namely, Data preparation, Rule Extraction and Rule Assessment. ANN's ability to classify patterns and estimate functions make its use in Data mining more significant. It is used to map complex input-output relationships and performs manipulation and cross fertilization to find patterns in data. This helps users to make more informed decisions. ANN is trained store, recognize, and retrieve patterns to filter noisy data, for combinatorial optimization problems and to control ill-defined problems [12, 41].

- **Servo Control**
ANN has emerged as a tool for controlling the response of parameters which are varying greatly in comparison to the PID controller due to its dynamic an robust control nature. Since ANN needs only input and output, it is easier to predict the parameters more correctly. This is extensively used in robotic actuators for motion control [3].

- **Cryptography**
Perceptron can used in cryptographic security. Based on the high-dimension Lorenz chaotic system and perceptron model within a neural network, a chaotic image encryption system with a perceptron model is proposed. The experimental results show that this algorithm has high security, and strong resistance to the existing attack methods [50].

- **Civil Engineering**
ANN is commonly used in various domains of civil engineering including structural engineering, construction engineering and management, environmental and

water resources engineering, traffic engineering, highway engineering, and geotechnical engineering. ANN is used in Structural Engineering in several ways. Pattern recognition and machine learning in structural analysis and design, Design automation and optimization, Structural system identification, Structural condition assessment and monitoring, Structural control and Structural material characterization and modeling are some of the areas. Environment Engineering has several ANN applications. It is used in river streamflow estimation, carbon dioxide concentration prediction from a gas furnace, feeding water control system in a boiling water reactor, multivariate forecasting problems encountered in the field of water resources engineering, long-term forecasting of potential energy inflows for hydropower operations planning and many more. Highway engineering uses ANN frequently. Radial-Gaussian based ANN is used for determining axle loads, axle spacing, and velocity from strain response of truck. BP algorithm is commonly used for condition rating of roadway pavements. A semi-virtual watershed model for small urban watersheds was developed in 2001. Traffic Engineering also uses ANN frequently for several applications. The identification of incident patterns in traffic data using Multilayer Perceptron, ART and SOM, Incident Detection Classifier uses field data to test a multilayer perceptron, function of steering and braking is also developed through ANN are common applications. Geotechnical Engineering involves using BP algorithm to develop a system for checking the concrete retaining walls, static capacity estimation of precast reinforced concrete piles from dynamic stress wave data and for checking the liquefaction potential of sandy soils [1].

- **Electrical Systems**

Adaline is used in power quality disturbances detection and frequency tracking. Adaline is advantageous of being simply calculated and easily implemented through hardware. Adaline is used to track voltage changes. When voltage quality changes, the voltage sudden change will cause the error raise of the adaline, and the weights of adaline vary with it. The simulating results of voltage quality disturbances detection and especially frequency tracking demonstrate that the new adaline and its algorithm can be applied to the precise analysis for power quality [2].

Synthesis of Adaptive Controllers for Dynamical Systems. First, the continuous state space model of the benchmark problem is stabilized. The overall system is then discretized and the proposed adaptive controller is applied. Perceptron training algorithm is used to analyze the effect of changing the parameters of the update algorithm on the performance of the closed-loop system [28].

- **Image Processing**

One of the popular technique for relevance feedback in content-based image retrieval (CBIR) [17] is formulated using SVM. It begins by learning a kernel function from a set of labeled and unlabeled data. Due to the use of unlabelled data it eliminates the problem of small sized training set. The kernel is utilized in a batch mode active learning method to recognize the informative and diverse examples via a min-max framework. Image processing applications are given in [42–44].

- **Compiler Design**
  Incremental Parsing Approach where parameters are estimated using a variant of the perceptron algorithm [9].
- **Finance**
  ANNs can be applied in several areas of finance too. They include stock price prediction, security trading systems, predicting bond ratings, modeling foreign exchange markets, evaluating risk for applications of loans, mortgages and credit cards and even prediction of financial distress.
  Due to the capability to deal with unstable relations, ANNs can be designed and trained to identify complex patterns and predict stock price trends. It can recognize the stocks that outperform the market more accurately than regression models.
  ANN for trading systems is commonly used to manage fidelity fund, pension funds and market returns. Bond Rating Prediction using limited data sets is commonly done. ANN applications for use in currency trading that are fee-based are not uncommon. They ensure services to a range of customers [36].

7   ANN Development Project

The development process of ANN consists of the six phases [5].
PHASE 1: Problem Definition and Formulation
The first phase relies heavily on proper understanding of the problem at hand which involves the following framework

- Define the Problem: The issue that we are trying to solve or the base idea that we want to implement.
- Specify the Objectives and Motivation: What is the aim of the project that can guide you throughout the project.
- Type of Problem: The advantage of ANN over other techniques should be evaluated and the modeling technique.

PHASE 2: System Design and Data Preparation
It is the first step in the actual ANN design. The designer determines the type of ANN, learning rule that fits the problem-supervised, unsupervised or reinforcement learning. This phase also includes data preparation. It has the following steps-data partitioning (partitions data into training set, validation set, and test set), data size (validation set should be large enough to detect differences between the models that are trained, test set should be large enough to indicate the overall performance of the final model and training set should be larger enough to ensure proper training), data representatives (choosing of the right data i.e data that closely correlates to the real world data), data de-duplication (ensures that the data training set, validation set, and test set is disjoint), and data pre-processing (Transform the raw data into a format such that ANN can best learn) [19].

PHASE 3: Metric Definition and System Training
It involves defining the performance metric and training of the network utilizing the training set, validation set, and test set, and simultaneously assessing the network performance be evaluating the error in prediction. Various structural parameters need to be considered such as network size, learning rate, number of iteration cycles for training, error etc. that have a considerable impact on the performance of the final network.

PHASE 4: System Verification and Validation
This phase verifies the generalization capability of the ANN with the use of the validation subset. It helps us in the confidence assessment of the ANN. Verification is intended to confirm the capability of the ANN. This phase also performs the comparison of the ANN-based model and its performance with other models of ANNs.

PHASE 5: System Implementation
Final code execution, testing and verification of the result to the desired output must be done before its released for use.

PHASE 6: System Maintenance
It involves updating the developed system as per the changes occurring in the cycle with time. It then goes back to Phase 2. This might involve training new samples and system variables involved in a new development cycle.

8 Issues with ANN

1. *Hardware Implementation*
   Neural nets are inspired from neurological architectures. They are increasingly used to mimic the biological counterparts. It is indeed imperative to focus on the fact that can such hardware and electronic equipment be developed to ensure high-speed implementation of ANN in digital computers. It can be quite challenging to implement the neural network using VLSI for real-time systems.

   - The input current received by and output current fed to a typical transistor is usually to 2–3 other transistors. The size of the neuro-network is typically scaled to 3000 which is a considerable quantitative difference.
   - Electronic circuits have time delays as well as the structure is fixed. This is unlike the brain where the time delays are negligible due to intensive network.
   - The circuit of biological network changes based on learning but the electronic circuit structure remain fixed.
   - Electronic circuits are mostly feed-forward network while the flow in biological network largely contains feedback loops and is very much a two-way.

- VLSI circuits are highly synchronized as they are governed by clock cycles which is not the case in neural-biology. It is a highly dynamic system.
- Since CMOS ANN have own activation function which is different from any activation function used in ANN, it is required to develop an approximation function.
- Training of the network using this activation function is another challenge for the designers.
- The discrete values of the channel length (L) and width (W) of MOS transistor architecture causes the Quantization effect.
- Power consumption of circuit to mimic the biological network is high [48, 54].

2. *Verification, Validation and Testing*
   The absence of a standard verification and validation procedure causes disparity [47]. Different development stages have varied verification and validation procedures. This leads to lack of trust and consistent assessment of the ANN used in different domains. It consists of issues such as-

   - Verification that the system learned the correct data and not something closely related to the data?
   - How many iterations to be performed on the network for training?
   - Has the network converged to the global minimum or a local minima?
   - How will the ANN handle the test data?
   - Can the network show the same result each time the same set is given as input?
   - Is there a quantifiable technique to describe the networks *memory* or data retention?
   - Are the correct set of input variables chosen for the network as per the problem domain?

3. *Over-Learning, Generalizing and Training Set*
   The aim of the training of ANN is to minimize the error generated in categorization of the sample data set. Lack of gathering or identifying the right (the data that can correlate with the outcomes that need to be predicted) and sufficient training data can hamper the training. This might lead to *Excessive learning or over-adaptation* i.e. ANN performs well on the sample data but not on test data. This is due to wrong noise data contained in training data. For example, if the data is for human face recognition but all photos contain cat faces predominantly the system learns to identify cat faces. Such a system is definitely not reliable. It is important to ensure performance of the sample and selection data(test data) should almost be similar. To avoid over-learning the number of hidden neurons can be reduced. If the learning is weak, new training data can be constantly added to contribute significantly. Thus, the goal of learning is to find an equilibrium between over-learning and generalization ability where the training set plays a key role which is challenging [49].

4. *Structure, Input Variables and Dimension Reduction of ANN*

   Based on the application there are several parameters that need to be focused on to make the suitable neural network. It requires various components. These include

   - Input variables should be carefully chosen based on the application.
   - The number of input nodes. They indicate the number of parameters or information given to identify the decision boundaries. This is difficult to determine since there are many parameters which are important but with varying degree. It is essential to prioritize the inputs to make the network simple and avoid negative impacts. It also improves performance.
   - Attention should be paid to the space dimension problem. Additional neurons can be added to the input layer but this increases the computation time from N to order of N^2 as well as increase the space dimension rapidly.
   - The type of network to be used among Multi-layer Perceptron, Convolution Network or Recurrent Network.
   - The number of hidden layers. They effect the generalization of the ANN.
   - The number of nodes in each layers and the connections between them. This is decided based on the f of data. The network can identify more complex training data if the number of nodes in hidden layers is more.
   - Magnitude of the weights need to be initialized.
   - Dependence among variables should be reduced so that each parameter can be independently evaluated for their impact on the network for easy extraction of information.
   - Redundant variables should be removed i.e the inputs that provide the same information should be eliminated. This will not only reduce the space dimension but also reduce the complexity of the network. Sample.

5. *Criteria of Convergence*

   Training can be stopped using three criteria namely, (i) training error (ii) gradient of error and (iii) cross-validation. The third criteria is the best but requires considerable amount of data and is the most computationally expensive. Hence most of the training takes place using the error function which is the difference between the predicted and the desired output for a given input. Some other methods used are training based on information theory developed in 1996, sum of squared error which is currently most commonly used. The criteria can even be the hit rate representing the percentage of correct classification. Thus based on the application the criteria needs to be effectively chosen.

6. *Parameter Optimization*

   Parameters such as learning rate, momentum coefficient, number of epochs and the training an test data size should be carefully selected since they have adverse impact on the training of the network. It might cause slow training, increased risk on local minima and reduce the generalizing capability of the network.

9 Recent Modification and Improvements

ANN has spread so much that now it is part and parcel of computer learning. Through filtration, this method has improvised itself to the extent that without Artificial Neural Network, no one can thing about training any machine learning, specially deep learning models.

9.1 Hardware Implementation

This issue is the main drawback of the limited calculation ability of computers. But this technology has been developed so much. The capability of the electronic components are increasing drastically and their size is shrinking day by day. However the basic component of a hardware like transistor, it's growing performance will reach it's limit due to the limitation of it's size. Their sizes can not be reduced below a certain level. To solve this issue, optical transistor are being developed. The optical logic gates are much faster than conventional electric gates. If they are completely developed, this will boos the processing power of hardware by a huge scale.

In recent times the cloud computing has grown so much. Google has developed **Tensor Processing Unit** (TPU) [51], that boosts the computation power of hardware by the factor no one has ever thought before. The TPU service is also provided by google to it's clients to perform tremendous amount of calculations in the cloud. So hardware problem is not going to be the issue with huge calculation task in near future. With each version of TPU, the speed is increasing tremendously. The peak performance of TPU v3 was up to 420 TFLOPS. US Department of Energy's Oak Ridge National Laboratory announced the top speeds of its Summit supercomputing machine, which nearly laps the previous record-holder, China's Sunway TaihuLight [55]. The Summit's theoretical peak speed is 200 petaflops, or 200,000 teraflops. So a single TPU chip capacity now can be compared with the world's most powerful supercomputer.

In the Table 3, a comparison between the performance between CPU and GPU can be seen. Various tasks were performed on these both, and their measure was compared. Multi Layer Perceptron(MLP), Long Short Term Memory(LSTM) and Convolutional Neural Network(CNN) model was trained for two times. Their performance are compared with when the same code was ran through TPU. The ratio implies the significant increase in the processing power of newly developed hardware.

9.2 Verification, Validation and Testing

With the recent development of web and cloud services, the data that are being created day by day is increasing by a very large amount. This large amount of data is

Table 3 K80 GPU die and TPU die performance relative to CPU for the NN workload. GM and WM are geometric and weighted mean [24]

Type	MLP0	MLP1	LSTM0	LSTM1	CNN0	CNN1	GM	WM
GPU	2.5	0.3	0.4	1.2	1.6	2.7	1.1	1.9
TPU	41.0	18.5	3.5	1.2	40.3	71.0	14.5	29.2
Ratio	16.7	60.0	8.0	1.0	25.4	26.3	13.2	15.3

that enables the ANN to perform significantly better than the early days. In present days, the collection of data for dataset is not a big thing, millions of images can be collected and trained on ANN, after which thousands of other images can be used to validate, verify and test.

9.3 Over-Fitting of Training Set and Model Structure

As mentioned above, unavailability of data for training a neural network model is not a issue now. Millions of images can be gathered from the web for any kind of model training. However the issue with over fitting is not a hardware issue, it's rather a software issue. The construction of models for neural network determines the accuracy that the model will give. This may be possible that even with millions of images, the model shows very poor performance due to the faulty design of the model. This is called under fitting problem.

Also there can be another scene that the model is built so well that this shows very high accuracy with the training data, however when tested with another data that the model has not seen before, the model will fail. This happens due to the factor called over fitting. The model adjusts it's neurons just to fit the data that it has been fetched in the training phase. The test data won't perform well without the model being altered.

9.4 Parameter Optimization and Model Training

While training the model, the parameters that are being fetched determines the quality of output the model will provide [8]. The new method of implementation are being used today for better output that is gained using precise parameters. In recent times, a model is also being developed that is used to tune the hyper parameters of a neural network model [4]. This is a giant leap in the designing of neural network models. The model does this with the help of recurrent neural networks that performs so well and updates itself after each iteration. So the hyper parameter selection is not a issue anymore, however this technology is not available for public at this time.

There are another models being developed that can program it's child neural network model. It can learn from it's experience gradually and implement it in the design of it's child neural network. These are the giant leaps that are around us today. The image classifier models are getting very accurate day by day. There has been numerous models for various computer vision task like pose estimation, text recognition etc. The technology is not going to be a barrier in near future.

10   Future of Artificial Neural Network

The designing method of ANN is developing day by day. Neural Network was made to mimic human brain functions. Despite of the recent exponential growth, the hardware is still not in the phase to take the load that a normal human brain takes throughout the day. Trillions of neurons fire simultaneously making the brain a large parallel processing device. A parallel processing of this scale is not possible today despite of the improvements in technology. The technology related to Artificial Narrow Intelligence has been growing rapidly day by day, but the construction of Artificial General Intelligence in not possible in the near future. A drastic boost is needed for this to happen. Researchers are doing their best each day to modify and update the computation procedure. The designing of Artificial General Intelligence will help humanity to reduce their effort in general day to day tasks and they can focus in the task where resources should be focused. Artificial Super Intelligence might be the last invention that humanity ever will invent. A super intelligent AI will outperform every existing computer, human even combined together by a very huge logarithmic scale. People are worried about the singularity phase that the Artificial Super Intelligence will posses that no one will understand. However this scenario does not seem possible in the far future. However development in technology might lead humanity to a super intelligent AI one day [40]. This is becoming world's most exciting technology. Almost every industry has already started investing on it. Various startups are actively associating themselves in this industry. The statistics is showing very good results.

As can be seen in recent studies, the annual investment in the field of Artificial Intelligence has started to grow exponentially. The Venture Capital invested on this field is currently way more than 3 Billion dollars. robotic Technology has also started evolving with a very good momentum. Companies like Boston Dynamics are producing both private as well as commercial robots with the help of artificial intelligence. This is the reason why this field has become a golden opportunity for the start-ups. They have also started to focus on this technology to invent something new or to improvise the existing ones. Recent job count in this field has started to increase significantly. Job opportunities in the field of data science, NLP, Computer Vision etc. are increasing day by day. This is a very good sign for the AI industry.

11   Conclusions

ANN has become remarkably important among efficient problem-solving paradigms. The increased use of ANN is due to several reasons such as the ability to learn and recognize by itself or by a teacher, non-linear and dynamic nature, and high tolerance to noisy data and measurement errors. In this paper we have attempted to draw together an overview of this widely used emerging technology. Inspired from neurons, the major constituents of human brain, Neural network first emerged around the mid 1900s. This paper gives a brief description of the time-line of its development over the century. It goes on to talk about the different methods of learning namely, Supervised, Unsupervised and Reinforcement Learning and various Neural Network models. Generally, the models vary according to different applications. This paper discusses the training algorithm and provide an insight as to which model can be used based on the application, complexity, topology and other variable parameters. ANNs are extensively consumed in implementation of various real-time applications such as biological modeling, pattern classification, time series prediction, civil engineering, servo control and even finance. ANN development cycle which involves six phases-problem definition, data preparation, metric definition and training, verification an validation, implementation and maintenance is commonly used methodology for solving ANN problems. However, ANN has several issues which should not be overlooked such as ANN's success rate depends on the quality and quantity of data as well as the number of epochs, lack of standard rules to implement, train and test the network and Determining the input, initializing the synaptic weights and finally, hardware implementation.

ANN thus cannot be the solution to all real-world problems but with the exposure and expanding interest in the domain the techniques will be modeled to larger domain of problems.

Acknowledgments This work was supported by Multimedia and Image Processing Laboratory, Department of Computer Science and Engineering National Institute of Technology Silchar, India.

References

1. Adeli, H.: Neural networks in civil engineering: 1989–2000. Comput. Aided Civil Infrastruct. Eng. **16**(2), 126–142 (2001)
2. Ai, Q., Zhou, Y., Xu, W.: Adaline and its application in power quality disturbances detection and frequency tracking. Electric Power Syst. Res. **77**(5–6), 462–469 (2007)
3. Anderson, D., McNeill, G.: Artificial neural networks technology. Kaman Sci. Corporation **258**(6), 1–83 (1992)
4. Bardenet, R., Brendel, M., Kégl, B., Sebag, M.: Collaborative hyperparameter tuning. In: Dasgupta, S., McAllester, D. (eds.) Proceedings of the 30th International Conference on Machine Learning, PMLR, Atlanta, Georgia, USA, Proceedings of Machine Learning Research, vol. 28, pp. 199–207 (2013)
5. Basheer, I.A., Hajmeer, M.: Artificial neural networks: fundamentals, computing, design, and application. J. Microbiol. Methods **43**(1), 3–31 (2000)

6. Basu, J.K., Bhattacharyya, D., Kim, Th: Use of artificial neural network in pattern recognition. Int. J. Softw. Eng. Appl. **4**(2) (2010)
7. Boser, B.E., Guyon, I.M., Vapnik, V.N.: A training algorithm for optimal margin classifiers. In: Proceedings of the Fifth Annual Workshop on Computational Learning Theory, pp. 144–152. ACM (1992)
8. Caruana, R., Lawrence, S., Giles, C.L.: Overfitting in neural nets: backpropagation, conjugate gradient, and early stopping. In: Advances in Neural Information Processing Systems, pp. 402–408 (2001)
9. Collins, M., Roark, B.: Incremental parsing with the perceptron algorithm. In: Proceedings of the 42nd Annual Meeting on Association for Computational Linguistics, Association for Computational Linguistics, p. 111 (2004)
10. Connally, P., Li, K., Irwin, G.W.: Prediction-and simulation-error based perceptron training: Solution space analysis and a novel combined training scheme. Neurocomputing **70**(4–6), 819–827 (2007)
11. Fukushima, K., Miyake, S., Ito, T.: Neocognitron: a neural network model for a mechanism of visual pattern recognition. IEEE Trans. Syst. Man Cybern. **5**, 826–834 (1983)
12. Gaur, P.: Neural networks in data mining. Int. J. Electron. Comput. Sci. Eng. (IJECSE, ISSN: 2277-1956) **1**(03), 1449–1453 (2012)
13. Grossberg, S.: Adaptive pattern classification and universal recoding: I. Parallel development and coding of neural feature detectors. Biol. Cybern. **23**(3), 121–134 (1976)
14. Handayani, N., Aindra, D., A, Wahyulis, D.F., Pathmantara, S, Asmara, R.A.: Application of adaline artificial neural network for classroom determination in elementary school. IOP Conference Series: Materials Science and Engineering **434**, 012030 (2018). https://doi.org/10.1088/1757-899X/434/1/012030
15. Haykin, S.: Neural Networks: A Comprehensive Foundation. International edition, Prentice Hall, URL https://books.google.co.in/books?id=M5abQgAACAAJ (1999)
16. Hebb, D.O.: The Organization of Behavior: A Neuropsychological Theory. Psychology Press (2005)
17. Hoi, S.C., Jin, R., Zhu, J., Lyu, M.R.: Semisupervised svm batch mode active learning with applications to image retrieval. ACM Trans. Inf. Syst. (TOIS) **27**(3), 16 (2009)
18. Hopfield, J.J.: Artificial neural networks. IEEE Circuits Dev. Mag. **4**(5), 3–10 (1988a)
19. Hopfield, J.J.: Artificial neural networks. IEEE Circuits Dev. Mag. **4**(5), 3–10 (1988b)
20. Huang, J., Lu, J., Ling, C.X.: Comparing naive bayes, decision trees, and svm with auc and accuracy. In: Third IEEE International Conference on Data Mining, IEEE, pp 553–556 (2003)
21. James, G., Witten, D., Hastie, T., Tibshirani, R.: An Introduction to Statistical Learning, vol. 112. Springer (2013)
22. Jha, G.K.: Artificial neural networks and its applications. IARI, New Delhi, girish_iasri@rediffmail com (2007)
23. Jha, R.K., Soni, B., Aizawa, K.: Logo extraction from audio signals by utilization of internal noise. IETE J. Res. **59**(3), 270–279 (2013)
24. Jouppi, N.P., Young, C., Patil, N., Patterson, D., Agrawal, G., Bajwa, R., Bates, S., Bhatia, S., Boden, N., Borchers, A., et al.: In-datacenter performance analysis of a tensor processing unit. In: 2017 ACM/IEEE 44th Annual International Symposium on Computer Architecture (ISCA), pp. 1–12. IEEE (2017)
25. Kohonen, T.: Correlation matrix memories. IEEE Trans. Comput. **100**(4), 353–359 (1972)
26. Kohonen, T.: Self-organized formation of topologically correct feature maps. Biol. Cybern. **43**(1), 59–69 (1982)
27. Kriesel, D.: A Brief Introduction to Neural Networks. URL available at http://www.dkriesel.com (2007)
28. Kuschewski, J.G,. Engelbrecht, R., Hui, S., Zak, S.H.: Application of adaline to the synthesis of adaptive controllers for dynamical systems. In: 1991 American Control Conference, pp. 1273–1278. IEEE (1991)
29. Li, S., Kwok, J.T., Zhu, H., Wang, Y.: Texture classification using the support vector machines. Pattern Recog. **36**(12), 2883–2893 (2003)

30. Maren, A.J., Harston, C.T., Pap, R.M.: Handbook of Neural Computing Applications. Academic Press (2014)
31. McCulloch, W.S., Pitts, W.: A logical calculus of the ideas immanent in nervous activity. Bull. Math. Biophys. 5(4), 115–133 (1943)
32. Minsky, M., Papert, S.: An Introduction to Computational Geometry. Cambridge tiass, HIT (1969)
33. Nilsson, N.J.: Learning Machines (1965)
34. Padhy, N., Simon, S.: Soft Computing: With MATLAB Programming. Oxford higher education, Oxford University Press, URL https://books.google.co.in/books?id=lKgdswEACAAJ (2015)
35. Polat, K., Güneş, S.: Breast cancer diagnosis using least square support vector machine. Digital Signal Process. 17(4), 694–701 (2007)
36. Rabuñal, J.R.: Artificial neural networks in real-life applications. IGI Global (2005)
37. Rojas, R.: Neural Networks: a Systematic Introduction. Springer Science & Business Media (2013)
38. Rosenblatt, F.: The perceptron: a probabilistic model for information storage and organization in the brain. Psychol. Rev. 65(6), 386 (1958)
39. Schrauwen, B., Verstraeten, D., Van Campenhout, J.: An overview of reservoir computing: theory, applications and implementations. In: Proceedings of the 15th European Symposium on Artificial Neural Networks, pp. 471–482, pp. 471–482 (2007)
40. Shabbir, J., Anwer, T.: Artificial intelligence and its role in near future. CoRR abs/1804.01396 (2018)
41. Singh, Y., Chauhan, A.S.: Neural networks in data mining. J. Theor. Appl. Inf. Technol. 5(1) (2009)
42. Soni, B., Debnath, S., Das, P.K.: Text-dependent speaker verification using classical lbg, adaptive lbg and fcm vector quantization. Int. J. Speech Technol. 19(3), 525–536 (2016)
43. Soni, B., Das, P.K., Thounaojam, D.M.: Improved block-based technique using surf and fast keypoints matching for copy-move attack detection. In: 2018 5th International Conference on Signal Processing and Integrated Networks (SPIN), pp. 197–202. IEEE (2018a)
44. Soni, B., Das, P.K., Thounaojam, D.M.: Keypoints based enhanced multiple copy-move forgeries detection system using density-based spatial clustering of application with noise clustering algorithm. IET Image Process. 12(11), 2092–2099 (2018b)
45. Soni, B., Das, P.K., Thounaojam, D.M.: Multicmfd: fast and efficient system for multiple copy-move forgeries detection in image. In: proceedings of the 2018 International Conference on Image and Graphics Processing, pp. 53–58 (2018c)
46. Steinbuch, K.: Die lernmatrix. Biol. Cybern. 1(1), 36–45 (1961)
47. Taylor, B.J., Darrah, M.A., Moats, C.D.: Verification and validation of neural networks: a sampling of research in progress. Intell. Comput.: Theory Appl. Int. Soc. Optics Photon. 5103, 8–17 (2003)
48. Tealab, A.: Time series forecasting using artificial neural networks methodologies: a systematic review. Future Comput. Inf. J. (2018)
49. Vemuri, V.R.: Main problems and issues in neural networks application. In: Proceedings 1993 The First New Zealand International Two-Stream Conference on Artificial Neural Networks and Expert Systems, p. 226. IEEE (1993)
50. Wang, X.Y., Yang, L., Liu, R., Kadir, A.: A chaotic image encryption algorithm based on perceptron model. Nonlinear Dyn. 62(3), 615–621 (2010)
51. Wei, G.Y., Brooks, D., et al.: Benchmarking tpu, gpu, and cpu platforms for deep learning. arXiv preprint arXiv:190710701 (2019)
52. Widrow, B., Hoff, M.E.: Adaptive Switching Circuits. Stanford Univ Ca Stanford Electronics Labs, Tech. rep. (1960)
53. Widrow, B., Lehr, M.A.: 30 years of adaptive neural networks: perceptron, madaline, and backpropagation. Proc. IEEE 78(9), 1415–1442 (1990)
54. Wilamowski, B.M., Binfet, J., Kaynak, M.O.: Vlsi implementation of neural networks. Int. J. Neural Syst. 10(03), 191–197 (2000)

55. Wolfson, E.: The US passed China with a supercomputer capable of as many calculations per second as 6.3 billion humans. https://qz.com/1301510/the-us-has-the-worlds-fastest-supercomputer-again-the-200-petaflop-summit/. Accessed 01 Dec 2019 (2018)
56. Wooldridge, M., Jennings, N.R.: Intelligent agents: theory and practice. Knowl. Eng. Rev. **10**(2), 115–152 (1995)
57. Yadav, N., Yadav, A., Kumar, M., et al.: An Introduction to Neural Network Methods for Differential Equations. Springer (2015)

Big Data and Deep Learning in Plant Leaf Diseases Classification for Agriculture

Mohamed Loey

1 Introduction

The artificial intelligent framework is a computer-based framework that can represent, reason about, and interpret big data. In doing so it can learn about the structure and analyze big data to extract meaning and patterns, derive new knowledge from big data, and identify strategies and behaviors to act on the outcomes of its analysis [1, 2]. Artificial intelligence is making an intelligent framework regarding being self-adaptive, self-awareness, creating, controlling a situation, and relating to the imposition [2]. Artificial intelligent frameworks come in many forms and have many intelligent applications, from processing big data sets to controlling drones. The concepts of the artificial framework are drawn from machine learning, and deep learning as a range of fields such as linguistics brain sciences, and psychology, forming many interdisciplinary relationships [1–3].

Machine learning is a field of computer science that gives a machine the ability to learn problems without specific computer programmed by using a training and testing set and a learning algorithm such as decision trees, naïve Bayes classifier, linear and logistic regression, support vector machines, artificial neural networks, and recently deep artificial neural networks [4, 5]. For an artificial intelligent image detection framework, the training database will contain different categorized images that will be used in a training big data process using a suitable computer algorithm to learn how to classify big data and predict their category [5].

Agriculture (Husbandry) is the first human activity and drives any economic framework for any given country [6, 7]. Husbandry and farming is also the vertebral column of any economic framework for any given nation. Husbandry and farming

M. Loey (✉)
Faculty of Computers and Artificial Intelligence, Department of Computer Science, Benha University, Benha 13511, Egypt
e-mail: mloey@fci.bu.edu.eg

© The Editor(s) (if applicable) and The Author(s), under exclusive license to Springer Nature Switzerland AG 2021
A.-E. Hassanien et al. (eds.), *Enabling AI Applications in Data Science*, Studies in Computational Intelligence 911, https://doi.org/10.1007/978-3-030-52067-0_8

185

helped humanity to advance and develop. Today, the most critical activities world-wide are the husbandry and nutrition industry, due to the increasing population and the increasing growth of their needs for eating for their life to continue. Husbandry and farming not only providing eating and raw material but also provides employment opportunities to a very large percentage of the population [8–11]. The increasing popularity of sensing mechanisms in the farm includes RGB imaging, spectral, thermal imaging, near-infrared imaging which can be ground-based or air-based on airborne drones to capture big data images [12].

Nutrition losses and waste in medium-income countries occur mainly at an early step of the nutrition value phase and can be affected by financial, managerial, and technical constraints in harvesting mechanisms as well as stockpiling and refrigeration facilities [13, 14]. The global nutrition supply is annually reduced demonstrating that our collective battle against leaf plant malady and pests in plants is not won. Leaf plant malady can be affected by different types of bacteria, viruses, fungi, pests, and other agents. Diseased plant symptoms can include leaf spots, blights, fruit spots, root, fruit rots, wilt, dieback, and decline. The major impact of the leaf plant malady is decreasing the nutrition ready-made to peoples. This can outcome in unsuitable nutrition to people or lead to hunger in some areas [15, 16].

Detecting and medicating different types of bacteria, viruses, fungi, pests were done by the unclad eye of the farmer by manually examining the plant leaf on-site and this stage is slow, and costly. The need for a partially or fully automated plant disease detection frameworks is a major growing research area to detect diseased plants. Plant Leaf disease detection is of extreme importance to recommend and choose the proper medication for diseased plants to prevent infections of uninfected. The plant leaf is the most common way to detect plant disease as it shows different symptoms for different plant diseases.

2   The Era of Deep Learning and Big Data

Traditional machine learning paradigms provided reasonable outcomes and performance regarding plant malady detection using plant leaf images. Deep learning (DL) is the latest technology that brought a big improvement in the area of artificial intelligence and machine learning in general. Deep learning is a machine learning function that mimics the operation of the human brain in processing data and classifying patterns. Deep learning is a subset field of machine learning in artificial intelligence (AI) that has artificial networks capable of learning from structured and unstructured data big data. Also known as deep artificial neural learning or deep artificial neural network as shown in Fig. 1 [17–19]. These days, DL is used at a large scale in the husbandry industry. Providing good big data labeled or unlabeled to a deep neural learning architecture yielded promising outcomes in various applications that comprise the base for automating the husbandry industry and using agricultural robots [20–22]. Besides their use in leaf plant malady recognition, DL is used in other areas in the application of image processing and computer vision in husbandry [23, 24].

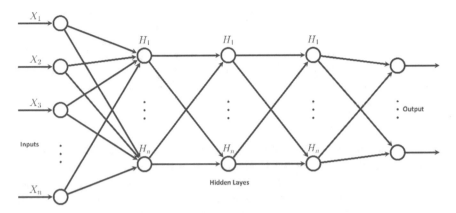

Fig. 1 Deep artificial neural learning

Convolution Neural Network (ConvNet) is a kind of deep neural learning that is commonly used in analyzing and classifying images. It learns features that are related spatially by treating an image as a volume as shown in Fig. 2. ConvNet has some specialized layers that transform the volume of the image in different ways. The convolutional layer does much of the computation for classifying an image. In Fig. 2, there is a sequence of kernels that slide or convolve, over an image volume within a convolutional layer [28]. One of the important benefits of ConvNets is that when the ConvNet's training increases, these kernels can identify textures, shapes, colors, and other features in the image [25–27].

In the following years, various improvements in deep ConvNets further increased the accuracy rate on the image detection/classification competition tasks. In 2012, Alex Krizhevsky designed AlexNet [26] to win the annual challenges of ImageNet Large Scale Visual Recognition Competition (ILSVRC). The AlexNet network

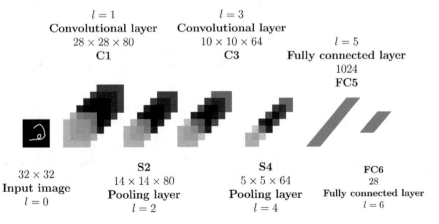

Fig. 2 Layers of the convolutional neural network

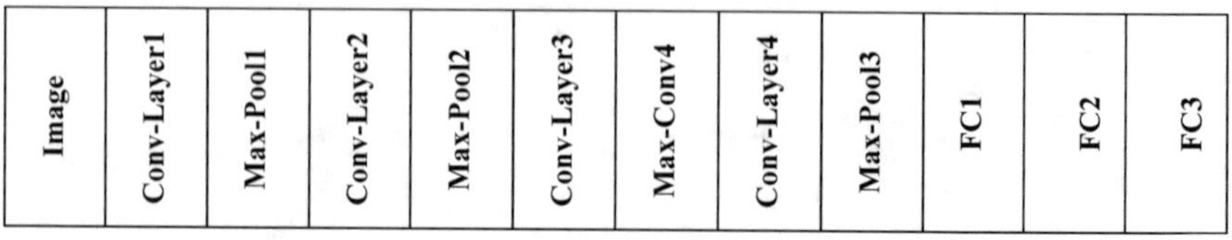

Image	Conv-Layer1	Max-Pool1	Conv-Layer2	Max-Pool2	Conv-Layer3	Max-Conv4	Conv-Layer4	Max-Pool3	FC1	FC2	FC3

Fig. 3 Layers of AlexNet

achieved a top-5 error of 15.3 in the ILSVRC Competition. AlexNet, as shown in Fig. 3, contained 8 layers; the first 5 were convolutional layers, some of them followed by max-pooling layers, and the last 3 layers were fully connected. Many pre-trained deep convolutional network paradigms were proposed like VGG-16, VGG-19 [29], GoogleNet [30], ResNet [31], X-ception [32], Inception-V3 [33] and DenseNet [34].

3 Deep Learning for Plant Diseases Detection

Leaf Plant malady caused by Abiotic (environmental impacts) or biotic (derived from living organisms) such as weather conditions, burning from chemicals, spring frosts, hail, etc. Some leaf plant malady is more dangerous and mostly not prevented because they are infectious and transmissible. Moreover, biotic leaf plant malady or infectious malady is too dangerous and risky that leads to the highest crop plant damages [35, 36]. biotic plant malady is categorized into:

3.1 Fungal Malady

Fungus or fungi lead to fungal leaf plant diseases. Fungi malady was responsible for most of leaf biotic plant diseases. Moreover, pathogenic fungi such as anthracnose, leaf spot, gall, canker, blight, rust, wilt, scab, coils, root rot, damping-off, mildew, and dieback lead to leaf plant diseases. Fungal spores are travel through the air by wind, water, insects, soil, and other invertebrates to infect other leaf plants [37].

3.2 Viral Malady

Viruses lead to viral leaf plant diseases. It is the rarest type of plant leaf Malady. Moreover, once plant infected, there are no chemical treatments to get rid of some viral leaf plant diseases.

3.3 Bacterial Malady

Leaf plant biotic bacteria malady cause many serious maladies of plants throughout the farms. Biotic bacteria malady is lead to more than 190 types of biotic bacteria. Bacteria can dispersal through sprinkle water, insects, other tools.

This part summarized the latest leaf plant malady detection papers for applying deep neural learning in the field of cultivation that shown in Table 1 that sum those researches.

Table 1 Survey of research papers based on deep learning in leaf plant diseases detection

Year	References	Database				Error rate (%)
		Crop type	Source	# of classes	No. of images	
2019	[62]	5 crops	Plant Village	9	15210	14
2019	[63]	2 crops	Own	2	3663	13
2019	[64]	19 crops	Plant Village	38	54,306	6.18
2018	[54]	Maize	Plant Village, websites	9	500	1.10
2018	[55]	14 crops	Plant Village	38	54300	0.25
2018	[56]	Tomato	Plant Village	7	13262	2.51
2018	[59]	Cucumber	Plant Village, websites	4	1184	6
2018	[60]	Wheat	Own	4	8178	13
2018	[57]	19 crops	Plant Village	38	56 k	8
2018	[58]	Tomato	Plant Village	5	500	14
2017	[38]	Tomato	Own	9	5 k	14
2017	[45]	Banana	Plant Village	3	3700	0.28
2017	[46]	Apple	Plant Village	4	2086	9.60
2017	[44]	Wheat	Own	7	9320	2.05
2017	[49]	Tomato	Plant Village	10	18 k	4.35
2017	[50]	Tomato	Plant Village	10	14828	0.82
2017	[47]	Rice	Own	10	500	4.52
2017	[48]	Apple	Own	4	1053	2.38
2017	[51]	Maize	Own	2	1796	3.30
2017	[52]	Cassava	Own	6	2756	7
2017	[53]	Olive	Plan tVillage, Own	3	299	1.40
2016	[41]	Cucumber	Own	7	7250	16.80
2016	[39]	5 crops	Internet	15	3 k	3.70
2016	[43]	14 crops	Plant Village	38	54300	0.66
2015	[40]	Cucumber	Own	3	800	5.10

Fig. 4 A representation of diseases and pests that affect tomato plants

   A deep transfer learning (DTL) detector has been introduced for tomato malady and their epidemic detection in [38] using a database of tomato leaf plant maladies that have 5 k pictures from filed cameras as shown in Fig. 4. The proposed detection framework used the ResNet50 with Region FCN. The framework achieved a minimum error rate accuracy of 14%. In [39] prosaists used a ConvNet to classify different types of leaf plant malady out of uninfected leaves. The database used was downloaded from a global computer network and depended on more than 3 k original images. The database representing infected leaves in different fields and two other classes for uninfected leaves and background images. Using data augmentation, the database enriched to more than 30 k images. Prosaists used a DTL Caffe-Net from ConvNet paradigm attaining precision between 91 and 98%, for separate class tests and an overall error rate of 3.7%.

   In [40] prosaists introduced a classification framework for viral leaf plant malady. Prosaists used their own convolutional neural network paradigm and improved the database by rotating the pictures more than 30 times every ten degrees. The database used to contain 8×100 pictures of cucumber leaf pictures representing 2 different malady and also well leaves. The proposed paradigm achieved a maximum detection error rate of 5.1% using a 4-fold cross-validation technique. In [41] Same prosaists introduced another research for classifying some types of viral cucumber malady. They have used a database of more than 7 k pictures including viral malady and healthy leaves. They have further divided the pictures into duo databases of uninfected and infected pictures. by data augmentation methods, 4-fold cross-validation strategy by training two ConvNet paradigms for each database, the classifiers achieved an average error rate of 17.7%.

   Plant-Village [42] is a large public database whose samples are shown in Fig. 5 of roughly more than 54 k pictures was used in [43] where prosaists used a ConvNet

Fig. 5 Some leaf images of Plant Village database

to identify more than ten crop species and 26 unhealthy leaves with a total of 38 labels. Prosaists used Alex-Net and GoogLe-Net paradigms with and without DTL gray and RGB pictures are segmented with different splitting ratios. Using GoogLe-Net with RGB pictures, the minimum achieved error was 0.7%. But when testing the introduced framework with pictures downloaded from the Internet the error was 68.6%.

In [44] prosaists used their wheat malady database that contains of more than 9 k pictures annotated at picture level based on husbandry experts. The database is split into seven wheat infirmity labels including an uninfected label. They have introduced a new model that uses a fully deep connected network (FDCN) to produce a spatial score that is collected by multiple instance learning (MIL) algorithm to recognize the infection, then localizing infected area of the leaf plant. The framework introduced 4 different paradigms; the best paradigm achieved a minimum error rate of 2.05%.

LeNet architecture used in [45] to classify 2 types of malady out of uninfected and infected banana leaf pictures collected from the Plant-Village database. The database consists of 3700 pictures and the paradigm was trained by trying different trainset/testset splits in gray and RGB scaled pictures. The introduced framework achieved a minimum error of 0.28% based on a 50% train/test split. In [46] used also PlantVillage database where prosaists used deep neural learning in classifying the seriousness of the apple black rot contagion. The portion of the database they used contains over nearly 160 pictures at 4 stages. Four different paradigms of ConvNet paradigms (VGG16, 19, Inception-V3, and ResNet50) used to achieve a minimum error rate of 9.6%.

Classification of ten common rice malady was introduced in [47] where prosaists created a custom deep ConvNet paradigm with different convolution filter sizes and pooling strategies. The proposed framework used a database consisted of 500 pictures of infected and uninfected rice leaves and stems captured from the experimental field. Another machine learning method such as standard backpropagation algorithm, particle swarm optimization, and support vector machines compared with the proposed framework outcomes. The proposed framework with stochastic pooling achieved an error rate 4.52% under the 10-fold cross-validation strategy. In [48] also used deep ConvNet for detecting 4 common types of apple leaf malady. In China, prosaists captured 1053 pictures from two apple experiment stations. The database was enriched using digital image processing techniques to produce more than 13 k pictures. The introduced ConvNet paradigm consisted of an adjusted Alex-Net paradigm followed by two max-pooling and Inception layers. Outcomes were compared with ConvNet transfer paradigms such as Alex-Net, GoogLe-Net, VGGNet-16, and ResNet-20. The best recognition error of nearly 2.38% has achieved using the adjusted proposed paradigm.

In [49] prosaists introduced a deep neural learning framework for malady classification on the leaves of the tomato leaf plant. Tomato pictures from the PlantVillage database used that comprised about 18 k pictures of ten different infected picture labels including an uninfected one. Prosaists used AlexNet and SqueezeNet paradigms and achieved error rate outcomes of 4.35, and 5.70% respectively. Comparing deep neural learning paradigms and traditional machine learning techniques was done in [50] where writers used both AlexNet and GoogLeNet to identify tomato infections. The proposed framework used leaf plant pictures from the PlantVillage database. Outcomes showed that pre-trained deep paradigms significantly perform better than shallow deep neural learning paradigms with a top error rate of 0.82% based on GoogLeNet.

The proposed ConvNets frameworks used in [51] to classify the northern-leaf blight (NLB) malady in maize crop as have or not the NLB lesions to labeled whether or not the entire picture contains malady leaf plant or not. The database used 1796 pictures, divided into more than 1 k unhealthy and 750 healthy leaves of maize pictures. The paradigm achieved a minimum error rate of 3.3% on a suggested test set. DTL paradigm was used in [52] to detect five different uninfected of the strategic crop of cassava which is considered in the world as the third-largest source of carbohydrates for human nutrition. The database taken from experimental fields consisted of 2756 pictures, that were manually cropped into individual leaflets to produce 15 k pictures of the cassava leaflets. Prosaists used a pre-trained paradigm

Inception v3 ConvNet with three different layers of classifiers applied to different trainset/testset splits. The Outcomes with 80/10 split and support vector machine as a classifier achieved a minimum 7% error rate.

Using a pre-train learning paradigm and limited size database, prosaists of [53] developed a vision framework to classify leaf on olive. The database contained RGB pictures divided into 3 classes: scorch, abiotic, and uninfected. The proposed paradigm was initially trained on the Plant-Village database then retrained on their database. The structure of the ConvNet was a simple LeNet paradigm with hand-crafted features inoculated at the FCs. The error rate result was 1.4%. In [54] prosaists worked on identifying eight infected out of uninfected one in maize leaf using the GoogLe-Net paradigm. The database used consisted of 500 different pictures as shown in Fig. 6 [54] that was collected from different origins such as web-sites and

Fig. 6 Samples of the maize leaves database

Plant-Village databases. The database enriched by applying different data augmentation techniques to increase the database size. The two proposed ConvNet paradigms achieved detection loss rate of 1.1 and 1.2% respectively based on trainset/testset.

Prosaists in [55] use of the Plant-Village database to classify plants leaves. The prosaists used six pre-train ConvNet architectures such as VGG16, Inception V4, ResNets-50, ResNets-101, and ResNets-152, and DenseNet-121 to calculate the best performance with less error rate. DenseNet 121was the best performing paradigm that achieving a test error rate of 0.25%. In [56] prosaists used Plant-Village database with only the tomato crop leaf pictures to classify 6 maladies out of uninfected tomato leaves. Prosaists have also shown the impact of changing the number of pictures, learning rates on both detection error rate and execution time. Using two pre-train ConvNet paradigms like Alex-Net and VGGNet-16 with hyper parameters tuning, the proposed paradigms achieved a minimum error of 2.51 and 2.71% respectively.

In [57], prosaists used generative adversarial networks (GANs) to augment their database and overcome the problem of the limited number of pictures. Prosaists used Plant-Village database and different ConvNet paradigms: a DTL Inception-v3 paradigm, and a DTL Mobile-Net paradigm. The trained paradigm was inserted inside a mobile application with an error rate of 11.4 and 8%. ConvNet was used as a feature extractor by [58], where prosaists developed a ConvNet paradigm to detect tomato leaf pictures. The output feature vector of the convolution layer is then fed to a Learning Vector Quantization (LVQ) classifier. The database used was Plant-Village database with 5 different labels split into 4 infected and one uninfected leaves. The proposed framework achieved an average miss-classification rate of 14% using the test-set.

In [59] prosaists classify four cucumber malady leaves. The proposed deep convolutional neural network ConvNet trained from scratch and DTL using Alex-Net paradigm. The database was created from the Plant-Village database and some web-sites. The proposed paradigm augmented with data augmentation methods and include only the leaf region. Outcomes showed that using DCNN achieved lowest miss-classification, and significantly CNN methods, with a minimum error rate of 6% using the DTL paradigm. In [60] prosaists introduced a paradigm to classify plant leaf malady in real-time using mobiles. Moreover, prosaists in [61] extended the work of that used an image processing algorithm to tackle plant malady classification. Enhanced deep residual neural network (Resnet) with pre-train paradigm used to classify three endemic wheat maladies with different levels. The database used contained more than 8 k wheat leaf pictures that were captured from Spain and Germany. Outcomes on real-time testing showed that balanced accuracy values increased from 0.78 on the traditional approach to 0.84 when using a Resnet. The whole paradigm was the development of mobile under real conditions.

In [62], prosaists was introduced a new architecture for the effective classification of plant malady. The database used here consists of several varieties of plants of both unhealthy/healthy, all these leaf plant pictures are collected from various sources and Plant-Village. Overall ConvNet paradigm was implemented from scratch and produces a miss-classification of 14%. Convolutional Neural Network paradigm used in [63]. The paradigm created and developed to perform plant malady classification using apple and tomato leaf pictures of uninfected and infected leaf plants. The paradigm consists of four convolutional layers each followed by pooling layers. The database contains 3663 of leaf pictures based on apple and tomato leaf picture. Training of the ConvNet paradigm was achieving a miss-classification of 13%. Finally, Based on NAS-Net a new paradigm for plant malady recognition [64]. The NAS-Net was a trainsct/testset using a publicly available Plant-Village project database that contains varied pictures of plant leaves. Using the paradigm, a miss-classification rate of 6.18% was achieved.

4 Hyperspectral Images for Plant Disease Detection

Modern hyperspectral imaging paradigms produce big databases conveying a large abundance of big data, however, they constitute many challenges in the analysis of these data. Hyperspectral imaging (HSI) is a technique in image processing that analyzes a wide spectrum of a large number of wavelengths instead of just assigning primary colors (red, green, blue) to each pixel of the picture. In the last years, HSI has gained importance and a central role in many areas of visual big data analysis, especially in Agriculture. Classical machine learning-based approaches to HSI data interpretation rely on the extraction of hand-crafted features on which to feed to a classifier. The advantages of using deep learning solutions lie in the automatic and deep neural learning process from big data itself which can build a paradigm with increasingly higher hieratical layers [65–67].

Hyperspectral data is too big, especially when multiple leaf plants are pictured for several days. A scan of a many leaves plant could easily be very big in data. If the whole hyper spectrum range is analyzed, then the process will take considerably more than selecting several wavelengths to analyses. However, there is a lot of information contained in the big hyperspectral data. In [68], introduced the outcomes of an experiment in the automatic detection of plant malady using hyperspectral imaging. A proposed model was developed to collect hyperspectral hyper-cubes of plant leaves in the VNIR and SWIR spectrum after inoculation to detect Tomato Spotted Wilt Virus (TSWV). The outcomes showed better than 90% accuracy based on SVM classifiers that uses three types of features. In Fig. 8, the hyperspectral imaging framework (Fig. 7) showing a leaf plant was pictured.

Deep convolutional neural networks (DConvNet) have been successfully used in many implementations such as speech recognition, object detection, document classification, and sentiment analysis. ConvNet used a 3D analog of the convolutional filter in HSI and such 3D- ConvNet frameworks have been used in the detection of hyperspectral pictures for some interesting computer science and engineering applications. In [65], prosaists introduced a novel 3D ConvNet called saliency map-based visualization that directly assimilates the hyperspectral big data. In Fig. 8, prosaists classify charcoal rot pictures using 3D convolutional neural network architecture. They evaluate the learned 3D-ConvNet model on 539 test pictures. The proposed model achieved a miss-classification rate of 4.27%.

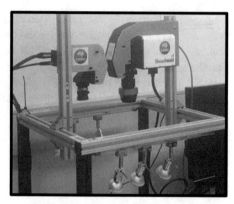

a) Hyperspectral imaging system

b) A leaf plant being imaged

Fig. 7 Hyperspectral imaging system [68]

5   Conclusion

Convolutional neural networks (CNN) and deep learning have revolutionized image processing and artificial intelligence research producing state of the art results, especially in the image detection field. After surveying different researches that have used DTL in precision husbandry especially in detecting plant leaf malady, it has been found that DTL has brought a massive advancement. It was also shown that DTL allows the growth of precision husbandry and independent husbandry robots for the cutting-edge of husbandry and nutrition production. Smart-phone frameworks are also a new tool for using DTL for helping farmers.

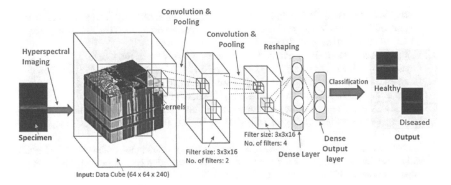

Fig. 8 3D convolutional neural network architecture

References

1. O'Leary, D.E.: Artificial intelligence and big data. IEEE Intell. Syst. **28**, 96–99 (2013). https://doi.org/10.1109/MIS.2013.39
2. Kibria, M.G., Nguyen, K., Villardi, G.P., et al.: Big data analytics, machine learning, and artificial intelligence in next-generation wireless networks. IEEE Access **6**, 32328–32338 (2018). https://doi.org/10.1109/ACCESS.2018.2837692
3. Allam, Z., Dhunny, Z.A.: On big data, artificial intelligence and smart cities. Cities **89**, 80–91 (2019). https://doi.org/10.1016/j.cities.2019.01.032
4. Shang, C., You, F.: Data analytics and machine learning for smart process manufacturing: recent advances and perspectives in the big data era. Engineering (2019). https://doi.org/10.1016/j.eng.2019.01.019
5. Ngiam, K.Y., Khor, I.W.: Big data and machine learning algorithms for health-care delivery. Lancet Oncol. **20**, e262–e273 (2019). https://doi.org/10.1016/S1470-2045(19)30149-4
6. Elijah, O., Rahman, T.A., Orikumhi, I., et al.: An overview of Internet of Things (IoT) and data analytics in agriculture: benefits and challenges. IEEE Internet Things J. **5**, 3758–3773 (2018). https://doi.org/10.1109/JIOT.2018.2844296
7. Aznar-Sánchez, J.A., Piquer-Rodríguez, M., Velasco-Muñoz, J.F., Manzano-Agugliaro, F.: Worldwide research trends on sustainable land use in agriculture. Land Use Policy **87**, 104069 (2019). https://doi.org/10.1016/j.landusepol.2019.104069
8. Ncube, B., Mupangwa, W., French, A.: Precision agriculture and food security in Africa BT. In: Katerere, D., Hachigonta, S., Roodt, A., Mensah, P. (eds.) Systems analysis approach for complex global challenges, pp. 159–178. Springer International Publishing, Cham (2018)
9. Gebbers, R., Adamchuk, V.I.: Precision agriculture and food security. Science (80-) **327**, 828–831 (2010). https://doi.org/10.1126/science.1183899
10. Erbaugh, J., Bierbaum, R., Castilleja, G., et al. Toward sustainable agriculture in the tropics. World Dev. **121**, 158–162 (2019). https://doi.org/10.1016/j.worlddev.2019.05.002
11. Liu, S., Guo, L., Webb, H., et al.: Internet of Things monitoring system of modern eco-agriculture based on cloud computing. IEEE Access **7**, 37050–37058 (2019). https://doi.org/10.1109/ACCESS.2019.2903720
12. Ip RHL, Ang L-M, Seng KP, et al (2018) Big data and machine learning for crop protection. Comput Electron Agric 151:376–383. https://doi.org/10.1016/j.compag.2018.06.008
13. Abu Hatab, A., Cavinato, M.E.R., Lindemer, A., Lagerkvist, C.-J.: Urban sprawl, food security and agricultural systems in developing countries: A systematic review of the literature. Cities **94**, 129–142 (2019). https://doi.org/10.1016/j.cities.2019.06.001

14. Pretty, J.N., Morison, J.I.L., Hine, R.E.: Reducing food poverty by increasing agricultural sustainability in developing countries. Agric. Ecosyst. Environ. **95**, 217–234 (2003). https:// doi.org/10.1016/S0167-8809(02)00087-7
15. Strange, R.N., Scott, P.R.: Plant disease: a threat to global food security. Annu. Rev. Phytopathol. **43**, 83–116 (2005). https://doi.org/10.1146/annurev.phyto.43.113004.133839
16. Loey, M., ElSawy, A., Afify, M.: Deep learning in plant diseases detection for agricultural crops: a survey. Int. J. Serv. Sci. Manag. Eng., Technol (2020)
17. Rong, D., Xie, L., Ying, Y.: Computer vision detection of foreign objects in walnuts using deep learning. Comput. Electron. Agric. **162**, 1001–1010 (2019). https://doi.org/10.1016/j.compag. 2019.05.019
18. Brunetti, A., Buongiorno, D., Trotta, G.F., Bevilacqua, V.: Computer vision and deep learning techniques for pedestrian detection and tracking: a survey. Neurocomputing **300**, 17–33 (2018). https://doi.org/10.1016/j.neucom.2018.01.092
19. Maitre, J., Bouchard, K., Bédard, L.P.: Mineral grains recognition using computer vision and machine learning. Comput. Geosci. (2019). https://doi.org/10.1016/j.cageo.2019.05.009
20. Gogul, I., Kumar, V.S.: Flower species recognition system using convolution neural networks and transfer learning. In: 2017 Fourth International Conference on Signal Processing, Communication and Networking (ICSCN), pp. 1–6 (2017)
21. Hedjazi, M.A., Kourbane, I., Genc, Y.: On identifying leaves: A comparison of CNN with classical ML methods. In: 2017 25th Signal Processing and Communications Applications Conference (SIU), pp. 1–4 (2017)
22. Dias, R.O.Q., Borges, D.L.: Recognizing plant species in the wild: deep learning results and a new database. In: 2016 IEEE International Symposium on Multimedia (ISM), pp. 197–202 (2016)
23. Abdullahi, H.S., Sheriff, R.E., Mahieddine, F.: Convolution neural network in precision agriculture for plant image recognition and classification. In: 2017 Seventh International Conference on Innovative Computing Technology (INTECH), pp. 1–3 (2017)
24. Gao, M., Lin, L., Sinnott, R.O.: A mobile application for plant recognition through deep learning. In: 2017 IEEE 13th International Conference on e-Science (e-Science), pp. 29–38 (2017)
25. Ciregan, D., Meier, U., Schmidhuber, J.: Multi-column deep neural networks for image classification. In: 2012 IEEE Conference on Computer Vision and Pattern Recognition, pp. 3642–3649 (2012)
26. Sutskever, I., Hinton, G.E., Krizhevsky, A.: A in neural information processing systems. Imagenet classification with deep convolutional neural networks. Adv. Neural Inf. Process Syst. 1097–1105 (2012)
27. Cireşan, D.C., Giusti, A., Gambardella, L.M., Schmidhuber, J.: Mitosis detection in breast cancer histology images with deep neural networks. In: Mori, K., Sakuma, I., Sato, Y., et al. (eds.) Medical Image Computing and Computer-Assisted Intervention–MICCAI 2013, pp. 411–418. Springer, Berlin Heidelberg (2013)
28. El-Sawy, A., EL-Bakry, H., Loey, M.: CNN for handwritten arabic digits recognition based on LeNet-5 BT. In: Hassanien, A.E., Shaalan, K., Gaber, T., et al. (eds.) Proceedings of the International Conference on Advanced Intelligent Systems and Informatics 2016, pp. 566–575. Springer International Publishing, Cham (2017)
29. Liu, S., Deng, W.: Very deep convolutional neural network based image classification using small training sample size. In: 2015 3rd IAPR Asian Conference on Pattern Recognition (ACPR), pp. 730–734 (2015)
30. Szegedy, C., Wei, L., Yangqing, J., et al.: Going deeper with convolutions. In: 2015 IEEE Conference on Computer Vision and Pattern Recognition (CVPR), pp. 1–9 (2015)
31. He, K., Zhang, X., Ren, S., Sun, J.: Deep residual learning for image recognition. In: 2016 IEEE Conference on Computer Vision and Pattern Recognition (CVPR), pp. 770–778 (2016)
32. Chollet, F.: Xception: deep learning with depthwise separable convolutions. In: 2017 IEEE Conference on Computer Vision and Pattern Recognition (CVPR), pp. 1800–1807 (2017)

33. Szegedy, C., Vanhoucke, V., Ioffe, S., et al.: Rethinking the inception architecture for computer vision. In: Proceedings of the IEEE Conference on Computer Vision and Pattern Recognition, pp. 2818–2826 (2016)
34. Huang, G., Liu, Z., Maaten, L.V.D., Weinberger, K.Q.: Densely connected convolutional networks. In: 2017 IEEE Conference on Computer Vision and Pattern Recognition (CVPR), pp. 2261–2269 (2017)
35. Sankaran, S., Mishra, A., Ehsani, R., Davis, C.: A review of advanced techniques for detecting plant diseases. Comput. Electron. Agric. 72, 1–13 (2010). https://doi.org/10.1016/j.compag.2010.02.007
36. Saleem, H.M., Potgieter, J., Arif, M.K.: Plant disease detection and classification by deep learning. Plants 8 (2019)
37. Jain, A., Sarsaiya, S., Wu, Q., et al.: A review of plant leaf fungal diseases and its environment speciation. Bioengineered 10, 409–424 (2019). https://doi.org/10.1080/21655979.2019.1649520
38. Fuentes, A., Yoon, S., Kim, S.C., Park, D.S.: A robust deep-learning-based detector for real-time tomato plant diseases and pests recognition. Sensors (Switzerland) 17 (2017). https://doi.org/10.3390/s17092022
39. Sladojevic, S., Arsenovic, M., Anderla, A., et al.: Deep Neural Networks Based Recognition of Plant Diseases by Leaf Image Classification. Comput. Intell. Neurosci. (2016). https://doi.org/10.1155/2016/3289801
40. Kawasaki, Y., Uga, H., Kagiwada, S., Iyatomi, H.: Basic study of automated diagnosis of viral plant diseases using convolutional neural networks. In: International Symposium on Visual Computing, pp. 638–645. Springer (2015)
41. Fujita, E., Kawasaki, Y., Uga, H., et al.: Basic investigation on a robust and practical plant diagnostic system. In: Proceedings of 2016 15th IEEE International Conference on Machine Learning and Applications ICMLA 2016, pp. 989–992 (2017). https://doi.org/10.1109/ICMLA.2016.56
42. Hughes, D.P., Salathe, M.: An open access repository of images on plant health to enable the development of mobile disease diagnostics (2015). https://doi.org/10.1111/1755-0998.12237
43. Mohanty, S.P., Hughes, D.P., Salathé, M.: Using deep learning for image-based plant disease detection. Front. Plant Sci. 7 (2016). https://doi.org/10.3389/fpls.2016.01419
44. Lu, J., Hu, J., Zhao, G., et al.: An in-field automatic wheat disease diagnosis system. Comput. Electron. Agric. 142, 369–379 (2017). https://doi.org/10.1016/j.compag.2017.09.012
45. Amara, J., Bouaziz, B., Algergawy, A.: A deep learning-based approach for banana leaf diseases classification. BTW, pp. 79–88 (2017)
46. Wang, G., Sun, Y., Wang, J.: Automatic image-based plant disease severity estimation using deep learning. Comput. Intell. Neurosci. (2017). https://doi.org/10.1155/2017/2917536
47. Lu, Y., Yi, S., Zeng, N., et al.: Identification of rice diseases using deep convolutional neural networks. Neurocomputing 267, 378–384 (2017)
48. Liu, B., Zhang, Y., He, D., Li, Y.: Identification of apple leaf diseases based on deep convolutional neural networks. Symmetry (Basel) 10, 11 (2017)
49. Durmus, H., Gunes, E.O., Kirci, M.: Disease detection on the leaves of the tomato plants by using deep learning. In: 2017 6th International Conference on Agro-Geoinformatics, Agro-Geoinformatics (2017). https://doi.org/10.1109/Agro-Geoinformatics.2017.8047016
50. Brahimi, M., Boukhalfa, K., Moussaoui, A.: Deep learning for tomato diseases: classification and symptoms visualization. Appl. Artif. Intell. 31, 299–315 (2017)
51. DeChant, C., Wiesner-Hanks, T., Chen, S., et al.: Automated identification of northern leaf blight-infected maize plants from field imagery using deep learning. Phytopathology 107, 1426–1432 (2017). https://doi.org/10.1094/PHYTO-11-16-0417-R
52. Ramcharan, A., Baranowski, K., McCloskey, P., et al.: Using transfer learning for image-based cassava disease detection 8, 1–7 (2017). https://doi.org/10.3389/fpls.2017.01852
53. Cruz, A.C., Luvisi, A., De Bellis, L., Ampatzidis, Y.: Vision-based plant disease detection system using transfer and deep learning, pp. 1–9 (2017). https://doi.org/10.13031/aim.201700241

54. Zhang, X., Qiao, Y., Meng, F., et al.: Identification of maize leaf diseases using improved deep convolutional neural networks. IEEE Access **6**, 30370–30377 (2018). https://doi.org/10.1109/ACCESS.2018.2844405

55. Too, E.C., Yujian, L., Njuki, S., Yingchun, L.: A comparative study of fine-tuning deep learning models for plant disease identification. Comput. Electron. Agric. 0–1 (2018). https://doi.org/10.1016/j.compag.2018.03.032

56. Rangarajan, A.K., Purushothaman, R., Ramesh, A.: Tomato crop disease classification using pre-trained deep learning algorithm. Proc. Comput. Sci. **133**, 1040–1047 (2018). https://doi.org/10.1016/j.procs.2018.07.070

57. Gandhi, R., Nimbalkar, S., Yelamanchili, N., Ponkshe, S.: Plant disease detection using CNNs and GANs as an augmentative approach. In: 2018 IEEE International Conference on Recent Research Development ICIRD 2018, pp. 1–5. (2018). https://doi.org/10.1109/ICIRD.2018.8376321

58. Sardogan, M., Tuncer, A., Ozen, Y.: Plant leaf disease detection and classification based on CNN with LVQ algorithm. In: 2018 3rd International Conference on Computer Science and Engineering (UBMK), pp 382–385 (2018)

59. Ma, J., Du, K., Zheng, F., et al.: Original papers a recognition method for cucumber diseases using leaf symptom images based on deep convolutional neural network. Comput. Electron. Agric. **154**, 18–24 (2018). https://doi.org/10.1016/j.compag.2018.08.048

60. Picon, A., Alvarez-Gila, A., Seitz, M., et al.: Deep convolutional neural networks for mobile capture device-based crop disease classification in the wild. Comput. Electron. Agric. 0–1 (2018). https://doi.org/10.1016/j.compag.2018.04.002

61. Johannes, A., Picon, A., Alvarez-Gila, A., et al.: Automatic plant disease diagnosis using mobile capture devices, applied on a wheat use case. Comput. Electron. Agric. **138**, 200–209 (2017). https://doi.org/10.1016/j.compag.2017.04.013

62. Hari, S.S., Sivakumar, M., Renuga, P., et al.: Detection of plant disease by leaf image using convolutional neural network. In: 2019 International Conference on Vision Towards Emerging Trends in Communication and Networking (ViTECoN), pp. 1–5 (2019)

63. Francis, M., Deisy, C.: Disease detection and classification in agricultural plants using convolutional neural networks—a visual understanding. In: 2019 6th International Conference on Signal Processing and Integrated Networks (SPIN), pp. 1063–1068 (2019)

64. Adedoja, A., Owolawi, P.A., Mapayi, T.: Deep learning based on NASNet for plant disease recognition using leave images. In: 2019 International Conference on Advances in Big Data, Computing and Data Communication Systems (icABCD). pp. 1–5 (2019)

65. Nagasubramanian, K., Jones, S., Singh, A.K., et al.: Plant disease identification using explainable 3D deep learning on hyperspectral images. Plant Methods **15**, 98 (2019). https://doi.org/10.1186/s13007-019-0479-8

66. Paoletti, M.E., Haut, J.M., Plaza, J., Plaza, A.: Deep learning classifiers for hyperspectral imaging: a review. ISPRS J. Photogramm. Remote Sens. **158**, 279–317 (2019). https://doi.org/10.1016/j.isprsjprs.2019.09.006

67. Venkatesan, R., Prabu, S.: Hyperspectral image features classification using deep learning recurrent neural networks. J. Med. Syst. **43**, 216 (2019). https://doi.org/10.1007/s10916-019-1347-9

68. Moghadam, P., Ward, D., Goan, E., et al.: Plant disease detection using hyperspectral imaging. In: 2017 International Conference on Digital Image Computing: Techniques and Applications (DICTA), pp. 1–8 (2017)

Machine Learning Cancer Diagnosis Based on Medical Image Size and Modalities

Walid Al-Dhabyani and Aly Fahmy

1 Introduction

Medical imaging is a valuable tool for diagnosing the existence of various diseases and the study of the experimental effects [1]. Large-scale medical images may support experts in medical fields with more details to increase diagnosis precision in pathological research [2–4]. Thus, the enhancement of the medical image is becoming very important. In addition, large-scale medical images can significantly help computer-aided automatic detection [5]. For instance, plurality of Computer Tomography (CT) [6] scanner sand Magnetic Resonance Imaging (MRI) [7] create medical images as practical as non-invasive examinations.

Bio-medical imaging is one of the foundations of intensive cancer treatment, diabetes, and bones, etc. It has many benefits including accessibility without destruction of tissue, Monitoring in real time, and minimally no invasiveness. In addition, it can function through a wide variety of size and time scales involved in pathological and biological processes. Time-scales are different from disease to another e.g., chemical reactions and protein binding need milliseconds while cancer needs years [8].

Early diagnosis was the most important factor in reducing the costs of any diseases such as cancer management and mortality. Biomedical imaging [9] presents an increasingly meaningful role in the cancer stages treatment [10]. These include screening [11], prediction [12], biopsy guidance [13], staging [14], therapy planning [15] and other diagnoses.

W. Al-Dhabyani (✉) · A. Fahmy
Faculty of Computers and Artificial Intelligence, Cairo University, Cairo, Egypt
e-mail: w.aldhabyani@grad.fci-cu.edu.eg

A. Fahmy
e-mail: a.fahmy@fci-cu.edu.eg

© The Editor(s) (if applicable) and The Author(s), under exclusive license
to Springer Nature Switzerland AG 2021
A.-E. Hassanien et al. (eds.), *Enabling AI Applications in Data Science*,
Studies in Computational Intelligence 911,
https://doi.org/10.1007/978-3-030-52067-0_9

In this chapter, the type and size of medical images are focused on. The chapter focuses on cancerous images when used with machine learning. And it is divided as follows: Sect. 2 introduces machine learning and transfer learning. Subsequently, Sect. 3 illustrates imaging and radiology. Section 4 explains medical images modalities, size, format and etc. Section 5 discusses a technique for manipulating medical images. And finally, Sect. 6 presents a conclusion and discussion.

2 Machine Learning

When medical images are available on the computer, a lot of applications have been built for automated analyzes. At First, from the seventieths to the nineties, medical image analysis was performed with the processing of the low-level pixel sequential application and mathematical modeling to establish systems of rule-based that answered to special tasks. In the same period, there is a relationship with skillful systems with many statements that were common in artificial intelligence. GOF-AI (good old-fashioned artificial intelligence) [16] is defined as an expert system that was usually vulnerable; Comparable to the processing systems of a rule-based image. In the late ninetieths, techniques under supervision, in which training dataset were applied to improve a system, became frequently common in the analysis of medical images. There are some examples such as the models of active shape (i.e., segmentation), atlas methods (i.e., the new data is matched with training data is called atlases) and the use of statistical classifiers and the idea of feature extraction (i.e., computer-aided diagnosis and detection). This machine learning (or pattern recognition) approach is yet popular and makes a lot of analysis system of medical images available that is the basis of several successful commercially. That is why we have seen a change from fully human-designed systems to computer-trained systems using sample data which is extracted from feature vectors. The algorithms of computer restrict the best decision boundary in the space of the characteristics of high-dimensional. Extracting discriminating image characteristics is a significant part of developing such systems. Human researchers are still carrying out this process and as such, we are referring to systems with handcrafted functionality.

The next reasonable step is to allow computers to discover the characteristics that optimally express the data for the specific problem. This definition is based on many deep learning algorithms: models (networks) consisting of several layers that change input data (i.e., photos) into outputs (i.e., positive/negative disease), while the learning is increased when moving to the level of higher features. Convolutional Neural Networks (CNNs) are the common powerful type of models for image analysis to date. CNNs include several layers that change their input with small-extent convolution filters. Work on CNN ran from the late seventieths [17] and already applied to the analysis of the medical images in 1995 by SCB Lo et al [18]. LeNet [19] was their first successful application in the real world for recognition of digit using hand-written. Notwithstanding these early achievements, the use of CNN networks did not gain traction until new technologies were created to efficiently train deep networks, and advances have been performed in computing systems. In challenge ImageNet, Krizhevsky et al. [20] contribution was the turning point in 2012. The proposal CNN, identified as AlexNet, achieved the best result in the competition by a big margin. New progress was made in the following years with the help of related but deeper architectures [21]. Deep convolution networks have become the preferred method in computer vision.

Those fundamental advances have been acknowledged by the medical image analysis community. However, the move to systems that learn features from handcrafted features systems is done gradually. Prior to AlexNet's breakthrough, there were several popular techniques for learning features. Bengio et al. [22] give a complete overview of these techniques. They cover image patches clustering, analysis of principal components, and much more. In addition, they define CNN which are trained end-to-end in their review. For a more comprehensive overview of how deep learning is applied in medical informatics, we are referring to Daniele Ravì et al. [23], where the analysis of medical images is concisely discussed. Moreover, Shen et al. [24] published a special review about deep learning applications for medical image analysis.

2.1 Machine Learning Methods

Machine learning has four methods: working under supervision, non-supervised, semi-supervised, and reinforcement learning. Medical images are part of supervised learning. The data must be labeled. Supervised learning is further subdivided into regression, classification and preferred relationships [25]. For example, if the predictable output is numeric to the price of a house, this is called regression. For example, if the predictable output is categorical spam or not, this is called classification. With medical images, classification is the most important part. And in some conditions, developers or researchers can use regressions.

2.2 Transfer Learning Model

Transfer learning (TL) [26, 27] is a process that provides a system for applying the information acquired from prior tasks to a new task domain somehow relevant to the previous domain. This theory is inspired by the concept that human can intuitively use their past knowledge to determine and resolve new problems. CNN pretrained model in ImageNet is used. ImageNet [28], actually the largest dataset for visual recognition and image classification, is a dataset of images with more than fourteen million images that are classified to a thousand categories of objects, constructed according to the hierarchy of WordNet.

TL is a similar trend in which a CNN may be trained on data from a particular domain and later reuse it to extract image features in another domain or as a first network to refine new data. Transfer learning shows that a suitable process for one task is applied to another task. To solve similar problems, it is the method of copying information from an already-trained network into a new network. Recently, TL approach is widely used in the researches of biomedical. It achieves great results.

3 Imaging and Radiology

In modern hospitals, a various amount of medical imaging data is obtained for diagnosis, treatment planning, and estimation of patient response. Such vast collections of images combined with other sources of images provide new possibility of using enormous image data to extract computerized resources for image-based diagnosis, teaching and biomedical research [29]. These applications are focused on identifying, retrieving and classifying patient data which reflect similar clinical outcomes [30], for example, images representing the same diagnosis.

3.1 Radiology Categorization

Imaging technology is used by radiology, which is a division of medicine, to treat and diagnose diseases. Radiologists are experts who are specialized in radiology. Radiology is divided into two distinct parts: diagnostic and interventional radiology [31].

3.1.1 Diagnostic Radiology

Diagnostic radiology helps healthcare professionals view structures in the internal body. Diagnostic radiologist is a specialist who is specialized in the translation of medical images. Diagnostic images are used by the physicians or radiologist to diagnose the reason of symptoms. They can also track how well the body is reacting to a treatment that a person is receiving for her or his disease and screen for various illnesses. These kinds of diseases or illnesses can be colon cancer, breast cancer, heart disease, or anything else. More types of radiology exams would be explained in Sect. 4.2.

3.1.2 Interventional Radiology

Interventional radiologist is a doctor that uses imaging such as fluoroscopy, ultrasound (US), MRI, and CT. For the doctor, the imaging is useful when putting catheters, wires, and little instruments into the body. This technology is used by doctors to diagnose any organs of the body. It does not necessitate looking through the body with open surgery or through a scope. Interventional radiologists are contributed in curing cancers, fibroids in the uterus, back pain, blockages in the veins and arteries, kidney problems and liver problems. Doctors make no incision or sometimes only a very small one. And people are not required to remain in the hospital after the procedure. But instead, moderate sedation is needed by most people (medicines for relaxing). Interventional radiology operation examples are angiography, cancer treatments, cryoablation, tumor ablation, placement of the catheter in the venous, vertebroplasty, breast biopsy, needle biopsies, uterine artery embolization, kyphoplasty, and microwave ablation.

4 Medical Images

A medical image is the depiction of an anatomical region's inner structure and its function in the form of a matrix of image components that are identified as voxels or pixels. It is a separated description arising from a method of reconstruction/sampling, designates numeral values to space locations. The pixels' number used to explain a given acquisition modality's field of view is a measure of the precision in which the feature or anatomy can be displayed. The pixel expresses numerical value depending on the protocol of retrieval, imaging modality, the reconstruction, and ultimately, the subsequent processing. This section explains the modalities of medical imaging and histopathology. In addition, size and type of medical images are discussed afterward.

4.1 Medical Image Modalities

Medical imaging introduces processes that present the human body visual information [32]. The goal of medical imaging is to help clinicians and radiologists to obtain the treatment and diagnostic process effectively. Medical imaging is a major portion of the diagnosis and treatment of illnesses. It describes a varity of imaging modalities, for example, CT, US, MRI, X-rays, and the hybrid modalities [33]. Examples of medical image modalities are shown in Fig. 2. They play an essential role in detecting anatomical and functional information for examination and treatment of the various organs of the body [34]. Figure 1 shows a typology of medical imaging modalities. Medical imaging is a fundamental tool in modern systems of medical care and computer-aided diagnosis (CAD). ML performs an essential role in CAD with its applications in detection and classification of tissues or cancer, medical image retrieval, medical image analysis, image-guided therapy, and the annotation of medical image [35–39]. The properties of medical imaging modalities and histopathology are explained in Sect. 4.4.

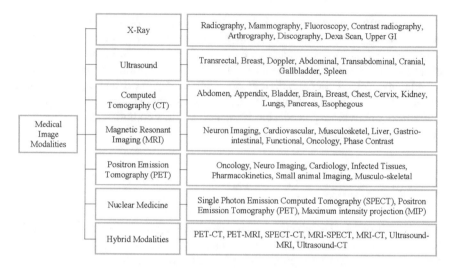

Fig. 1 Typology of medical imaging modalities

4.2 *Taxonomy of Medical Image Modalities and Characteristics*

There are several types of medical Image modalities. Each medical image has its own organ and properties [40]. Modalities operate an essential function in the discovery of functional and anatomical knowledge about several organs of the body for research and diagnosis [34]. Some examples of medical images are displayed in Fig. 2. Medical imaging is an important help in machine learning algorithms and modern health-care.

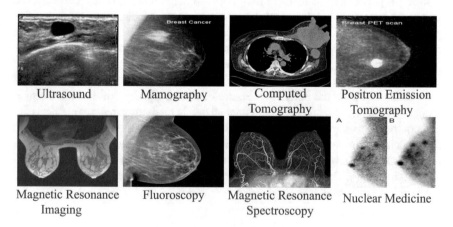

Fig. 2 Example of medical image modalities types and their labels under each image

Digital image size of the radiograph relies on the dimension of the Detector Element (from 0.2 mm to 0.1 mm (measured in millimeters)), Field of View of 18 × 24 cm to 35 × 43 cm, and the depth of bit of 10 to 14 bits for each pixel. This outcomes in standard image sizes of (8 to 32) MB for each single projection image. For CAD algorithms that are always utilizied with images of digital radiology, it is necessary to "process" the image data (raw). Communication protocols used in medicine such as Digital Imaging and Communication in Medicine (DICOM) [41] third-party interface box [41] are required for the linking of legacy equipment to capture and encapsulate images for archiving.

There are two sources of power which are radiation and non-radiation. Radiation is used with X-rays, Positron Emission Tomographies (PETs), and CTs. PETs use radiotracers to read through the body. X-rays use ionizing radiation. CTs always use ionizing radiation. The other one is the non-radiation which is used with MRIs and Ultrasounds. USs use sound waves. MRIs use magnetic waves. Of course, each scan can achieve optimal scan result. Some common types of medical image modalities and their properties are as follows:

- **X-ray**. It is the most popular scan. The most common subtypes of its radiology are radiography, Mammography, Upper GI, Fluoroscopy, Dexa Scan, Discography, Discography, Arthrography, and Contrast radiography. The average size is 5.1 MB [42]. The screens size monitor of radiography is generally suitable for 35 cm × 43 cm radiographs. The resolution displayed in picture mode, for example, is 2, 048 × 2, 560 pixels (i.e., 5 MP (megapixels)) or even 3-megapixel displays are commonly in use. An x-ray takes images that are useful to look at foreign objects in tissues and bones. The Conditions diagnosed with X-rays are general infection or injuries, fractured bones, breast tumor, osteoporosis, gastrointestinal problems, bone cancer, and arthritis,. The most famous x-ray types are:

  – **Mammography Scan**. Digital mammography (DM) is an imaging technique for x-ray projection and is primarily used for breast imaging. The average size is 83.5 MB [42]. The size of DM image varies in the detector element from 0.01 to 0.05 mm and data per pixel from 12 to 16 bits to make images of 8–50 Megabyte (MB) for a FoV of (18 × 24 cm) and (24 × 30 cm).
  – **Fluoroscope Scan**. It is an image of an x-ray projection in real-time obtaining arrangement utilized for dynamic assessment of a lot of patient procedures in interventional and diagnostic radiology. Video sequences of x-ray images of (1 to 60) frames per second are obtained to view and evaluate the anatomy of the patient. Standard sizes of the images vary from 512 × 512 × 8 bit to 1024 × 1024 × 12 bit. 2048 × 2048 × 12 Bit arrays are used for spot image applications.

- **Computed Tomography (CT)**. The most common form of organs that this kind of radiology is used for are brain, breast, chest, cervix, lungs, kidney, pancreas, abdomen, appendix, bladder, and esophagus. As a general rule, a CT image should have about as many pixels in each dimension as in detector channels that provide data for a view. The average size is 153.4 MB [42]. For example, an array of 1024-

channel linear detectors justifies a reconstructed image of 1024 × 1024 pixels; If an offset scan mode is used, an image of up to 2048 × 2048 pixels can be justified. The most common image size is 512 × 512 × 12 bits.

- **Magnetic Resonance Imaging (MRI)**. The most common form of organs that this kind of radiology is used for are neuroImaging, cardiovascular, musculoskeletal, liver, gastrointestinal, functional, oncology, and phase contrast. The average size is 83.5 MB [42]. This yields a temporary resolution of 20–30 ms for images with a resolution in the plane of 1.5-2.0 mm.

- **Ultrasound Scan**. The most common form of organs that this kind of radiology is used for are transrectal, breast, doppler, abdominal, transabdominal, cranial, gallbladder, and spleen. This offers clear images of soft-tissues and their motions, and shows blood images moving through vessels. The average size is 69.2 MB [42]. The size of the image is typical images of gray-scale: 640 × 480 × 8 bits or 512 × 512 × 8 bits. The color images size is similar as the size of the matrix with 24 bits (RGB - 8 bits). The RGB stands for Red, Green, and Blue colors.

- **Positron Emission Tomography (PET) Scan**. The most common form of organs that this kind of radiology is used for are infected tissues, pharmacokinetics, oncology, cardiology, neuroimaging, and musculoskeletal. The average size is 365.9 MB [42]. The matrix size of the image is 1005 × 1005 with 0.1 mm of pixel size. They are employed for tracking cancer, diagnosis, and disease of the coronary arteries. The conditions diagnosed with PET are epilepsy of Alzheimer, cancers, disease, disease of Parkinson, cardiac, issues of memory, and problems with previous therapy.

- **Hybrid Modalities**. Hybrid modalities merge the strength of two imaging technologies into one, offering an even wider and more accurate look inside the body [32]. For example, US-CT, US-MRI, MRI-CT, PET-MRI, MRI-CT, SPECT-CT, MRI-SPECT, and SPECT-CT, etc are some types of hybrid modalities.

- **Nuclear Medicine**. Molecular imaging utilizes markers that are tagged with radioactive substances that are inserted into the patient and redistributed according to metabolism. It uses a matrix of 128 × 128, 128 or 120 views of 360 degrees (a cardiac camera is unable to rotate 360 degrees, so it will rotate 180 degrees) And it adjusts the zoom factor to perform a pixel size of around 3 mm as desired. The standard zoom factors for large FoV camera, American College of Radiology (ACR), are 1.33 to 1.46 mm and 2.7 to 3.3 mm. It uses a rotation radius as similar to 20 cm as possible (helpful for an elliptical orbit). The protocol of ACR has a total count of 32 M. Diagnosis determines the severity or treats a diversity of illnesses including gastrointestinal disorders, many cancers, neurological, endocrine, heart disease, and additional deformities in the body.

4.2.1 Other Medical Images

Image obsession includes pathology (complex workflow and the largest sizes of the image), cardiology, dermatology, ophthalmology, and many other images. For dental imaging, Information Object Definition (IOD) of intraoral X-ray images reflects the need to describe the characteristics of a specific acquisition and knowledge required for the practice of dentistry [43]. The Visible light IOD is highly related for a large part of the data generated in dermatology and pathology [44].

4.3 Histopathology Images

Pathology reviews of biopsied tissues are often considered the gold standard. However, pathology slides reviewed is difficult even for professional pathologists [45], as shown in Fig. 3. A slide of digital pathology at 40X magnification always has millions or billions of pixels and can use up several gigabytes of disk space. Pathologists are often forced to look for micrometastases in these large archives, small groups of tumor cells that contain an early sign of cancer. The size of these groups of tumor cells are less than a thousand pixels in diameter Fig. 3. This makes reviewing the slides of pathology without losing one of these tiny ones but clinically actionable evidence very hard and time-consuming. In histopathology, Whole-slide images (WSI) are very big and usually every WSI has a full $80,000 \times 80,000$ pixels of locative resolution and about twenty gigabytes in the size of storage at 40× magnification [46].

Due to the high spatial resolution used in digital pathology, the dimensions of WSI is very large, which generally exceed nine hundred million pixels. In addition, there are several types of histopathology images (biopsy types) that are shown in Fig. 4. Properties of histopathology images are explained in Sect. 4.4.

4.4 Medical Images and Histopathology Sizes

These days, the resolution and size of medical images become larger and sharper. However, they use a large format in storage and need more computer resources when

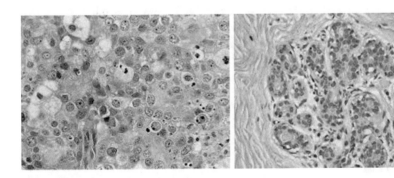

Fig. 3 Example of histopathology images Scan

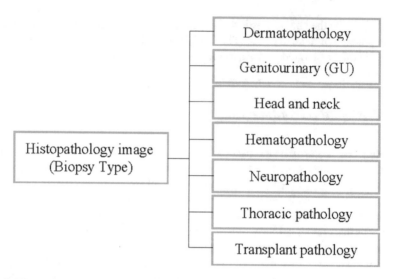

Fig. 4 Histopathology image (Biopsy Types)

they are used in machine learning. The resolution and size of the medical images are explained in this subsection.

This is necessary because the applications of AI in radiology may be sensitive to the changes in the equipment (e.g., many manufacturers and several scanners), the study population features (e.g., distribution and prevalence of disease rainbow in the chosen population), selection of reference standard (e.g., pathology, radiologist interpretation, clinical results) and image protocols with various image attributes (e.g., spatial resolution, "signal-to-noise" ratio, contrast, and temporal resolution) [47]. Thus, there will be a need for consistent, perfect communication, and transparency of patient cohort requirements to ensure that training data is generalized to target hospital sites and that AI algorithms are guaranteed to be implemented safely and responsibly.

Medical images and histopathology sizes differ from each other. Of course, histopathology has the largest image size. They are different in image size, image resolution and image type. Properties of radiology and histopathology images are as follows:

4.4.1 Medical Images

Medical image size is illustrated in Table 1. We can see big differences in image size among each modality. Medical image modalities have average sizes that range from 5.1 to 365.9 MB.

Table 1 Average size per modality of radiology [42]

Modality of radiology	Average size in (MB)	Bit-depth
Magnetic resonance angiography	55.6	8 or 16
Breast imaging	38.8	
Panoramic X-ray	5.1	
Radiotherapy	30.6	12
Endoscopy image	6.8	16
Radiographic image "conventional plain films"	36.7	8 to 24
Nuclear medicine	7.0	8
Radiofluoroscopy	30.1	8
Digital radiography	17.4	14 or 16
PET	365.9	8 or 16
Mammography (MG)	83.5	14
MRI	98.6	16
Computerized radiography	14.1	12
X-ray angiography	157.5	14
US	69.2	8 to 24
CT scan	153.4	16

4.4.2 Histopathology

The size of histopathology is illustrated in Table 2. There are big differences in medical images size among different type of histopathology magnifications. Dealing with these kinds of images sizes are more difficult. In general, the image of histopathology must be subdivided into tiles (small images) to interact with machine learning. It is a powerful method to defeat the curse of high dimension in this kind of images.

An important challenge with digital pathology is the image study the average size of files. The file size will vary from fifty MB on average for a WSI to more than six GB, and this relies on the slide scan magnification. Furthermore, a pathology imaging study has a number of slides that can range from two to sixty slides. Accordingly, Unlike other specialties such as radiology, where the average size of an image is usually tens or hundreds of MB, WSI studies can have a magnitude of thousands of MB [48].

The WSIs are saved in a compressed format for efficient storage. The compression technique usually used is the encoding of JPEG2000 or JPEG, which has a factor of high-quality to maintain the image information. A common WSI captured with an objective of $\times 20$ could theoretically reflect more than 20 Gb of storage if it is not compressed, but the size decreases to an average range of 200 to 650 Mb after compression [48].

Table 2 Average size through the study type of pathology [42]

Histopathology (Biopsy type)	Slides	Average size (MB)	
		×20 compressed resolution	×40 compressed resolution
Transplant pathology (wedge)	2–3	600–900	1,500–2,250
Transplant pathology (needle)	2–3	250–375	700–1050
Thoracic pathology	12	3,240	
Neuro-pathology	9	1,872	
Hemato-pathology	31		40,300
Neck and head	3	1,965	
GU	9	1,701	
Dermato-pathology	6	1,392	

Whole Slide Images: Modern scanning systems that produce WSI are formed of hardware platforms of software-driven robotic that operate on principles related to the compound of common microscopes. A glass slide of pathology is placed into the scanner and constantly ran under an objective microscope lens at a constant high speed. Because the histopathological sample has a tissue section with a recognized density, algorithm-defined real-time focus settings are used to keep centered on the resulting images. The objective lens defines the image output resolution and the total recording speed with a greater resolution, resulting in smaller fields of view and longer scanning times [48]. The resulting image is usually displayed in 24 bit/pixel which is full color in both dimensions, with a size of several thousand pixels. In general, the image is saved in a structure due to the size of the data that resembles a pyramid arranged in layers, each with a different resolution, to allow efficient display through specialized software given by a provider. A typical case of pathology can hold several WSI images, each of which can be in storage in hundreds or thousands of MB [48].

4.5 Medical Image Formats Type

The image file formats present a regulated method to save information specifying an image in a file in computer. A dataset of medical image usually depends on an image or more images which describe the projection of an anatomic volume on an imaging surface (planar/projection images), a series of images which describe fine sections by volume (multislice 2D images or tomographic), data from a volume (3D images), or based on volume overtime to perform a dynamic list of acquisitions or duplicated acquisition of the related tomographic image (4D images). The format of the file explains how image data is arranged in the image file. Furthermore, the way that the software must interpret the pixel data for accurate loading and display. On the other hand, the formats of the image files are regulated means for arranging and saving digital images. The format of the image file can save data in compressed or uncompressed. They are as follows:

4.5.1 File Format

The file formats of a medical image can be distributed into two divisions [49]. The first is the format designed to regulate the images which use diagnostic modalities to produced, e.g., Digital Imaging and Communications in Medicine (DICOM) [41]. The second is the format created for the purpose of facilitating and strengthening analysis after post-processing, e.g., Nifti [50], Analyze [51] and Minc [52]. Medical image files are usually saved with one of two possible arrangements that follow. The first one is that both image data and metadata are included in an individual file, with the metadata saved at file initialization. This paradigm is utilized by the Minc, Nifti, and Dicom file formats, even though other formats allow it. The second arrangements saves the image data in a file and the metadata in another. A two-file model is used by analyze file format which are (.img and.hdr).

4.5.2 Image Format

Most digital image formats are classified by their compression method into two divisions. If no information is lost from the digital image file, principally when a compressed algorithm is used, it is called lossless [53]. It contains some type of images such as TIFF, PNG, BMP, and RAW. All RGB digital images are saved in those image formats without information loss. This arises with the advantages of high-quality reproductions. Though, it needs a large memory size to store those files in it. The second has losses, resulting in the loss of certain image information to obtain a smaller file size including GIF and JPEG. The size and resolution of all images vary with regard to dots per inch (dpi) and bit depth.

4.6 Grayscale Images and Color Images

Images generally represent natural objects. The photometric interpretation explains how to represent the pixel data for the accurate display of the monochrome image or color image. To define whether the color information is saved in the pixel values of the image or not, samples per pixel (also known as channel numbers) is the concept to specify this case. Monochrome images, which color data is not saved in the image, have one example per pixel. The image is displayed using a scale of gray shades from white to black. The gray color shades number definitely depends on the bits number applied to save the sample that synchronizes with the pixel-depth. The color information in color image is stored in the sample per pixel which is three. The images are displayed with an RGB. The number of color density is clearly dependent on the bits number applied to save the samples which corresponds to the depth of the pixel in this case.

For example, color is used to represent the velocity and direction of blood flow in ultrasound (Doppler US), to display additional details in an anatomical image in shades of gray as colored overlays, as at the activation sites of fMRI, to concurrently display anatomical and functional images such as in PET/MRI or PET/CT, and occasionally instead of shades of gray to emphasize signal variations.

The use of 2D, 3D, and 4D in machine learning is as follows. Machine learning uses a gray-scale image with an array of 3 matrices (height, width, and depth) where the depth is equal to number one. The 2D color image uses an array of 3 matrices (height, width, and depth). The Gray-scale volumetric image uses an array of 4 matrices (height, width, depth, and slices of images) where the depth is equal to number one. The 3D color image uses an array of 4 matrices (height, width, depth, and slices of images). More information about grayscale and color image are as follows:

4.6.1   Grayscale Images

Grayscale images (or black & white) images are images that have one color which is gray. It arranges from 0 to 255. It is placed with a number of one in the third dimension in the image matrix. Clinical-radiological images, for example, MRI, CT, and X-ray have a grayscale photometric interpretation.

4.6.2   Color Images

Since color is a primary distinguishing feature for the various biological structures, every pixel contains three color components. Color image (also called RGB) is a colorful image. The image is containing three images above each other. Color map are typically used to display some medical images e.g., PET, Nuclear medicine images, and etc. Within a predefined color map, every pixel of the image is kinked to a color, though the color refers only to the screen and is associated knowledge, and is not actually saved in the pixel values. The images yet contain a sample/pixel and are usually labeled in pseudo-color. They typically need multiple samples per pixel to translate the color data into pixels. In addition, they choose a model of color which shows how colors can be obtained by merging the samples [54]. Eight bits are usually maintained for every part of the color component or sample. The depth of pixels is estimated by multiplying the number of samples per pixel by the depth of the sample . Most medical images are usually saved using the model of RGB color. The pixel should be understood in this case as a mixture of the three prime colors and three samples are stored for each pixel.

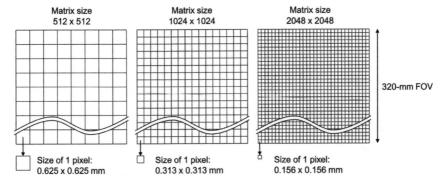

Fig. 5 The relation between the size of the matrix and the size of one pixel. If the field of view is 320 mm, the theoretical size of one pixel in an image matrix is: 0.156 mm for 2048 × 2048; 0.313 mm for 1024 × 1024; and 0.625 mm for 512 × 512 [55]

5 Manipulating Medical Images

To let the image interacts with machine learning, it should be manipulated. There are a lot of methods to change the properties of the images such as re-sizing, augmentation, image compression and other techniques. We will focus on re-sizing and compression.

5.1 Enhancing Medical Image Size and Resolution

There are several methods and algorithms for re-sizing and compressing images. They are as follows:

5.1.1 Traditional Re-Sizing

There are some common types of functions and library that are used to resize medical images e.g., python library and OpenCV library. This kind of re-sizing change the resolution of the image. In addition, deep learning algorithms will produce not accurate results because medical image loses some of its information. The relationship between size and pixel are explained in Fig. 5.

5.1.2 Super Resolution Re-Sizing

With the revolution of Generative Adverisal Network (GAN) [56], we can synthesize and produce great images. Super-resolution GAN (SRGAN) [57] uses a combination of a deep network with an adversary network to create higher resolution images.

Table 3 The Royal College of Radiologists' guidelines on the Lossy compression ratio [74]

Modality	Compression Ratio
Radiotherapy CT	(No compression)
CT (all fields)	5:1
Digital Angiography	10:1
Ultrasound (US)	10:1
Magnetic Resonance Images (MRI)	5:1
Skeletal Radiography	10:1
Chest Radiography	10:1
Mammography (MG)	20:1

SRGAN is attractive to a human with more details compared with the related design without GAN e.g., SRResNet [57]. A HR image is downsampled to a LR image. The generator of SRGAN upsamples LR images to SR images. They use a discriminator to discriminate the images of HR and backpropagate the loss function of GAN to train the discriminator and the generator. Another algorithm is used in enhancing images such as Enhanced Super-Resolution Generative Adversarial Networks (ESRGAN) [58]. Dealing with super resolution images can be found in [59].

5.2 Image Compression

Modern medical imaging has created vast quantities of data so that the storage and transmission systems can be overfilled rapidly [60, 61]. Data compression has generally been key to overcome this dilemma [62, 63]. Various compression techniques of medical image, in this context, have been improved to decrease image volume, such as genetic algorithms [64], Region of Interest Coding Techniques [65, 66], low-complexity compression [67], Fractal coding methods [68], and Lossless dynamic and adaptive compression [69]. However, JPEG2000 [70, 71], JPEG [72], JPEG-LS(JLS) [73] and TIFF are still one of the widespread methods used as standards because they provide greater performance when compressing images.

We perceive that these methods are based on algorithms that are coded to multi-resolution domains using modified images [75, 76]. In addition, several medical applications allow the employ of loss compression unless the compressed image leads to the wrong diagnosis [77].

There are several advantages of image compression that can be summarized as follows:

- It offers a significant saving of bandwidth with regard to sending a smaller amount of data via the internet.
- Significant saving on storage space.
- Offers a higher level of immunity and security against illegal monitoring.

To save a gray image with a size of 1024 × 1024, approximately 1Mb of disk space and for color images, this will be multiplied by 3. The process of size reduction is the exclusion of unnecessary data. If the image is decreased with a compression ratio of 10:1, the storage needed is reduced to 300 KB [78]. Compression ratios for medical images are shown in Table 3. Thus, the possibilities to reorganize compressed data (e.g., interactive multi-resolution transmission with JPIP [79], tiled DICOM [80], or tiling in BigTIFF [81]) are predominant to allow a fine viewing experience. More information about medical image compression can be found in [82].

5.3 Whole Slide Images and Tile Images

5.3.1 Very Large Image Size.

When using deep learning to classify images e.g., houses or dogs, an image of a small size like 128 × 128 is frequently applied as an input. The large size image also need to be re-sized to a smaller size that is suitable for adequate distinguishing, because an expansion in the magnitude of the input image results in an expansion in the parameter to be measured, the computing power needed and memory.

In opposition, WSI [83] has several cells, and the image composed of billions or more of pixels, which typically makes it difficult to analyze. Nevertheless , re-sizing the whole image to a smaller size e.g., 128 × 128 will drive to information loss at the cellular level, ending in a marked reduction in the accuracy of the identifier. Accordingly, the whole WSI is regularly separated into fractional regions of approximately 128 × 128 pixels or ("patches"), and each patch is independently evaluated, for example, RoIs detection. Because of advances in memory and computing power, patch size is expanding (e.g., 900 × 900), which is expected to lead to improved accuracy [84]. The process of incorporating the outcome from every patch is an opportunity to get better yet. Because the WSI, for example, can contain a huge number of patches (hundreds of thousands), it is very likely that false positives will occur, even though the individual patches are minutely classed. The regional average of every decision is a potential answer to this, such that the regions are marked as ROI even if the ROI is spread over several patches.

Though, this methodology can suffer from false-negatives, leading to a lack of small ROIs, for example, isolated tumor cells [85]. Scoring in a few applications such as Immunohisto Chemistry (IHC), staging the metastasis of lymph nodes in specimens or patients, and staging the diagnosis of prostate cancer with Glisson multi-region score inside one slide, more complicated algorithms are incorporate to integrate patch or object-level choices [85–89].

However, pathological images can precisely be labeled only by pathologists, and it takes a lot of work to label it at the territorial level in an immense WSI. reusing ready-to-analyze data is conceivable in machine learning as training data, for example, natural images [90] in the macroscopic diagnosis of skin and ImageNet [21] in International Skin-Imaging Collaboration (ISIC).

5.4 Algorithms Used with Medical Images

There are numbers of common algorithms used in machine learning with medical imaging. Must of them are discussed by [91], they explore some algorithms such as Naive Bayes Algorithm, Support Vector Machines, Decision trees, Neural Networks, and k-Nearest Neighbors. Some other algorithms are used in deep learning such as CNN [20], Deep Convolutional Neural Network (DCNN) [20], U-net [92], Fully Convolutional Networks (FCN) [93], LeNet [94], etc. However, deep learning gives a great result when acting with medical images and image segmentation. [95–97], etc., are examples of deep learning algorithms that are used with medical images. Many open source medical images datasets are available nowadays for researchers such as [98–100], etc.

6 Conclusions and Discussion

Medical images play a glory rule when determine to investigate human body. Machine learning algorithms produce great results with medical images. Medical image properties such as resolution, size, and type, effect machine learning result. In addition, the best image type is PNG for all scan type because it contains all the details about the image. besides, it is time-consuming and takes larger data than other compressed formats. Furthermore, Medical imaging modalities contain many scanning methods for diagnostic and care purposes for visualizing the body of human. Such modalities are also very beneficial for patient follow-up on the severity of the already diagnosed disease state and/or a treatment plan is underway. Such modalities of medical imaging involve themselves in all aspects of hospital care. The key aim is to get the right diagnosis.

Acknowledgements The authors would like to express their gratitude to Mohammed Almurisi for his support.

References

1. Fowler, J.F., Hall, E.J.: Radiobiology for the Radiologist. Radiat. Res. **116**, 175 (1988)
2. Mifflin, J.: Visual archives in perspective: enlarging on historical medical photographs. Am. Arch. **70**(1), 32–69 (2007)
3. Cosman, P.C., Gray, R.M., Olshen, R.A.: Evaluating quality of compressed medical images: SNR, subjective rating, and diagnostic accuracy. Proc. IEEE **82**(6), 919–32 (1994)
4. Kayser, K., Görtler, J., Goldmann, T., Vollmer, E., Hufnagl, P., Kayser, G.: Image standards in tissue-based diagnosis (diagnostic surgical pathology). Diagn. Pathol. **3**(1), 17 (2008)
5. Ramakrishna, B., Liu, W., Saiprasad, G., Safdar, N., Chang, C.I., Siddiqui, K., Kim, W., Siegel, E., Chai, J.W., Chen, C.C., Lee, S.K.: An automatic computer-aided detection system for meniscal tears on magnetic resonance images. IEEE Trans. Med. Imaging **28**(8), 1308–1316 (2009)

6. Brenner, D.J., Hall, E.J.: Computed tomography-an increasing source of radiation exposure. New Engl. J. Med. **357**(22), 2277–2284 (2007)
7. Foltz, W.D., Jaffray, D.A.: Principles of magnetic resonance imaging. Radiat. Res. **177**(4), 331–348 (2012)
8. Fass, L.: Imaging and cancer: a review. Molecular oncology. **2**(2), 115–52 (2008)
9. Ehman, R.L., Hendee, W.R., Welch, M.J., Dunnick, N.R., Bresolin, L.B., Arenson, R.L., Baum, S., Hricak, H., Thrall, J.H.: Blueprint for imaging in biomedical research. Radiology **244**(1), 12–27 (2007)
10. Hillman, B.J.: Introduction to the special issue on medical imaging in oncology. J. Clin. Oncol. **24**(20), 3223–3224 (2006)
11. Lehman, C.D., Isaacs, C., Schnall, M.D., Pisano, E.D., Ascher, S.M., Weatherall, P.T., Bluemke, D.A., Bowen, D.J., Marcom, P.K., Armstrong, D.K., Domchek, S.M.: Cancer yield of mammography, MR, and US in high-risk women: prospective multi-institution breast cancer screening study. Radiology **244**(2), 381–388 (2007)
12. de Torres, J.P., Bastarrika, G., Wisnivesky, J.P., Alcaide, A.B., Campo, A., Seijo, L.M., Pueyo, J.C., Villanueva, A., Lozano, M.D., Montes, U., Montuenga, L.: Assessing the relationship between lung cancer risk and emphysema detected on low-dose CT of the chest. Chest **132**(6), 1932–1938 (2007)
13. Nelson, E.D., Slotoroff, C.B., Gomella, L.G., Halpern, E.J.: Targeted biopsy of the prostate: the impact of color Doppler imaging and elastography on prostate cancer detection and Gleason score. Urology **70**(6), 1136–1140 (2007)
14. Kent, M.S., Port, J.L., Altorki, N.K.: Current state of imaging for lung cancer staging. Thorac. Surg. Clin. **14**(1), 1–3 (2004)
15. Fermé, C., Vanel, D., Ribrag, V., Girinski, T.: Role of imaging to choose treatment: Wednesday 5 October 2005, 08:30–10:00. Cancer Imaging. 2005;5(Spec No A):S113
16. Haugeland J. Artificial Intelligence: The Very Idea. MIT Press (1989)
17. Fukushima, K.: Neocognitron: a self-organizing neural network model for a mechanism of pattern recognition unaffected by shift in position. Biol. Cybern. **36**(4), 193–202 (1980)
18. Lo, S.C., Lou, S.L., Lin, J.S., Freedman, M.T., Chien, M.V., Mun, S.K.: Artificial convolution neural network techniques and applications for lung nodule detection. IEEE Trans. Med. Imaging **14**(4), 711–718 (1995)
19. LeCun, Y., Bottou, L., Bengio, Y., Haffner, P.: Gradient-based learning applied to document recognition. Proc. IEEE **86**(11), 2278–2324 (1998)
20. Krizhevsky, A., Sutskever, I., Hinton, G.E.: Imagenet classification with deep convolutional neural networks. In: Advances in Neural Information Processing Systems, pp. 1097–1105 (2012)
21. Russakovsky, O., Deng, J., Su, H., Krause, J., Satheesh, S., Ma, S., Huang, Z., Karpathy, A., Khosla, A., Bernstein, M., Berg, A.C.: Imagenet large scale visual recognition challenge. Int. J. Comput. Vis. **115**(3), 211–252 (2015)
22. Bengio, Y., Courville, A., Vincent, P.: Representation learning: a review and new perspectives. IEEE Trans. Pattern Anal. Mach. Intell. **35**(8), 1798–1828 (2013)
23. Ravì, D., Wong, C., Deligianni, F., Berthelot, M., Andreu-Perez, J., Lo, B., Yang, G.Z.: Deep learning for health informatics. IEEE J. Biomed. Health Inf. **21**(1), 4–21 (2016)
24. Shen, D., Wu, G., Suk, H.I.: Deep learning in medical image analysis. Ann. Rev. Biomed. Eng. **21**(19), 221–248 (2017)
25. Smola, A., Vishwanathan, S.V.: Introduction to Machine Learning, vol. 32, pp. 34. Cambridge University, UK (2008)
26. Caruana, R.: Multitask learning. Mach. Learn. **28**(1), 41–75 (1997)
27. Thrun, S.: Is learning the n-th thing any easier than learning the first? In: Advances in Neural Information Processing Systems, pp. 640–646 (1996)
28. House, D., Walker, M.L., Wu, Z., Wong, J.Y., Betke, M.: IEEE Computer Society Conference on Computer Vision and Pattern Recognition Workshops, 2009. CVPR Workshops 2009, pp. 186–193. IEEE (2009)

29. Kumar, A., Kim, J., Cai, W., Fulham, M., Feng, D.: Content-based medical image retrieval: a survey of applications to multidimensional and multimodality data. J. Digit. Imaging **26**(6), 1025–1039 (2013)
30. Sedghi, S., Sanderson, M., Clough, P.: How do health care professionals select medical images they need? In: Aslib Proceedings. Emerald Group Publishing Limited (29 Jul 2012)
31. Freeny, P.C., Lawson, T.L.: Radiology of the Pancreas. Springer Science & Business Media (6 Dec 2012)
32. Anwar, S.M., Majid, M., Qayyum, A., Awais, M., Alnowami, M., Khan, M.K.: Medical image analysis using convolutional neural networks: a review. J. Med. Syst. **42**(11), 226 (2018)
33. Heidenreich, A., Desgrandschamps, F., Terrier, F.: Modern approach of diagnosis and management of acute flank pain: review of all imaging modalities. Eur. Urol. **41**(4), 351–362 (2002)
34. Rahman, M.M., Desai, B.C., Bhattacharya, P.: Medical image retrieval with probabilistic multi-class support vector machine classifiers and adaptive similarity fusion. Comput. Med. Imaging Graph. **32**(2), 95–108 (2008)
35. Sánchez Monedero, J., Saez Manzano, A., Gutiérrez Peña, P.A., Hervas Martínez, C.: Machine learning methods for binary and multiclass classification of melanoma thickness from dermoscopic images. IEEE Trans. Knowl. Data Eng. 2016 (ONLINE)
36. Miri, M.S., Lee, K., Niemeijer, M., Abràmoff, M.D., Kwon, Y.H., Garvin, M.K.: Multimodal segmentation of optic disc and cup from stereo fundus and SD-OCT images. In: Medical Imaging 2013: Image Processing. International Society for Optics and Photonics, vol. 8669, p. 86690O (13 Mar 2013)
37. Gao, Y., Zhan, Y., Shen, D.: Incremental learning with selective memory (ILSM): towards fast prostate localization for image guided radiotherapy. IEEE Trans. Med. Imaging **33**(2), 518–534 (2013)
38. Tao, Y., Peng, Z., Krishnan, A., Zhou, X.S.: Robust learning-based parsing and annotation of medical radiographs. IEEE Trans. Med. Imaging **30**(2), 338–350 (2010)
39. Camlica, Z., Tizhoosh, H.R., Khalvati, F.: Autoencoding the retrieval relevance of medical images. In: 2015 International Conference on Image Processing Theory, Tools and Applications (IPTA), pp. 550–555. IEEE (10 Nov 2015)
40. Branstetter, B.F.: Practical Imaging Informatics: Foundations and Applications for PACS Professionals. Springer, New York (2009)
41. Bidgood Jr., W.D., Horii, S.C., Prior, F.W., Van Syckle, D.E.: Understanding and using DICOM, the data interchange standard for biomedical imaging. J. Am. Med. Inf. Assoc. **4**(3), 199–212 (1997)
42. Lauro, G.R., Cable, W., Lesniak, A., Tseytlin, E., McHugh, J., Parwani, A., Pantanowitz, L.: Digital pathology consultations-a new era in digital imaging, challenges and practical applications. J. Digit. Imaging **26**(4), 668–677 (2013)
43. Tirado-Ramos, A., Hu, J., Lee, K.P.: Information object definition-based unified modeling language representation of DICOM structured Reporting: A Case Study of Transcoding DICOM to XML. J. Am. Med. Inf. Assoc. **9**(1), 63–72 (2002)
44. Seibert, J.A.: Modalities and data acquisition. In: Practical Imaging Informatics, pp. 49–66. Springer, New York, NY (2009)
45. Li, Y., Ping, W.: Cancer metastasis detection with neural conditional random field (19 Jun 2018). arXiv:1806.07064
46. Cruz-Roa, A., Gilmore, H., Basavanhally, A., Feldman, M., Ganesan, S., Shih, N., Tomaszewski, J., Madabhushi, A., González, F.: High-throughput adaptive sampling for whole-slide histopathology image analysis (HASHI) via convolutional neural networks: application to invasive breast cancer detection. PloS one **13**(5) (2018)
47. Kamnitsas, K., Baumgartner, C., Ledig, C., Newcombe, V., Simpson, J., Kane, A., Menon, D., Nori, A., Criminisi, A., Rueckert, D., Glocker, B.: Unsupervised domain adaptation in brain lesion segmentation with adversarial networks. In: International Conference on Information Processing in Medical Imaging, vol. 25, pp. 597–609. Springer, Cham (2017)

48. Park, S., Pantanowitz, L., Parwani, A.V.: Digital imaging in pathology. Clin. Lab. Med. **32**(4), 557–584 (2012)
49. Larobina, M., Murino, L.: Medical image file formats. J. Digit. Imaging **27**(2), 200–206 (2014)
50. NIFTI documentation, (Available via website, 2018). https://nifti.nimh.nih.gov/nifti-1/ documentation (Cited May 18, 2018)
51. Robb, R.A., Hanson, D.P., Karwoski, R.A., Larson, A.G., Workman, E.L., Stacy, M.C.: Analyze: a comprehensive, operator-interactive software package for multidimensional medical image display and analysis. Comput. Med. Imaging Graph. **13**(6), 433–454 (1989)
52. MINC software library and tools, (Available via website, 2018). http://www.bic.mni.mcgill. ca/ServicesSoftware/MINC (Cited May 18, 2018)
53. Ukrit, M.F., Umamageswari, A., Suresh, G.R.: A survey on lossless compression for medical images. Int. J. Comput. Appl. **31**(8), 47–50 (2011)
54. Wikipedia: Encyclopedia of Graphics File Formats, (Available via website, 2019). https://en. wikipedia.org/wiki/Machine-learning (25 March 2019)
55. Hata, A., Yanagawa, M., Honda, O., Kikuchi, N., Miyata, T., Tsukagoshi, S., Uranishi, A., Tomiyama, N.: Effect of matrix size on the image quality of ultra-high-resolution CT of the lung: comparison of 512x512, 1024x1024, and 2048x2048. Acad. Radiol. **25**(7), 869–876 (2018)
56. Goodfellow, I., Pouget-Abadie, J., Mirza, M., Xu, B., Warde-Farley, D., Ozair, S., Courville, A., Bengio, Y.: Generative adversarial nets. In: Advances in Neural Information Processing Systems, pp. 2672–2680 (2014)
57. Ledig, C., Theis, L., Huszár, F., Caballero, J., Cunningham, A., Acosta, A., Aitken, A., Tejani, A., Totz, J., Wang, Z., Shi, W.: Photo-realistic single image super-resolution using a generative adversarial network. In: Proceedings of the IEEE Conference on Computer Vision and Pattern Recognition, pp. 4681–4690 (2017)
58. Wang, X., Yu, K., Wu, S., Gu, J., Liu, Y., Dong, C., Qiao, Y., Change Loy, C.: Esrgan: enhanced super-resolution generative adversarial networks. In: Proceedings of the European Conference on Computer Vision (ECCV), pp. 0–0. 2018
59. Wang, Z., Chen, J., Hoi, S.C.: Deep learning for image super-resolution: a survey (16 Feb 2019). arXiv:1902.06068
60. Xia, Q., Ni, J., Kanpogninge, A.J., Gee, J.C.: Searchable public-key encryption with data sharing in dynamic groups for mobile cloud storage. J. UCS **21**(3), 440–453 (2015)
61. Chaabouni, I., Fourati, W., Bouhlel, M.S.: Using ROI with ISOM compression to medical image. Int. J. Comput. Vis. Robot. **6**(1–2), 65–76 (2016)
62. Suruliandi, A., Raja, S.P.: Empirical evaluation of EZW and other encoding techniques in the wavelet-based image compression domain. Int. J. Wavelets, Multiresolution Inf. Process. **13**(02), 1550012 (2015)
63. Ang, B.H., Sheikh, U.U.: Marsono MN. 2-D DWT system architecture for image compression. J. Signal Process. Syst. **78**(2), 131–137 (2015)
64. Shih, F.Y., Wu, Y.T.: Robust watermarking and compression for medical images based on genetic algorithms. Inf. Sci. **175**(3), 200–216 (2005)
65. Doukas, C., Maglogiannis, I.: Region of interest coding techniques for medical image compression. IEEE Eng. Med. Biol. Mag. **26**(5), 29–35 (2007)
66. Hernandez-Cabronero, M., Blanes, I., Pinho, A.J., Marcellin, M.W., Serra-Sagristà, J.: Progressive lossy-to-lossless compression of DNA microarray images. IEEE Signal Process. Lett. **23**(5), 698–702 (2016)
67. Pizzolante, R., Carpentieri, B., Castiglione, A.: A secure low complexity approach for compression and transmission of 3-D medical images. In: 2013 Eighth International Conference on Broadband and Wireless Computing, Communication and Applications, pp. 387–392. IEEE (28 Oct 2013)
68. Bhavani, S., Thanushkodi, K.G.: Comparison of fractal coding methods for medical image compression. IET Image Process. **7**(7), 686–693 (2013)
69. Castiglione, A., Pizzolante, R., De Santis, A., Carpentieri, B., Castiglione, A., Palmieri, F.: Cloud-based adaptive compression and secure management services for 3D healthcare data. Future Gener. Comput. Syst. **1**(43), 120–134 (2015)

70. Ciznicki, M., Kurowski, K., Plaza, A.J.: Graphics processing unit implementation of JPEG2000 for hyperspectral image compression. J. Appl. Remote Sens. **6**(1), 061507 (2012)
71. Bruylants, T., Munteanu, A., Schelkens, P.: Wavelet based volumetric medical image compression. Signal Process. Image Commun. **1**(31), 112–133 (2015)
72. Pu, L., Marcellin, M.W., Bilgin, A., Ashok, A.: Compression based on a joint task-specific information metric. In: 2015 Data compression conference. IEEE, pp. 467–467 (7 Apr 2015)
73. Starosolski, R.: New simple and efficient color space transformations for lossless image compression. J. Visual Commun. Image Represent. **25**(5), 1056–1063 (2014)
74. The Adoption of Lossy Image Data Compression for the Purpose of Clinical Interpretation, (Available via website, 2017). https://www.rcr.ac.uk/sites/default/files/docs/radiology/pdf/IT-guidance-LossyApr08.pdf (Cited 15 October 2017)
75. Wu, X., Li, Y., Liu, K., Wang, K., Wang, L.: Massive parallel implementation of JPEG2000 decoding algorithm with multi-GPUs. In: Satellite Data Compression, Communications, and Processing X. International Society for Optics and Photonics, vol. 9124, pp. 91240S (22 May 2014)
76. Blinder, D., Bruylants, T., Ottevaere, H., Munteanu, A., Schelkens, P.: JPEG 2000-based compression of fringe patterns for digital holographic microscopy. Opt. Eng. **53**(12), 123102 (2014)
77. Chemak, C., Bouhlel, M.S., Lapayre, J.C.: Neurology diagnostics security and terminal adaptation for PocketNeuro project. Telemed. e-Health. **14**(7), 671–678 (2008)
78. Dewan, M.A., Islam, R., Sharif, M.A., Islam, M.A.: An Approach to Improve JPEG for Lossy Still Image Compression. Computer Science & Engineering Discipline, Khulna University, Khulna 9208
79. Hara, J.: An implementation of JPEG 2000 interactive image communication system. In: 2005 IEEE International Symposium on Circuits and Systems, pp. 5922–5925. IEEE (23 May 2005)
80. Supplement 145: Whole Slide Microscopic Image IOD and SOP Classes, (Available via website, 2019). ftp://medical.nema.org/MEDICAL/Dicom/Final/sup145-ft.pdf (Cited 10 March 2019)
81. BigTIFF: BigTIFF Library, (Available via website, 2019). http://bigtiff.org/ (Cited 12 March 2019)
82. Liu, F., Hernandez-Cabronero, M., Sanchez, V., Marcellin, M.W., Bilgin, A.: The current role of image compression standards in medical imaging. Information **8**(4), 131 (2017)
83. Farahani, N., Parwani, A.V., Pantanowitz, L.: Whole slide imaging in pathology: advantages, limitations, and emerging perspectives. Pathol. Lab. Med. Int. **7**(23–33), 4321 (2015)
84. Komura, D., Ishikawa, S.: Machine learning methods for histopathological image analysis. Comput. Struct. Biotechnol. J. **1**(16), 34–42 (2018)
85. Liu, Y., Gadepalli, K., Norouzi, M., Dahl, G.E., Kohlberger, T., Boyko, A., Venugopalan, S., Timofeev, A., Nelson, P.Q., Corrado, G.S., Hipp, J.D.: Detecting cancer metastases on gigapixel pathology images (3 Mar 2017). arXiv:1703.02442
86. Wang, D., Khosla, A., Gargeya, R., Irshad, H., Beck, A.H.: Deep learning for identifying metastatic breast cancer (18 Jun 2016). arXiv:1606.05718
87. Mungle, T., Tewary, S., Das, D.K., Arun, I., Basak, B., Agarwal, S., Ahmed, R., Chatterjee, S., Chakraborty, C.: MRF ANN: a machine learning approach for automated ER scoring of breast cancer immunohistochemical images. J. Microsc. **267**(2), 117–129 (2017)
88. Wang, D., Foran, D.J., Ren, J., Zhong, H., Kim, I.Y., Qi, X.: Exploring automatic prostate histopathology image gleason grading via local structure modeling. In: 2015 37th Annual International Conference of the IEEE Engineering in Medicine and Biology Society (EMBC), pp. 2649–2652. IEEE (25 Aug 2015)
89. Wollmann, T., Rohr, K.: Automatic breast cancer grading in lymph nodes using a deep neural network (24 Jul 2017). arXiv:1707.07565
90. Gutman, D., Codella, N.C., Celebi, E., Helba, B., Marchetti, M., Mishra, N., Halpern, A.: Skin lesion analysis toward melanoma detection: a challenge at the international symposium on biomedical imaging (ISBI) 2016, hosted by the international skin imaging collaboration (ISIC) (4 May 2016). arXiv:1605.01397

91. Erickson, B.J., Korfiatis, P., Akkus, Z., Kline, T.L.: Machine learning for medical imaging. Radiographics **37**(2), 505–515 (2017)
92. Ronneberger, O., Fischer, P., Brox, T.: U-net: convolutional networks for biomedical image segmentation. In: International Conference on Medical Image Computing and Computer-assisted Intervention, vol. 5, pp. 234–241. Springer, Cham (2015)
93. Long, J., Shelhamer, E., Darrell, T.: Fully convolutional networks for semantic segmentation. In: Proceedings of the IEEE Conference on Computer Vision and Pattern Recognition, pp. 3431–3440 (2015)
94. LeCun, Y., Bengio, Y.: Convolutional networks for images, speech, and time series. In: The Handbook of Brain Theory and Neural Networks, vol. 3361, no. 10 (1995)
95. Al-Dhabyani, W., Gomaa, M., Khaled, H., Aly, F.: Deep learning approaches for data augmentation and classification of breast masses using ultrasound images. Int. J. Adv. Comput. Sci. Appl. **10**(5) (2019)
96. Khalifa, N.E., Taha, M.H., Hassanien, A.E., Hemedan, A.A.: Deep bacteria: robust deep learning data augmentation design for limited bacterial colony dataset. Int. J. Reason. Based Intell. Syst. **11**(3), 256–64 (2019)
97. Khalifa, N.E., Taha, M.H., Hassanien, A.E., Mohamed, H.N.: Deep iris: deep learning for gender classification through iris patterns. Acta Informatica Medica **27**(2), 96 (2019)
98. Al-Dhabyani, W., Gomaa, M., Khaled, H., Fahmy, A.: Dataset of breast ultrasound images. Data Brief **1**(28), 104863 (2020)
99. Cancer Imaging Archive, (Available via website, 2018). http://www.cancerimagingarchive. net (Cited 20 October 2018)
100. National Cancer Institute, Genomic data commons data portal (legacy archive), (Available via website, 2018). https://portal.gdc.cancer.gov/legacy-archive/ (Cited 18 October 2018)

Edge Detector-Based Hybrid Artificial Neural Network Models for Urinary Bladder Cancer Diagnosis

Ivan Lorencin, Nikola Anđelić, Sandi Baressi Šegota, Jelena Musulin, Daniel Štifanić, Vedran Mrzljak, Josip Španjol, and Zlatan Car

1 Optical Biopsy and ANN Utilization

Bladder cancer is one of the most common cancers of the urinary tract and is a result of uncontrolled growth of bladder mucous cells. Some of the potential risk factors for urinary bladder cancer are [1]: smoking, previous cases of disease in family members, exposure to chemicals and prior radiation therapy. The symptoms that most commonly suggest bladder cancer are: lower back pain, painful urination and blood in urine. When some or all of those symptoms are present, the diagnostic procedure for urinary bladder cancer must be performed. As one of the key procedures, cystoscopy is nowadays frequently performed examination during the process of urinary bladder cancer diagnosis. It can be defined as an endoscopic method for visual evaluation of urethra and bladder via insertion of the probe called cystoscope

I. Lorencin · N. Anđelić (✉) · S. Baressi Šegota · J. Musulin · D. Štifanić · V. Mrzljak · Z. Car
Faculty of Engineering, University of Rijeka, Vukovarska 58, 51000 Rijeka, Croatia
e-mail: nandelic@riteh.hr

I. Lorencin
e-mail: ilorencin@riteh.hr

S. Baressi Šegota
e-mail: sbaressisegota@riteh.hr

J. Musulin
e-mail: jmusulin@riteh.hr

D. Štifanić
e-mail: dstifanic@riteh.hr

V. Mrzljak
e-mail: vmrzljak@riteh.hr

Z. Car
e-mail: car@riteh.hr

J. Španjol
Faculty of Medicine, University of Rijeka, Braće Branchetta 20/1, 51000 Rijeka, Croatia
e-mail: josip.spanjol@medri.uniri.hr

© The Editor(s) (if applicable) and The Author(s), under exclusive license to Springer Nature Switzerland AG 2021
A.-E. Hassanien et al. (eds.), *Enabling AI Applications in Data Science*, Studies in Computational Intelligence 911,
https://doi.org/10.1007/978-3-030-52067-0_10

225

[2]. In clinical practice, there are two basic types of cystoscopy, and these are flexible cystoscopy and rigid cystoscopy [3]. Difference between two aforementioned methods is that in flexible cystoscopy, the procedure is performed by using flexible fiber-optic, while in the case of rigid cystoscopy procedure is performed by using rigid fiber-optic. Modern cystoscopes are equipped with digital camera that is used to broadcast cystoscopy images on a computer screen. Such approach offers a possibility for optical evaluation of lesions on bladder mucosa. This evaluation is performed by a specialist urologist in vivo. By utilizing this method, papillary lesions of transitional epithelium are detected in high number of cases. Negative side of this method is the fact that is hard to distinguish inflammatory and scarring changes in bladder mucosa from Carcinoma in-situ (CIS) or bladder cancer that does not exceed 75% [4, 5]. For these reasons, confocal laser endomicroscopy (CLE) is utilized for lesions observation. Such approach offers higher accuracy of urinary bladder cancer diagnosis and it can be called optical biopsy. By utilizing this procedure, malignant changes in urinary bladder mucosa can be detected according to micro-structural changes. Images obtained with CLE can roughly be divided in four classes, and these are:

- High-grade carcinoma,
- Low-grade carcinoma,
- Carcinoma in-situ (CIS) and
- Healthy tissue.

CIS represents a group of abnormal cells that may become cancer, while low-grade cancer cells represent slower growing and spreading cancer cells. On the other hand, high-grade carcinoma represents a term describing faster growing and spreading cancer cells. A right diagnosis in regard of cancer gradation can be crucial in treatment planning [6]. An example of all four classes is given in Fig. 1.

Aforementioned micro-structural changes for all three carcinoma classes are detected with higher accuracy if AI methods, mainly ANNs are utilized for classification of images collected by CLE. Such approach can offer a significant increase in bladder cancer diagnosis accuracy, especially in the case of CIS. Principle of ANN utilization in optical biopsy is presented with a dataflow diagram in Fig. 2.

A standard procedure of ANN utilization in tasks of image recognition and classification includes use of convolutional neural networks (CNN). Such procedure represents a standard approach with applications in various fields of science and technology [7, 8]. Described trend can also be noticed in medical applications [9–11]. Such approach can be costly from a standpoint of computational resources [12]. For these reasons, the utilization of simpler algorithms can be of significant importance. This feature is particularly emphasized in clinical practice where fewer computing resources are available. Here lies a motive for utilization of hybrid ANN models. In this chapter, an idea of edge detector-based ANN models utilization in diagnosis of urinary bladder cancer is presented.

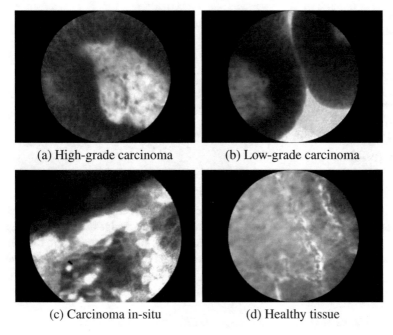

(a) High-grade carcinoma   (b) Low-grade carcinoma

(c) Carcinoma in-situ   (d) Healthy tissue

Fig. 1 Overview of bladder tissue images obtained with CLE

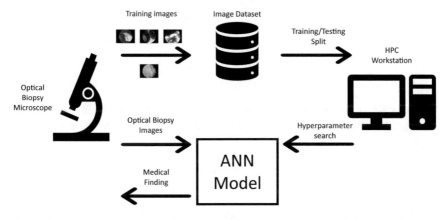

Fig. 2 Dataflow diagram of proposed ANN utilization for optical biopsy of urinary bladder

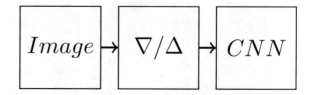

2 Representation of Edge Detector-Based Hybrid ANN Models

The idea of edge detector-based hybrid ANN model utilization in urinary bladder cancer diagnosis is presented in [13], where Laplacian edge detector is utilized for designing the hybrid ANN model. The main principle of edge detector-based hybrid ANN model is application of edge detector on all images contained in the dataset. Such operation can be defined as convolution of image with an edge detector kernel as:

$$B = A * K, \tag{1}$$

where A represents the original image and K represents a kernel:

$$K = \begin{bmatrix} K_{0,0} & K_{1,0} & \dots & K_{m,0} \\ K_{0,1} & K_{1,1} & \dots & K_{m,1} \\ \cdot & \cdot & \cdot & \cdot \\ \cdot & \cdot & \cdot & \cdot \\ \cdot & \cdot & \cdot & \cdot \\ K_{0,n} & K_{1,n} & \dots & K_{m,n} \end{bmatrix}, \tag{2}$$

whose elements are defined depending on the edge detector operator. Convolution of a digital image with edge detector kernel is defined with its convolutional sum as:

$$B = \sum_{i=0}^{m} \sum_{j=0}^{n} A_{m-i,n-j} K_{i+1,j+1}, \tag{3}$$

where m represents image height and n represents image width in number of pixels. According to ANN type, hybrid models can be divided to:

- hybrid models designed with MLP and
- hybrid models designed with CNN.

Dataflow diagram of MLP-based hybrid models is shown in Fig. 3.

The idea behind MLP-based hybrid models is to utilize edge detectors for feature extraction. Such procedure is equivalent to convolutional layers in CNN. This step is combined with image scaling, that is equivalent to pooling layers in CNN. The detailed procedure of image downsizing is presented in [13]. Such approach allows

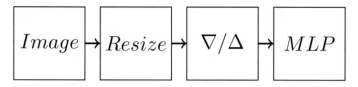

Fig. 4 Dataflow diagram for hybrid models designed with MLP

urinary bladder cancer diagnosis by using simpler ANN architecture, in comparison with commonly used CNN. On the other hand, the idea behind hybrid CNN models is to emphasize image edges in order to increase the accuracy of urinary bladder cancer recognition and, at the same time, to achieve shorter training procedures. The dataflow diagram that represents a CNN-based model is presented in Fig. 4.

According to edge detector type used, hybrid models can be divided to:

- Gradient-based hybrid models and
- Laplacian-based hybrid models.

Description of aforementioned hybrid models is given in the following sections.

3 Gradient-Based Hybrid Models

As the name suggests, gradient-based hybrid models are designed by using gradient-based edge detectors. The idea behind gradient-based edge detectors is utilization of the gradient operator on an image. Digital image can be represented with a discrete function with two variables as:

$$A = f(x, y), \tag{4}$$

where x represents width and y represents height dimension of an image. Gradient of image A can be defined as a 2-D vector of partial derivations in both dimensions:

$$\nabla A = \left(\frac{\partial A}{\partial x}, \frac{\partial A}{\partial y} \right). \tag{5}$$

From above-given statements and Eq. 5 it can be concluded that ∇A will be obtained by utilizing two different kernels, one for each image dimension. In other words, in order to obtain $\frac{\partial A}{\partial x}$, the kernel K_x will be used and in order to obtain $\frac{\partial A}{\partial y}$, kernel K_y is utilized. Aforementioned kernels are defined as equivalents to partial derivatives, or:

$$K_x \iff \frac{\partial A}{\partial x} \tag{6}$$

and:

$$K_y \iff \frac{\partial A}{\partial y}. \tag{7}$$

It can be seen that by utilizing above presented kernels, partial derivatives of image are achieved as convolutions of image with kernels:

$$\frac{\partial A}{\partial x} = A * K_x \tag{8}$$

and:

$$\frac{\partial A}{\partial y} = A * K_y. \tag{9}$$

Since gradient of an image is a 2-D vector it can be defined with an associated magnitude:

$$B = \sqrt{\left(\frac{\partial A}{\partial x}\right)^2 + \left(\frac{\partial A}{\partial y}\right)^2} \tag{10}$$

and phase:

$$\Phi = \arctan\left(\frac{\frac{\partial A}{\partial y}}{\frac{\partial A}{\partial x}}\right). \tag{11}$$

For purposes of edge detection, gradient magnitude is used to represent output image. Partial derivatives of images are achieved by using different methodologies. Most common approaches to edge detection using gradient magnitude are:

- Roberts edge detector,
- Prewitt edge detector and
- Sobel edge detector.

Above listed methods are described in next subsections, where a detailed description of kernel construction procedure is given for each detector, respectively.

3.1 Roberts Hybrid Model

Roberts hybrid model is based on the utilization of Roberts edge detector or Roberts cross. This approach represents a simple method for image partial derivative approximation [14]. Using numerical formulations, image gradient can be approximated with:

$$\nabla A \approx ((A_{i,j} - A_{i+1,j+1}), (A_{i+1,j} - A_{i,j+1})), \tag{12}$$

where the first difference represents a partial derivative on x and the second difference a partial derivative on y coordinate. If partial derivatives are approximated around interpolated pixel $P(i + 0.5, j + 0.5)$ [14], kernels:

$$R_x = \begin{bmatrix} 1 & 0 \\ 0 & -1 \end{bmatrix} \tag{13}$$

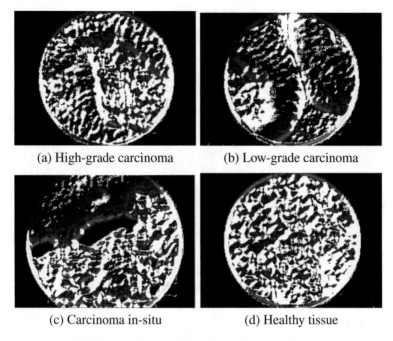

(a) High-grade carcinoma   (b) Low-grade carcinoma

(c) Carcinoma in-situ   (d) Healthy tissue

Fig. 5 Overview of bladder tissue images with applied Roberts edge detector

and:

$$R_y = \begin{bmatrix} 0 & 1 \\ -1 & 0 \end{bmatrix}, \tag{14}$$

are constructed. According to kernels presented in Eqs. (13) and (14) gradient magnitude can be calculated to represent a new image with extracted edges. Examples of all four class members with applied Roberts edge detector are presented in Fig. 5.

3.2 Prewitt Hybrid Model

Another approach to gradient-based edge detection is to use Prewitt edge detector that is based on finite difference approximation. For the case of 1-D function, such approximation can be written as:

$$f'(x) \approx \frac{f(x+1) - f(x-1)}{2}. \tag{15}$$

In this way, approximations of partial derivatives for 2-D image can be expressed as:

$$\frac{\partial A}{\partial x} \approx \frac{A_{i+1,j} - A_{i-1,j}}{2}, \tag{16}$$

for x coordinate, and:

$$\frac{\partial A}{\partial y} \approx \frac{A_{i,j+1} - A_{i,j-1}}{2}. \tag{17}$$

for y coordinate. Following the same logic as in the case of Roberts edge detector, edge detector kernels can be constructed around pixel $A_{i,j}$ as:

$$P_x = \frac{1}{2}\begin{bmatrix} -1 & 0 & 1 \end{bmatrix}, \tag{18}$$

for x coordinate, and:

$$P_y = \frac{1}{2}\begin{bmatrix} -1 \\ 0 \\ 1 \end{bmatrix}, \tag{19}$$

for y coordinate. Due to mathematical formalism, coefficient $\frac{1}{2}$ is added before kernels, but it can be eliminated during realization of the detector [14]. The presented kernels are quite sensitive to noise, so they are modified in order to eliminate the noise. Such modifications were made by extending the original kernels into 3×3 kernels [15], yielding:

$$P_x = \begin{bmatrix} -1 & 0 & 1 \\ -1 & 0 & 1 \\ -1 & 0 & 1 \end{bmatrix}, \tag{20}$$

for partial derivative on x coordinate and:

$$P_y = \begin{bmatrix} -1 & -1 & -1 \\ 0 & 0 & 0 \\ 1 & 1 & 1 \end{bmatrix}, \tag{21}$$

for partial derivative on y coordinate. Results of gradient magnitude for the case of Prewitt edge detector are shown in Fig. 6 for all four classes.

3.3 Sobel Hybrid Model

As another gradient-based method, Sobel edge detector is utilized. This edge detector is based on Prewitt edge detector with the difference in the middle row in the case of approximation of partial derivative in x coordinate:

$$S_x = \begin{bmatrix} -1 & 0 & 1 \\ -2 & 0 & 2 \\ -1 & 0 & 1 \end{bmatrix}, \tag{22}$$

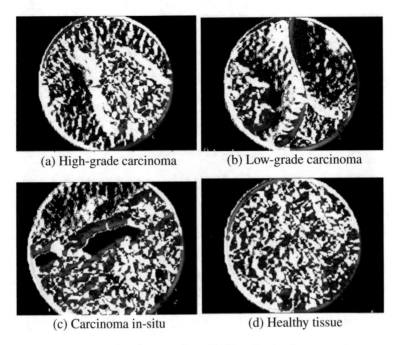

(a) High-grade carcinoma   (b) Low-grade carcinoma

(c) Carcinoma in-situ   (d) Healthy tissue

Fig. 6 Overview of bladder tissue images with applied Prewitt edge detector

and with the difference in the middle column in the case of the approximation of the partial derivative in y coordinate:

$$S_y = \begin{bmatrix} -1 & -2 & -1 \\ 0 & 0 & 0 \\ 1 & 2 & 1 \end{bmatrix}. \tag{23}$$

As it can be noticed, in both cases a modification is based on multiplication with coefficient 2. By utilization of this approach, edge pixels are more emphasized than in the case of Prewitt edge detector [16]. An overview of achieved results is given in Fig. 7 for all four classes.

4 Laplacian-Based Hybrid Model

Another approach to edge detection is utilization of Laplacian edge detector. As the name suggests, this edge detector is based on Laplacian of an image that can be defined as equivalent to second derivative. For the case of an image, Laplacian edge detector represents a gradient of a gradient, defined as [13]:

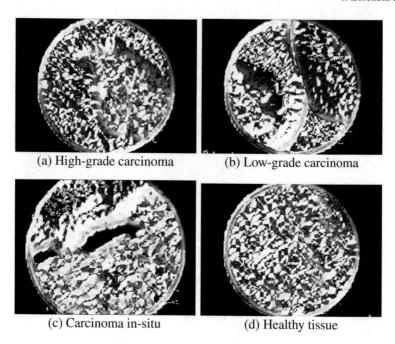

(a) High-grade carcinoma   (b) Low-grade carcinoma

(c) Carcinoma in-situ   (d) Healthy tissue

Fig. 7 Overview of bladder tissue images with applied Sobel edge detector

$$\Delta A = \nabla \cdot \nabla A = \frac{\partial^2 A}{\partial x^2} + \frac{\partial^2 A}{\partial y^2}, \tag{24}$$

where Δ represents the Laplace operator. Second partial derivative of an image on x coordinate can be approximate as a finite difference of two first partial derivatives:

$$\left(\frac{\partial^2 A}{\partial x^2}\right)_{i,j} \approx \left(\frac{\partial A}{\partial x}\right)_{i,j} - \left(\frac{\partial A}{\partial x}\right)_{i-1,j}. \tag{25}$$

Both first partial derivatives can be approximated with:

$$\left(\frac{\partial A}{\partial x}\right)_{i,j} \approx A_{i+1,j} - A_{i,j} \tag{26}$$

and:

$$\left(\frac{\partial A}{\partial x}\right)_{i-1,j} \approx A_{i,j} - A_{i-1,j}. \tag{27}$$

By combining Eqs. (25), (26) and (27), second partial derivative of an image on x coordinate can be approximated with:

$$\left(\frac{\partial^2 A}{\partial x^2}\right)_{i,j} \approx (A_{i+1,j} - A_{i,j}) - (A_{i,j} - A_{i-1,j}), \tag{28}$$

which yields:

$$\left(\frac{\partial^2 A}{\partial x^2}\right)_{i,j} \approx A_{i+1,j} - 2 \cdot A_{i,j} + A_{i-1,j}. \tag{29}$$

For the case of the second partial derivative on y coordinate, the procedure is similar. Second partial derivative can be approximated with the finite difference of first partial derivatives as:

$$\left(\frac{\partial^2 A}{\partial y^2}\right)_{i,j} \approx \left(\frac{\partial A}{\partial y_{i,j}}\right)_{i,j} - \left(\frac{\partial A}{\partial x}\right)_{i,j-1}. \tag{30}$$

First partial derivatives can also be approximated with finite differences giving:

$$\left(\frac{\partial^2 A}{\partial y^2}\right)_{i,j} \approx A_{i,j+1} - 2 \cdot A_{i,j} + A_{i,j-1}. \tag{31}$$

Combination of two approximations of second derivatives gives an approximation of Laplacian of an image:

$$\Delta A \approx A_{i+1,j} + A_{i,j+1} - 4 \cdot A_{i,j} + A_{i-1,j} + A_{i,j-1}. \tag{32}$$

By using above presented equation, Laplacian kernel can be construed as:

$$L = \begin{bmatrix} 0 & 1 & 0 \\ 1 & -4 & 1 \\ 0 & 1 & 0 \end{bmatrix}. \tag{33}$$

An overview of the results achieved with Laplacian edge detector is given in Fig. 8 for all four classes.

It is important to notice that Laplacian edge detector is designed with just one kernel, in difference with gradient-based edge detectors that are designed with two kernels. This difference comes from the fact that gradient is a vector, while Laplacian is a scalar. The result of this property is a difference in execution times of these two types of edge detectors.

5 Model Evaluation

In order to determine the optimal ANN model, it is necessary to define model evaluation criteria. The standard procedure for classifier evaluation is receiver operating characteristic (ROC) analysis that is used for binary classifier evaluation [17]. ROC

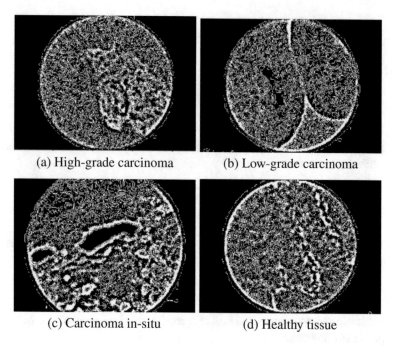

(a) High-grade carcinoma   (b) Low-grade carcinoma

(c) Carcinoma in-situ   (d) Healthy tissue

Fig. 8 Overview of bladder tissue images with applied Laplacian edge detector

analysis is based on a graphical representation of classification performances that is called ROC curve. In most cases ROC curve is constructed in such manner that $x-$axis represents true positive rate and $y-$axis represents false positive rate. As a single scalar value for measuring classification performances area under the ROC curve (AUC) is introduced. It is interesting to notice that in this case, both binary classes are taken into consideration for classifier evaluation. For these reasons, AUC is a dominant classification performance measure [17]. For cases of multi-class classification, modified ROC analysis is performed. In the particular case of urinary bladder cancer diagnosis, the data set is divided in four classes. Classification performances could be visually interpreted by using confusion matrix. In the case of four-class classification, confusion matrix C can be defined as:

$$M = \begin{bmatrix} M_{11} & M_{12} & M_{13} & M_{14} \\ M_{21} & M_{22} & M_{23} & M_{24} \\ M_{31} & M_{32} & M_{33} & M_{34} \\ M_{41} & M_{42} & M_{43} & M_{44} \end{bmatrix}, \tag{34}$$

where elements on a main diagonal (M_{11}, M_{22}, M_{33} and M_{44}) represent a fraction of the correct classification in a particular class, while the other elements represent fractions of incorrect classifications in all other classes. An example of a confusion matrix for the case of urinary bladder cancer diagnosis is presented in Fig. 9.

Fig. 9 An example of four class confusion matrix used for ANN model evaluation

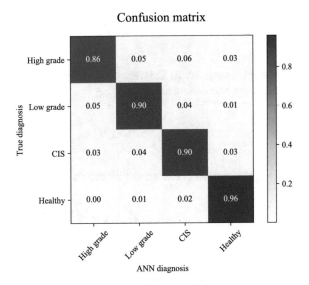

It can be noticed that, in this case, construction of a ROC curve is not straightforward as it is in the case of binary classifier. For these reasons, method for AUC value averaging must be introduced. In the case of a binary classifier, the horizontal axis of a ROC curve represents false positive rate (FPR), while vertical axis represents true positive rate (TPR). TPR can be calculated as:

$$TPR = \frac{P_C}{P_C + P_U},\qquad(35)$$

where P_C represents the number of correct classifications into the positive class and P_U represents the number of incorrect classifications into the negative class. On the other hand, FPR can be calculated as:

$$FPR = \frac{N_U}{N_C + N_U},\qquad(36)$$

where N_U represents the number of incorrect classifications into the positive class and N_C represents the number of correct classifications into the negative class. Due to inability for standard binary ROC AUC approach to evaluation of non-binary classifiers, some variations must be applied. For these reasons average AUC values are introduced. For purposes of this chapter, two types of average AUC value are introduced, and these are:

- micro-average AUC and
- macro-average AUC.

As it is in the case of a binary ROC curve, for the case of average values, ROC curves are designed by using TPR and FPR. For the case of micro-average, TPR

can be calculated as:

$$TPR_{micro} = \frac{tr(M)}{G(M)},\tag{37}$$

where $tr(M)$ represents the trace of a confusion matrix that can be defined as a sum of main diagonal elements, or:

$$tr(M) = \sum_{m=1}^{N} M_{mm}\tag{38}$$

and $G(M)$ represents the sum of all matrix elements, or:

$$G(M) = \sum_{m=1}^{N} \sum_{n=1}^{N} M_{mn}.\tag{39}$$

FPR can be expressed as a fraction of all false classifications and a sum of all matrix elements. Such fraction can be written as:

$$FPR_{micro} = \frac{G(M) - tr(M)}{G(M)}.\tag{40}$$

For the case of macro-average ROC analysis, TPR can be calculated as an average of TPR values of individual classes:

$$TPR_{macro} = \frac{1}{N} \sum_{n=1}^{N} TPR_n.\tag{41}$$

For example, TPR for the case of the first class can be defined as a fraction between a number of correctly classified images and a total number of images that are members of the first class. Such expression can be written as:

$$TPR_1 = \frac{M_{11}}{M_{11} + M_{12} + M_{13} + M_{14}}.\tag{42}$$

On the other hand, macro-average FPR can be defined as an average of individual FPR values:

$$FPR_{macro} = \frac{1}{N} \sum_{n=1}^{N} FPR_n.\tag{43}$$

For example, FPR for the case of the first class can be defined as a fraction of a number of images that are incorrectly classified as members of the first class and total number of images that are classified as members of the first class. Such expression can be written as:

$$FPR_1 = \frac{M_{21} + M_{31} + M_{41}}{G(M) - (M_{11} + M_{21} + M_{31} + M_{41})}. \tag{44}$$

Both $\overline{AUC}_{micro}$ and $\overline{AUC}_{macro}$ will be used for evaluation of various architectures of hybrid ANN models for urinary bladder cancer diagnosis.

6 Definition of Hyperparemeter Space

In order to determine the optimal ANN architecture, the grid-search procedure is performed. Grid search takes all possible combinations of hyperparameters and creates a model for each of those hyperparameter combinations, which is then trained and evaluated. Such procedure is based on extensive search through the defined hyperparameter space, that is defined according to theoretical knowledge of model selection [18, 19]. The constraints are particularly introduced for the number of neurons and layers. An overview of possible hyperparameters for the case of CNN-based hybrid models is given in Table 1. Hyperparameters are the values which define the architecture of the ANN/CNN. Choice of hyperparameters has a great influence on the performance of the ANN [18, 20].

By using above presented hyperparameters, a grid search procedure will be performed for cases of all four hybrid CNN models. Alongside that, six different configurations of CNN-based models will be utilized. These configurations are presented in Table 2.

From architectures presented above, one with best performances will be determined for the case of all four hybrid models. For the case of MLP-based hybrid models, the hyperparameter space is defined in Table 3.

Alongside hyperparameter search, four different sizes of input images will be used, and these are:

- 30×30,
- 50×50,
- 100×100 and
- 200×200.

Table 1 Definition of hyperparameter space for the case of CNN-based hybrid models

Hyperparameter	Values
Activation function per layer	ReLU, Tanh, Logistic Sigmoid
Number of feature maps per layer	2, 5, 10, 20
Kernel size per layer	[3, 3] [5, 5] [10, 10]
Pooling size per layer	[2, 2] [5, 5]
Number of neurons per layer	10, 20, 40, 80
Number of epochs	1, 10, 20, 50
Batch size	4, 8, 16, 32, 64

Table 2 Used CNN configurations with presented numbers of convolutional, pooling and fully connected layers

CNN	Convolutional layers	Pooling layers	Fully connected layers
1.	1	1	1
2.	1	1	2
3.	1	1	3
4.	2	2	1
5.	2	2	2
6.	2	2	3

Table 3 Definition of hyperparameter space for the case of MLP-based hybrid models

Hyperparameter	Values
Activation functions per layer	ReLU, Tanh, Logistic Sigmoid
Number of neurons per layer	10, 20, 40, 80
Number of hidden layers	1, 2, 3, 4, 5
Number of epochs	1, 10, 20, 50
Batch size	16, 32, 64

As it is in the case of CNN-based models, the best models for each edge detector will be presented.

7 Hybrid Models Comparison

In order to determine the optimal configuration of hybrid ANN model, it is necessary to compare results achieved with each of the presented models. As presented in previous sections, the grid-search procedure is used for determination of the model with the highest classification performances. Such procedure is performed for each of eight described hybrid models. As evaluation measures, $\overline{AUC}_{micro}$, $\overline{AUC}_{macro}$ and average edge detector execution time are used. When grid search procedure is performed for determination of CNN-based hybrid models, architectures presented in Table 4 are obtained.

It can be noticed that all achieved models are designed with some combination of layers that is one of lower complexity. In other words, grid-search algorithm is trending towards the use of simpler models. From $\overline{AUC}_{micro}$ and $\overline{AUC}_{macro}$ values presented in Table 5 similar classification performances are noticeable. Furthermore, it can be seen that higher values are achieved if $\overline{AUC}_{macro}$ is used as a classification measure.

When grid-search algorithm is performed for determining hyperparameters of MLP-based hybrid models, architectures presented in Table 6 are obtained. As it is

Table 4 Overview of hyperparameters used for design of hybrid CNN models for all four edge detectors

Hyperparameter	Hybrid model			
	Roberts	Prewitt	Sobel	Laplacian
Feature maps (first layer)	5	2	2	2
Activation function (first layer)	Tanh	Tanh	ReLU	ReLU
Kernel size (first layer)	[5, 5]	[3, 3]	[3, 3]	[3, 3]
Pooling size (first layer)	[2, 2]	[2, 2]	[2, 2]	[2, 2]
Feature maps (second layer)	10	–	–	2
Activation function (second layer)	ReLU	–	–	Sigmoid
Kernel size (second layer)	[5, 5]	–	–	[5, 5]
Pooling size (second layer)	[2, 2]	–	–	[2, 2]
Number of neurons (first fully connected layer)	40	10	20	40
Activation function (first fully conected layer)	Tanh	Tanh	ReLU	Tanh
Number of neurons (second fully connected layer)	–	10	10	20
Activation function (second fully conected layer)	–	Sigmoid	Sigmoid	Sigmoid
Number of neurons (third fully connected layer)	–	10	–	–
Activation function (third fully connected layer)	–	ReLU	–	–
Solver	Adam	SGD	SGD	Adam
Number of epochs	10	10	20	10
Batch size	32	16	32	32

Table 5 Overview of achieved results with CNN-based hybrid models

Hybrid model	AUC_{micro}	$\overline{AUC}_{macro}$
Roberts	0.990	0.995
Prewitt	0.987	0.994
Sobel	0.990	0.997
Laplacian	0.992	0.997

Table 6 Overview of hyperparameters of optimal hybrid MLP models for all four edge detectors

Hyperparameter	Hybrid model			
	Roberts	Prewitt	Sobel	Laplacian
Number of neurons (first hidden layer)	10	80	10	10
Activation function (first hidden layer)	ReLU	ReLU	ReLU	ReLU
Number of neurons (second hidden layer)	80	80	80	80
Activation function (second hidden layer)	Sigmoid	Sigmoid	Tanh	Sigmoid
Number of neurons (third hidden layer)	40	–	10	–
Activation function (third hidden layer)	Sigmoid	–	ReLU	–
Solver	Adam	SGD	Adam	Adam
Number of epochs	10	20	10	10
Batch size	16	32	4	4
Image size	100×100	50×50	50×50	100×100

in the case of CNN-based models, MLP-based models are also converging to the models of intermediate complexity. Such conclusion could be driven from the fact that all obtained models are constructed with two or three hidden layers.

As another interesting property, the fact of algorithm converging into the intermediate image size can be noticed. Lower classification performances are achieved if larger input images are used. By using above presented models, classification performances represented with $\overline{AUC}_{micro}$ and $\overline{AUC}_{macro}$ are achieved. These values are presented in Table 7.

Table 7 Overview of achieved results with MLP-based hybrid models

Hybrid model	$\overline{AUC}_{micro}$	$\overline{AUC}_{macro}$
Roberts	0.987	0.996
Prewitt	0.986	0.994
Sobel	0.987	0.996
Laplacian	0.973	0.977

Fig. 10 Comparison of $\overline{AUC}_{micro}$ values achieved with CNN-based and MLP-based hybrid models

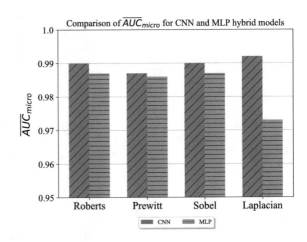

When achieved results are compared, it can be noticed that CNN-based models are performing better from the $\overline{AUC}_{micro}$ standpoint. It can also be noticed that the highest $\overline{AUC}_{micro}$ value is achieved if Laplacian-based CNN model is utilized. On the other hand, Laplacian-based MLP models are showing the lowest performances from $\overline{AUC}_{micro}$ standpoint. Such comparison is presented in Fig. 10.

When $\overline{AUC}_{macro}$ is used for model comparison, it can be noticed that the same results are achieved in both cases of Sobel and Laplacian utilization. It is interesting to notice that if $\overline{AUC}_{macro}$ is used as a classification measure, Roberts-based MLP model is showing higher performances that its CNN version. The gap between CNN and MLP model is still noticeable for the case of Laplacian-based models. Such property is shown in Fig. 11.

As the last measure for model comparison, the average edge detector execution time can be used. In this case, it can be noticed that Laplacian-based model achieved significantly shorter execution times, in comparison to gradient-based edge detectors. Such property is logical because the fact of two-kernel nature of gradient and one-kernel nature of Laplacian edge detectors. It is also interesting to notice that Roberts edge detector requires shorter time to execute, in comparison to other gradient detectors, as presented in Fig. 12. This characteristic is the result of simpler kernels.

Fig. 11 Comparison of
$\overline{AUC}_{macro}$ values achieved
with CNN-based and
MLP-based hybrid models

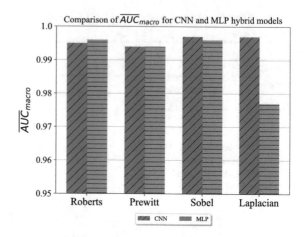

Fig. 12 Comparison of
average execution times for
all four edge detectors

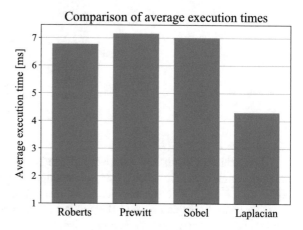

When all presented facts are summed up, it can be concluded that the lowest
requirements in regard of computing resources are achieved if Laplacian-based
hybrid models are utilized. This conclusion is supported by the fact that this detector requires the shortest execution times. On the other hand, Laplacian-based CNN
model is also achieving the highest classification performances. A limiting fact for
Laplacian-based CNN model is higher model complexity, in comparison to other
CNN-based models. A simpler CNN model could be used if Sobel hybrid model is
used. Such approach, despite lower model complexity, requires more complex image
pre-processing.

Acknowledgments This research has been (partly) supported by the CEEPUS network CIII-HR-
0108, European Regional Development Fund under the grant KK.01.1.1.01.0009 (DATACROSS),
project CEKOM under the grant KK.01.2.2.03.0004 and University of Rijeka scientific grant uniri-
tehnic-18-275-1447.

References

1. Janković, Slavenka, Radosavljević, Vladan: Risk factors for bladder cancer. Tumori J. **93**(1), 4–12 (2007)
2. Al Bahili, H.: General Surgery & Urology: key principles and clinical surgery in one book. Saudi Med. J. **36** (2015)
3. Hashim, H. Abrams, P., Dmochowski, R.: The Handbook of Office Urological Procedures. Springer (2008)
4. Duty, B.D., Conlin, M.J.: Principles of urologic endoscopy. In: Campbell-Walsh Urology, 11th edn. Elsevier, Philadelphia, PA (2016)
5. Lerner, S.P., Liu, H., Wu, M.-F., Thomas, Y.K., Witjes, J.A.: Fluorescence and white light cystoscopy for detection of carcinoma in situ of the urinary bladder. In: Urologic Oncology: Seminars and Original Investigations, vol. 30, pp. 285–289. Elsevier (2012)
6. Babjuk, M., Böhle, A., Burger, M., Capoun, O., Cohen, D., Compérat, E.M., Hernández, V., Kaasinen, E., Palou, J., Rouprêt, M., et al.: Eau guidelines on non–muscle-invasive urothelial carcinoma of the bladder: update 2016. Eur. Urol. **71**(3), 447–461 (2017)
7. Lorencin, I., Anđelić, N., Mrzljak, V., Car, Z.: Marine objects recognition using convolutional neural networks. NAŠE MORE: znanstveno-stručni časopis za more i pomorstvo **66**(3), 112–119 (2019)
8. Dong, C., Loy, C.C., Tang, X.: Accelerating the super-resolution convolutional neural network. In European Conference on Computer Vision, pp. 391–407. Springer, Berlin (2016)
9. Kang, E., Min, J., Ye, J.C.: A deep convolutional neural network using directional wavelets for low-dose X-ray CT reconstruction. Med. Phys. **44**(10), e360–e375 (2017)
10. Han, Xiao: MR-based synthetic CT generation using a deep convolutional neural network method. Med. Phys. **44**(4), 1408–1419 (2017)
11. Anwar, S.M., Majid, M., Qayyum, A., Awais, M., Alnowami, M., Khan, M.K.: Medical image analysis using convolutional neural networks: a review. J. Med. Syst. **42**(11), 226 (2018)
12. Zhang, X., Zhou, X., Lin, M., Sun, J.: Shufflenet: an extremely efficient convolutional neural network for mobile devices. In: Proceedings of the IEEE Conference on Computer Vision and Pattern Recognition, pp. 6848–6856 (2018)
13. Lorencin, I., Anđelić, N., Španjol, J., Car, Z.: Using multi-layer perceptron with laplacian edge detector for bladder cancer diagnosis. Artif. Intell. Med. **102**, 101746 (2020)
14. Muthukrishnan, R., Radha, M.: Edge detection techniques for image segmentation. Int. J. Comput. Sci. Inf. Technol. **3**(6), 259 (2011)
15. Morse, B.S.: Lectures in image processing and computer vision. Department of Computer Science, Brigham Young University (1995)
16. Gao, W., Zhang, X., Yang, L., Liu, H.: An improved sobel edge detection. In: 2010 3rd International Conference on Computer Science and Information Technology, vol. 5, pp. 67–71. IEEE (2010)
17. Fawcett, Tom: An introduction to ROC analysis. Pattern Recogn. Lett. **27**(8), 861–874 (2006)
18. Hastie, T., Tibshirani, R., Friedman, J.: The Elements of Statistical Learning: Data Mining, Inference, and Prediction. Springer Science & Business Media (2009)
19. Bishop, C.M.: Pattern Recognition and Machine Learning. Springer, Berlin (2006)
20. Lorencin, I., Anđelić, N., Mrzljak, V., Car, Z.: Genetic algorithm approach to design of multi-layer perceptron for combined cycle power plant electrical power output estimation. Energies **12**(22), 4352 (2019)

Predicting Building-Related Carbon Emissions: A Test of Machine Learning Models

Emmanuel B. Boateng, Emmanuella A. Twumasi, Amos Darko,
Mershack O. Tetteh, and Albert P. C. Chan

1 Introduction

Global warming, caused by the growing concentration of greenhouse gases in the atmosphere, has severe implications for both economic and human development [2]. A significant driver of global warming is CO_2 emission. Generally, CO_2 emission has progressively increased since 2009, from 31.89 billion tons to 36.14 billion tons in 2014 [33]. According to the International Energy Agency and the United Nations Environment Programme [20], the buildings and construction sector contributes almost 40% of energy and process-related emissions. As such, the buildings and construction sector produce one-third of global energy-related emissions. In China, buildings accounted for 28% of national total emissions in 2011, and this figure is

E. B. Boateng
School of Health Sciences, The University of Newcastle, Newcastle, Australia
e-mail: emmanuel.boateng@uon.edu.au

E. A. Twumasi
School of Surveying and Construction Management, College of Engineering and Built Environment, Technological University Dublin – City Campus, Dublin, Ireland
e-mail: D17127590@mytudublin.ie

A. Darko (✉) · M. O. Tetteh · A. P. C. Chan
Department of Building and Real Estate, The Hong Kong Polytechnic University, Hung Hom, Kowloon, Hong Kong, China
e-mail: amos.darko@connect.polyu.hk

M. O. Tetteh
e-mail: mershack-opoku.tetteh@connect.polyu.hk

A. P. C. Chan
e-mail: albert.chan@polyu.edu.hk

A.-E. Hassanien et al. (eds.), *Enabling AI Applications in Data Science*, Studies in Computational Intelligence 911, https://doi.org/10.1007/978-3-030-52067-0_11

estimated to reach 35% by 2020 [9]. With a rapidly growing urban population, energy consumption and CO_2 emissions are expected to rise.

Considering these, reducing CO_2 emissions in the buildings and construction sector is key to attaining the Paris Agreement commitment and the United Nations (UN) Sustainable Development Goals (SDGs) [20]. However, recent reports by the International Energy Agency and the United Nations Environment Programme [20] suggest that climate actions by the sector geared towards reducing building emissions to the desired level are unenviable. As a result, most countries and their institutions are making efforts to reduce building-related CO_2 emissions and develop low-carbon cities by introducing carbon mitigation schemes such as incentives for adopting decarbonising measures. Also, researchers have made significant efforts, however, according to a recent review [23], much of the investigations have been focussed on energy use from buildings while studies on carbon emissions from buildings are rare.

Accurate estimation of building emissions is of prime importance for the attainment of more sustainable built environment outcomes [26]. A reliable method for predicting emissions could serve as a tool for international organisations and environmental policymakers to design and implement sound climate change mitigation policies [2]. However, studies have shown that carbon emissions from construction activities have been underestimated [25]. Likewise, there have been many simulation studies on buildings [23]. Nevertheless, simulated applications are not effective in estimating energy consumption from buildings [3]. An alternative to building energy simulations are statistical and machine learning (ML) algorithms [24].

On the other hand, classical statistical approaches are not appropriate for modelling the complex (e.g., non-linear, non-parametric, chaotic, etc.) behaviour of carbon emission variables [2, 15]. ML approaches have been known to yield promising results in numerous industrial settings. Whereas linear regression and ML techniques such as artificial neural networks (ANNs) and support vector machines (SVMs) are popular in estimating building energy consumption, tree-based models are seldomly used. However, recent advancement in ML algorithms has resulted in the development of robust ML approaches for modelling and forecasting. As such, different and novel ML algorithms need to be continuously tested in the quest to attain high-quality forecasts for providing realistic foresights and making better-informed decisions concerning sustainability in the built environment.

There is limited research on robust ML algorithms that can accurately predict building-related CO_2 emissions in a timely manner. This chapter evaluates and compares the accuracy and computational costs in predicting building-related CO_2 emissions, using China's emissions as illustration. Six ML algorithms: decision tree (DT), random forest (RF), support vector regression (SVR), extreme gradient boosting (XGBoost), k-nearest neighbour (KNN) and adaptive boosting (AdaBoost) were evaluated. The rest of the chapter is organised as follows. Section 2 describes the data and methodology used in this study, Sect. 3 presents the results and discussions. Concluding remarks are given in Sect. 4.

2 Data and Methodology

2.1 Data Description and Pre-processing

Time series data spanning over the period 1971–2014 on building-related CO_2 emissions (CO_2 emissions from residential buildings and commercial services, % of total fuel combustion), population size (total), R&D (Trademark applications, total), urbanisation (urban population), GDP (GDP per capita, current US$), and energy use (kg of oil equivalent per capita) were sourced from [32] and International Energy Agency [18, 19] repositories. However, to develop accurate models, the study follows Shahbaz et al. [27], Bannor and Acheampong [5] to use quadratic-sum method to transform the annual data from low-frequency to high-frequency data. Hence, this study applied quarterly data between 1971Q1 and 2014Q4, representing 176 quarters. Literature on fundamental variables that influence building emissions at the macro-level was used in informing variable selection. See Acheampong and Boateng [2] for extensive literature on the variables.

Multivariate imputation by chained equations technique was used in treating missing data [8]. Similar approach was used by Bannor and Acheampong [5] as it has been known to give reasonable replacements than mean, median, and modal imputations. Upon prior experimentations, the input data was standardised while the output data was normalised. This pre-processing of data was to eliminate instances of one variable dominating the other [6], as the variables used in this study have different units [2]. Moreover, this process eases the severe computations performed by ML models. 80% (140 observations) of the data was used in training each model, while the remaining 20% (36 observations) were used for validation. Same data ratios were applied in previous studies (e.g. [1, 2]). Table 1 presents the descriptive statistics for all variables used in this study.

Table 1 Descriptive statistics

Variables	Mean	SD	Min	Max
Building emissions	13.61253	5.937167	5.19005	21.64584
Urbanisation	3.73e+08	1.84e+08	1.45e+08	7.48e+08
R&D	373691.7	504102.1	5627.75	2206486
Population size	1.14e+09	1.60e+08	8.33e+08	1.37e+09
GDP	1675.802	1668.072	237.2819	6253.393
Energy use	986.5826	522.8532	459.1008	2238.488

2.2 Machine Learning Estimation Algorithms

2.2.1 Adaptive Boosting (AdaBoost)

AdaBoost was introduced in the ninety's by Freund and Schapire [14]. The algorithm uses an ensemble learning technique to combine several "weak learners" to form a "strong learner" in yielding high-quality and accurate predictions. This combination of weak learners is essential as a single learner is governed by rules that are not strong enough to make predictions. These weak learners are mostly termed as decision tree stumps. Larger stumps have much influence in making decisions than smaller ones based on the "majority rule". A key advantage of AdaBoost is that it is more resistant to overfitting than many ML algorithms.

Upon the initial iteration, equal weights are assigned to individual samples by the base leaner or first decision stump. For each subsequent iteration, false predictions are assigned to the subsequent learner with greater weights while there is a reduction in weights allocated to correct predictions. This process of assigning bigger weights suggest a more significant value is placed on engaging false predictions to improve generalisability. The procedure is repeated sequentially until the AdaBoost algorithm deems that all observations have been rightly predicted. Due to their efficiency, these algorithms are one of the most powerful ML tools and have been used in diverse engineering and industrial fields. Despite its accolades in the ML arena, it has often been used in classification rather than regression problems. The final hypothesis of the AdaBoost regression model is generated in Eq. (1) as:

$$h_f(x) = \inf \left\{ y \in Y : \sum_{t:h_t(x) \leq y} \log\left(\frac{1}{\beta_t}\right) \geq \frac{1}{2} \sum_t \log\left(\frac{1}{\beta_t}\right) \right\} \quad (1)$$

where β_t is a measure of the confidence in the predictor, which is computed using $\beta_t = \varepsilon_t/(1 - \varepsilon_t)$. ε_t is the error rate of $h_t(x)$. Time $t = 1, ..T$, where T is the number of rounds/iterations, and y consists of the label space. The loss in the ensemble could be computed using a linear, exponential, or square loss function [29]. In summary, Ada-Boost adjusts adaptively to the misclassified or prediction errors re-turned by the weak leaner [14].

2.2.2 Decision Tree (DT)

Decision trees (DTs) are non-parametric ML tools that are used for both regression and classification tasks. DTs are simple and understandable, as the final tree can explain how a prediction was yielded. They comprise of a supervised learning algorithm and are advantageous in handling smaller data sets than deep learning models. These algorithms have been applied and attained successes in many fields due to their efficiency and interpretability [30, 34]. Generally, a decision tree consists of branches (splits) and nodes, where a node could be a leaf, internal, or root node. The branches influence the complexity of the model; simpler models are less likely to face fit problems. Through an iterative process, a tree is built from the root to the leaves. The root node is the first node, the internal node (non-terminal) relates to questions asked about features of variable X, and the leaf node also termed as the terminal node contains the prediction or output.

A cost function is used in splitting a node, where the lowest cost is determined as the best cost. Specifically, for regression, on each subset of data, the DT algorithm computes a mean squared error (MSE), and the tree with the least MSE is designated as the point of split. Whereas, metrics such as cross-entropy or Gini index are used to evaluate splits for classification problems. For cross-entropy (CE) and Gini index (GI), partitioning the dataset into subsets continues at each internal node based on an assessment of the functions [17]:

$$CE = -\sum_{k=1}^{K} P_m(k) log P_m(k) \tag{2}$$

$$GI = \sum_{k \neq k'} P_m(k) P_m(k') = \sum_{k=1}^{K} P_m(k)(1 - P_m(k)) \tag{3}$$

where $P_m(k)$ is the ratio of class k observations in node m. The process of splitting continues until each node reaches a specified minimum number of training samples or a maximum tree depth.

2.2.3 Extreme Gradient Boosting (XGBoost)

XGBoost is a scalable end-to-end tree-boosting algorithm [10]. It can be used to solve classification, regression, and ranking problems. This algorithm performs parallel tree learning using a novel sparsity-aware system [10]. XGBoost is fast, performs out-of-core computations, and has many tunable parameters, hence can be tweaked for improving model fit. Given these benefits, XGBoost has been the go-to model for most winning teams in ML hackathons and competitions such as Kaggle when dealing with structured data. The fundamental factor behind the triumphs of XGBoost is its scalability in all circumstances [10]. Because the algorithm yields state-of-the-art results when solving a wide range of problems by using an optimal amount of resources to train the model. XGBoost uses the gradient boosting framework at its core; as such, there are several similarities between the gradient boosting algorithm and XGBoost.

A leaf node in an XGBoost model signifies a decision. The predicted outcome is the sum of the predicted scores by K trees [11] and is expressed in Eq. (4).

$$\hat{y}_i = \sum_{k=1}^{K} f_k(x_i), f_k \in F \tag{4}$$

where F is the space of function containing all possible regression trees, f is the function in the functional space, x_i is the i-th training sample, and $f_k(x_i)$ is the prediction from the k-th decision tree. The objective function to be optimised is expressed in Eq. (5).

$$obj(\theta) = \sum_{i}^{n} l(y_i \hat{y}_i) + \sum_{k=1}^{K} \Omega(f_k) \tag{5}$$

where $\sum_{i}^{n} l(y_i, \hat{y}_i)$ is a differential loss function that controls the predictive accuracy, $\sum_{k=1}^{k} \Omega(f_x)$ and penalises the complexity of the model. Striving for simplicity prevents over-fitting. The function that controls the complexity is formulated in Eq. (6).

$$\Omega(f) = \gamma^T + \frac{1}{2}\lambda \sum_{j=1}^{T} \omega_j^2 \tag{6}$$

where γ and λ are user-configurable parameters, T is the number of leaves, and ω_j^2 is the score on the j-th leaf. The leaf is split and gains a score using Eq. (7).

$$gain = \frac{1}{2}\left[\frac{\left(\sum_{i \in I_L} g_i\right)^2}{\sum_{i \in I_L} h_i + \lambda} + \frac{\left(\sum_{i \in I_R} g_i\right)^2}{\sum_{i \in I_R} h_i + \lambda} - \frac{\left(\sum_{i \in I} g_i\right)^2}{\sum_{i \in I} h_i + \lambda} \right] - \gamma \tag{7}$$

where I_L and I_R are the instance sets of left and right nodes after the split, and $g_i = \partial_{\hat{y}(t-1)} l(y_i, \hat{y}^{(t-1)})$ and $h_i = \partial_{\hat{y}(t-1)}^2 l(y_i, \hat{y}^{(t-1)})$ are the first and second-order gradient statistics on the loss function.

2.2.4   K-Nearest Neighbour (KNN)

The KNN is a simple and non-parametric tool applied to regression and classification problems. The algorithm is often referred to as the lazy learning technique. The theory of KNN assumes that input observations in local neighbourhoods will tend to have similar outcomes. The algorithm performs reasonably well with low-dimensional data [17]. For classification, the intuition behind the algorithm is to locate the distances between the query and all the data points, choosing a fixed number (k) of its closest neighbours in the feature space, and then selects the most frequent label. On the other hand, KNN regression estimates the value of the target $\hat{y}$ of x as the average of the labels of its nearest neighbours:

$$\hat{y} = \frac{1}{k} \sum_{i=1}^{k} y_i(x) \tag{8}$$

Distance measures such as Euclidean, Manhattan, Minkowski, and Hamming can also be implemented. With the KNN algorithm, selecting the number of k is important since it greatly affects the fit of the model. Cross-validation approaches are highly recommended in determining the optimal number of k.

2.2.5 Random Forest (RF)

RF is an ensemble learning technique that combines numerous DTs to efficiently classify or predict an outcome [7]. This process of combination also referred to as bagging involves training each DT with a random sample by replacement from the calibration dataset. Bagging minimises the variance of the base learner, however, has minimal influence on the bias. Each random sample grows a tree; with the best split of each node selected among a random subset of m_{try}. The best split among the selected predictors is chosen based on the lowest MSE. A clear distinguishing element of RF from DT is the random sampling of training observations and random subsets of candidate variables for splitting nodes. The final prediction is derived by averaging the predictions of previous K trees as:

$$f(x) = \frac{1}{K} \sum_{k=1}^{K} T(x) \tag{9}$$

Considering the average of numerous DT predictions, there is a significantly lower risk of overfitting. RFs are well known for their capacity to handle huge numbers of variables with few samples and are advantageous in evaluating variable importance [16].

2.2.6 Support Vector Regression (SVR)

SVR is a kind of support vector machine (SVM) that supports linear and non-linear regression in a higher-dimensional space. This kernel-based model has advantages in high dimensional spaces since its optimisation does not depend on the dimensionality of the input area [12]. Based on Vapnik [31] concept of support vectors, Drucker et al. [12] developed the SVR technique. The support vector algorithm is contemporarily used in industrial settings due to its sound inclination towards real-world applications [28]. Using a supervised learning approach, the SVR trains by employing a symmetrical loss function that similarly penalises misestimates [4]. In SVC (support vector classification), model complexity is minimised with the heuristic that all observations are correctly classified, and for SVR, minimal deviation of the predicted value from the actual value is expected. The decision function of an SVR is expressed as:

$$f(x) = \sum_{i=1}^{n} (\alpha_i - \alpha_j^*) K(x_i, x) + \rho \tag{10}$$

where α_i and α_i^* are the Lagrange multipliers, ρ is an independent term, and $K(x_i, x)$ is defined as the kernel function. ρ can be estimated by using the Karush-Kühn-Tucker (KKT) criteria [13, 21, 22, 28]. The kernel function could be linear (x, x'), polynomial $([\gamma(x, x') + r]^d)$, sigmoid $(\tanh[\gamma(x, x') + r])$, or radial basis function (RBF) $(\exp(-\gamma||x - x'||^2))$.

2.3 Hyperparameter Optimisation and Setup Environment

A grid-search with 10-fold cross-validation is performed to select the best parameters for each algorithm, suitable for comparison purposes. This approach shuffles and resamples the training data into 10 equal folds, fits the model on a combination of one set of hyperparameters on nine folds and tests the model on the remaining fold [5]. A carefully tuned model is at lower risks of fit problems. The best score function returns the combination of hyperparameters and associated arguments suitable to develop the model. For all models, their mean cross-validation (MCV) scores are evaluated where the highest MCV score is used as the basis to select the optimal combination of hyperparameters.

Upon prior experimentations, certain hyperparameters were deemed to influence the performance of the models considerably and hence used in tuning the ML algorithms. In the case of the AdaBoost algorithm, the default base estimator, i.e., "DecisionTreeRegressor" at a maximum depth of 3 is maintained. Five estimators/trees (50, 100, 200, 300, and 350), two loss functions (linear and exponential), and ten learning rates (0.01, 0.02, 0.03, 0.04, 0.05, 0.06, 0.07, 0.1, 0.2, and 0.3) were tested over 10-folds of training data. This yielded 1000 models. For the DT algorithm, five minimum sample splits (2, 3, 5, 7, and 9), five maximum tree depths (3, 5, 7, 9, and 11), and four minimum sample leafs (1, 3, 5, and 7) were also experimented on 10-folds of training data; totalling 1000 models. The default splitter strategy, i.e., "best" is maintained to choose to best split at each node, while the maximum leaf node is set at "None". The "None" argument suggests an infinite number of leaf nodes.

For the XGBoost algorithm, five estimators (100, 200, 300, 400, and 500), five maximum tree depths (1, 3, 5, 7, and 9), two learning rates (0.1 and 0.2), and two boosters (gbtree and dart) were evaluated on 10-folds of training data, resulting in 1000 models. Similarly, 1000 models were developed for the KNN algorithm by experimenting on four number of neighbours (1, 3, 5, and 7), five number of the power parameter "p" (2, 3, 4, 5, and 7), and five leaf sizes (1, 10, 20, 30, and 40) with 10-folds of training data. The algorithm used to compute the nearest neighbours was set to "auto" to identify the most appropriate algorithm. In the case of the RF algorithm, five trees (5, 10, 20, 30, and 100), five maximum depths (1, 3, 5, 7, and 9), four minimum samples leaf (1, 3, 5, and 7) were assessed on 10-folds of calibration data, resulting in 1000 models. The maximum features parameter was maintained with the default of "auto", which equals the number of features.

For the SVR algorithm, a too-small C (regularisation parameter) coefficient will under-fit the training data, while a too-large C coefficient may over-fit the training data. As such, the value of C should be carefully selected to produce an optimal model. The gamma determines the influence of a single training sample, as well as the capacity to capture the shape of the data. The kernel function determines how a lower-dimensional data is mapped into a higher dimensional space. Ten C arguments (0.5, 0.8, 0.9, 1.0, 2.0, 3.0, 4.0, 5.0, 10.0, and 50.0), five gammas' (0.001, 0.01, 0.1, 1.0, and 2.0), and two kernels (RBF and polynomial function) were assessed on 10-folds of training data, yielding 1000 models. For each algorithm, the ideal combination of parameters and associated arguments is selected based on the model with the peak MCV score.

For hardware and software environment, we used an Intel i5-2520 M [4] at 3.2 GHz CPU, 8 GB memory operating on Ubuntu 18.04.2 LTS. Two GPUs were used, Intel 2nd Generation Core Proce, and NVIDIA GeForce GTX 1050Ti. All processing units were involved to perform the severe parallel computations in rapidly generating numerous models. Spyder (Python 3.7) was used to write and execute the programming codes.

2.4 Performance Evaluation Metrics

The accuracies of each algorithm in predicting the 36 building-related CO_2 emissions in the test data set is assessed, by evaluating the deviations between the predicted and actual emissions, models with lower errors were ranked high in terms of accuracy. The root-mean-square error (RMSE), coefficient of determination (R^2), and mean absolute percentage error (MAPE) and mean absolute error (MAE) were used in comparing the levels of deviations among the six ML algorithms. The computational efficiency of the models, specifically, the elapsed time taken during the grid-search with 10-fold cross-validation process on each algorithm was measured.

3 Results and Discussion

3.1 Development and Validation of the AdaBoost Model

Grid-search with 10-fold cross-validation process suggested that the AdaBoost model with a learning rate of 0.05, loss function as linear, and the number of estimators as 300 had the highest MCV score of 0.992434. This combination of parameters was used to configure, train, and validate the AdaBoost model. The correlation between the three parameters and their arguments can be observed in Fig. 1. Figure 2 shows that the trained AdaBoost model has its best-fit line pass through majority of the test data. This indicates that the AdaBoost model has developed good generalising capabilities. The variables X_0 to X_4 within the internal node's tallies with features of

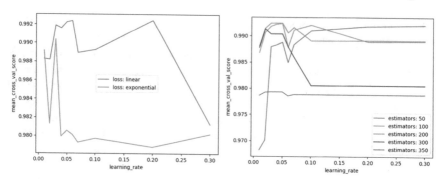

Fig. 1 Grid-search with 10-fold cross-validation results for AdaBoost models

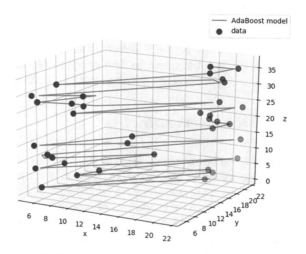

Fig. 2 3D regression of AdaBoost model

the input data set; X_0 is "energy use", X_1 is "GDP", X_2 is "population", X_3 is R&D and X_4 is "urbanisation".

3.2 Development and Validation of the DT Model

For the DT algorithm, the combination of a maximum tree depth of 9, minimum samples on a leaf node of 1, and minimum samples split of 3, produced the highest MCV of 0.997685. These parameters and their arguments were used in developing the DT model. The relationship between hyperparameters can be seen in Fig. 3. Section of the validated DT model is shown in Fig. 4.

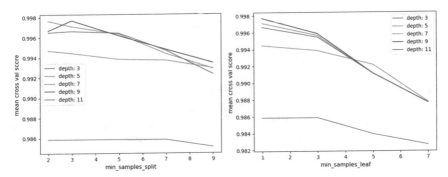

Fig. 3 Grid-search with 10-fold cross-validation results for DT models

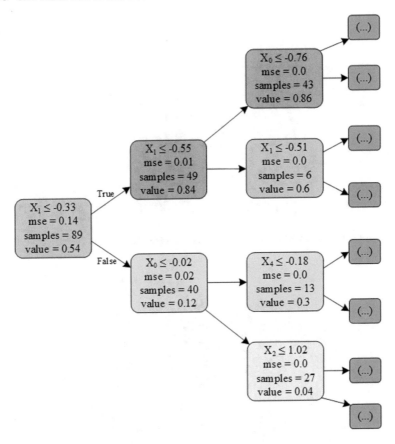

Fig. 4 Sectional plot of DT model

3.3 Development and Validation of the KNN Model

The KNN model with the parameter combination of 1 leaf size, 1 number of neighbours, and a p of 5 resulted in the highest MCV score of 0.998302. The relationship among these parameter arguments and the corresponding MCV score can be seen in Fig. 5. The leaf sizes are laid on each other in the left illustration, with the peak MCV score at a p of 5. The KNN model in Fig. 6 shows promising results, as its best-fit line accurately passes through/near the test data points.

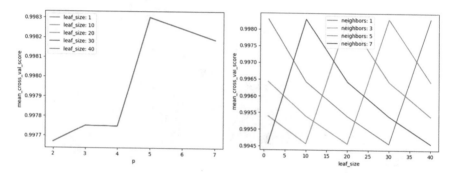

Fig. 5 Grid-search with 10-fold cross-validation results for KNN models

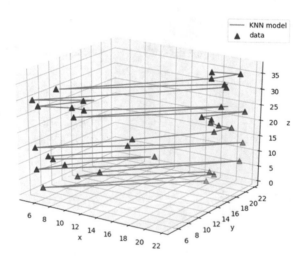

Fig. 6 3D regression of KNN model

3.4   Development and Validation of SVR Model

The SVR model with the parameter combination of a C of 1.0, gamma of 0.1, and an RBF kernel attained the highest MCV score of 0.968532. These parameters were deemed optimal to develop the SVR model. As shown in Fig. 7, the RBF-SVR models tend to have superior MCV scores, implying that the poorest performing RBF-SVR model was better than the highest performing POLY-SVR model. The developed RBF-SVR model is presented in Fig. 8, however, some of its predictions tend to slightly deviate from the test data points.

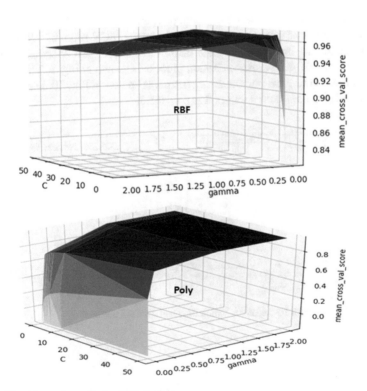

Fig. 7 3D grid-search results for SVR models

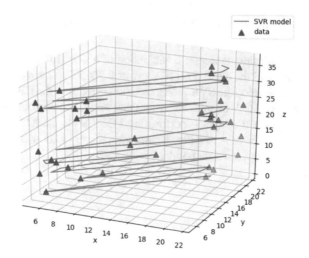

Fig. 8 3D regression of SVR model with RBF kernel

3.5 Development and Validation of the RF Model

The RF model with a maximum tree depth of 9, minimum samples on a leaf node as 1, and the number of trees as 10 attained the highest MCV score of 0.997515. The relationship among the parameters and their corresponding MCV scores during the grid-search with 10-fold cross-validation process is shown in Fig. 9. The RF model configured with the ideal parameter arguments is shown in Fig. 10. Due to brevity, the 2nd tree at a maximum depth of 2 is presented.

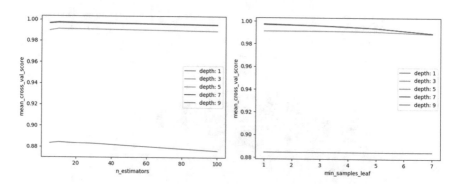

Fig. 9 Grid-search with 10-fold cross-validation results for RF models

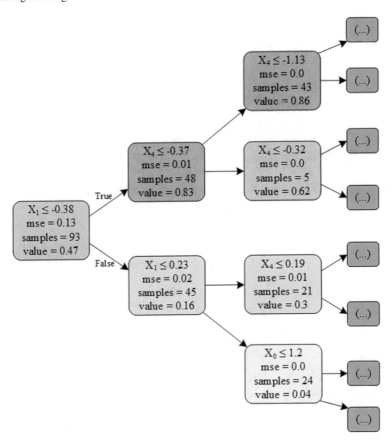

Fig. 10 RF model plot of 2nd tree at a maximum depth of 2

3.6 Development and Validation of the XGBoost Model

After the grid-search with 10-fold cross-validation procedure, the XGBoost model with the parameter combination of a "dart" booster, 0.2 learning rate, maximum tree depth of 3 and 500 trees yielded the highest MCV score of 0.997011. As shown in Fig. 11, irrespective of the kind of booster, higher MCV scores was positively associated with increasing learning rates in this study. The ideal parameters were used to develop the XGBoost model (Fig. 12). The variables X_0 to X_4 in the AdaBoost, RF, and DT models correspond to f0 to f4 in the XGBoost model. The scores on the leaf nodes are also known as weights.

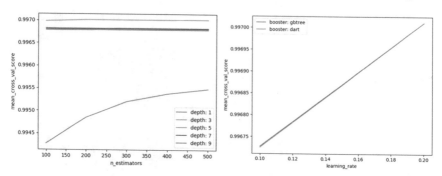

Fig. 11 Grid-search with 10-fold cross-validation results for XGBoost models

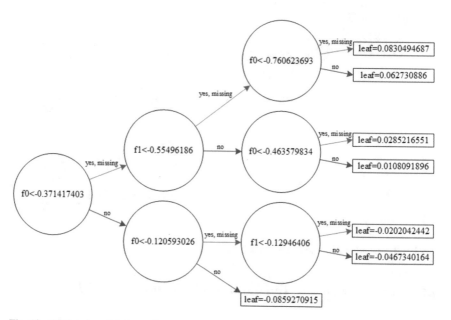

Fig. 12 XGBoost model plot at 1st tree

3.7 Performance Evaluation and Comparison of ML Models

The observed variance that each validated model explained was evaluated as shown in Fig. 13. The RF model explained almost 100% (99.88%) of the observed variance. In theory, a model explaining 100% of the variance would always have the actual values equal to the predicted values. Though the SVR model had the lowest R^2 in this study, its capacity to account for 97.67% of the observed variance only leaves less than 3% unaccounted-for.

The actual and predicted building-related CO_2 emissions from the six ML models on the test sample were plotted (Fig. 14). It can be observed that predictions from

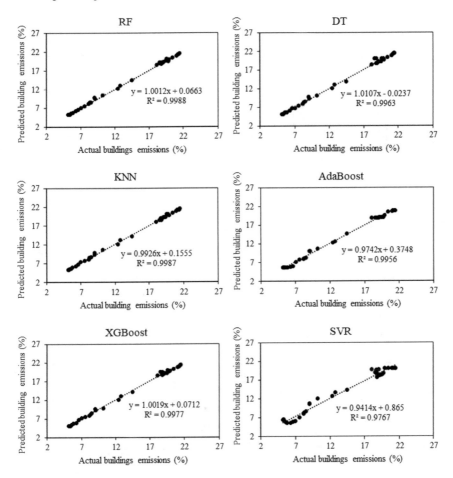

Fig. 13 Scatter plot of actual and predicted building emissions

the SVR slightly deviates from the actual building-related CO_2 emissions while that of the RF model tend to nearly-perfectly lay on the actual emissions. Further metrics on errors are presented in Table 2, where the RF model shows superior accuracy, closely followed by the KNN model. Time considerations are essential in industrial and problem-solving scenarios, as such, the elapsed time taking by the six algorithms in generating 1000 models each during parameter optimisation is further evaluated. The DT model outperformed all other models in generating 1000 models in just 2.2 s, followed by the KNN model in 3.2 s. Based on these results, it can be concluded that the RF algorithm is the best performing ML algorithm in accurately predicting building-related CO_2 emissions, whereas the best algorithm in terms of time efficiency is the DT algorithm. The KNN model is highly recommended when practitioners want to have accurate predictions in a timely manner.

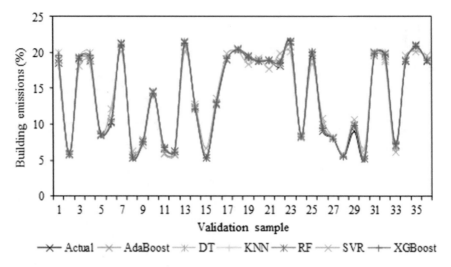

Fig. 14 Actual versus predicted building emissions

Table 2 Performance evaluation of ML models

Model	MAE[a]	MAPE[b]	MSE[a]	RMSE[a]	R^2	Elapsed time	Ranking Accuracy	Time
RF	0.14972	1.182106	0.051918	0.227854	0.9988	18.3 s	1	4
KNN	0.164794	1.391133	0.0525	0.22913	0.9987	3.2 s	2	2
XGBoost	0.201254	1.658298	0.096684	0.31094	0.9977	1.0 min	3	5
DT	0.243482	1.848901	0.158474	0.398088	0.9963	2.2 s	4	1
AdaBoost	0.336348	3.173315	0.177449	0.421247	0.9956	1.5 min	5	6
SVR	0.795811	7.200395	0.906077	0.951881	0.9767	4.6 s	6	3

[a]CO_2 emissions from residential buildings and commercial and public services (% of total fuel combustion), [b]%

4 Conclusions

This chapter evaluated and compared the performance of six ML algorithms in predicting building-related CO_2 emissions. Using China's emissions as an illustration, the models took into account five input parameters influencing building-related CO_2 emissions: urbanisation, R&D, population size, GDP, and energy use. Data on these variables throughout 1971–2014 were sourced from World Bank Open Data and IEA repositories. Each model was developed using 140 observations and validated on 36 observations. The findings indicated that the RF model attained the highest R^2 of 99.88%, followed by the KNN model (99.87%), XGBoost (99.77%), DT (99.63%), AdaBoost (99.56%), and the SVR model (97.67%). Overall, the RF algorithm is the best performing ML algorithm in accurately predicting building-related CO_2 emissions, whereas the best algorithm in terms of time efficiency is the DT algorithm. The KNN model is highly recommended when practitioners want to have accurate predictions in a timely manner.

Some algorithms could have performed better in terms of their computational efficiency; however, the combination and choice of certain parameters for such algorithms during the grid-search process can be computationally laborious and hence to a marginal extent, could limit our results in those aspects. The implication of this chapter is that depending on the trade-off, i.e. speed over accuracy or vice versa, RF, KNN, and DT models could be added to the toolkits of environmental policymakers to provide high-quality forecasts and patterns of building-related CO_2 emissions that are more accurate promptly. Researchers could also provide more comparative empirical evidence by replicating the methods applied in this chapter for other countries. Other robust and hybrid ML algorithms could be tested to explain 100% of the observed variance.

References

1. Abidoye, R.B., Chan, A.P.: Improving property valuation accuracy: a comparison of hedonic pricing model and artificial neural network. Pacific Rim Prop. Res. J. **24**(1), 71–83 (2018)
2. Acheampong, A.O., Boateng, E.B.: Modelling carbon emission intensity: application of artificial neural network. J. Clean. Prod. **225**, 833–856 (2019). https://doi.org/10.1016/j.jclepro.2019.03.352
3. Ahmad, M.W., Mourshed, M., Rezgui, Y.: Trees vs Neurons: comparison between random forest and ANN for high-resolution prediction of building energy consumption. Energy Build. **147**, 77–89 (2017)
4. Awad, M., Khanna, R.: Support vector regression. In: Efficient Learning Machines: Theories, Concepts, and Applications for Engineers and System Designers, pp. 67–80. Apress, Berkeley, CA (2015). https://doi.org/10.1007/978-1-4302-5990-9_4
5. Bannor, B.E., Acheampong, A.O.: Deploying artificial neural networks for modeling energy demand: international evidence. Int. J. Energy Sect. Manag. ahead-of-print (ahead-of-print) (2019). https://doi.org/10.1108/ijesm-06-2019-0008
6. Boateng, E.B., Pillay, M., Davis, P.: Predicting the level of safety performance using an artificial neural network. In: Ahram T, Karwowski W, Taiar R (eds) Human Systems Engineering and Design. Human Systems Engineering and Design, vol. 876, pp. 705–710. Springer International Publishing, Cham (2019). https://doi.org/10.1007/978-3-030-02053-8
7. Breiman, L.: Random forests. Mach. Learn. **45**(1), 5–32 (2001)
8. Buuren, S.V., Groothuis-Oudshoorn, K.: mice: Multivariate imputation by chained equations in R. J. Stat. Softw. 1–68 (2010)
9. Chen, H., Lee, W., Wang, X.: Energy assessment of office buildings in China using China building energy codes and LEED 2.2. Energy Build. **86**, 514–524 (2015)
10. Chen, T., Guestrin, C.: XGBoost: A scalable tree boosting system. In: Association for Computing Machinery, pp. 785–794 (2016). https://doi.org/10.1145/2939672.2939785
11. Chen, T., Guestrin, C.: Xgboost: A scalable tree boosting system. In: Proceedings of the 22nd ACM SIGKDD International Conference on Knowledge Discovery and Data Mining, ACM, pp. 785–794 (2016)
12. Drucker, H., Burges, C.J., Kaufman, L., Smola, A.J., Vapnik, V.: Support vector regression machines. In: Advances in Neural Information Processing Systems, pp. 155–161 (1997)

13. Fletcher, R.: Practical Methods of Optimization. Wiley (2013)
14. Freund, Y., Schapire, R.E.: A decision-theoretic generalization of on-line learning and an application to boosting. J. Comput. Syst. Sci. **55**(1), 119–139 (1997)
15. Gallo, C., Conto, F., Fiore, M.: A neural network model for forecasting CO_2 emission. AGRIS on-line Papers in Economics and Informatics 6 (665-2016-45020), pp. 31–36 (2014)
16. Grömping, U.: Variable importance assessment in regression: linear regression versus random forest. Am. Stat. **63**(4), 308–319 (2009)
17. Hastie, T., Tibshirani, R., Friedman, J.: The Elements of Statistical Learning: Data Mining, Inference, and Prediction. Springer Science & Business Media (2009)
18. IEA: CO_2 Emissions from Fuel Combustion. All rights reserved (2019)
19. IEA: World Energy Balances. All rights reserved (2019)
20. International Energy Agency and the United Nations Environment Programme: 2019 global status report for buildings and construction: towards a zero-emission, efficient and resilient buildings and construction sector (2019)
21. Karush, W.: Minima of functions of several variables with inequalities as side constraints. M Sc Dissertation Department of Mathematics, University of Chicago (1939)
22. Kuhn, H.W., Tucker, A.W.: Nonlinear programming. In: Neyman, J. (ed.) Proceedings of the Second Berkeley Symposium on Mathematical Statistics and Probability. University of California Press, Berkeley (1951)
23. Lu, M., Lai, J.: Review on carbon emissions of commercial buildings. Renew. Sustain. Energy Rev. **119**, 109545 (2020)
24. Pedersen, L.: Use of different methodologies for thermal load and energy estimations in buildings including meteorological and sociological input parameters. Renew. Sustain. Energy Rev. **11**(5), 998–1007 (2007)
25. Ren, Z., Chrysostomou, V., Price, T.: The measurement of carbon performance of construction activities. In: Smart and Sustainable Built Environment (2012)
26. Seo, S., Hwang, Y.: Estimation of CO_2 emissions in life cycle of residential buildings. J. Constr. Eng. Manag. **127**(5), 414–418 (2001)
27. Shahbaz, M., Van Hoang, T.H., Mahalik, M.K., Roubaud, D.: Energy consumption, financial development and economic growth in India: new evidence from a nonlinear and asymmetric analysis. Energy Econ. **63**, 199–212 (2017)
28. Smola, A.J., Schölkopf, B.: A tutorial on support vector regression. Stat. Comput. **14**(3), 199–222 (2004)
29. Solomatine, D.P., Shrestha, D.L., AdaBoost, R.T.: A boosting algorithm for regression problems. In: 2004 IEEE International Joint Conference on Neural Networks (IEEE Cat. No. 04CH37541), pp. 1163–1168. IEEE (2004)
30. Tsai, C.-F., Chiou, Y.-J.: Earnings management prediction: A pilot study of combining neural networks and decision trees. Expert Syst. Appl. **36**(3), 7183–7191 (2009)
31. Vapnik, V.: The nature of statistical learning theory. Springer, New York (1995)
32. World Bank: World Bank Open Data. The World Bank Group https://data.worldbank.org/ (2019). Accessed 25 Dec 2019
33. World Bank: Carbon emissions data. https://data.worldbank.org/indicator/EN.ATM.CO2E. KT?end=2014&start=2000 (2020). Accessed 1 Mar 2020
34. Wu, D.: Supplier selection: A hybrid model using DEA, decision tree and neural network. Expert Syst. Appl. **36**(5), 9105–9112 (2009)

Artificial Intelligence System for Breast Cancer Screening Based on Malignancy-Associated Changes in Buccal Epithelium

A. V. Andreichuk, N. V. Boroday, K. M. Golubeva, and D. A. Klyushin

1 Introduction

Breast cancer is a common and dangerous disease. Survival of patients with this diagnosis depends on the accuracy of early diagnosis. A combination of methods for detecting early signs of cancer and artificial intelligence, in particular, statistical machine learning, provides a solution to this problem. A wide review of works on this topic is given in [8].

Now, fractal analysis is one the most promising field of investigation of cell heterogeneity in healthy and tumor tissues. Fractal dimension (FD) is a power diagnostic tool used for defining heterogeneity of cells of complex endometrial hyperplasia and well-differentiated endometrioid carcinoma [4]. Also, it is an independent prognostic factor for survival in melanoma [3], leucemia [1, 14], and other diseases [12, 17, 18]. The nuclear patterns of human breast cancer cells attracted a special attention [6, 11, 19–21, 26, 29]. These investigations have demonstrated the significant potential role of fractal analysis in assessing the morphological information. Nevertheless, all these investigations were aimed only on tumor cells but not cells that are distant from the tumor.

The effect of malignancy-associated changes (MAC) in cells distant from a tumor was considered in numerous papers, see for example [2, 5, 31, 32]. The authors of these papers have demonstrated that the analysis of MAC in buccal epithelium is one

A. V. Andreichuk · K. M. Golubeva · D. A. Klyushin (✉)
Faculty of Computer Science and Cybernetics, Taras Shevchenko National University of Kyiv, Akademika Glushkova Avenu 4D, Kiev 03680, Ukraine
e-mail: dokmed5@gmail.com

N. V. Boroday
Institute for Problems of Cryobiology and Cryomedicine, National Academy of Sciences of Ukraine, Pereyaslavskaya str., 23, Kharkiv 61015, Ukraine
e-mail: boroday1@ukr.net

© The Editor(s) (if applicable) and The Author(s), under exclusive license to Springer Nature Switzerland AG 2021
A.-E. Hassanien et al. (eds.), *Enabling AI Applications in Data Science*, Studies in Computational Intelligence 911, https://doi.org/10.1007/978-3-030-52067-0_12

of the perspective noninvasive methods for the effective screening of cancer. Such methods can be divided in two groups: methods involving the analysis of MAC in non-tumor cells located near a tumor [31], Us-Krasovec [32] and methods involving the analysis of MAC in non-tumor cells located far from a tumor, in particular, in buccal epithelium (oral mucosa) [2, 5].

As far as now there are no any results on the fractal properties of nuclear patterns of buccal epitheliocytes in tumor and healthy conditions, the aim of the paper is to present an AI system that estimates the distinguishes between the distributions of the fractal dimension of chromatin in feulgen-stained nuclei in buccal epithelium of patients with breast cancer, fibroadenomatosis and healthy people and implement a screening test for breast cancer.

2 Input Data

The study was conducted on samples from 130 patients: 68 patients with breast cancer, 33 patients with fibroadenomatosis and 29 healthy individuals. Every sample contains in average 52 nuclei of buccal epithelium. The input data set for the study consists of 20,256 images of interphase nuclei of buccal epithelium (6752 nuclei were scanned without filter, through a yellow filter and through a violet filter). Each image consists of three channels: red, green, blue, as well as gray halftones.

At the first step, we had obtained the scrapes from various depths of the spinous layer (conventionally they were denoted as median and deep) after gargling and taking down of the superficial cell layer of buccal mucous. The smears were dried out under the room temperature and fixed during 30 min in the Nikiforov mixture and we made Feulgen reaction with the cold hydrolysis in 5 n. HCl during 15 min under the temperature t = 21–22 °C. The DNA content stained by Feulgen was estimated using the Olympus computer analyzer, consisting of the Olympus BX microscope, Camedia C-5050 digital zoom camera and a computer. We investigated 52 cells in average in every preparation. The DNA-fuchsine content in the nuclei of the epitheliocytes was defined as a product of the optical density on area. Thus under investigation of the interphase nucleus we obtained a scanogram of the DNA distribution which is rectangular table (matrix) 160 × 160 pixels.

3 General Scheme of the Screening

According to the results of the diagnosis, the patient may be assigned to one of two groups—either breast cancer patients or non-breast cancer patients (healthy and patients with fibroadenomatosis).

For each patient, we receive a scrape from the oral mucosa, which is used to distinguish between 23 and 81 cells. On the basis of cytospectrophotometry data, we obtain a DNA scanogram of the inter-phase nuclei of these cells in two filters (yellow and violet), as well as without it.

For each image, there are three channels—red, green, blue—and a gray version. We perform pre-processing of images for binarization and denoising. For each filter/channel pair, we use a special diagnostic test. The blue channel with the yellow filter gives dark grey verging on black, so such pairs were not considered.

The main stages of the diagnostic test

1. By each scanogram of a cell nucleus we calculate its fractal dimension.
2. For each patient we form a vector of the fractal dimensions of his cell nuclei.
3. For both classes we form training and control samples.
4. We evaluate the performance of the classification using cross-validation with various sample sizes.

Diagnostic samples. Cross-validation

At the stage of cross-validation data are broken down into p parts. Next, using 1NN method we train a classification model on $(p - 1)$ training parts and evaluate it on $p - 1$th control part of the data. The procedure is repeated p times. By using this method, each part is used for testing, so the classification quality assessment is objective because all data is used evenly. As a similarity measure we used p-statistics [9].

At this stage, results are obtained for all 11 filter/channel pairs. Since the number of pairs is odd, we can use the voting method: each patient belongs to the group to which she is assigned by most filter/channel pairs.

We also apply the voting method to evaluate the filters: for each filter we take the 3 most accurate individual results, i.e. we drop out one of the channels (in any case, for the yellow filter the blue channel does not matter), and take into account their voices.

4 Image-Preprocessing: Segmentation, Binarization, and Denoising

Microscopic images in pure form are usually unacceptable for analysis due to defects, digital noise caused by the need to increase the photo sensitivity of the material, as well as foreign objects. For further analysis of the incoming image pre-processing and segmentation are required. To reduce the noise level of the microscope images, a median filter was used, one of the types of digital filters described in [23], which is widely used in digital signal and image processing.

The next step is processing is separating the background pixels from the pixels of the objects in the image (cell nuclei or third-party objects). The image is binarized, that is, the background pixels are white and the nucleus pixels are black. To achieve this goal, the Otsu algorithm [22] is appropriate.

The binarized image is getting rid of residual noise of salt and pepper type (separate small groups of white or black dots, respectively). The binary morphological operations [28] were used, namely: opening to get rid of black points, and then closing to fill white points.

Then we calculate the fractal dimension for pre-processed images. Thus, a set of fractal characteristics will be constructed for each patient. As a measure of fractal dimension we use the Minkowski dimension computed by the modified box-counting algorithm [10].

Each patient has a set of fractal characteristics, the size of which is not constant (each cell number is different). To compare the set of characteristics of different patients, it is necessary to use the measure of proximity between samples, and for this purpose p-statistics is used [9].

For classification of patients the closest neighbor method was used (to classify an element x, we select from the training sample an element closest to x and relate x to the class of this element).

Let us consider these stages in detail.

Images obtained directly from the microscope are often subject to noise and defects. They are caused by the increased photo sensitivity of used photographic materials. This means that raw images are not usable and require further processing, otherwise there is a high risk of incorrect results. Therefore, for further analysis, we will apply the already processed image sets.

The blurring process can be seen as reducing the sharpness of the image. Blurring makes the image details less clear and often used for smoothing.

Images that are perceived as too sharp can be softened by using various blurring methods and levels of intensity. Often, images are smoothed out to remove or reduce noise on the image. When selecting contours, the best results are often achieved with preceded noise reduction.

Consider the effect of different blur filters on the image of the cell nucleus (Fig. 1). We prefer the median filter [23] because it is a widely used digital signal and image processing for noise reduction and well-preserving image boundaries. The median filter is classified as a nonlinear filter. Unlike many other image blurring methods, the implementation of this filter does not involve a convolution or a predefined kernel.

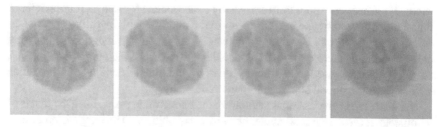

Fig. 1 Initial image, image after Gaussian blurring with the kernel 5 × 5, image after medial filtering with the kernel 5 × 5, and image after "moving blurring" with the kernel 5 × 5, respectively

Pre-processing algorithm. For each pixel of the input image, do the following:

1. Add to the array all pixels of the image that are in the window with the center in the current pixel.
2. Sort this array in ascending order.
3. The value in the middle of the sorted array of neighbors is a median of window centered in the current pixel. Replace all pixel values of the image with the value of the median.

To separate nucleus from the background on the image we use the Otsu method [22] that is intended for automatic computation of threshold value and construction of binary images from grey-level images where the pixel values vary from 0 to 255. Thus, we have the range of pixels of an image in 256 gray levels. Let us denote the number of pixels at level l by n_l and the total number of pixels by $N = \sum_{i=0}^{255} n_l = 160 \times 160 = 25,600$. Compute a probability distribution of pixel values:

$$p_l = \frac{n_l}{N}, \ p_l \geq 0, \ \sum_{l=0}^{255} p_i = 1. \tag{1}$$

The Otsu method separates pixels into two classes by minimizing the within-class variance at the threshold T:

$$v_w^2(T) = p_1(T)v_1^2(T) + p_2(T)v_2^2(T), \tag{2}$$

where $p_1(T) = \sum_{l=0}^{T} p_l$ and $p_2(T) = \sum_{l=T+1}^{255} p_l$ are the probabilities of the first and second classes, respectively, separated by a threshold T, $v_1^2(T) = \sum_{l=0}^{T}(l - \mu_1(T))^2 p_l / p_1(T)$ is the variance of the first class, $v_2^2(T) = \sum_{l=T+1}^{255}(l - \mu_2(T))^2 p_l / p_2(T)$ is the variance of the second class, $\mu_1(T) = \sum_{l=1}^{T} l p_l / p_1(T)$ is the mean of the first class, $\mu_2(T) = \sum_{l=T+1}^{255} l p_l / p_2(T)$ is the mean of the second class.

Otsu proved that minimizing the within-class variance at the threshold T is equivalent to finding the optimal threshold that maximizes the between-class variance σ_B^2:

$$T^* = \arg\max_{0 \leq T \leq 255} \sigma_B^2(T), \tag{3}$$

where $\sigma_B^2(T) = p_1(T)p_2(T)(\mu_2(T) - \mu_1(T))^2$.

Otsu algorithm

1. Input the grey-level image $G(i, j)$, $i = \overline{1, 160}$, $j = \overline{1, 160}$.
2. Initialize a counter $k = 0$.
3. Compute an image histogram $p(l)$ and a frequency $N(l)$ for every level $l = \overline{0, 255}$ of the pixel values.
4. Compute the initial values of $p_1(0)$, $p_2(0)$, $\mu_1(0)$ and $\mu_2(0)$.

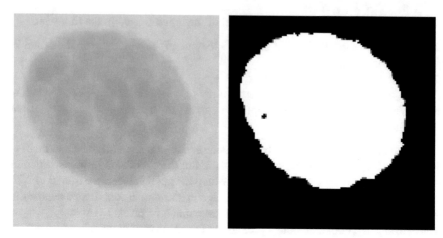

Fig. 2 Before and after binarization

5. Update $p_1(T)$, $p_2(T)$, $\mu_1(T)$, $\mu_2(T)$ for every possible value of a threshold $T = 0, 255$:

$$N_T = \sum_{i=0}^{255} p(i), \quad p_1(T) = \frac{1}{N_T} \sum_{i=0}^{T-1} p(i) = \sum_{i=0}^{T-1} N_i, \quad p_2(T) = 1 - p_1(T),$$

$$\mu_T = \frac{1}{N_T} \sum_{i=0}^{255} i p_i = \sum_{i=0}^{255} i N_i, \quad \mu_1(T) = \frac{1}{N_T p_1(t)} \sum_{i=0}^{255} i p_i = \frac{1}{p_1(t)} \sum_{i=0}^{255} i N_i, \quad \mu_2(T) = \frac{\mu_T - \mu_1(T) p_1(T)}{p_2(T)}.$$

6. Compute $\sigma_B^2(T) = p_1(T) p_2(T) [\mu_1(T) - \mu_2(T)]^2$.
7. If $\sigma_B^2(T)$ is greater than current value T, save $\sigma_B^2(T)$ and the value T.
8. The desired threshold maximizes $\sigma_B^2(T)$.

The result is shown at Fig. 2.

Morphological operations. Morphological operations [28] are performed on two images: an input image and a special one, which depends on the operation and the type of out-of-task performed. Such a special image in mathematical morphology is called a *structural element*, or a *primitive*. Any image is regarded as a subset of Euclidean space R^2 or an integer grid Z^2, where n—is the dimension of space. The structural element is a binary image (geometric shape). It can be of any size and structure, but as a rule the size of such an image is 3×3, 4×4, 5×5 pixels, that is, much smaller than the input image. In each element there is a special point, which is called *origin*. It can be selected anywhere in the image, but most often it is a central pixel.

Consider the basic operations of mathematical morphology from space R^2. The *transfer* of multiple pixels A to a given vector s is defined as:

$$A_s = \{a + s | a \in A\} \; \forall s \in R^2. \tag{4}$$

The transfer can be determined by an ordered pair of numbers (x, y), where x is the number of pixels offset along the axis X, and y is the offset along the axis y.

For two sets A and B from space, the *erosion* (narrowing) of the set by the structural element is defined as:

$$A \ominus B = \left\{ z \in Z^2 \middle| B_z \subseteq A \right\}. \tag{5}$$

In other words, the erosion of the set A by a structural element B is such a geometric location of points for all such positions of points of the center z, at which the set B is completely contained in A.

Dilation (expansion) of the set is defined as:

$$A \oplus B = \left\{ z \in Z^2 \middle| \left(B^s \right)_z \cap A \neq 0 \right\}. \tag{6}$$

In this case, the dilation of a set A by a structural element B is the set of all such displacements z, in which the sets A and B coincide with at least one element.

The dilation is a commutative function, meaning the following expression:

$$A \oplus B = B \oplus A = \bigcap_{a \in A} B_a. \tag{7}$$

The *opening* of the set A by structural element B is

$$A \odot B = (A \ominus B) \oplus B. \tag{8}$$

Thus, opening of the set A by a structural element B is computed as erosion A by a structural element B, and the result of which is dilated over the same structural element B. In the general case, the opening smoothies out the contours of the object, eliminates narrow isthmus and eliminates the projections of small width.

The *closing* of the set A by a structural element B is obtained by performing the operation of dilation of the set A by a structural element B, which is followed by the operation of erosion of the resulting set by the structural element B. Closing is defined by the following expression:

$$A \odot B = (A \ominus B) \oplus B. \tag{9}$$

As a result of the closing operation, the contour segments are smoothed, but, unlike the opening, in general, small breaks and long valleys of small width are filled, as well as small holes are opened and the gaps of the contour are filled.

Salt and Pepper Noise Reduction. The "salt and pepper" type noise erases black dots called "pepper" and fills the holes in the image called "salt". First, open A by B, it will remove all the black points (pepper), then close A by B, and all the holes will be filled (salt).

We will use an octagon-type structural element with $R = 2$:

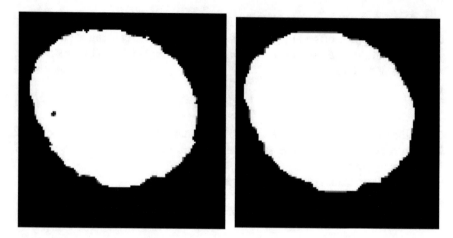

Fig. 3 Before and after "salt-and-pepper" denoising

$B = \{(-2, -1), (-2, 0), (-2, 1), (-1, -2), (1, -2), (0, -2), (-1, -1), (-1, 0), (-1, 1),$
$(0, -1), (0, 0), (0, 1), (1, -1), (1, 0), (1, 1), (2, -1), (2, 0), (2, 1), (-1, 2), (1, 2), (0, 2)\}$

Let's look at an example of the effect of "salt and pepper" denoising with an octagon structural element on the image of a cell nucleus (Fig. 3).

Contour detection. Roberts Operator. For further calculations, it is necessary to select the contours of the image, because on the basis of them the fractal dimension will be calculated. To do this, we use the Roberts operator, a discrete differential operator used in image processing to distinguish boundaries.

Roberts [25] proposed the following equations:

$$y_{ij} = \sqrt{x_{ij}}, \quad z_{ij} = \sqrt{\left(y_{ij} - y_{i+1,j+1}\right)^2 + \left(y_{i+1j} - y_{i,j+1}\right)^2}, \qquad (10)$$

where x is the value of the pixel intensity of the image, z is the approximated derivative, i, j are the coordinates of the pixel.

The resulting image highlights the changes in intensity in diagonal direction. The transformation of each pixel by the Roberts cross-operator can tell a derivative image along a non-zero diagonal, and the combination of these transformed images can also be considered as a gradient from the top two pixels to the bottom two.

The difference vector module can also be calculated in the Manhattan metric. The Roberts operator uses the module of this total vector, which shows the largest difference between the four points covered. The direction of this vector corresponds to the direction of the largest gradient between the points.

Convolution algorithm. We convolute the image with the kernels $\begin{pmatrix} 1 & 0 \\ 0 & -1 \end{pmatrix}$ and $\begin{pmatrix} 0 & 1 \\ -1 & 0 \end{pmatrix}$. If $I(x, y)$ is the point of the initial image, $G_x(x, y)$ is the point of the image convoluted using the first kernel, and $G_x(x, y)$ is the point of the image convoluted using the first kernel, then the gradient can be defined as:

$$\nabla I(x, y) = G(x, y) = \sqrt{(G_x(x, y))^2 + (G_y(x, y))^2}. \tag{11}$$

The gradient direction can be defined as:

$$\Theta(x, y) = \tan^{-1}\left(\frac{G_y(x, y)}{G_x(x, y)}\right). \tag{12}$$

Let's look at an example of the action of Roberts operator on the image of the cell nucleus (Fig. 4).

Fractal dimension. In a broad sense, a fractal means a figure whose small parts are at random magnification similar to itself. Let Ω be a finite set in metric space X.

Definition Let $\varepsilon > 0$. A set of not more than a counted family of subsets $\{\omega_i\}_{i \in I}$ of space X is called an ε-*coverage* of the set Ω if the following conditions holds:

1. $\Omega \subset \bigcup_{i \in I} \omega_i$,
2. $\forall i \in I \ diam \ \omega_i < \varepsilon$.

Let $\alpha > 0$ and $\Theta = \{\omega_i\}_{i \in I}$ be a coverage of the set Ω. Let's define the following function, which in some sense determines the size of this coverage:

$$F_\alpha(\Theta) = \sum_{i \in I} (diam \ \omega_i)^\alpha. \tag{13}$$

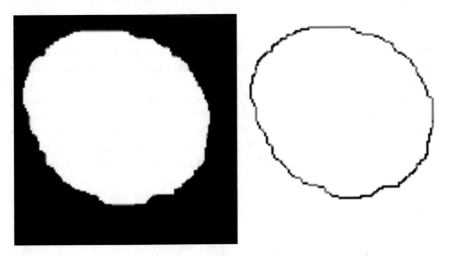

Fig. 4 Before and after contour detection

Denote by $M_\alpha^\varepsilon(\Omega)$ the minimum size of ε-coverage of the set:

$$M_\alpha^\varepsilon(\Omega) = \inf_\alpha F_\alpha(\Theta). \tag{14}$$

Here the infimum is taken over all the coverages of the set Ω. Obviously, the function $M_\alpha^\varepsilon(\Omega)$ does not decrease with decreasing ε, since with decreasing ε we also narrow down the set of possible coverages. Therefore, it has a finite or infinite boundary at:

$$M_\alpha(\Omega) = \lim_{\varepsilon \to +0} M_\alpha^\varepsilon(\Omega).$$

Definition The value $M_\alpha(\Omega)$ is called the *Hausdorff α-measure* of the set Ω.

From the properties of the Hausdorff α-measure it is follows that $M_\alpha(\Omega)$ is a decreasing function. Moreover, for any set Ω there is a critical value α_0 such that:

$$M_\alpha(\Omega) = 0 \quad \forall \alpha > \alpha_0,$$

$$M_\alpha(\Omega) = +\infty \quad \forall \alpha < \alpha_0.$$

The value $M_{\alpha_0}(\Omega)$ can be null, finite positive, or infinite.

Definition The number α_0 is called the *Hausdorff dimension of the set* Ω.

Mandelbrot [13] proposed the following definition of a fractal.

Definition A set is called a *fractal* if its Hausdorff dimension strictly exceeds its topological dimension.

The Minkowski dimension. Fractal sets usually have a complex geometric structure, and its main property is self-similarity. The fractal dimension is the characteristic that describes this property [33].

When measuring the fractal dimension of various natural and artificial objects, we deal with problems associated with the fact that there are several definitions of fractal dimension. The fundamental concept is the Hausdorff dimension, but its computation is often a difficult task. Therefore, other dimensions are more commonly used in practice, such as the *Minkowski dimension*, which is calculated using the box-counting algorithm [27].

For an arbitrary positive δ some function $M_\delta(\Omega)$ is calculated. If $M_\delta(\Omega) \propto \delta^{-D}$, then the set Ω has a fractal dimension D. Thus,

$$\dim_M \Omega = D = \lim_{\delta \to 0} \frac{\ln(M_\delta(\Omega))}{-\ln(\delta)}, \tag{15}$$

where the value $M_\delta(\Omega)$ is equal to the number of n-dimensional cubes with sides δ required to cover the set Ω.

Definition The number $\dim_M \Omega$ is called the *Minkowski dimension of the set* Ω.

Note that it may not always exist. The relation between the Minkowski dimension and the Hausdorff dimension is expressed by the following theorem:

Theorem *Schroeder [30].* $\dim_M \Omega \leq \dim_H \Omega$.

Box-counting algorithm. To calculate the Minkowski dimension of a binary image we use the version of box-counting algorithm described in [10].

Let $M(r) = \sum_{\omega_r \in W_r} \bar{\omega}_r$, $r \in R$, where R is a number of values r that does not exceed the length of the sides of the image, $W_r = \{\omega_r | \omega_r \in S\}$ is a partitioning of the image into squares ω_r that have sides r, $\bar{\omega}_r$ is the number of pixels filled in ω_r.

Consider the graph of the dependence of $\ln(M(r))$ from $\ln(r^{-1})$ and the tangent of the angle of the slope of this graph. It is an approximation of the Minkowski's dimension $\dim_M \Omega$. The box-counting algorithm requires to calculate a regression line to find the angle of its slope. We will look for direct regression using the least squares method. We need to find such a linear function $f(x)$ that minimize $\sum_{r \in W_r} (y - f(x_r))^2$, where $x_r = \ln(r^{-1})$, $y_r = \ln(M(r))$.

Let's look at an example of calculating Minkowski's fractal dimension on the image of a cell nucleus (Fig. 5).

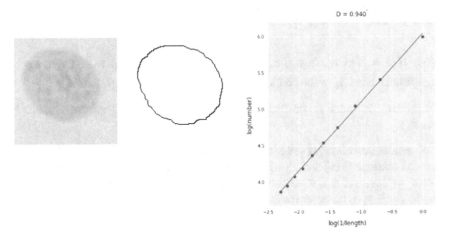

Fig. 5 Cell nucleus, its contour and the graph of linear regression for computation of the Minkowski dimension of the nucleus

5 Similarity Measure

Let $x = (x_1, x_2, \ldots, x_n)$ and $y = (y_1, y_2, \ldots, y_m)$ be two samples from the general populations G_1 and G_2, respectively, obtained by the sample sampling. Let us construct a similarity measure $\mu(x, y)$ between x and y and two-sided confidence interval (μ_1, μ_2) with given significance level, which does not depend on whether the null hypothesis that $G_1 = G_2$ is true or false. Further, using this similarity measure we construct the non-parametrical test for sample heterogeneity. First, introduce necessary notions and theorems [9].

Definition Intervals $(a_k, b_k) = (a_k(x_1, \ldots, x_k), b_k(x_1, \ldots, x_k))$, $k = 1, 2, \ldots$ are refereed to as *asymptotic intervals* for random values p_k corresponding to the significance level β, if

$$\lim_{n \to \infty} P(p_k \in (a_k(x_1, \ldots, x_k), b_k(x_1, \ldots, x_k))) = 1 - \beta. \tag{16}$$

Definition The ends of these intervals are called *asymptotic confidence limits*.

Definition The value β is called an *asymptotic significance level* of the sequence (a_k, b_k).

Definition In particular, when all values $p_1, p_2, \ldots, p_k, \ldots$ are the same, i.e. $p_k = p \; \forall k = 1, 2, \ldots$, the interval (a_k, b_k) is called an *asymptotic confidence interval* for value p, and the value β is called an *asymptotic significance level* of the interval (a_k, b_k).

Note If $\beta = \lim_{n \to \infty} \beta_k$, then β can be considered as *approximate value of the true significance level* β_k of the confidence intervals (a_k, b_k): $\beta_k \approx \beta$.

Definition The value β is called an *approximate significance level* of the intervals (a_k, b_k).

Consider the following criterion for the test of hypothesis H about equality distribution functions $F_1(u)$ and $F_2(u)$ of the general population G_1 and G_2, respectively. Let $x = (x_1, \ldots, x_n) \in G_1$ and $y = (y_1, \ldots, y_m) \in G_1$, $x_{(1)} \leq \ldots \leq x_{(n)}$, $y_{(1)} \leq \ldots \leq y_{(m)}$, be order statistics. Suppose that $G_1 = G_2$. Then [7]

$$p_{ij}^{(n)} = p\left(y_k \in (x_{(i)}, x_{(j)})\right) = \frac{j - i}{n + 1}. \tag{17}$$

We can calculate a frequency h_{ij} of the random event A_{ij} and compute confidence limits $p_{ij}^{(1)}$ and $p_{ij}^{(2)}$ for the probability $p_{ij}^{(n)}$ corresponding to given significance level $\beta : B = \left\{ p_{ij}^{(n)} \in \left(p_{ij}^{(1)}, p_{ij}^{(2)} \right) \right\}$, $p(B) = 1 - \beta$. These limits have been calculated by more that 20 formulae [24]. For definiteness, let us take the Wilson interval

$$p_{ij}^{(1,2)} = \frac{h_{ij}m + g^2/2 \mp g\sqrt{h_{ij}(1-h_{ij})m + g^2/4}}{m + g^2}. \tag{18}$$

The parameter g we may put be equal to 3. Denote by N a number of all confidence intervals $I_{ij}^{(n)} = \left(p_{ij}^{(1)}, p_{ij}^{(2)}\right)$ and L a number of intervals I_{ij} such that $p_{ij} \in I_{ij}^{(n)}$. Put $h = \mu(x, y) = \frac{L}{N}$. As far as h is a frequency of the random event $B = \left\{p_{ij}^{(n)} \in I_{ij}^{(n)}\right\}$ having the probability $p(B) = 1 - \beta$, then by putting $h_{ij} = h$, $m = N$ and $g = 3$ in formulae [18] we get a confidence interval $I = \left(p^{(1)}, p^{(2)}\right)$ for the probability $p(B)$ which have confidence level approximately equal to 0.95. A criterion for the test of hypothesis $G_1 = G_2$ with significance level approximately equal to 0.05 may be formulated by the following way: if the confidence interval $I = \left(p^{(1)}, p^{(2)}\right)$ contains the probability $p(B) = 1 - \beta$, then hypothesis $G_1 = G_2$ is accepted, otherwise it is rejected. The statistics h is called *p-statistics*.

Hereinafter we shall suppose that the samples x and y have the same size n. Let us compute the significance levels for confidence intervals $I_{ij}^{(n)}$ and $I^{(n)}$. Let us find the estimation of the asymptotic significance level of the confidence intervals $I_{ij}^{(n)}$.

Theorem [15] *If (1) $n = m$; (2) $0 < \lim_{n\to\infty} p_q^{(n)} = p_0 < 1$ and (3) $0 < \lim_{n\to\infty} \frac{i}{n+1} = p^* < 1$, then the asymptotic level β of the sequence of confidence intervals $I_{ij}^{(n)}$ for the probabilities $p_{ij}^{(n)} = p\left(y_k \in \left(x_{(i)}, x_{(j)}\right)\right)$ does not exceed 0.05.*

Theorem [15] *If in a strong random experiment E samples $x = (x_1, \ldots, x_n) \in G_1$ and $y = (y_1, \ldots, y_m) \in G_2$ have the same size, then the asymptotic significance level of the interval $I^{(n)} = \left(p^{(1)}, p^{(2)}\right)$ constructed by the 3σ-rule when $g = 3$ with the help of formulas [3], does not exceed 0.05.*

6 Results

We call the group of breast cancer patients "positive" and those with fibroadenomatosis and healthy "negative". So, the total number of positive examples is 68, and the negative ones are $33 + 29 = 62$.

We used a 1NN voting procedure among the pairs channel-filter and among the filters separately. Also, we used one-fold cross-validation with several sizes of control samples (5, 10, 20% and leave-one-out). The data were broken into p parts, so that $p - 1$ were used for model evaluation and one for testing. This procedure was repeated p times. The number p depends on the size of a control sample. As a similarity measure, we used p-statistics.

The results of testing are provided in Tables 1, 2, 3, 4, 5, 6, 7, 8, 9 and 10. These tables contain the sensitivity, specificity and accuracy of the cross-validation with different channels (red, green, blue, and grey) and filters (violet, yellow, and without filters) depending on size of control set.

Table 1 Results of cross-validation on control samples (20% of the all samples) using 1NN voting among the channels

N	Pair filter/channel	Sensitivity	Specificity	Accuracy
1	Blue	0.8889	0.5728	0.7309
2	Grey	0.8955	0.8730	0.8843
3	Green	0.8649	0.9286	0.8967
4	Red	0.6018	1.0000	0.8009
5	(Violet filter) blue	0.6029	0.5645	0.5837
6	(Violet filter) grey	0.9494	0.8310	0.8900
7	(Violet filter) green	0.8272	0.9796	0.9034
8	(Violet filter) red	0.6735	0.9375	0.8055
9	(Yellow filter) grey	0.8333	0.9423	0.8878
10	(Yellow filter) green	0.8250	0.9600	0.8925
11	(Yellow filter) red	0.8462	0.9615	0.9038
	Mean	0.8008	0.8683	0.8345

Table 2 Results of cross-validation on control samples (10% of the all samples) using 1NN voting among the channels

N	Pair filter/channel	Sensitivity	Specificity	Accuracy
1	Blue	0.9545	0.5648	0.7597
2	Grey	0.9286	0.9500	0.9393
3	Green	0.9296	0.9661	0.9478
4	Red	0.5913	1.0000	0.7956
5	(Violet filter) blue	0.5645	0.5147	0.5396
6	(Violet filter) grey	0.9688	0.9091	0.9389
7	(Violet filter) green	0.9067	1.0000	0.9533
8	(Violet filter) red	0.6733	1.0000	0.8366
9	(Yellow filter) grey	0.9178	0.9825	0.9501
10	(Yellow filter) green	0.8816	0.9815	0.9315
11	(Yellow filter) red	0.9315	1.0000	0.9658
	Mean	0.8407	0.8976	0.8689

From Tables 1, 2, 3 and 4, we can see that the highest average and individual (i.e. separately for filter/channel pairs) accuracy is achieved using the one-on-all principle—when one patient in the control sample compares with all others. This is natural, because in this way the comparison is made with as many training samples as possible. The highest accuracy obtained was 99.28% for grey channel in yellow and violet filters. This result was achieved by using the principle of one against all (Table 4). Specificity is 100% in all acceptable results, meaning that cancer patients

Table 3 Results of cross-validation on control samples (5% of the all samples) using 1NN voting among the channels

N	Pair filter/channel	Sensitivity	Specificity	Accuracy
1	Blue	0.9545	0.5648	0.7597
2	Grey	0.9437	0.9830	0.9634
3	Green	0.9571	0.9833	0.9702
4	Red	0.5965	1.0000	0.7982
5	(Violet filter) blue	0.5500	0.5000	0.5250
6	(Violet filter) grey	0.9851	0.9683	0.9767
7	(Violet filter) green	0.9315	1.0000	0.9658
8	(Violet filter) red	0.6800	1.0000	0.8400
9	(Yellow filter) grey	0.9577	1.0000	0.9789
10	(Yellow filter) green	0.9189	1.0000	0.9595
11	(Yellow filter) red	0.9577	1.0000	0.9789
	Mean	0.8575	0.9090	0.8833

Table 4 Results of cross-validation on control samples (leave one out) using 1NN voting among the channels

N	Pair filter/channel	Sensitivity	Specificity	Accuracy
1	Blue	0.9525	0.5596	0.7560
2	Grey	0.9444	1.0000	0.9722
3	Green	0.9710	0.9836	0.9773
4	Red	0.5913	1.0000	0.7957
5	(Violet filter) blue	0.5500	0.5000	0.5250
6	(Violet filter) grey	0.9855	1.0000	0.9928
7	(Violet filter) green	0.9315	1.0000	0.9658
8	(Violet filter) red	0.6476	1.0000	0.8238
9	(Yellow filter) grey	0.9855	1.0000	0.9928
10	(Yellow filter) green	0.9315	1.0000	0.9658
11	(Yellow filter) red	0.9714	1.0000	0.9857
	Mean	0.8602	0.9130	0.8866

Table 5 Results of 1NN voting among filters (20% of all samples)

3 best performance channels	Sensitivity	Specificity	Accuracy
Without filter	0.8701	0.9811	0.9256
Yellow filter	0.8333	0.9423	0.8878
Violet filter	0.8947	1.0000	0.9474

Table 6 Results of 1NN voting among filters (10% of all samples)

3 best performance channels	Sensitivity	Specificity	Accuracy
Without filter	0.9189	1.0000	0.9595
Yellow filter	0.9178	0.9825	0.9501
Violet filter	0.9444	1.0000	0.9722

Table 7 Results of 1NN voting among filters (5% of all samples)

3 best performance channels	Sensitivity	Specificity	Accuracy
Without filter	0.9315	1.0000	0.9658
Yellow filter	0.9577	1.0000	0.9789
Violet filter	0.9714	1.0000	0.9857

Table 8 Results of 1NN voting among filters (leave-one-out)

3 best performance channels	Sensitivity	Specificity	Accuracy
Without filter	0.9444	1.0000	0.9722
Yellow filter	0.9714	1.0000	0.9857
Violet filter	0.9714	1.0000	0.9857

Table 9 Results of 1NN voting among filters (leave-one-out) after medial filtering only

3 best performance	Sensitivity	Specificity	Accuracy
Without filter	0.7303	0.9268	0.8286
Yellow filter	0.7500	0.9524	0.8512
Violet filter	0.6939	1.0000	0.8469

Table 10 Results of 1NN voting among channels (leave-one-out) after medial filtering only

N	Pair filter/channel	Sensitivity	Specificity	Accuracy
1	Blue	0.6129	0.5051	0.5590
2	Grey	0.8125	0.9400	0.8763
3	Green	0.6355	1.0000	0.8178
4	Red	0.6579	0.6667	0.6623
5	(Violet filter) blue	0.6951	0.7708	0.7330
6	(Violet filter) grey	0.6250	0.5541	0.5895
7	(Violet filter) green	0.8072	0.9787	0.8930
8	(Violet filter) red	0.6569	0.9643	0.8106
9	(Yellow filter) grey	0.7647	0.9333	0.8490
10	(Yellow filter) green	0.7363	0.9744	0.8553
11	(Yellow filter) red	0.7037	0.7755	0.7396
	Mean	0.7007	0.8239	0.7623

will not be accidentally classified into a group of non-cancer patients, which is very important.

If you can use images that were taken with only one filter, we can see in Tables 5, 6, 7 and 8 that it is appropriate to take violet filter. At the same time, accuracy is increased compared to the maximum individual accuracy everywhere, except for one against all.

To discover the effect of median filtering we compared Tables 1 and 10 and 8 and 9 respectively. In Tables 9 and 10 demonstrating the results without morphological operations we see a significant decrease in accuracy both on average and individually for all characteristics. Therefore, the use of morphological operations is highly recommended for pre-processing the images used in cancer diagnosis. This also means that the noise on the images that appeared during cytospectrophotometry is of a type similar to that of salt and pepper.

7 Conclusion

The fractal dimension of feulgen-stained nuclei of buccal epithelium is a highly informative feature for the screening of breast cancer. In combination with the nearest neighbor method and similarity measure the fractal dimension of nuclear pattern of feulgen-stained buccal epitheliocytes suggests its usability as a diagnostic breast cancer marker. The high accuracy achieved in our AI system of breast cancer screening in an "old-school manner" of machine learning are comparable with a novel AI systems based on deep learning [16], and at the same time it does not require high-end computers and huge volume of data for training.

References

1. Adam, R., Silva, R., Pereira, F., et al.: The fractal dimension of nuclear chromatin as a prognostic factor in acute precursor B lymphoblastic leukemia. Cell Oncol. **28**, 55–59 (2006)
2. Andrushkiw, R.I., Boroday, N.V., Klyushin, D.A., Petunin, YuI.: Computer-Aided Cytogenetic Method of Cancer Diagnosis. Nova Publishers, New York (2007)
3. Bedin, V., et al.: Fractal dimension of chromatin is an independent prognostic factor for survival in melanoma. BMC Cancer **10**, 260 (2010)
4. Bikou, O., et al.: Fractal dimension as a diagnostic tool of complex endometrial hyperplasia and well-differentiated endometrioid carcinoma. In Vivo **30**, 681–690 (2016)
5. Boroday, N., Chekhun, V., Golubeva, E., Klyushin, D.: In vitro and in vivo densitometric analysis of DNA content and chromatin texture in nuclei of tumor cells under the influence of a nano composite and magnetic field. Adv. Cancer Res. Treat. **2016**, 1–12 (2016)
6. Einstein, A., Wu, H., Sanchez, M., Gil, J.: Fractal characterization of chromatin appearance for diagnosis in breast cytology. J. Pathol. **185**, 366–381 (1998)
7. Hill, B.: Posterior distribution of percentiles: Bayes' theorem for sampling from a population. J. Am. Stat. Assoc. **63**(322), 677–691 (1968)

8. Huanga, S., Yang, J., Fong, S., Zhaoa, Q.: Artificial intelligence in cancer diagnosis and prognosis: opportunities and challenges. Cancer Lett. **471**, 61–71 (2020). https://doi.org/10.1016/j. canlet.2019.12.007

9. Klyushin, D., Petunin, Yu.: A nonparametric test for the equivalence of populations based on a measure of proximity of samples. Ukrainian Math. J. **55**(2), 181–198 (2003)

10. Li, J., Du, Q., Sun, C.: An improved box-counting method for image fractal dimension estimation. Pattern Recognit. **42**(11), 2460–2469 (2009)

11. Losa, G., Castelli, C.: Nuclear patterns of human breast cancer cells during apoptosis: characterization by fractal dimension and (GLCM) co-occurrence matrix statistics. Cell Tissue Res. **322**, 257–267 (2005)

12. Losa, G.: Fractals and their contribution to biology and medicine. Medicographia **34**, 365–374 (2012)

13. Mandelbrot, B.: The Fractal Geometry of Nature. W. H. Freeman and Co., San Francisco (1982)

14. Mashiah, A., Wolach, O., Sandbank, J., et al.: Lymphoma and leukemia cells possess fractal dimensions that correlate with their interpretation in terms of fractal biological features. Acta Haematol. **119**, 142–150 (2008)

15. Matveichuk, S., Petunin, Yu.: A generalization of the Bernoulli model occurring in order statistics. I. Ukrainian Math. J. Ukr. Math. J. **42**, 459–466 (1990). https://doi.org/10.1007/bf0 1071335

16. McKinney, S.M., Sieniek, M., Godbole, V., et al.: International evaluation of an AI system for breast cancer screening. Nature **577**, 89–94 (2020). https://doi.org/10.1038/s41586-019-1799-6

17. Metze, K.: Fractal dimension of chromatin and cancer prognosis. Epigenomics **2**(5), 601–604 (2010)

18. Metze, K.: Fractal dimension of chromatin: potential molecular diagnostic applications for cancer prognosis. Expert Rev. Mol. Diagn. **13**(7), 719–735 (2013)

19. Muniandy, S., Stanlas, J.: Modelling of chromatin morphologies in breast cancer cells undergoing apoptosis using generalized Cauchy field. Comput. Med. Imaging Graph. **32**, 631–637 (2008)

20. Nikolaou, N., Papamarkos, N.: Color image retrieval using a fractal signature extraction technique. Eng. Appl. Artif. Intell. **15**(1), 81–96 (2002)

21. Ohri, S., Dey, P., Nijhawan, R.: Fractal dimension in aspiration cytology smears of breast and cervical lesions. Anal. Quant. Cytol. Histol. **26**, 109–112 (2004)

22. Otsu, N.: A threshold selection method from gray-level histograms. Automatica **11**, 23–27 (1975)

23. Perreault, S., Hébert, P.: Median filtering in constant time. IEEE Trans. Image Process. **16**(9), 2389–2394 (2007)

24. Pires, A.: Interval estimators for a binomial proportion: comparison of twenty methods. REVSTAT Stat. J. **6**, 165–197 (2008)

25. Roberts, L.: Machine perception of 3-D solids. In: Tippett, J.T., et al. (eds.) Optical and Electro-optical Information Processing. MIT Press, Cambridge (1965)

26. Russo, J., Linch, H., Russo, J.: Mammary gland architecture as a determining factor in the susceptibility of the human breast to cancer. Breast J. **7**, 278–291 (2001)

27. Sarker, N., Chaudhuri, B.B.: An efficient differential box-counting approach to compute fractal dimension of image. IEEE Trans. Syst. Man Cybern. **24**, 115–120 (1994)

28. Serra, J.: Image Analysis and Mathematical Morphology. Academic Press Inc., London (1982)

29. Sharifi-Salamatian, V., Pesquet-Popescu, B., Simony-Lafontaine, J., Rigaut, J.P.: Index for spatial heterogeneity in breast cancer. J Microsc. **216**(2), 110–122 (2004)

30. Schroeder, M.: Fractals, Chaos, Power Laws: Minutes from an Infinite Paradise. W. H. Freeman, New York (1991)

31. Susnik, B., Worth, A., LeRiche, J., Palcic, B.: Malignancy-associated changes in the breast: changes in chromatin distribution in epithelial cells in normal-appearing tissue adjacent to carcinoma. Anal. Quant. Cytol. Histol. **17**(1), 62–68 (1995)

32. Us-Krasovec, M., Erzen, J., Zganec, M., et al.: Malignancy associated changes in epithelial cells of buccal mucosa: a potential cancer detection test. Anal. Quant. Cytol. Histol. **27**(5), 254–262 (2005)
33. Voss, R.F.: Random fractal forgeries. In: Earnshaw, R. (ed.) Fundamental Algorithms in Computer Graphics, pp. 805–835. Springer, Berlin (1985)

The Role of Artificial Intelligence in Company's Decision Making

Djamel Saba, Youcef Sahli, and Abdelkader Hadidi

1 Introduction

Decision making is an act that is often performed by humans [1]. Indeed, people are faced with situations in which decision-making is necessary. Often decisions do not require complex thinking processes because generally, it only takes a heuristic experience or sometimes even intuition to decide. But in other situations, decision making is more difficult. This is due to several factors [2]:

- Structural complexity of decisions;
- The impact of the decision taken can be significant, it can be economic, political, organizational, environmental, etc.;
- The need for speed in decision-making, this is the case of medical or military emergencies or even the diagnosis of industrial installations.

Computerized Decision Support Systems (CDSS) have been developed to assist the decision-maker in his task. Emery et al. [3], have introduced the notion of Management Decision Systems. CDSS also designate Interactive Decision Support Systems, since interaction with the decision-maker occupies a prominent place. The purpose of the CDSS is for assisting to the decision-maker rather than to replace it. Several researchers have taken an interest in this field [4–6] which has led to CDSS which are

D. Saba (✉) · Y. Sahli · A. Hadidi
Unité de Recherche en Energies Renouvelables en Milieu Saharien, URER-MS, Centre de
Développement des Energies Renouvelables, CDER, 01000 Bouzareah, Adrar, Algeria
e-mail: saba_djamel@yahoo.fr

Y. Sahli
e-mail: sahli.sofc@gmail.com

A. Hadidi
e-mail: hadidiabdelkader@gmail.com

© The Editor(s) (if applicable) and The Author(s), under exclusive license
to Springer Nature Switzerland AG 2021
A.-E. Hassanien et al. (eds.), *Enabling AI Applications in Data Science*,
Studies in Computational Intelligence 911,
https://doi.org/10.1007/978-3-030-52067-0_13

first interactive and then intelligent. Then, the CDSS is dedicated to semi-structured tasks, the interaction is of considerable contribution because it is based on the involvement of humans in conjunction with the processing of information by the computer, unlike structured situations where solutions and procedures are fully automatable.

To make a system intelligent, one or more knowledge bases were introduced with inference engines [7]. Intelligence can relate to reasoning in decision-making, or in the interaction between man and machine [8]. As a result, the machines became more efficient and therefore could support cooperation with the decision-maker. A system is said to be cooperative if it has additional capacities to cooperate with its user [9]. Cooperation, in this case, consists of sharing the tasks to be carried out between the user and the system [10]. In addition, certain situations require that the decision not be taken by a single individual, but rather within the framework of consultation and collective solicitation [11]. When the problem to be solved is divided into sub-tasks which will be assigned separately to participants, decision-making, in this case, takes on a so-called man-man cooperation aspect between the participants. However, this will not prevent each decision-maker from being helped by an individual cooperative CDSS (human/machine cooperation) in the task assigned to him. When, a priori, there is no division of tasks between the decision-makers who collectively participate in each step of the decision-making process, and, when the latter will lead the group towards a common or consensual decision, the decision, in this case, is taken in a context of collaboration between decision-makers [12]. The collective decision-making process involves a group of people who may be in one place or spread across several places, at the same time or at different times. The current trend is that decision-makers are geographically distant. Indeed, globalization has brought about a change in organizational structures as well as in the attitudes of managers who find them facing new challenges. This new situation is characterized by the following points:

- The evolution of information technology which has enabled geographically distant individuals to share data and ideas;
- The distribution of organizations across the planet;
- Stronger competition;
- The opening of the international market.

Consequently, decision-making requires that decision-making tools support collective decision-making processes where group members will be involved in a cooperative or collaborative decision.

This work fits into the context of collaborative decision support systems. The latter is considered in two dimensions: the collective dimension and the individual dimension. The collective dimension concerns the collaborative aspect because it consists of providing collective decision support where each participant is involved in each step of the decision-making process. There is no sharing of tasks between participants and the group of decision-makers who engage in decision making may be geographically distant [13]. They follow a collective decision-making process which is guided by a facilitator. Such a system is characterized by a strong interactional component that requires rigorous coordination between the facilitator and the decision-makers.

However, the individual dimension concerns the cooperative aspect of the decision because it consists of providing decision support to a decision-maker who is expert in a given field and who is proposed to solve a particular problem. Problem-solving follows a pre-established decision support process which is based on the breakdown of the problem into tasks and sub-tasks. The cooperation, in this case, is of the man/machine type. Indeed, the decision-maker has knowledge and skills, but the machine can also be endowed with knowledge and skills which allow task sharing between the two actors and thus lead to the implementation of cooperation.

In this work, AI technology is used as a basic tool for the decision process. Indeed, AI offers many advantages for its design and development. They consist of distributing complex treatments on entities and intelligent objects of less complexity. The problem consists of finding a good breakdown of the treatments than in implementing good coordination which can ensure the cooperation of the entities (objects) whose common objective is to accomplish the execution of the process of assistance to the collective decision.

The objective of this document is to provide an information platform for the role of AI in decision support in companies. Hence, the remainder of this paper is organized as follows. Section 2 presents the decision making. Section 3 explains the models and decision-making process. Section 4 explains the Interactive Decision Support System. Section 5 details the Cooperative decision support systems and group decision making. Section 6 clarifies the algorithms, software, and AI data. Section 7 provides sensors and connected objects and AI hardware. Then, the Machine Learning, the AI response in assisting decision making, in Sect. 8. Finally, Sect. 9 concludes the paper.

2 Decision Making

Human beings are often faced with problems for which decision-making is necessary. The latter is often taken based on our intuitions or according to our past experiences [14]. However, this decision is not always easy to make because it can be a complex problem for which the decision sought can have relatively significant consequences. Therefore, a bad decision can be expensive and sometimes even fatal. For example, when it comes to managing technological or natural risks, it becomes necessary to take into account a large amount of data and knowledge of different natures and qualities, and for this, managers have more and more use of computers to acquire powerful tools for the decision. However, it becomes necessary to formalize these kinds of problems to guarantee better decision-making while being effective. Efficiency means making decisions quickly by exploring all the possibilities. In addition, the criterion of cost of execution of the decision must also be taken into account in the choice of the decision [15]. Also, it is inappropriate to adopt a trial-and-error strategy to manage an organization and use decision support systems that assess the situation and provide the various alternatives and their impacts [16].

2.1 Decision and Decision Making

Decision support was based on individual experience, knowledge, the experience of advisers to decision-makers, as well as historical analysis [17]. However, opinion and subjectivity were of great importance. In the 20th century, mathematical tools were introduced in Decision support [4]. These models and their algorithms are based on concepts and theories such as probability, decision analysis, graph theory, or operational research. Also, computerized decision support systems have appeared and have taken an increasing place in certain decision-making processes, sometimes to the point of replacing humans with automatic processes. The GIS (Geographic Information Systems) has developed a lot since the 1970s, with the advantage of presenting visually and cartographically [18]. In this context, agencies and design offices have specialized in advice and decision support for banks, companies, governments, communities. Subsequently, business intelligence and the arrival of new methods such as OLAP have been installed [19].

Decision-making is a complex cognitive process that is based on rational and/or irrational assumptions of origin. This process is activated when we feel the need to act when there is a choice for us from a set of alternatives. Preferring one action over another, choosing at random can be the result of decision-making. But in some areas, rational decision-making is required because the consequences can be life-threatening. For example, in the medical field, decision-making intervenes in establishing the diagnosis on which the prescription of treatment and patient monitoring depends. However, in other situations where rapid action is compulsory or when all the information is not available, experts may favor their intuition. Moreover, according to the context and "environmental decision-making, it may concern a single decision-maker or a group. In the latter case, we speak of collective or group decision-making. In this type of situation, the decision process must, in addition, ensure coordination between the actions of the different decision-makers, and to reach a collective decision, it must also provide functionalities for integrating the different proposals of the decision-makers. In general, the decision marks the beginning or ends a situation, in other words, it requires a change of state. In other words, it can be defined as "an act by which one or more decision-makers choose between several options allowing bringing a satisfactory solution to a given problem" [1]. This notion of decision has evolved as the decision-making procedures are transformed and made more complex. In the classical sense of the term, we equate the decision with the act by which an individual (having the power to decide), takes the measures favoring the creation and the distribution of wealth in a company by relying on a set of information to its availability in the market. In the modern approach, decision-making appears more like "a process of progressive engagement, connected to others, marked by the recognized existence of several paths to reach the same and only goal". However, the evolution of the concept of decision is indicative of a certain number of evolutions in the way of apprehending the process of decision-making. The decision is no longer a single, constant act based on the pursuit of profit but is based on a successive set of decisions of lesser significance. The decision is no longer based on the search for a single objective but incorporates a larger number of variables. The decision comes in a more random context in the sense that the manner of achieving the objective pursued can go through different types of actions. These changes are understandable because they only underline the changes in the productive system [20]. The business environment has become more complex, more uncertain, and decision-making is no longer the responsibility of a single individual but can be shared among a large number of actors acting within the company. This increase in the number of decision-makers also reflects the diversity of decisions that must be taken in a company.

There are several definitions of the decision, the first definition is as follows: "A decision is an action that is taken to face difficulty or respond to a change in the environment, that is to say, to solve a problem that arises for the individual (or organization)" [21]. A second definition is as follows: "A decision is the result of a mental process which chooses one among several mutually exclusive alternatives" [22]. A decision is also defined as "a choice between several alternatives", or by "the fact that it also concerns the process of selecting goals and alternatives" [23]. A decision is often seen as the fact of an isolated individual ("the decision-maker") exercising a choice between several possibilities of actions at a given time. In general, the decision can be defined as the act of deciding after deliberation, and that the actor exercises an important role. It is therefore not a simple act, but rather the outcome of a whole decision process. Then, decision-making can appear in any context of daily life, whether professional, sentimental, or familiar, etc. The process, in its essence, solves the various challenges that must be overcome. When it comes to making a decision, several factors are frustrated [1]. Ideally, use your reasoning skills to be on the right track and lead to a new stage or, at least, to resolve an actual or potential conflict. However, all decision-making must include great knowledge of the problem. By analyzing and understanding it, it is then possible to find a solution. Of course, when faced with simple questions, decision-making takes place practically on its own and without any complex or profound reasoning. On the other hand, in the face of decisions that are more transcendent for life, the process must be thought through and treated. When a young person has to choose which studies to pursue after high school, he must make a reasoned decision, since this choice will have important consequences. In the business field, each transcendent decision for a company involves extensive research and study, as well as collaboration between multidisciplinary teams. Finally, the decision-making process is a complex process, the study of which can be facilitated by reference to theoretical models. The model of limited rationality or IMCC proposed by Herbert Simon has four phases [24]: intelligence, modeling, choice, and control.

- **Intelligence**: the decision-maker identifies situations in his environment for which he will have to make decisions.
- **Modeling**: the decision-maker identifies the information, the structures to have possible solutions.
- **The choice**: from the evaluation of each solution, the decision-maker chooses the best of them.
- **Control**: confirms the question made or questions it.

If all the problems that managers had to solve were well defined, they could easily find a solution by following a simple procedure. However, managers often face more complex problems since they frequently have to assess several options, each with advantages and disadvantages. Before even making a final decision, managers must have a thorough understanding of all the conditions and events related to the problem to be resolved. The conditions surrounding decision-making show almost unlimited variations. However, it is generally possible to divide them according to three types of situations: decisions taken in a state of certainty, those made in a state of risk and those made in a state of uncertainty.

2.2 Categories of Problems and Decision-Making and Their Foundations

To improve the quality of decision-making, it is necessary to be able to categorize the problems and thus determine how they will be treated (Table 1).

Intuition-based decisions are those where the factors related to the problem are unknown or those where he is unfamiliar with the adopted solutions. He will choose a solution without having carried out an in-depth analysis. Ultimately, this empirical approach is based on the manager's experience. Then, decisions based on the manager's judgment are also very common in a job. The knowledge and experience accumulated in the position he holds or even within the company can be of great help to him in decision-making on many routine problems. His experience has taught him to use a set of routine rules that have proven to work. Her common-sense allows her to make a multitude of daily decisions without making a detailed analysis of the situation since she is familiar with it. It is difficult to conclude that decision-making is objective, rational, and systematic exercise. Decision-making is a mental process carried out by a human being and therefore linked to all the negative and positive aspects characterizing the individual. The search for the ideal solution

Table 1 Problems categories

Problem category	Nature of intervention	Description of the intervention
Urgent problems	Immediate response	Strike declarations, negative consumer reaction to the launch of a new clothing collection, are typical examples of these problems
Non-urgent problems	Without presenting immediate response	Without presenting the importance of this problem type. The vast majority of management problems fall into this category
Opportunity problems	Appropriate actions are taken on time	It's all about seizing the opportunity to use new ideas rather than struggling to resolve

comprising maximum yield and minimum consumption of resources is not the daily characteristic of all managers. Hence, the manager must be creative in choosing the solution to try to obtain the maximum benefits while minimizing the costs related to his decision. In addition to dealing with different categories of problems, managers find themselves in different decision-making situations. Decisions can be routine, adaptive, innovative, programmed and unscheduled (Table 2).

The policy and procedures explained in writing help individuals to make a choice quickly, as they make detailed analysis unnecessary. However, since most everyday

Table 2 Categories of decision-making

Category of decision-making	Description
Innovative or unscheduled decisions	• Unique, exceptional, and unstructured, these decisions are rarely made • Innovative decisions involve a great deal of uncertainty and do not result from the application of specific rules • To make good, innovative decisions, you need to have sound judgment, creative imagination, and an analytical mind and use quantitative methods to arrive at a better choice
Adaptive decisions	• These decisions represent choices that managers make in response to changes in the market • Unlike those who have to make an innovative decision, executives who have to make an adaptive choice may have already found themselves in similar circumstances • They then know to some extent the conditions that apply as well as the potential outcome of their decision • The risk involved remains significant, however, and several executives generally participate in the decision-making process • This type of decision made by managers is in reaction to decisions made by competitors
Current decisions or scheduled decisions	• Decisions of this nature provide a solution to simple and repetitive problems that arise frequently, such as choosing a meal for dinner or clothes to wear depending on the circumstances. In the business world, hundreds of repetitive decisions are made, most often by following a set of rules, procedures, or established measures • An administrative policy can indeed facilitate routine decision-making. We can thus adopt a rule stipulating for example: — Anyone who applies for the corporate credit card must have a score of 60 points or more to obtain it — That an employee who arrives constantly late will receive a disciplinary sanction, i.e. 2 days of suspension without pay

Table 3 Decision-making methods

Method	Traditional	Modern
Programmable decision	• Habit • Routine • Standardized operating procedures	• Operational research (The models, Mathematical analysis, Computer simulation) • Computer processing of data by programs (algorithms)
Non-programmable decision	• Judgment • Intuition, creativity • Emperies rules • Selection and training of decision makers	• Heuristic problem solving techniques and their computerization (artificial intelligence, expert systems, programming under constraints, …) • Computer processing of knowledge extraction from data (warehouse and data mining)

decisions are repeated frequently and the responsibility of different people, it is important to write guidelines and procedures and then make sure that everyone understands them.

Decisions can be classified into several levels. The SIMON classification proposes two types of decisions, programmed decisions, and unscheduled decisions also called "well-structured decisions", and "weakly structured decisions" (Table 3) [25]. For the first type, it concerns repetitive and routine decisions, for which and a procedure can be defined to carry them out, which avoids having to reconsider them each time they arise. Optimization techniques via mathematical programming and linear programming are used. Their models consist of having several variables linked by a mathematical equation, and it is a question of maximizing one of them. In this class of decisions, the machine takes advantage of the man.

Many problems in organizations can be analyzed in terms of a well-structured decision. For example, the allocation of resources (agent, time, people, space, equipment, assignment of employees or equipment to work, etc.). In this case, a decision-maker must allocate scarce resources for various activities to optimize a measurable objective. The solution is also the best possible, it follows after application of the previously mentioned models. Then, the little (or not) structured decisions (Non-programmable decisions), this second type, these are decisions for which no specific procedure is defined to carry them out because they are new, unstructured or very unusual. Then, solving an unstructured decision problem requires the intuition and know-how of the decision-maker, who is the predominant element of the man/machine couple. These kinds of problems have the following characteristics:

• Their resolution is highly dependent on preferences, judgments, intuition, and the experience of the decision-maker.

Table 4 Classification of decisions according to managerial level

Category of decision-making	Description
Regulatory activity	• These are well-defined problems, fairly easily quantifiable and often of a technical nature • These problems have been solved by classical operational research whose linear programming is the typical example of methods used. Example of problems: inventory management, weekly planning, and optimization of production, scheduling of tasks, etc.
Steering activity	• Concerns problems of intermediate complexity such as the development of the planning of a new project, the methods of setting up a new production unit, etc. • These problems reveal conflicting objectives, difficulties in collecting precise and quantifiable data, aspects of uncertainty and risks, etc.
Strategic planning activity	• Deals with complex problems where the decision is directed towards the achievement of a general goal, often in the long term and some aspects of which are often poorly defined and poorly quantified • This is the highest level of the decision concerning the first manager of the company, for example, the creation and establishment of a new factory, or the establishment of a policy for the creation of jobs by a senior manager

- The objectives pursued during decision-making are numerous, conflicting, and highly dependent on the user's perception.

The search for a solution to these kinds of problems involves a mixture of information search, problem formulation, calculation, and data manipulation. In this class of decisions, one of the most important aspects is that man takes advantage of the machine, unlike structured problems. The decision-maker can adopt a progressive strategy with backtracking. Also, it is necessary to provide him with methods and tools to assist him in solving the decision problem.

The decisions can also be classified according to the managerial level. Indeed, the levels are distinguished by the nature of the activities that take place there. This influences the type of decisions that decision-makers are led to make. There are three types of activities (Table 4).

3   Models and Decision-Making Process

Decision models are generally based on assumptions of the rationality of decision-makers and possible solutions. There are two types of models: normative models and descriptive models [26]. For normative models, the proposed solution is the best and it is possible to prove it. While for the other, only fairly good or satisfactory solutions are provided.

- **Normative models**: in this type of model, there are three categories of normative models, complete enumeration which seeks the best solution among a relatively small set of alternatives. The main methods are tables, decision trees, and multi-criteria analysis. Then, the models for optimization via algorithms, which seek to find the best solution among a large or even infinite set of alternatives, using a step-by-step improvement process. The main methods are linear programming, integer linear programming, convex programming, and multi-objective programming which is a variant of linear programming for several functions (criteria) to be optimized simultaneously. Finally, for models for optimization using analytical formulas, is the find the best solution in one step using an analytical formula.
- **Descriptive models**: give a satisfactory solution by exploring part of the solutions. Among the descriptive models, the simulation concerns a technique that makes it possible to carry out decision making by observing the characteristics of a given system under different configurations. This technique makes it possible to solve by choosing the best among the alternatives evaluated. The second model concerns prediction, which makes it possible to predict the consequences of the different alternatives according to prediction models. Markov models are among the best-known methods in this category. Prediction provides a fairly good solution or a satisfactory solution. Then, heuristic models allow reaching a satisfactory solution at lower cost by using heuristic programming techniques and knowledge-based systems. Rather, these methods are used for complex and poorly structured problems where finding solutions can result in high cost and time.

The decision process consists of determining the steps to be followed by a decision-maker to arrive at opting for a decision as a solution to a problem posed. In 1960 Simon has proposed the IDC model breaking down the decision-making process into three stages (intelligence—design—choice). In 1977, this same researcher revised his model by adding a fourth step (review or evaluation) [27]. The latter can be seen as a process evaluation step to validate or not the decision to be applied (Fig. 1). This model remains to this day a reference for decision modeling.

- **Study of the existing (discover the problem and collect the data)**: this is a phase of identifying the problem. It involves identifying the objectives or goals of the decision-maker and defining the problem to be solved. For this, it is necessary to seek the relevant information according to the concerns of the decision-maker.
- **Design (formulation and modeling of the problem)**: this is a modeling phase proper. The decision-maker builds solutions and imagines scenarios. This phase leads to different possible paths to solving the problem.
- **Calculations (display of results)**: the models formulated in the previous step are used to perform the calculations associated with the resolution of the problem addressed. The display of the results is done through output devices (screens).
- **Choice (choose from the alternatives)**: this is a phase of selecting a particular mode of action, that is to say making a decision.

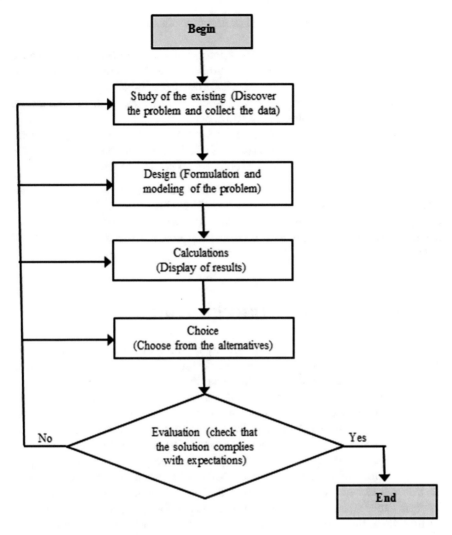

Fig. 1 Modeling of the decision-making process

- **Evaluation (check that the solution complies with expectations)**: this phase makes it possible to evaluate the solution was chosen (the decision is taken). It can lead to backtracking to one of the three previous phases or, on the contrary, to the validation of the solution.

Decisions are not made after posing the problem and collecting all the information, but gradually during a long process of action and planning [28]. Indeed, for any decision-making process, we develop a decision model. The latter generally consists of five elements:

- The decision-maker: an individual or group responsible for making a particular decision.
- A set of inputs to the decision-making process: these are numerical or qualitative data models for interpreting this data, experience with similar data sets, similar decision situations, constraints associated with decision-making, etc.
- The decision process itself: a set of steps, more or less well understood, to transform inputs into outputs in the form of decisions.
- A set of outputs from the decision-making process including the decisions themselves, and ideally a set of criteria to assess the decisions produced by the process in the relation to the set of needs, problems or objectives of the decision.

Generally speaking, decision-making is considered to be any mental process after which any individual, faced with several mutually exclusive alternatives, chooses one of them. Some decisions are easy to make. However, there are complex decisions that require the monitoring of a decision-making process associated with assistance to the decision-maker during this process. This aid is provided by CDSS.

4 Interactive Decision Support System (IDSS)

Decision support is defined as "The activity of someone who, based on models, clearly explained but not necessarily completely formalized, helps to obtain elements of response to questions posed by an actor in a process of decision. Elements contributing to clarify the decision and normally to recommend, or simply to encourage, a behavior likely to increase the coherence between the evolution of the process, on the one hand, the objectives and the system of values at the service of which these actors are placed on the other hand" [29]. The first interactive decision support system was released in 1971 and when it comes to Big Data, it is simply a logical evolution of what has been done before [30]. Michael Scott Morton based his research on how computers and analytical models would help managers make strategic decisions, using the machines, communications systems, and databases of the time. So he managed a pioneer project in which three executives of a large organization (the production manager, marketing director and the president of the division) used the ISSD (interactive support system decision) or MDS (management decision system), to plan the monthly production of washing and drying machines [30]. His research gave birth to the first interactive decision support system in June 1967. At the time, there was no microcomputer and we don't talk about smartphones like today, however, Michael Scott Morton created this system which he describes himself as an expensive invention but beneficial for the decision-making process taking into account the characteristics of the computers of the time, far from being "individual and portable" [31]. This machine consisted of a central unit, a keyboard and a special screen because at that time the screens were not able to display graphics. It was, therefore, necessary to develop a screen to display the curves calculated by the system. This system was connected by a telephone line with the company to retrieve the data on the central computer. Undoubtedly the challenge for companies, in the context of their decision-making support projects, is the processing of their data to extract their value. And since these concepts and systems have been developed for more than 40 years, the excuse for not carrying out these projects cannot be in technology, as Michael Scott Morton declares, "The general unresolved issue I see is one of understanding the management of change. Without a better understanding of this, it is hard to implement and learn from DSS applications. As an engineer trained in the technology it took me a while to understand that the hard problems lie in the 'soft' domains of management and human behavior, not in the hardware and software".

In general, an IDSS is a computer system that must help a decision-maker to follow a decision support process. It is thanks to the interactivity which is the basis of the cooperation of the two partners (the system and the user) that the IDSS will fulfill the role for which it was designed. These are systems that use computers to:

- Assist decision-makers during the decision-making process concerning semi-structured tasks,
- Help rather than replace the judgment of decision-makers,
- Improve the quality of decision-making rather than efficiency.

In addition, various functions are assigned to the IDSS:

- They should mainly help with poorly structured or poorly structured problems by connecting human judgments and calculated information.
- They must have a simple and user-friendly interface to avoid the user being lost in front of the complexity of the system.
- They must provide help for different categories of users or different groups of users.
- They must support interdependent or sequential processes.
- They must be adaptive over time. The decision-maker must be able to withstand rapidly changing conditions and adapt the IDSS to deal with new situations. An IDSS must be flexible enough for the decision-maker to add, destroy, combine, change and rearrange the variables in the decision process as well as the various calculations, thus providing a rapid response to unexpected situations.
- They must leave control of all stages of the decision-making process to the decision-maker so that the decision-maker can question the recommendations made by the IDSS at any time. An IDSS should help the decision-maker and not replace him.
- The most advanced IDSS uses a knowledge-based system that provides effective and efficient assistance in particular in problems requiring expertise.
- They must allow the heuristic search.
- They must not be black box type tools. The functioning of an IDSS must be done in such a way that the decision-maker understands and accepts it.
- They must use templates. Modeling makes it possible to experiment with different strategies under different conditions. These experiences can provide new insights and learning.

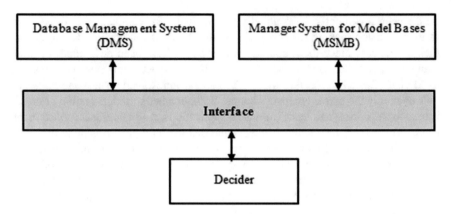

Fig. 2 IDSS structure

Since there are several definitions for IDSS, there are also several architectures. Previously, researchers defined IDSS in terms of data and models for solving poorly structured problems. They proposed the following IDSS architecture (Fig. 2).

This architecture is composed of a Human/Machine interface, a Database Management System (DMS) including the Database as well as a Manager System for Model Bases (MSMB) including a Model Database. In addition, a system based on this architecture has interactive capabilities that allow the user to be involved in solving unscheduled, poorly structured, or semi-structured problems. In the same context, a conceptual architecture is proposed, and a prototype system which validates the technical feasibility of the proposed architecture is presented by Szu-Yuan Wang et al. The relational data model is used in the proposed approach, and how its tabular structure can be exploited to provide a unified logical view of data management and model management [32]. In addition, the concept of the macro is used to extend the conventional meaning of the model to include the composite requests that often occur in a DSS environment. A primitive knowledge base of macros, data, and associated dictionaries for both is included in the architecture. Organized in a three-level hierarchy, the knowledge base consists of macros and system-wide, group, and individual data. Also included in the architecture as a language system and modules to monitor system performance, maintain data and macro security, and link the DSS to other parts of the local computing environment.

In the development of enterprise applications, a critical success factor is to make good architectural decisions. There are text templates and support for tools to capture architectural decisions. One of the inhibitors on large-scale industrial projects is that the capture of architectural decisions is considered as a retrospective documentation task and therefore undesirable which brings no advantage during the design work. A major problem with such a retrospective approach is that the justification for the decision is not available to decision-makers when they identify, make, and apply decisions. Often, a large community of decision-makers, possibly distributed, is involved in these three stages. Zimmermann et al. proposed a new conceptual framework for the proactive identification of decisions, the collaboration of decision-makers, and the implementation of decisions [33]. Based on a meta-model explicitly capturing an aspect of reuse and collaboration, these models capture the knowledge acquired on other projects using the same architectural style. An exemplary application of these concepts to a service-oriented architecture, reusable architectural decision models can accelerate the identification of the decision and improve the quality of decision-making. Reusable architectural decision models can also simplify the exchange of architecture design logic and expose decisions made as model transformation parameters in model-driven software development. In addition, enterprise architecture modeling languages holistically capture the structure of an enterprise. They represent how the services and business processes of an organization are supported by IT infrastructure and applications. However, the reasoning behind the selection of specific design decisions in architecture remains generally implicit. The work of Plataniotis et al. proposed the EA Anamnesis approach which captures the information of rationalization of the design in the solution space of the architecture of the business [34]. This contribution concerns a formal metamodel that captures the reasoning behind design decisions and the relationships between them. Then, they proposed an approach with concepts from the domain of problematic space of enterprise architecture, such as objectives, principles, requirements. In addition, we provide a bridge to existing EA Anamnesis concepts that are part of the solution space. In doing so, they presented the extent to which EE design decisions, which define the design of EE, are consistent with the given objectives, principles, and requirements. The extension is assessed with a real-world case study within a research and technology organization. In the same context, visualizations have become a de facto standard as a means of decision-making in the discipline of enterprise architecture management (EA). As a result, these visualizations are often created manually, so that they quickly become obsolete because of the underlying data changes frequently. Therefore, EA management tools require mechanisms to generate visualizations. In this perspective, a major challenge consists in adapting current EA visualizations to an organization-specific metamodel. At the same time, end-users want to interact with the visualization in terms of changing data immediately in the visualization for strategic planning of an environmental assessment. To date, there is no standard, framework, or reference model for the generation of such an interactive EA visualization. Schaub et al. they presented a framework for the interactions of different models to produce interactive visualizations, they described the requirements for interactive EA management visualizations referring to the concepts of the framework, then they applied the framework to a prototypical implementation detailing the models used as an example [35]. Then, the contribution of Marín et al. is a conceptual architecture based on intelligent services for the manufacture of support systems [36]. Current approaches cannot cope with the dynamism inherent in the manufacturing field, where unexpected and drastic changes occur at all times and affect cross-border systems. This approach alleviates these problems and focuses on the need to support real-time automatic negotiation, planning, planning, and optimization, inside and between factories. They presented an example of the implementation of this approach in the aerospace industry, instantiating the conceptual architecture using different technologies.

There are several classifications of IDSS, a classification according to the quantity of information handled, a classification according to the level of the decision involved, a classification according to the scope of the decision, and a classification according to the conceptual level of the system [37]. In the classification which is based on the amount of information handled an enterprise IDSS is linked to large data warehouses and is used by several managers in the enterprise. Then, a simple user or desktop IDSS is a small system that resides in an individual manager's ordinator. So in the classification which is based on the level of decision, researchers distinguish four types of decision-making systems according to the level of the decision involved in an IDSS [37]:

- **Executive Information System (EIS)**: is a system specially designed to meet the needs of a company's senior management and which is exclusively reserved for it. The interest of EIS is to facilitate access to relevant information by allowing navigation from the summary to the detail. The aim is to compare this information with the objectives sought.
- **Executive Support System (ESS)**: an ESS not only provides access to critical data but also integrates the analysis of this data. This means that for an EIS to be considered an ESS, it must offer the possibility of also supporting the design and choice phases. An ESS allows, in addition to navigation and multidimensional crossing, to process large volumes, to ensure the integration of information of various origins and to manage hierarchies and aggregates.
- **Decision Support System (DSS)**: it is an interactive system that helps the decision-maker to use data and models to find a solution to an unstructured problem and to analyze the effect of possible changes in the environment on the organization. The goal of the DSS is to assist decision making, not to replace the decision-maker. It should allow for strategic planning, as well as long-term budgeting.
- **Planning Support System (PSS)**: it allows an analysis of the feasibility of the procedures or decisions adopted. It, therefore, provides intelligent assistance to the decision-maker.

In the classification which is according to the scope of the decision. When only the size of the decision is taken into account, the IDSS can be classified into three categories:

- **The operational IDSS**: avoids the mental overload of the operator by proposing solutions allowing him to deal quickly with complex situations. This automaton is part of expert systems, only equips those of operators who may have to solve very difficult problems under the constraint of emergency.
- **The management IDSS**: it presents operational managers with the daily indicators and alarms useful for monitoring the work of operators (compliance with quality standards, the workload of resources). A management SIAD should equip any production process.

- **The strategic IDSS**: it presents managers with a periodic time series. It provides the management committee with a shared and early assessment of the essential indicators.

Finally, for the classification which is according to the conceptual level of the system using the assistance mode as the criterion, there are four generic types of decision support systems:

- A data-centric IDSS highlights access to and manipulation of a time series of data internal to the organization and sometimes of external data;
- A model-oriented IDSS highlights access to and manipulation of simulation, optimization, financial, and statistical model. A Model-Oriented IDSS uses data and parameter provided by users to help decision-makers analyze a situation, but is not necessarily data-centric;
- Knowledge-driven DSS providing the problem-solving expertise stored as facts, rules or procedures;
- A document-oriented IDSS provides expertise in solving specialized problems, stored as facts, rules, procedures or in similar structures;
- IDSS oriented communication supports more than a person working on a shared task.

5 Cooperative Decision Support Systems and Group Decision Making

Cooperation can be located at different levels when implementing cooperative systems:

- **Human-machine cooperation**: cooperative systems at this level have additional capacities to guide the user in his problem-solving process. One of the most important aspects of decision-making is that humans take advantage of the machine, unlike structured problems. To make a SIAD cooperative is to endow it with additional capacities to cooperate with its user and guide him in his problem-solving process [38]. Cooperating in this case, in particular, means distributing the tasks to be carried out between the user and the system. The task that is the subject of cooperation must be broken down into coherent sub-tasks. The distribution of tasks between the two agents (user, system) is done dynamically, according to the performance of each of the two agents and the workload of the operator. In addition, these systems are characterized by human/machine interactions through intelligent adaptive interfaces.
- **Mediated human-human cooperation**: cooperative systems of this level have in themselves functionalities capable of supporting cooperation. In a general framework, a decision is said to be cooperative when it involves several actors. This process of collective construction of the decision must be managed as a project with real meetings between the different decision-makers and return dates for

sub-tasks. These so-called CDSS (Collaborative Decision Support Systems) or GDSS (Group Decision Support Systems) systems must support the concepts of communication, coordination, and cooperation between the various players concerned [39].

- **An interpersonal communications tool**: an infrastructure such as the Internet makes it possible to implement communication tools (messaging, discussion forums, etc.).
- **A task management tool**: It ensures the distribution of tasks into sub-tasks as well as the assignment of roles or agents to these tasks. It is this entity that allows cooperation between various users or between the machine and the users.
- **A knowledge capitalization tool**: capitalize on the knowledge of the decision-makers involved in the decision-making process so that everyone can refer to it if necessary. Decision-makers involved in distributed and asynchronous decision-making processes can be supported by this tool by reusing existing solutions for example or simply parts of already established solutions.
- **A dynamic Human/Machine Interaction tool**: this tool must be more advanced than that of traditional SIADs. It must take into account constraints relating to the cognitive load of the user during the execution of a task, constraints relating to the resolution of the global problem.

In some situations, decisions are not made by a single individual. Even if the responsibility for decision-making rests with a single individual, the decision can be taken within the framework of consultation and collective solicitation. The collective decision-making process involves a group of people who may be in one place or spread across several places, at the same time or at different times. Group decision making is a process used by most organizations and at different hierarchical levels. It involves discussing the possible choices with a group to determine a final solution. The decision, in this case, is the outcome of the decision process. When the problem to be solved is divided into different sub-tasks which will be assigned to individual participants, decision-making, in this case, is not only collective but also includes an aspect of cooperation between participants. However, in the research area dedicated to IT and collaborative tools, the collaborative decision is associated with an "argumentative" process where each participant must take into account the other collaborators to understand the constraints and solutions to the problem posed, the interests and everyone's priorities.

In the field of artificial intelligence, collaborative decision-making is defined as being associated with a group of distributed entities that cooperate to achieve objectives that exceed the individual capacities of the entities [40]. Collaborative decision-making is generally associated with a distributed reasoning method according to which a group of entities work collaboratively via a common research space. Finally, collective decision-making can be defined as the engagement of a group of decision-makers in a decision-making process. This group can be divided. The actors can have different points of view that they can argue. They may have a private workspace but must have the means to communicate to share a public workspace.

A group of people who come together and engage in a collaborative decision-making process. But getting together doesn't mean that group members have to be in one place. Indeed, they can be dispersed geographically but they have the means to communicate. In addition, there are three ways in which group work can be organized:

- **Interactive group**: the members of an interactive group communicate with each other and try to continue their task. In a face-to-face meeting, only one person in the group can come up with their idea at any given time, because group members can only pay attention to one person at a time. The interactive group can take advantage of social synergies; however, negotiations between members can cause the process to be lost.
- **Nominal group**: in a nominal group, the members of the group work separately on the same task and one of the results is chosen as the product of the group. The members of a nominal group do not benefit from the social synergies of large groups, but on the other hand, they are not affected by undesirable effects of interactive work such as production stoppage or apprehension of evaluation. A nominal group can also be used to provide an anonymity process for group members.
- **Teamwork**: teamwork combines aspects of both interactive and nominal group work. The working group is divided into teams (typically 2-5 people), who work separately. The teams are small enough not to undergo the process losses of large groups but large enough to take advantage of social synergies. However, the teams can undergo the negotiations between members as in the interactive working groups.
- **Group size**: Some research has shown that the optimal size of a group without technological support is 3–5 people. Indeed, this size of the group favors the number of ideas generated per person and better supports informal discussions. However, in certain cases, when the problem requires a great deal of expertise, a larger number of people becomes necessary. When a GDSS is used, it will be necessary to test the system to define the optimal group size because it depends on the nature of the problem and also on the way in which the participants react with the system.
- **Anonymity**: anonymity is not necessary for a collective decision-making meeting. However, in some cases, it may be beneficial to allow participants to express their opinions freely without being influenced by other group members (their leaders, for example). However, this can have the disadvantage of being a source of demotivation on the part of the participants or even creating conflicts insofar as the participants can use derogatory expressions.
- **Composition of the group**: there are two types of actors involved in a group decision-making process: participants or experts in decision-making and the facilitator also called the mediator or initiator.
  - **The facilitator**: he is a member of the group. It is the actor who has the role of overseeing the collective decision-making process. First, it launches and prepares the different phases of the process. He then defines the problem of the decision and organizes the group of decision-makers. He is responsible

for making the decision-making process converge. He is also responsible for disseminating the results to participants at the end of the decision-making session. He is therefore responsible for the entire process and its completion, namely the decision.

- **Participants**: these are the other actors in the group other than the facilitator. They represent the experts in decision-making. Participants are first invited to the meeting. During the meeting, they receive the problem from the facilitator. Their roles are to participate in the collective decision-making session. For this, they produce ideas and comments which they share with the other participants. The activity of the participants is coordinated by the facilitator.

6 Algorithms, Software and AI Data

To describe the universe of algorithms, software, and tools for AI development, it is important to use a simplified segmentation of the domain with the tools of automatic reasoning, those of machine learning which integrate neural networks and deep learning. Finally, the AI bricks integration layer that constitutes the notions of agents and multi-agent networks. This segmentation covers the main current uses of AI since most of the image processing, natural language, and perception solutions are deep learning, those that manage structured data and in particular forecasting are machine basic learning and those which relate to reasoning and planning relate to different variations of rule engines, solvers and associated tools. However, many parts of the AI exploit the different bricks among them:

- Evolutionary or genetic algorithms which can be based on deep learning and which test several versions of solutions to keep only the best.
- The representation of knowledge which extracts it from unstructured textual data, for example, via deep learning, and then most often uses it in automatic reasoning with the tools of symbolic AI [41]. This is the direction of the side arrow in the diagram. But this representation of knowledge, used in particular in conversational agents can be fully exploited in neural networks.
- Affective AI which exploits a wide range of machine learning and deep learning tools and even automatic reasoning to capture and classify external elements of human emotions and act accordingly.
- Transfer learning, or transfer learning, is a variant of deep learning that allows you to train a neural network from a neural network already trained to complete, update or use it in a neighboring field of the initial domain.
- The building blocks of reasoning and planning that solve logic problems where there are elements of factual information in the form of rules, facts and constraints, which must be available beforehand to exploit them [42]. These problems can relate to maintenance, how to manage the driving of an autonomous vehicle, the operation of a robot or a numerically controlled machine in a factory or the optimization of the exploitation of multiple resources to accomplish tasks. The best known tools in this area are expert systems built with solvers which

exploit the bases of formal rules and facts. Rule engines are now called BRMS for Business Rules Management Systems and are often integrated into DMS for Decision Management Systems. These systems can incorporate fuzzy logic to handle imprecise facts and rules.

- **Machine learning or automatic learning**: is used to make forecasts, classification and automatic segmentation by exploiting generally multidimensional data, such as a customer database or an Internet server log. Machine learning is based on a probabilistic approach. Machine learning tools are used to exploit large volumes of corporate data, in other words "big data". Machine learning can rely on simple neural networks for complex tasks involving multidimensional data.
- **Neural networks**: are a sub-domain of machine learning for performing identical tasks, but when the probabilistic space managed is more complex. This elementary bio mimicry is exploited when the dimension of the problem to be managed is reasonable. Otherwise, we quickly move on to deep learning, especially for image and language processing.
- **Deep learning**: allows you to go further than machine learning to recognize complex objects such as images, handwriting, speech and language. Deep learning uses multilayer neural networks, knowing that there are many variations. However, this is not the solution to all of the problems that AI seeks to address. Deep learning can also generate content or improve existing content, such as automatically coloring black and white images. Deep means deep. But deep learning does not think. It is deep because it taps into neural networks with many layers of filters. That's all! Deep learning, however, is not exclusively dedicated to image and language processing. It can be used in other complex environments such as genomics. It is also used in so-called multimodal approaches which integrate different senses such as vision and language.
- **Agent networks**: or multi-agent systems are a little-known area that covers the science of orchestrating the technical building blocks of AI to create complete solutions. A Chatbot like a robot is always a motley assembly of the bricks below with rules engines, machine learning and several deep learning techniques [43]. Agent networks are both conceptual objects and tools for assembling AI software bricks. The principle of an agent is that it is conceptually autonomous, with inputs and outputs. The assembly of agents in multi-agent networks is a "macro" version of the creation of AI solutions.

The common point of machine learning and deep learning is to use data for the training of probabilistic models. Along with algorithms/software and hardware, data is the third key component of most AI today. There are generally three types of data to train a machine learning and deep learning system: training data, test data and production data.

In machine and supervised deep learning, training and test data contains their label, that is, the information that must be generated by the system to be trained. It is a test game with a good distribution of the possible space of the application. Firstly, it is arbitrarily divided into two subsets, one for training and the other for qualification tests of the trained neural network which determine a recognition error rate. In general, the share of the tagged base dedicated to training is greater than that which is dedicated to tests and in a 3/4 and 1/4 ratio. However, training and test data is essential for the vast majority of machine learning-based AI systems, whether for supervised or unsupervised learning. Then, reinforcement learning uses a smaller amount of data but is generally based on models already trained beforehand, followed by the addition of reinforcement data which are used for reinforcement learning. Finally, these are new training datasets that allow you to adjust that of an already trained neural network.

- **Training data**: these are the data sets that will be used to train a machine learning or deep learning model to adjust the parameters. In the case of image recognition, it will be a database of images with their corresponding tags that describe their content. The larger the base, the better the system will train, but the longer it will take. The image training bases have a size which depends on the diversity of the objects to be detected. In addition, in medical imaging, training bases for specialized pathologies can be satisfied with a few hundred thousand images to detect a few tens or hundreds of pathologies. At the other extreme of complexity, Google Search's image training base relies on hundreds of millions of images and allows the detection of more than 20,000 different objects. However, training a 50,000 image system takes about a quarter of an hour in cloud resources, but it depends on the resources allocated on that side. When you go to hundreds of millions of images, it will take thousands of servers and up to several weeks for training. In practice, the training games for deep learning solutions are limited in size by the computing power required. You can also perform incremental training as you add data using neural network transfer techniques. It is necessary to have quality training data, which often requires a great deal of filtering, cleaning and reduplication before data ingestion, a task already existing in the context of big data applications.
- **Test data**: this is the data, also tagged, which will be used to check the quality of the training of a system. These data must have a statistical distribution close to training data, in the sense that they must be well representative of the diversity of data that we find in the training base and that we will have in production data. However, test data is a subset of a starter set, part of which is used for training and another, more limited part, which is used for testing. They will be injected into the trained system and the resulting tags will be compared with the base tags. This will identify the system error rate. We will go to the next step when the error rate is considered acceptable for the production of the solution. Finally, the level of acceptable error rate depends on the application. Its generally accepted maximum is the error rate of human recognition. But as we are generally more demanding with machines, the rate truly accepted is much lower than that of humans.
- **Production data**: this is untagged data that will feed the system when it is used in production to forecast missing tags. While training data is normally anonymized for system training, production data can be nominative as well as the associated forecast generated by the solution.

- **Reinforcement data**: the use of this expression to describe the data used for reinforcement learning. In a Chatbot, this will be, for example, the reactivity data of users to Chatbots' responses allowing them to identify which are the most appropriate. In a way, these are results of A/B testing performed on the behavior of AI-based agents. Anything that can be captured from the real-world reaction to the actions of an AI-based agent will potentially adjust behavior through retraining. However, reinforcement learning is ultimately a kind of incremental supervised learning because it is used to evolve already trained systems in small impressionist touches.

The data that feeds the AI systems comes from inside and/or outside the company. They come from all kinds of various sensors: connected objects, from the simplest (connected thermometer) to the most sophisticated (machine tool, smartphone, personal computer). As with traditional big data applications, data sources must be reliable and data well extracted and prepared before being injected into machine-based systems like deep learning. However, the most advanced solutions jointly exploit external open data and cross it with data that only the company can control. It's a good way to create differentiated solutions.

7 Sensors and Connected Objects and AI Hardware

Sensors and connected objects play a key role in many artificial intelligence applications. Microphones and cameras power speech recognition and artificial vision systems. Smartphones and Internet access tools in general are creating data drums on user behavior. The smart city and autonomous vehicles are also powered by all kinds of sensors. However, one of the ways to get closer to and even go beyond humans is to multiply sensory sensors. The main difference between humans and machines is the range of these sensors. For humans, the range is immediate and concerns only its surroundings. For machines, it may be distant and global. We see around us, we feel the temperature, we can touch, etc. The machines can capture environmental data on a very large scale. This is the advantage of large-scale connected object networks, such as in "smart cities". And the volumes of data generated by connected objects are growing, creating both a technological challenge and an opportunity for their exploitation. In addition, the brain has an unknown characteristic: it does not include sensory cells. This explains why you can do open brain surgery on someone awake. The pain is only noticeable at the periphery of the brain. When you have a migraine, it is generally linked to peripheral pain in the brain, which does not come from inside. The computer is in the same case: it does not have its sensory sensors. He doesn't feel anything if he's not connected outside. AI without sensors or data is useless. Indeed, convolutional neural networks use low resolution source images to take into account current hardware constraints. They are rare to operate in 3D with stereoscopic vision.

The sensor market has experienced strong development since the late 2000s thanks to the emergence of the smartphone market, powered by the iPhone and Android smartphones. There are currently about 1.5 billion units sold per year, and they are renewed approximately every two years by consumers. As an example, connected home management platforms also take advantage of many room sensors to optimize comfort [7, 44]. They play on the integration of disparate data: outdoor and indoor temperature, humidity, brightness and the movements of users, captured with their smartphones. This allows for example to anticipate the temperature of the accommodation in anticipation of the return home of its occupants [8, 13]. This orchestration is going more and more often through deep learning to identify user behaviors and adapt system responses.

Innovation in photo and video sensors is also ceaseless, if only by the miniaturization of those that equip smartphones and are now equipped with 3D vision. The American Rambus is working on its side on a photo sensor that does not need optics! Vibration sensors and microphones have unsuspected industrial applications revealed by AI: the detection of anomalies. Thus, sensors placed in industrial vehicles or machines generate a signal which is analyzed by deep learning systems capable of identifying and characterizing anomalies.

One of the key tools of AI is the deep learning training servers. If it gives very good results, as in image recognition, it consumes a lot of resources during its training phase. It takes easily 1000–100,000 times more machine power to train an image recognition model than to run it next. This explains why, for example, GPUs and other TPUs (Tensor Processing Units) have a computing capacity of around 100 Tflops/s while the neural bricks of Huawei's latest Kirin 980s and the A11 Bionic are content with 1 at 4 Tflops/s. And again, training the largest neural networks requires hundreds or even thousands of servers using these GPUs and TPUs. In addition, GPUs are the most widely deployed hardware solution to accelerate neural networks and deep learning. They are the ones who made deep learning possible, especially for image processing and from 2012.

Quantum computers are used to solve so-called exponential mathematical problems, whose complexity and size grow exponentially with their size. They are based on qubits, information management units handling 0 s and 1 s, but in an overlapping state, and arranged in registers of several qubits. A system based on n qubits is capable of simultaneously representing 2n states to which various operations can be applied simultaneously.

8 The Machine Learning, the AI Response in Assisting Decision Making

Machine Learning is the most promising area of application for Artificial Intelligence. It uses specific algorithms, created jointly by analysts and developers, which consolidate and categorize a variety of raw data, combining the processing and analysis of information [45]. Machine Learning continues to develop its understanding of information to provide an increasingly relevant and useful result to the user. It's also learning that is done with or without human intervention. This ability to adapt automatically ensures that the database of the "machine" is constantly updated and adapts to the requirements. This allows him to deal with a request from a holistic point of view, providing a qualitative and targeted analysis of the need. It is interesting to observe the correlation between Big Data and Machine Learning, namely, to offer a "machine" that can grasp and process a large amount of raw data. This opens up the field of possibilities, especially since whatever the field of study, the applications is multiple. Take for example the case of companies' financial and administrative departments: imagine being able to identify improvements within their service on various subjects such as fraud detection, decision-making, reporting, etc.

Machine Learning should be seen as both a tool that optimizes processes and improves employee engagement, but also as a great lever for customer satisfaction.

The absence of human intervention in the analysis and processing of data makes Machine Learning the most avant-garde technology concerning the optimization of business processes (reduction of the workload of third parties or human interaction). For example, it can process a supplier's order, taking into account the current stock level of the warehouse and that of the requesting point of sale. In parallel, it can deliver a relevant analysis of the most trendy products (based on the level of sales of a set of stores), to assist the logistics manager when making his replenishment choices. The added value of Machine Learning lies here in its real-time qualitative analysis, which the company can use as decision support. Machine Learning also represents a lever in controlling fraud risk, by creating algorithms capable of:

- Analyze information on all types of existing fraud;
- Analyze the information to determine the new types of fraud possible;
- And assess the associated financial impacts.

These helps support organizations and make their anti-fraud policy more effective. For a single person, this requires time as well as meticulous research work. If Machine Learning is capable of carrying out this detection work, organizations only have to put their efforts into resolving disputes. Organizations can also create new detection policies based on machine learning analyzes.

9   Conclusion

AI is a discipline of computer science closely linked to other sciences: mathematics, logic and statistics which serve as a theoretical basis, the human sciences (cognitive sciences, psychology, philosophy, linguistics, etc.) and neurobiology which help to reproduce components of human intelligence by bio mimicry, and finally, the hardware technologies which serve as physical support for the execution of AI software. AI represents at the same time a whole section of IT with its diversity, its technological bricks, its methods, its assemblies and solutions of all kinds, and a set of technologies which are now embedded in almost all aspects of digital. It's a veritable motley ecosystem. The vast majority of AI commercial solutions are made up of odds and ends, depending on specific needs. It's far from having generic AI solutions, at least in business. In the general public, AI bricks are already used in everyday life without users noticing it. This is for example the case of face tracking systems in the development of photos and videos in smartphones. The most generic technological building blocks of AI are software libraries and development tools such as TensorFlow, PyTorch, scikit-learn or Keras. Since 2015, artificial intelligence (AI) has become one of the top priorities of major players in the digital economy and businesses. The wave of AI is starting to match that of the Internet and mobility in terms of impact. It is rather difficult to escape! It has become a tool for business competitiveness both in the digital sector and in the rest of the economy.

In the environment, decision makers are faced with difficult decisions. Many of these decisions are made through either intuition, experience or ineffective traditional approaches. Making appropriate decisions usually involves controlling and managing risk. Although decision makers have some control over the levels of risk they are exposed to, risk reduction must be pursued by experts to reduce costs and use resources efficiently.

Improving added value is the major challenge for companies to be competitive. However, an Interactive Decision Support System (IDSS) is one that determines good production practices and the product that maximizes added value.

Businesses are increasingly confronted with problems of such complexity and size that they cannot be solved by a human, no matter how expert. This is the case for extracting knowledge from a large mass of documents, integrating data from heterogeneous sources, detecting breakdowns or anomalies, re-planning in "real time" in the event incidents in the fields of transport or production chains, etc. In addition, the company tends to delegate certain dangerous or tedious activities to robots or software agents: drones for air combat, demining robots, exploration rovers of the planet Mars… However, performing these activities requires autonomy, complete or partial, based on learning and adaptation capacities in the event of an unforeseen situation. Finally, humans and machines (computers, robots, etc.) are increasingly called upon to communicate using natural language, speech, or even images. It is precisely the aim of artificial intelligence and decision support to deal with these problems which are too complex for humans or to take an interest in autonomy, learning or even human- machine communication. Demand, in commercial terms, therefore exists and is likely to increase over the next few years.

References

1. Kvist, T.: Decision making. In: Apical Periodontitis in Root-Filled Teeth: Endodontic Retreatment and Alternative Approaches (2018)
2. Landrø, M., Pfuhl, G., Engeset, R., et al.: Avalanche decision-making frameworks: classification and description of underlying factors. Cold Reg. Sci. Technol. (2020)
3. Emery, J.C., Morton, M.S.S.: Management decision systems: computer-based support for decision making. Adm. Sci. Q. (1972). https://doi.org/10.2307/2392104

4. Druzdzel, M.J., Flynn, R.R.: Decision support systems. In: Understanding Information Retrieval Systems: Management, Types, and Standards (2011)
5. Adhikari, N.K.J., Beyene, J., Sam, J., et al.: Effects of computerized clinical decision support systems on practitioner performance. JAMA J. Am. Med. Assoc. (2005)
6. Turban, E., Watkins, P.R.: Integrating expert systems and decision support systems. MIS Q. Manag. Inf. Syst. (1986). https://doi.org/10.2307/249031
7. Saba, D., Sahli, Y., Abanda, F.H., et al.: Development of new ontological solution for an energy intelligent management in Adrar city. Sustain. Comput. Inform. Syst. **21**, 189–203 (2019). https://doi.org/10.1016/J.SUSCOM.2019.01.009
8. Saba, D., Laallam, F.Z., Degha, H.E., et al.: Design and development of an intelligent ontology-based solution for energy management in the home. In: Hassanien, A.E. (ed.) Studies in Computational Intelligence, 801st edn, pp. 135–167. Springer, Cham, Switzerland (2019)
9. Saba, D., Berbaoui, B., Degha, H.E., Laallam, F.Z.: A Generic optimization solution for hybrid energy systems based on agent coordination. In: Hassanien, A.E., Shaalan, K., Gaber, T., Tolba, M.F. (eds.) Advances in Intelligent Systems and Computing, pp. 527–536. Springer, Cham, Cairo, Egypte (2018)
10. Saba, D., Degha, H.E., Berbaoui, B., et al.: Contribution to the modeling and simulation of multiagent systems for energy saving in the habitat. In: Proceedings of the 2017 International Conference on Mathematics and Information Technology, ICMIT 2017 (2018)
11. Saba, D., Laallam, F.Z., Berbaoui, B., Fonbeyin, H.A.: An energy management approach in hybrid energy system based on agent's coordination. In: The 2nd International Conference on Advanced Intelligent Systems and Informatics (AISI' 16). Advances in Intelligent Systems and Computing, Cairo, Egypt (2016)
12. Saba, D., Laallam, F.Z., Hadidi, A.E., Berbaoui, B.: Contribution to the management of energy in the systems multi renewable sources with energy by the application of the multi agents systems "MAS". Energy Proc. **74**, 616–623 (2015). https://doi.org/10.1016/J.EGYPRO.2015.07.792
13. Saba, D., Maouedj, R., Berbaoui, B.: Contribution to the development of an energy management solution in a green smart home (EMSGSH). In: Proceedings of the 7th International Conference on Software Engineering and New Technologies—ICSENT 2018. ACM Press, New York, New York, pp. 1–7 (2018)
14. Miller, C.C., Ireland, R.D.: Intuition in strategic decision making: friend or foe in the fast-paced 21st century?. Acad. Manage., Exec (2005)
15. Mukhopadhyay, A., Chatterjee, S., Saha, D., et al.: Cyber-risk decision models: to insure IT or not? Decis. Support Syst. (2013). https://doi.org/10.1016/j.dss.2013.04.004
16. Uusitalo, L., Lehikoinen, A., Helle, I., Myrberg, K.: An overview of methods to evaluate uncertainty of deterministic models in decision support. Environ. Model, Softw (2015)
17. Burstein, F., Holsapple C, Power DJ (2008) Decision support systems: a historical overview. In: Handbook on Decision Support Systems, 1
18. Ai, F., Dong, Y., Znati, T.: A dynamic decision support system based on geographical information and mobile social networks: a model for tsunami risk mitigation in Padang, Indonesia. Saf. Sci. **90**, 62–74 (2016). https://doi.org/10.1016/J.SSCI.2015.09.022
19. Scotch, M., Parmanto, B., Monaco, V.: Evaluation of SOVAT: an OLAP-GIS decision support system for community health assessment data analysis. BMC Med. Inform. Decis. Mak. (2008). https://doi.org/10.1186/1472-6947-8-22
20. Lerner, J.S., Li, Y., Valdesolo, P., Kassam, K.S.: Emotion and decision making. Annu. Rev. Psychol. (2015). https://doi.org/10.1146/annurev-psych-010213-115043
21. Yamashige S (2017) Introduction to decision theories. In: Advances in Japanese Business and Economics
22. Lands, Ii S.: Better Choices : You Can Make Better Choices. Authorhouse (2015)
23. Schwarz, N.: Emotion, cognition, and decision making. Cogn. Emot. (2000)
24. Simon, H.A.: A behavioral model of rational choice. Q. J. Econ. (1955). https://doi.org/10.2307/1884852

25. Rabin, J., Jackowski, E.M.: Handbook of Information Resource Management. Dekker, M (1988)
26. Banning, M.: A review of clinical decision making: models and current research. J. Clin. Nurs. (2008). https://doi.org/10.1111/j.1365-2702.2006.01791.x
27. Schweizer, R., Johanson, J.: Internationalization as an entrepreneurial process. Artic. J. Int. Entrep. (2010). https://doi.org/10.1007/s10843-010-0064-8
28. de Witte, B.: The decision-making process. In: A Companion to European Union Law and International Law (2016)
29. Zarte, M., Pechmann, A., Nunes, I.L.: Decision support systems for sustainable manufacturing surrounding the product and production life cycle—a literature review. J. Clean. Prod. (2019)
30. Scott Morton, M.S.: Management decision systems; computer-based support for decision making. Division of Research, Graduate School of Business Administration, Harvard University (1971)
31. Jones, J.W., McCosh, A.M., Morton, M.S.S., Keen, P.G.: Management decision support systems. decision support systems: an organizational perspective. Adm. Sci. Q. (1980). https://doi.org/10.2307/2392463
32. Wang, M.S.-Y., Courtney, J.F.: A conceptual architecture for generalized decision support system software. IEEE Trans. Syst. Man Cybern. SMC **14**, 701–711 (1984). https://doi.org/10.1109/TSMC.1984.6313290
33. Zimmermann, O., Gschwind, T., Küster, J., et al.: Reusable Architectural Decision Models for Enterprise Application Development, pp. 15–32. Springer, Berlin, Heidelberg (2007)
34. Plataniotis, G., de Kinderen, S., Ma, Q., Proper, E.: A conceptual model for compliance checking support of enterprise architecture decisions. In: 2015 IEEE 17th Conference on Business Informatics. IEEE, pp. 191–198 (2015)
35. Schaub, M., Matthes, F., Roth, S.: Towards a conceptual framework for interactive enterprise architecture management visualizations. In: Lecture Notes in Informatics (LNI), Proceedings—Series of the Gesellschaft fur Informatik (GI) (2012)
36. Marin, C.A., Monch, L., Leitao, P., et al.: A conceptual architecture based on intelligent services for manufacturing support systems. In: 2013 IEEE International Conference on Systems, Man, and Cybernetics. IEEE, pp. 4749–4754 (2013)
37. Harper, P.R.: A review and comparison of classification algorithms for medical decision making. Health Policy (New York) (2005). https://doi.org/10.1016/j.healthpol.2004.05.002
38. Zaraté, P.: Outils pour la décision coopérative. Hermès Science (2013)
39. Nof SY: Collaborative control theory and decision support systems. Comput. Sci. J. Mold (2017)
40. Jarrahi, M.H.: Artificial intelligence and the future of work: human-AI symbiosis in organizational decision making. Bus. Horiz. (2018). https://doi.org/10.1016/j.bushor.2018.03.007
41. Saba, D., Degha, H.E., Berbaoui, B., Maouedj, R.: Development of an Ontology Based Solution for Energy Saving Through a Smart Home in the City of Adrar in Algeria, pp. 531–541. Springer, Cham (2018)
42. Saba, D., Zohra Laallam, F., Belmili, H., et al.: Development of an ontology-based generic optimisation tool for the design of hybrid energy systems. Development of an ontology-based generic optimisation tool for the design of hybrid energy systems. Int. J. Comput. Appl. Technol. **55**, 232–243 (2017). https://doi.org/10.1504/IJCAT.2017.084773
43. Bollweg, L., Bollweg, L., Kurzke, M., et al.: When robots talk—improving the scalability of practical assignments in moocs using chatbots. EdMedia + Innov. Learn. (2018)
44. Saba, D., Sahli, Y., Berbaoui, B., Maouedj, R.: Towards smart cities: challenges, components, and architectures. In: Hassanien, A.E., Bhatnagar, R., Khalifa, N.E.M., Taha, M.H.N. (eds.) Studies in Computational Intelligence: Toward Social Internet of Things (SIoT): Enabling Technologies, Architectures and Applications. Springer, Cham, pp. 249–286 (2020)
45. Lemley, J., Bazrafkan, S., Corcoran, P.: Deep learning for consumer devices and services: pushing the limits for machine learning, artificial intelligence, and computer vision. IEEE Consum. Electron. Mag. (2017). https://doi.org/10.1109/MCE.2016.2640698

The Virtues and Challenges of Implementing Intelligent 'Student-Success-Centered' System

Fatema Abdulrasool

1 Introduction

Universities exist to guarantee student success. There are different definitions and facets of student academic success. The most prevailing definition is linked to student retention and graduation rates [41]. "Institutions of higher education have been concerned about the quality of education and use different means to analyze and improve the understanding of student success, progress, and retention" [9]. Student success is a complicated endeavor that requires a comprehensive framework that put in mind the technical, social, organizational, and cultural aspects [14, 41]. Information management directly aid the accomplishment of university mission [10].

In the last year (2019) DUCAUSE Top 10 IT issues list, the issues were categorized into three main themes: Empowered Students, Trusted Data, and Business Strategies. Student success was ranked the second most problematic issue in the higher education intuitions while student-centered institution occupied the fourth position [41].

Forty years ago, teaching and learning processes were conducted in person and without any use of technology. The introduction of distance learning in 1998 signals the beginning of technology utilization to support higher education operations. Technology has affected not only student academic experience, but also the whole operations of higher education institutions (HEIs). Currently, the whole student life-cycle is managed and reformed by advanced technologies. The technologies used can positively or negatively impact the institution ability to attract, admit, enroll, and retain

F. Abdulrasool (✉)
University of Tsukuba, Ibaraki 305-8577, Japan
e-mail: fabdulrasool.85@gmail.com

© The Editor(s) (if applicable) and The Author(s), under exclusive license to Springer Nature Switzerland AG 2021
A.-E. Hassanien et al. (eds.), *Enabling AI Applications in Data Science*, Studies in Computational Intelligence 911, https://doi.org/10.1007/978-3-030-52067-0_14

315

students as well as other stakeholders and partners. Despite the fact that the advancement in technologies made e-campus initiative possible, it intensifies the need for technology protection and monitoring [23, 39].

Business model ensures business continuity which can be thrive in the 21st century through cost-efficient use of technology. In higher education, understanding data is the key to achieve the most pressing priority: student success. Universities produce, process, transmit, store, and collect streams of information. The focus is on organizing, standardizing, and safeguarding data and then apply it to advance their mission [23, 41].

Implementing AI techniques in higher education environment is in its early stages. Analyzing higher education data started later than other sectors because institutional data was not stored electronically until recently. Further, the organizational structure challenged the adoption of intelligent knowledge systems [6].

The use of AI to analyze educational data has been increased due to different factors which are: enhancement in AI techniques; atomization of student data; and increasing global competition which requires more innovative ways to deal with the budgetary constraints and increase return on investments [5, 40].

In the strive to improve student outcomes, the focus is not only on student academic journey but also on their life circumstances. To do so, trusted and meaningful data will help in monitoring student achievement and guiding the decision-making process. The extensive use of IT in all university operations (teaching and learning, research, and business functions) and the increasing funding challenges necessitate the integration of IT in the overall institutional strategy and business model [23, 41].

"HEIs are facing a growing governance crisis which might cause a decrease in staff and students retention rates. 'Good' university governance does not simply happen. It is the product of continuous efforts to find the most appropriate governance structures, protocols and processes. It is also about timing and judgement: it requires boards of governors to recognize when a governance model is not working, why and how to repair it. Ultimately, governance models are created by people to govern people. They are only as good as they who devise and apply them, as well as those who live by them" [52].

In this study, we will discuss the virtues and challenges of embedding AI techniques in universities processes to improve student success rates. We will also discuss the role of a good governance system in ensuring the achievement of the intended goals. The paper is arranged as follows.

Section 2 describes the meaning of student success; Sect. 3 explains business intelligence in universities; Sect. 4 describes the strategic role of IT in universities; Sect. 5 details the virtues of AI in university; Sect. 6 lists the challenges facing the adoption of AI in universities; Sect. 7 provides a full description of IT Governance concept; Sect. 8 discusses the differences between agility and IT Governance; Sect. 9 discusses the intelligent "student-success-centered" system; finally, the conclusion and the references.

2 Student Success

"Student success is a complex endeavor. Universities embrace the aspiration, yet they are still grappling with big questions about how to define, measure, and structure student success, all while keeping the student at the center" [41].

There are different definitions and facets of student academic success. The most prevailing definition is linked to student graduation and retention rates. The different preventive and corrective techniques implemented by universities to guarantee students success falls into three categories which are:

1. Lagging (backward): retention and graduation rates are considered as lagging indicators which investigate events that have already occur and fail to give an explanation for these events as of why and how it happened.
2. Leading (predictive): focus on analyzing student ongoing performance in different areas such as: course performance, and full-time continuous enrollment, accumulated GPA, number of registered courses per semester, attendance, and e-learning activities. These indicators are preventative measures that can help in identifying at risk student and direct universities resources towards supporting them before they quit their study.
3. Actionable (behavioral): study student behavior such as timely registration, participating in advising sessions, active participation and engagement in social networks and university events [41].

Many studies have been conducted to understand the factors that could elevate student success rates. Some are related the student profile and previous educational background while the others are related to the student current academic progress. Progression reports extracted from the institutional Learning Management System (LMS) can be utilized to develop proactive programs to mark and support at risk students [7, 12].

Students retention is a very complex problem that requires a comprehensive system to solve it. Researches directed toward understanding this phenomenon has been intensified [21]. Reports show that one fourth of students failed to continue their study. Universities are leveraging technology to identify at risk students and effectively support them to ensure their success [21, 33].

There are many indicators that can predict student's performance even before enrollment. AI can help in analyzing and finding the hidden pattern and create knowledge that helps in guiding students during program selection phase. College entry exams results, previous academic history, and students profile information (gender, age, country, etc.) are examples of data analyzed to create that knowledge [9, 31, 41].

Data-driven technology brings new opportunities to help student success even after graduation by connecting them with companies that offer jobs or internship positions [29]. Further, it helps students in developing employability skills which are:

1. Communication skills (verbal and non-verbal).
2. Teamwork (working in a group to achieve a common objective).
3. Problem solving (using logical process to overcome occurring problem).
4. Initiative and enterprise (creativity and self-directed).
5. Planning and organizing (creating task timelines and meeting deadlines).
6. Learning (self-learning and development).
7. Technology (using technology effectively and efficiently) [1].

3   Business Intelligence

Nowadays universities are abandoning their obsolete scattered systems and implementing university-wide ERP system that integrates all administrative functions in one comprehensive system with centralized data warehouse. If implemented effectively, the system could help the institution to achieve its strategic objectives as stated in the strategic plan [42, 44].

Student Information Systems (SIS) are used to automate processes related to admissions, course scheduling, exams scheduling, curricula follow up and development, study plans, students transfer from one program to another from within the university or transfers from other universities, graduation, student extra-curriculum activities, student complaints and enquiries, academic advising, counselling, etc.

Some universities might take their SIS to new levels and include students' hobbies, faculty research interests and publications, faculty conference attendance, etc., and thus moving from a basic student information system to an academic information system.

SIS play a major role in achieving university objectives by providing the required information to all university stakeholders especially students, faculty, deans, chairpersons, vice presidents, the president and the board of directors.

Some of the university systems are closely interconnected like grading and registration systems, whereas other systems such as alumni and complaint systems require less integration due to the static nature of data. Unfortunately, software vendors failed in developing a comprehensive software solution that covers end to end university business processes [50].

The daily and accumulated information creates a wealth of knowledge which can be inquired in different ways to provide the different stakeholders with different reports and information to be used for their purposes. For example, student information systems can be used to monitor student progress and identify difficulties and solve them and thus improve students' retention and completion rates. Another example of student information system utilization is to study course taking patterns. The university senior management can use the student information system to do their planning, e.g., future recruitment of faculty or for identifying sources for future students' recruitment or training opportunities for the students' internships. Therefore, it is evident that keeping good student information system and having proper

IT governance for such systems will have a direct influence of achieving university objectives.

The increasing adoption of enterprise management systems such as ERP, data warehousing, supply chain management systems, and customer relationship management systems have expanded the definition of IT infrastructure. As a result, organizations transformed their fundamental mission from applications development towards platform building and solutions delivery. The traditional style of managing IT infrastructures for cost-effectiveness and efficiency has been extended to incorporate issues related to global reach and range, flexibility, scalability, and openness to emerging technologies [47].

The processes of selecting, implementing and evaluation universities ERP systems have been considered as challenging tasks due to factors related to the decentralized organizational structure of universities which weaken the communication and collaboration between IT and business functions. As an outcome of the aforementioned issue, risks increased while benefits are not fully optimized, and resources are wasted. Institutions who are able to successfully implement high quality, efficient, effective, and widely accepted ERP system; may gain tremendous amount of benefits and will surely enhance the university competitiveness capabilities in the turbulent marketplace [42, 44].

Universities are building an increasing attention towards enterprise wide business intelligence to fulfill the growing demand for performance management. Business intelligence defined as "an umbrella term that includes the applications, infrastructure and tools, and best practices that enable access to and analysis of information to improve and optimize decisions and performance" [50].

The implementation of business intelligence tools that enhance the analytical capabilities are very complicated due to the complex system landscapes of universities, autonomy of departments as well as the decentralization of information. Although the access to the university scattered information has been improved due to the adoption of document management tools; the data maintenance issue has been raised [50].

Classification is an important aspect of data mining (DM) and machine learning (ML) which are used to forecast the value of a variable based on previously known historical variables. Statistical classification, decision tree, rule induction, fuzzy rule induction, and neutral networks are classification algorithms used to predict student performance. In this study we are not going into the specifics of these algorithm since it is beyond our scope. Figure 1 shows the subcategories of these algorithms. Some of these algorithms proved to be better than others in predicting student success [45].

Zacharis [54] forecasted student course performance by examining the data stored in the Moodle server using neural network algorithm. The researcher studied student interaction with the system in terms of number of emails, content created with wiki, number of accessed materials, and online quizzes scores. The prediction model shows an accuracy level up to 98.3%.

In a previous study, neural networks, logistic regression, discriminant analysis, and structural equation modelling were compared to determine the most accurate method in predicting student performance. Results show that the neutral network method is

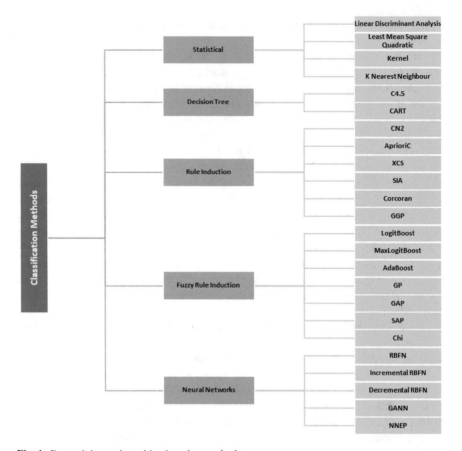

Fig. 1 Data mining and machine learning methods

the most accurate method followed by discriminant analysis [32]. Other study found that the decision tree is the easiest to be understood and implemented [45].

4 The Strategic Role of Technology in Universities

Online technologies that may threaten traditional universities may also expand their capacities. Traditional universities have natural advantages in delivering online learning by simply utilizing the available resources. They already have the required assets, experience, and human resources [15].

Management of universities realized the importance of utilizing intelligent systems to improve the overall performance including improving student retention and graduation rates. These systems can help in creating student's admission policies, predict the number of students to be enrolled in the forthcoming semester, attract prospective students, manage resources, update course offering, estimate hiring needs, or make financial decisions [9, 22, 55].

Internationalization of Higher Education Institutions is growing rapidly. The number of students studying cross national borders are increasing. Universities are opening branch campuses in other countries which are mostly operated by local partners while some institutions have established their own premises. E-learning distance education has not grown as expected, partially because of the limited ICT capacities especially in the third world countries [35]. The special characteristics of Higher Education Institutions business process brings unique opportunities and risks which requires special attention from top management. There is a shift in the perception of IT function in universities. Currently, IT is considered as strategic value creator instead of operational cost center. Innovative technologies enhance universities teaching and learning, research, and administrative performance [44].

There are different layers for implementing AI in higher education ranging from simple basic conversional interfacing capable of answering simple questions, to more complex layer that analyze and relate hundreds of factors and produce knowledge [7]. Some of AI applications in universities are:

1. "Jil Watson": Teaching assistant chatbot which was developed at Georgia Tech Uni-versity by a team headed by Ashok Goel.
2. Declara: Machine learning tool able to create personalized learning pathways.
3. Assessment and Learning in Knowledge Spaces (ALEKS): Measure student success and suggest specific content that suit each student.
4. REVEL: e-learning platform.
5. Cognitive computing is also being used across the campus.
6. Retention360: Includes an AI persona named "RENEE" that automatically launches interventions by analyzing different variables such as student profiles and best practices.
7. AdmitHub: AI chatbots help incoming students during admission process [35].

The real strength of traditional universities is their ability to blend online and on class learning experiences which has proven more effective than adopting one method. Hybrid instruction has the potential to take traditional universities to new levels by allowing them not only to respond to competition but also to serve more students with their existing resources. Universities that fail to employ online learning technology will loss the opportunity to grow and may also lose students as the cost disparity between the traditional model and the technology-enabled model increases [15].

The University of South Florida (USF) implemented variety of programs, processes, and policies that integrates Artificial Intelligence and human intelligence to promote student persistence and retention. As a result, the university achieved tremendous benefits. The retention rate has been improved from 86% in 2008 to 91% in 2017. The four years graduation rate has been improved from 38% in 2011–2012 (2008 cohort), to 61% in 2016–2017 (2014 cohort) [36].

5   Virtues of AI in Universities

Computing capacities and the level of technology utilization defined the future of HEIs. The advances in Artificial Intelligence brings new possibilities and challenges for teaching and learning, research, and governance of HEIs. AI techniques can be used in many aspects of universities administrative functions such as: helpdesk, automated timetable, and decision making. It also helps in the university core missions; teaching and learning, and research; by means like: AI data analysis tools and personalized online learning [43].

The following sections illustrates the benefits of implementing AI in universities.

5.1   Admission

AI advancements can be utilized to identify prospected student and direct university resources to serve this cohort. It can also facilitate personalized application process without any extra cost [22].

Pre-college data can give an accurate prediction of students' college performance. Effective AI model that analyze these early available factors can be valuable to improve student retention [32].

Abdul Aziz et al. [2] conducted a study to measure the impact of gender, race, hometown location, family income, and university entry mode on student academic performance. The results show that the race is the most influential factor followed by family income, and gender, while the hometown location came in the last position.

5.2   Registration

Artificial Intelligence can improve student experience and therefore, student satisfaction. AI techniques that utilize student education history, profile, e-learning performance, and historical data about other students can help students in planning their study plan and registration process. It can also initiate early warnings if the pattern shows that the student is at risk of not completing the program or found a mismatch between the student skills, and the academic program. Many universities recognized the importance of AI in helping students with registration processes, Elon University is an example. It provides students with a tool for tracking registration process and helps them in planning the upcoming semester courses [7].

5.3 Teaching and Learning

E-learning has been defined as "a system which is able to electronically transfer, manage, support, and supervise teaching and learning materials" [53]. Implementing AI algorithms in teaching and learning models is useful due to their accuracy and efficiency in imitating the human decision-making process [5].

Implementing intelligent LMS not only help universities in identifying at risk students, it also provides each student with personalized learning which means, based on student progress in LMS, the students' strength and weakness areas will be identified. The system will suggest studying materials based on this analysis. As a result, student retention and graduation rates will be improved [12].

Research shows that one-to-one teaching level up student success rates. Thought it is irrational to adopt it in traditional classrooms, it is attainable with the advancement of AI and e-learning techniques. The adaptive educational systems are LMS capable of designing personalized educational model for each and every individual student. These systems integrate students learning needs with the pedagogical expertise to improve learning outcomes. The first step in constructing these models is to design an accurate students' profile by analyzing the information stored in the SIS such as gender, sex, educational background and also personality traits and skills; and then link these variables using AI algorithms with the students' performance in the LMS. The success of these systems relies on the accuracy of data as well as the efficiency of the applied AI algorithms. Educators are provided with tools to monitor and evaluate students online performance which will help them in identifying students who needs extra attention, furthermore, it will give them an insight about the most challenging study area so they can focus more on them [5, 26, 29, 39].

While the access to LMS is restricted to the university enrolled students, Massive Online Open Courses (MOOC) systems are open to public with no cost. Stanford and Harvard universities are examples of universities offering this service to support lifelong learning. The challenge for MOOC lies in its capability to analyze the enormous data generated by the massive number of users. The completion rates for the students registered in these courses is really low (less than 13%). Robust AI techniques are needed to enhance the performance of these systems to the level of LMS, hence improve the retention and completion rates [5].

Although e-learning systems are implemented to improve efficiency, effectiveness, satisfaction and motivation of students; many systems failed to achieve the intended objectives. The spotted problems are poor management; neglecting important phases of system development lifecycle; inappropriate allocation of recourses; poor data quality; and compliance issues with rules and legislations [53].

5.4   Research

Research computing deals with the governance of a set of software, hardware and human expertise that support research practices. Many years ago, the access to high-performance research facilities was restricted to elite researchers since it was very expensive. As the cost declined and technologies advances, the access for all researchers has been assured [19].

There are many software packages dedicated to help researchers in interpreting and relating data. The new packages included AI techniques which proved to be more accurate than human in analyzing data and selecting the best regression model.

5.5   Awareness and Engagement

Student engagement has been defined as "the time and effort students devote to activities that are empirically linked to desired outcomes of college and what institutions do to induce students to participate in these activities".

Many studies have been conducted to understand the impact of deploying intelligent technologies on student engagement, Beeland [8] study is an example. The aforementioned study aims to evaluate whether the use of interactive whiteboard will affect student engagement. The results show a positive relationship between the use of interactive whiteboard in classrooms and student engagement [8].

AI can proactively help in improving students-faculty-university interaction which in turns improve student success [14, 41].

5.6   Advising

Universities are implementing AI techniques in identifying students with high probabilities of dropping out and quitting their study programs (at risk students). Universities are designing special care programs tailored to these students to help them in completing their degree [12, 13].

The efficiency of intelligent advising systems that can automatically identify, guide and advice students reached up to 93% compared to traditional advising process performed by human advisors [3].

5.7   Summary

In previous sections, we explained the benefits that can be achieved by implanting AI techniques in university processes. Table 1 summarizes our findings.

Table 1 Benefits of implementing AI techniques in universities

Function	Students	University management and educators
Admission	Choose the best-fit academic program	Predict the forthcoming semester admission rates Automatic responses to inquiries Personalized care and support
Registration	Curriculum design Selecting courses	Plan timetable Predict the needed resources (classes, equipments, and lecturers)
Teaching and learning	Customized learning	Assisting teachers in teaching and learning
Research	Help with finding the best research model	Control research quality
Awareness and engagement	Personalized communication	Plan communication and reporting schedules
Advising	Progress reports Recommend studying materials	Highlight at risk students and provide support for them

6 Challenges of AI in Universities

Although the interest in using AI technologies to support student success has been increasing drastically, it did not receive the full attention from the university top management. Successful implementation of a comprehensive student-successes-centered AI system cannot be achieved without an active engagement of the top management who can influence the system and university culture [48].

Information quality (accuracy, completeness, and integrity) and algorithm bias can affect the outcomes of AI function. Ethical and privacy issues pertaining to the use of students' personal data must be addressed. Transparency, security, and accountability are major concerns of information privacy. Noncompliance with privacy rules may jeopardize institution reputation and existence [4, 40, 55].

Many factors that affect the accuracy of the AI function can be avoided through the implementation of a good IT Governance system that ensure the quality of data, Business/IT alignment, clear communication between business and IT function, and compliance with internal and external laws, rules, and regulation [55].

The governance and management of high-performance computing and campus cyberinfrastructure deemed to be inappropriate [19].

7 IT Governance

Despite the huge amount of money that universities invest in IT, it's never seeming to achieve its full potential. Nowadays, universities recognized the importance of IT Governance in supporting their mission in attaining the full potential of their IT spending [25, 30]. The concept of IT Governance focuses on the sustainability in controlling, managing and monitoring IT activities through five driven mechanisms which are strategic IT-business alignment, value delivery, IT resource management, IT risk management, and performance measurement [49]. IT Governance is often the weakest link in a corporations' overall governance structure. It represents one of the fundamental functional governance models receiving a significant increase in attention by business management [11].

IT Governance concept is defined as "a set of relationships and processes designed to ensure that the organizations' IT sustains and extends the organization's strategies and objectives, delivering benefits and maintaining risks at an acceptable level" [24, 34].

IT Governance can be generally applied to any organization, e.g., banks, schools, universities, industrial companies, etc. However, we believe that each domain or type of business may necessitate its own style of IT Governance which makes the IT Governance model or methodology more applicable and able to yield better outcomes.

IT Governance can be deployed using a mixture of structures (strategies and mechanisms for connecting IT with business), processes (IT monitoring procedure) and relational mechanisms (participation and collaboration between management) [11, 38].

Data quality is a complex research area. The most important data quality dimensions are accuracy, completeness, and security/privacy/accessibility. Figure 2 which was created by COBIT detailed data quality dimensions [17].

Data Governance is a subset of the overall IT Governance. Data Management International (DAMA) defines Data Governance as "the exercise of authority, control and shared decision-making (planning, monitoring and enforcement) over the management of data assets. Data Governance is high-level planning and control over data management" [37].

As the proliferation of technology and communication occurs, the more all-embracing the governance issues have become. Data and information are what enables decision making, therefore, Data Governance is the starting point in any discussion concerning governance frameworks" [37].

Data Governance can't be separated from IT Governance and it needs a holistic approach. The university top management including the president and the board must be deeply involved in the Data Governance endeavors since it can be used as a strategic tool to enhance the organization ability to compete in the highly diverse marketplace [23].

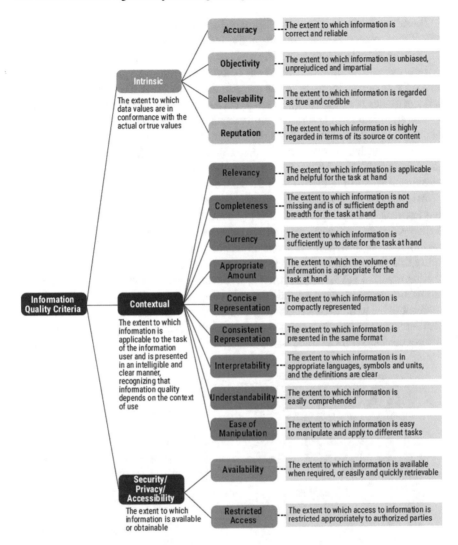

Fig. 2 Quality criteria for information [17]

8   IT Governance and Agility

While the implementation of IT Governance has been viewed by many researchers as a crucial aspect to enhance business performance and achieve organizational objectives, several studies argue that agility plays an equally vital role in this regard.

Agility has been defined as the process whereby an organization and its executive boards is able to answer the market emerging opportunities and demands both effectively and efficiently.

Based on Couto, E., et al. the IT Governance practices involving the organizations procedures, supervision, monitoring and control are proven to have positive impact on the organizational performance. The main challenge, however, arises from the ability of the presumed rigidness and relatively static IT Governance approach in coping with a rapidly changing industry needs and requirements. This issue of IT Governance and agility has been deemed as a stress between adaptation and anticipation [20].

The significance of enterprise agility has been emphasized by Tallon and Pinsonneault [51] where they were able to find through an enquiry to 241 organizations that there is a positive correlation between alignment and agility and between agility and organizational performance. The study was able to particularly confirm the substantial effect of agility in volatile business environments [51].

Adapting IT to a rapidly changing environment has been perceived as a challenging task in key aspects of development, design, testing and implementation. Agility concept has been misunderstood and confused with other terminologies leading to more difficulties in scope understanding and implementation. Unlike flexibility that deals with known changing factors, agility is more viewed as the ability to adhere with a yet unknown context. Ultimately, these challenging variables raise further questions on whether the current well-established IT Governance structures are responsive to unknown regulatory and economic changes in any presumed business setting including the educational sector and universities. Figure 3 illustrates the differences between IT Governance and agility.

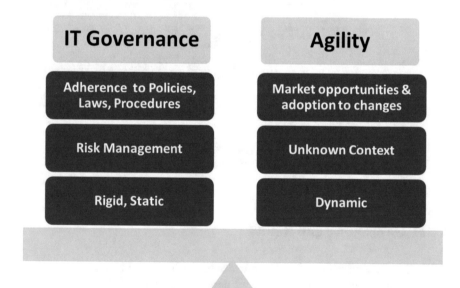

Fig. 3 IT governance versus agility

9 Intelligent "Student-Success-Centered" System

University resources, policies, and processes affects students 'success. All universities functions must be strategically planned and orchestrated towards achieving its core mission [31].

A conceptual framework that focuses on intelligent governance of AI in universities was developed (Fig. 4). The kernel of this model is student success which is an ongoing process that requires regular evaluating, directing and monitoring. Any change in the university internal policies or the environment surrounding the educational institution must be examined.

Away from the traditional perception of student success, the model defines student success in terms of its ability to proactively support and guide students during their entire academic life to which will surely improve retention and completion rates.

Students success can be forecasted in early stages, even before the student enrollment. With the availability of reliable data, AI can suggest the appropriate academic program for each student. It also helps student even after graduation by connecting them with the marketplace.

A successful framework must bear in mind the needs of all stakeholders. Students, parents, faculty, administrative staff, and top management are the key stakeholders and they may have not only different needs but may be also conflicting with the

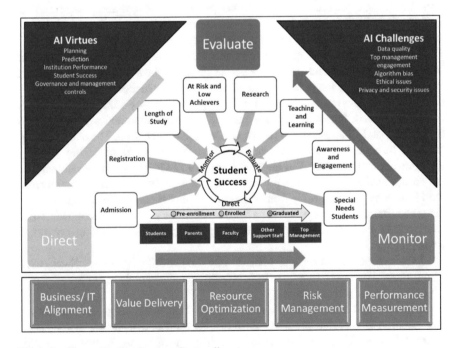

Fig. 4 Intelligent 'Student-Success-Centered' system

needs of each other. The tripartite IT Governance has been introduced as a solution to avoid these conflicts.

King [29] specified 5 steps that must be embedded in the institution strategy to become an intelligent campus. The top management involvement in the goals setting plans executing is crucial. The advocated five steps are:

1. IT/Business alignment.
2. A portfolio of personalized services.
3. Data strategy.
4. Resource Planning.
5. Agility [29].

Like student success governance, IT Governance is also a continuous process which focuses on three main area; evaluating, directing, and monitoring IT initiatives. Policies, laws, and procedures guides the planning, implementation and monitoring of IT projects. Best practices are used to enlighten the management about the best way to maximize their profits from their IT investments while risks are controlled, and resources are optimized.

Universities are facing many issues with regards to the control, management, and governance of IT. These issues result from the decentralized structure of universities which makes it very difficult to trace, evaluate, and monitor the interwind systems with business processes scattered throughout the different departments. The New technologies that are proliferating brings new challenges and possibilities for teaching, learning and research, as well as university administration processes; hence, models to justify additional investment and long-term sustainability plans are needed. The increasing legislation intended to protect and secure the information assets are increasing drastically and universities must develop plans and policies to assure compliance which is the core to its sustainability and existence. The level of information technologies awareness increased among the university stakeholders specially students; this leads to growing demand for more advance technologies acquisition. As the reliance on IT increased, the risk of system failure increased. The clash between the need to centralize IT and the resistance from scholars is increasing [18, 56]. Research shows that successful implementation of IT Governance relies not only on the design of the IT Governance framework, but also on how well-communicated the IT strategy and IT policy from the board is. In the context of universities, for the sake of minimizing capital risk and failure of trial and error; universities have to analyze their current status of IT governance performance before taking any further steps [27].

IT Governance structure is spearheaded by the Chief Information Officer (CIO) who supposed to report directly to the university president and should have a seat in the university council committee.

To govern and manage universities initiative that focus on student success, a new position in the vice president level has been created with "student success" notion (VPSS). This move proved to be fruitful and it helped in lifting retention and completion rates. VPSS is accountable for evaluating, directing, and monitoring the university operation by ensuring the availability of well-communicated policies, procedures, and plan that orchestrate student success. Although VPSS is accountable for the student success, the whole institution is responsible for the management and implementation of the student success system. The system covers the university-wide practices including advising, institutional research (IR), business intelligence, and IT [41].

Communication and collaboration strategies and plans must be in place to grantee a successful alignment between student success operations and IT Governances processes. The relationship between the VPPS and CIO can positively or negatively affect the overall performance of the university.

IT Governance focuses on IS and their risk management and performance. IT Governance will help those responsible for university governance and IT Management to understand, direct and manage the IS and IS security. Implementing university governance in general and IT Governance in particular, will put the university in compliance with the needs key legislation such as Sarbanes-Oxley in the USA.

Risk is an integral part of business process. Universities must be proactive in identifying and managing risk by developing a risk management plan which will contribute to its survival. Understanding and managing risk is an inherent part of the business process. The plan aims to control, mitigate and confront risk that await the university before they become a threat. The plan contains a clear repository of risks and controls and how they are related to the business process management along the dimensions of time and ownership. Reporting findings of audits helps in attaining compliance with more efficiency. Compliance initiatives such as the Sarbanes-Oxley forces university models to conduct more transparent audit trail. Security plan must be developed for information systems with technical and management requirement to assure data and communications integrity, confidentiality, accountability, access controls, data availability, as well as disaster recovery contingency measures [46].

IT Governance helps higher education in attaining regulatory compliance specially when it is facing higher scrutiny from the public. Even though the scholars advocated top-down management as an effective technique to enforce Information Security Policy (ISP); the unique structure of HEIs necessitate the need for another approach to inspire compliance. Participatory leadership and shared governance that involves employees in decision-making might be the key to foster compliance in HEIs. ISP enforcement which is achieved using coercive force does not blend well with higher education culture [28]. Performance measurement of IT processes/IT controls is one of the critical operational aspects of IT governance. IT governance process starts with planning phase where the IT and business are aligned. The planning phase has a guiding impact on the succeeding operating phase where the monitoring of IT resources, risks, and management is taking place. The last stage is the evaluation stage where the KPIs are measured and evaluated (benefits, costs, opportunities, and risks) [38].

Many countries have liked the financial incentives to institutional performance, Japan and United States are examples [41]. The adoption of an independent role of an internal auditing function in enhancing governance by a university has not been examined thoroughly. The multi-theoretical approach to university governance and the views of their chief executive officers to examine the extent to which internal auditing can be beneficial as a control mechanism which was adopted by Australian public universities under an environment of change management was studied by Christopher [16]. The findings highlight negative consequences of change [16].

10 Conclusion

Student success is a very challenging endeavor and it is receiving increasing attention. It is a continuous practice that drain resources. Intelligent planning and implementation of student success programs supported by AI technologies can brings opportunities and challenges for the universities. Universities are facing many issues with regards to the control, management, and governance of intelligent IT. Like student success governance, IT Governance is also a continuous process which focuses on three main area; evaluating, directing, and monitoring IT initiatives. In this study, we discussed the virtues and challenges of implementing Intelligent 'Student-Success-Centered' System. We also constructed a conceptual framework for this system. The aim of study is to increase the awareness about the importance of IT Governance in helping the universities in achieving its ultimate goal: student success.

References

1. 8 job skills you should have: Youth Central: https://www.youthcentral.vic.gov.au/jobs-and-car eers/plan-your-career/8-job-skills-you-should-have, 12 Jan 2019
2. Abdul Aziz, A., Ismail, N. H., Ahmad, F.: First semester computer science students' academic performances analysis by using data mining classification algorithms. In: Proceeding of the International Conference on Artificial Intelligence and Computer Science (AICS 2014), Bandung, Indonesia, pp. 100–109 (2014)
3. Al Ahmar, A.: A prototype student advising expert system supported with an object-oriented database. Int. J. Adv. Comput. Sci. Appl. Spec Issue Artif. Intell., 100–105 (2011)
4. Alexander, B.: 5 AIs in search of a campus. EDUCAUSE Rev., 11–18 (2019)
5. Almohammadi, K., Hagras, H., Alghazzawi, D., Aldabbagh, G.: A survey of artificial intelligence techniques employed for adaptive educational systems within e-learning platforms. JAISCR, 47–64 (2017)
6. Baker, R.S.: Educational data mining: an advance for intelligent systems in education. AI and Education, pp. 78–82 (2014)
7. Barret, M., Branson, L., Carter, S., DeLeon, F., Ellis, J., Gundlach, C., Lee, D.: Using artificial intelligence to enhance educational opportunities and student services in higher education. Inq. J. Virg. Community Coll. 22 (2019)
8. Beeland, W.D.: Student Engagement, Visual Learning and Technology: Can Interactive Whiteboards Help? (2002)
9. Bin Mat, U.: An overview of using academic analytics to predict and improve students' achievement: a proposed proactive intelligent intervention. In: IEEE 5th Conference on Engineering Education (ICEED), pp. 233–237. IEEE (2013)
10. Blaschke, S., Frost, J., Hattke, F.: Towards a micro foundation of leadership, governance, and management in universities. Higher Educ. 711–732 (2014)
11. Brown, A.E., Grant, G.G.: Framing the frameworks: a review of IT governance research. Commun. Assoc. Inform. Syst. 15, 696–712 (2005)

12. Bryant, T.: Everything depends on the data. EDUCAUSE Rev. https://er.educause.edu/articles/2017/1/everything-depends-on-the-data
13. Castillo, J.R.: Predictive analytics for student dropout reduction at Pontificia Universidad Javeriana Cali. EDUCAUSE Rev. (2019). https://er.educause.edu/articles/2019/12/predictive-analytics-for-student-dropout-reduction-at-pontificia-universidad-javeriana-cali
14. Cheston, A., Stock, D.: The AI-first student experience. EDUCAUSE Rev. (2017). https://er.educause.edu/articles/2017/6/the-ai-first-student-experience
15. Christensen, C., Eyring, H.J.: The Innovative University: Changing the DNA of Higher Education. Forum for the Future of Higher Education, pp. 47–53 (2011)
16. Christopher, J.: The adoption of internal audit as a governance control mechanism in Australian public universities–views from the CEOs. J. Higher Educ. Policy Manag. **34**(5), 529–541 (2012)
17. COBIT® 2019: Framework: Introduction & Methodology. ISACA (2019)
18. Coen, M., Kelly, U.: Information management and governance in UK higher education institutions: bringing IT in from the cold. Perspect. Policy Pract. Higher Educ., 7–11 (2007)
19. Collins, S.: What Technology? Reflections on Evolving Services. Educause, pp. 68–88, 30 Oct 2009
20. Couto, E.S., Lopes, F.C., Sousa, R.D.: Can IS/IT governance contribute for business agility? Procedia Comput. Sci. (2015). http://hdl.handle.net/11328/1311
21. De Carvalho Martinho, V.R., Nunes, C., Minussi, C.R.: Intelligent system for prediction of school dropout risk group in higher education classroom based on artificial neural networks. In: IEEE 25th International Conference on Tools with Artificial Intelligence. IEEE (2013). https://doi.org/10.1109/ictai.2013.33
22. Dennis, M.J.: Artificial Intelligence and recruitment, admission, progression, and retention **22**(9), 2–3 (2018)
23. Grajek, S.: Top 10 IT issues, 2019: the student genome project. EDUCAUSE Rev., 4–41 (2019)
24. Gunawan, W., Kalensun, E.P., Fajar, A.N.: Applying COBIT 5 in higher education. Mater. Sci. Eng. **420**, 12108 (2018)
25. Hicks, M., Pervan, G., & Perrin, B.: A Study of the review and improvement of IT governance in Australian universities. In: CONF-IRM 2012 Proceedings. AIS Electronic Library (AISeL) (2012)
26. Humble, N., Mozelius, P.: Artificial intelligence in education-a promise, a threat or a hype? In: Proceedings of the European Conference on the Impact of Artificial Intelligence and Robotics, pp. 149–156. EM-Normandie Business School Oxford, UK, England (2019)
27. Jairak, K., Praneetpolgrang, P.: Applying IT governance balanced scorecard and importance-performance analysis for providing IT governance strategy in university. Inform. Manag. Comput. Sec., 228–249 (2013)
28. Kam, H.-J., Katerattanakul, P., Hong, S.: IT governance framework: one size fits all? In: AMCIS 2016: Surfing the IT Innovation Wave-22nd Americas Conference on Information, San Diego (2016)
29. King, M.: The AI revolution on campus. EDUCAUSE Rev, 10–22 (2017)
30. Ko, D., Fink, D.: Information technology governance: an evaluation of the theory-practice gap. Corp. Gov. Int. J. Bus. Soc., 662–674 (2010)
31. Kuh, G.D., Bridges, B. K., Hayek, J.C.: What Matters to Student Success: A Review of the Literature. America: National Postsecondary Education Cooperative (2006)
32. Lin, J.J., Imbrie, P., Reid, K.J.: Student retention modelling: an evaluation of different methods and their impact on prediction results. In: Proceedings of the Research in Engineering Education Symposium, Palm Cove, QLD (2009)
33. Lin, S.-H.: Data Mining for Student Retention Management. CCSC: Southwestern Conference, pp. 92–99. The Consortium for Computing Sciences in Colleges (2012)
34. Iliescu, F.-M.: Auditing IT governance. Informatica Economică **14**(1), 93–102 (2010)
35. Marginson, S.: Dynamics of national and global competition in higher education. Higher Educ., 1–39 (2006)
36. Miller, T., Irvin, M.: Using artificial intelligence with human intelligence for student success. EDUCAUSE Rev. (2019). https://er.educause.edu/articles/2019/12/using-artificial-intelligence-with-human-intelligence-for-student-success

37. Mosley, M.: DAMA-DMBOK Functional Framework. Data management Association (2008)
38. Nicho, M., Khan, S.: IT governance measurement tools and its application in IT-business alignment. J. Int. Technol. Inform. Manag. 26(1), 81–111 (2017)
39. Norris, A.: The catch 22 of technology in higher education. EDUCAUSE Rev. (2018). https://er.educause.edu/blogs/sponsored/2018/8/the-catch-22-of-technology-in-higher-education
40. Pardo, A., Siemens, G.: Ethical and privacy principles for learning analytics. B. J. Educ. Technol., 438–450 (2014)
41. Pelletier, K.: Student success: 3 big questions. EDUCAUSE Rev. (2019). https://er.educause.edu/articles/2019/10/student-success–3-big-questions
42. Pollock, N., Cornford, J.: ERP systems and the university as a 'Unique' organisation. Inform. Technol. People 17(1), 31–52 (2004)
43. Popenici, S.A., Kerr, S.: Exploring the impact of artificial intelligence on teaching and learning in higher education. Res. Pract. Technol. Enhanc. Learn. (2017). https://doi.org/10.1186/s41039-017-0062-8
44. Rabaa'i, A.A., Bandara, W., Gable, G.G.: ERP systems in the higher education sector: a descriptive case study. In: 20th Australian Conference on Information Systems, pp. 456–470. ACIS, Melbourne (2009)
45. Romero, C., Ventura, S., Espejo, P.G., Hervás, C.: Data mining algorithms to classify students. In: The 1st International Conference on Educational Data Mining, Montréal, Québec, Canada, pp. 8–17 (2008)
46. Rosca, I.G., Nastase, P., Mihai, F.: Information systems audit for university governance in bucharest academy of economic studies. Informatica Economica 14(1), 21–31 (2010)
47. Sambamurthy, V., Zmud, R.W.: Research commentary: the organizing logic for an enterprise's IT activities in the digital era-a prognosis of practice and a call for research. Inform. Syst. Res. 11(2), 105–114 (2000)
48. Shulman, D.: Personalized learning: toward a grand unifying theory. EDUCAUSE Rev., 10–11 (2016)
49. Subsermsri, P., Jairak, K., Praneetpolgrang, P.: Information technology governance practices based on sufficiency economy philosophy in the thai university sector. Inform. Technol. People, 195–223 (2015)
50. Svensson, C., Hvolby, H.-H.: Establishing a business process reference model for universities. Procedia Technol, 635–642 (2012)
51. Tallon, P.P., Pinsonneault, A.: Competing perspectives on the link between strategic information technology alignment and organizational agility: insights from a mediation model. MIS Q. 35, 463–486 (2011)
52. Trakman, L.: Modelling university governance. Higher Educ. Q., 63–83 (2008)
53. Urh, M., Vukovic, G., Jereb, E., Pintar, R.: The model for Introduction of Gamification into E-learning in higher education. In: 7th World Conference on Educational Sciences, (WCES-2015), pp. 388–397. Procedia-Social and Behavioral Sciences, Athens, Greece (2015)
54. Zacharis, N.Z.: Predicting student academic performance in blended learning using artificial neural networks. Int. J. Artif. Intell. Appl. (IJAIA), 17–29 (2016)
55. Zeide, E.: Artificial intelligence in higher education: applications, promise and perils, and ethical questions. EDUCAUSE Rev., 31–39 (2019)
56. Zhen, W., Xin-yu, Z.: An ITIL-based IT service management model for Chinese universities. In: International Conference on Software Engineering Research, pp. 493–497 (2007)

Boosting Traditional Healthcare-Analytics with Deep Learning AI: Techniques, Frameworks and Challenges

Prabha Susy Mathew and Anitha S. Pillai

1 Introduction

Artificial Intelligence (AI), Machine Learning (ML) and Deep Learning (DL) has gained a lot of traction in the recent past especially in the healthcare sector. The amalgamation of EHR systems with these technologies are advancing healthcare sector into a new realm. The analytics platforms and ML algorithms running in the background of a clinical setup, uses the data stored in EHR, making the treatment option and outcomes more precise and predictable than ever before. Google has developed a ML algorithm to help identify cancerous tumors on mammograms. To improve radiation treatment Google's DeepMind Health is working with University College London Hospital (UCLH) to develop ML algorithms capable of distinguishing healthy and cancerous tissues. To identify skin cancer, Stanford is using a DL algorithm. The use of new AI solutions in healthcare is limitless ranging from simplifying administrative process, treating diseases, predicting possibilities of a patient contracting a certain disease and to giving personalized care to patients [1].

The purpose of this chapter is to provide valuable insights on Deep Learning techniques used in clinical setup. First, quick overview about Artificial Intelligence, Machine Learning and Deep Learning are presented. The next sections will discuss about the DL tools, framework and techniques followed by its limitations and opportunities. Use cases of DL techniques in general and finally its contributions in the healthcare sector will be highlighted to emphasize its significance for providing better healthcare outcomes.

P. S. Mathew (✉)
Bishop Cotton Women's Christian College, Bengaluru, India
e-mail: prabha.susan102@gmail.com

A. S. Pillai
School of Computing Sciences, Hindustan Institute of Technology and Science, Chennai, India
e-mail: anithasp@hindustanuniv.ac.in

A.-E. Hassanien et al. (eds.), *Enabling AI Applications in Data Science*,
Studies in Computational Intelligence 911,
https://doi.org/10.1007/978-3-030-52067-0_15

2 Background

While the terms Artificial Intelligence (AI), Machine Learning (ML) and Deep Learning (DL) are often used interchangeably, they are different. The concept of AI came first, followed by ML and lastly, ML's advanced sub-branch DL. DL is gaining a lot of traction and popularity due to its ability to solve complex problems.

2.1 Artificial Intelligence

John McCarthy, widely recognized as the father of Artificial Intelligence, in mid-1950s coined the term "Artificial Intelligence" which he defined as "the science and engineering of making intelligent machines". AI enabled machines can be classified as either weak AI also known as Artificial Narrow Intelligence (ANI), Strong AI also known as Artificial General Intelligence (AGI) or Artificial Super Intelligence (ASI) [2].

Artificial Narrow Intelligence: Artificial Narrow Intelligence is programmed explicitly to focus on a specific or singular task. ANI is today used in many applications such as playing chess, self-driving cars, Sophia the humanoid, recommendation systems, smart phone assistants, generating weather reports, Expert systems etc.

Artificial General Intelligence: Artificial General Intelligence refers to machines that exhibit human intelligence and can-do task that a human can perform. There are no real examples for AGI except for the human like AI seen in movies.

Artificial Super Intelligence: Artificial Super Intelligence refers to machines that will surpass intelligence of humans. Machines will exhibit intelligence, that one has not seen before.

2.2 Machine Learning

Machine Learning makes use of algorithms that detect patterns and learn directly from the data and experience, to make predictions without being explicitly programmed. The three important classes of ML algorithms are:

Supervised Learning: In supervised learning, a system is trained with labelled data. The system then uses these training data to learn relationship of a given inputs to a given output. It is used to predict category of a given data.

Unsupervised Learning: In unsupervised learning, system is trained with unlabeled data. Based on the similarity of the data, clusters are formed, and data is assigned to it.

Reinforcement Learning: It focuses on the principle of learning by experience. The systems are trained by means of virtual rewards or punishments. So, the training of machine models is done by allowing it to make a sequence of decisions to maximize the rewards it gains for the optimal decisions taken.

2.3 Deep Learning

Deep learning is a subset of ML. DL algorithms uses a huge amount of unsupervised data to extract complex representations with more accurate results than traditional machine learning approaches [3]. It achieves this by emulating the hierarchical learning approach of human brain with the help of multi-layer artificial neural network also known as deep neural network. Each of these layers progressively extract features from the raw input [4]. Deep learning algorithm performance scales with the increase in the size of data. As these DL algorithms use huge amount of data and does complex processing it requires high-end machines contrary to traditional ML algorithms. For this reason, Graphics Processing Unit (GPU) is an integral part for deep learning. Feature extraction process in traditional machine learning requires the domain expert to select the features, in order to reduce complexity of data and to improve the efficiency. On the other hand, deep learning algorithms learns high-level features without the domain expert's manual intervention. Usually deep learning algorithm takes long time in training phase but takes much lesser time in the testing phase giving a much accurate result when compared to machine learning algorithms [5, 6].

3 Deep Learning Frameworks/Libraries

A deep learning framework is a library or a toolkit for defining deep learning models without having to worry about the specifics of the background algorithms. There are several deep learning frameworks available and each of it is built to cater different objectives. Some of the deep learning frameworks [7–15] are.

3.1 Google's TensorFlow

It is an open source google project. It is one of the most widely adopted and most popular deep learning frameworks. This library does numerical computation using data flow graphs. Its comprehensive and flexible architecture allows deploying of computation to one or more CPUs or GPUs without modifying code [9]. It is available both in desktop and mobile. It can be used for Text based detection/summarization, Image/Sound recognition and Time series/Video analysis [12].

3.2 Keras

It is a python-based library, capable of running on TensorFlow, Theano or CNTK supporting both convolutional and recurrent networks. It was developed with an intension to enable fast experimentation. Keras is easy to use and primarily used in classification, summarization, tagging and translation. It runs seamlessly on CPUs as well as GPUs but does not support multi-GPU environments for training a network in parallel. It supports neural network layers such as convolution layers and recurrent layers [11].

3.3 Caffe

It is another popular open source deep learning framework for visual recognition and classification. Caffe stands for Convolutional Architecture for Fast Feature Embedding. It is supported with interfaces like C, C++, Python and MATLAB. It offers fantastic speed for processing images and modularity, so for modelling CNN (Convolutional Neural Network) or image processing issues it is a right choice. It works also works with CUDA deep neural networks (cuDNN) [7]. Caffe provides meagre support for recurrent networks, language modelling while it does not support fine granular network layers. Facebook open sources a version of Caffe known as Caffe2 to help researchers and developers train huge machine learning models. It is now a part of PyTorch [8, 12].

3.4 CNTK

The Microsoft Cognitive Toolkit (CNTK) is an open source DL library that consists of all building blocks which describes neural networks as a series of computational steps via a directed graph. It is created to deal with huge datasets and facilitates efficient training for handwriting, voice and facial recognition. It supports both CNN (Convolutional Neural Network) and RNN (Recurrent Neural Network) [8].

3.5 Apache MXNet

It is a deep learning framework designed for flexibility, efficiency and productivity. It supports languages like Python, Julia, C++, R or JavaScript. Due to its scalability it is used mainly for handwriting/speech recognition, forecasting and natural language processing (NLP). MXNet offers great scalability and can work with variety of GPUs as its backend written in C++ and CUDA (Compute Unified Device Architecture).

It supports long short-term memory (LTSM) networks along with both RNNs and CNNs [11, 12].

3.6 Facebook's PyTorch

It is open source framework which runs on python. It uses CUDA along with C/C++ libraries for processing, to scale production of building models and to provide overall flexibility. PyTorch uses dynamic computational graphs. It works best for small projects and prototyping purposes. while for cross-platform solution TensorFlow is a better option than PyTorch. It can be used in deep learning applications such as image detection/classification, NLP and Reinforcement Learning [7, 8].

3.7 Theano

It is a deep learning framework written in python enabling its users to use python libraries to define, enhance and evaluate mathematical expressions with multidimensional array. It is tightly integrated with NumPy for data computations and uses GPUs for performing data intensive computations [7, 13].

3.8 Deeplearning4j

It is a distributed deep learning framework written mainly in Java and Scala. DL4J supports different neural networks, like CNN (Convolutional Neural Network), RNN (Recurrent Neural Network), RNTNs (recursive neural tensor networks) and LSTM (Long Short-Term Memory). It integrates with Kafka, Hadoop and Spark using any number of GPUs or CPUs. MapReduce is used to train the network, while large matrix operations are executed depending on other libraries such as ND4J (TensorFlow library). This framework can be used for image recognition, fraud detection, text mining, parts-of-speech tagging, and NLP. This framework as it is implemented in java it is more efficient compared to other frameworks developed using python. For image recognition using multiple GPUs it is as fast as Caffe [12].

3.9 Chainer

It is a Python-based deep learning framework designed with an aim to provide flexibility. It runs on top of NumPy and CuPy Python libraries and provides other extended libraries also. It is a neural network framework for dynamic computation graphs for

NLP tasks. It supports CUDA computation along with multi-GPU. It provides better GPU performance than TensorFlow. It is faster than other python-based frameworks. It supports both RNN and CNN. It is mainly used for sentiment analysis, machine translation, speech recognition, etc. ChainerMN outperformed MXNet, CNTK and TensorFlow for ImageNet classification using ResNet-50 giving the best performance in a multi-node setting [10].

Some other deep learning frameworks are Gluon, Sonnet, Paddle, Fast.ai, nolearn, BigDL, Lasagne, Elephas, Deeplearning.scala, Apache SINGA etc. [7, 8] (Table 1).

A single framework cannot be chosen to be the best. There are several factors that are investigated before zeroing in a library such as modularity, ease of use, speed, documentation, community support, programming language, popularity, application, easy to deploy etc. Theano is one of the oldest and a stable framework. Chainer provides excellent documentation support. With multiple GPUs and CPUs, MXNet offers decent computational scalability which is great for enterprises. Keras and PyTorch are easy to use while TensorFlow is the most popular deep learning framework. Caffe is known for its image processing speed. Deeplearning4j is best for those comfortable with Java. CNTK provides better flexibility and scalability while operating on multiple machines. Choice of frameworks can be made based on its application and factors that offers best solution to the business challenge at hand.

4 Deep Learning-Techniques/Architecture

ML and DL algorithms have been used to find solutions for complex health problems and for providing long term quality care to patients. Medical images are valuable data available in healthcare that when combined with other demographic information can give valuable insights. Typical deep learning techniques are composed of algorithms like deep neural network (DNN), deep feed-forward neural network (DFFNN), recurrent neural networks (RNNs), long short-term memory (LSTM)/gated recurrent unit (GRU), convolutional neural networks (CNNs), deep belief networks (DBN), deep stacking networks (DSNs), generative adversarial networks (GAN), Autoencoders and many more.

4.1 Feed Forward Neural Network (FFNN)

Deep feedforward neural networks/feedforward neural networks also known as multilayer perceptron's (MLP) are the foundation of many deep learning networks like CNNs and RNNs. It represents the first generation of neural networks [16]. In this architecture, information flows only in the forward direction which start at the input layer, goes to the hidden layers, and then the output layer. Due to its simplicity this architecture is popular in all engineering domains although it has achieved major success in computer vision and speech recognition applications [13].

Table 1 Comparison of deep learning frameworks

Framework	Release year	Core language	CUDA support	Support models	Multi GPU	Advantages	Disadvantages
TensorFlow	2015	Python, C++, CUDA	✓	CNNs, RBMs, DBNs, and RNNs	✓	1. Powerful supporting services 2. During graph creation it can optimize all operations that are defined 3. Good documentation and guidelines with community support 4. It supports desktop, mobile, embedded devices and distributed training 5. Flexible system architecture 6. Supported by cloud computing platforms	1. Struggles with poor results for speed when compared to CNTK, MXNet 2. "define and run" mode makes debugging difficult 3. Static graphs
Keras	2015	Python	✓	CNN, RNN	✓	1. Quick and easy prototyping possible 2. Has built-in support for training on multiple GPUs 3. Provides simpler API for developers 4. Light weight architecture	1. Not always easy to customize 2. Backend support is restricted to engines such as TensorFlow, Theano, or CNTK

(continued)

Table 1 (continued)

Framework	Release year	Core language	CUDA support	Support models	Multi GPU	Advantages	Disadvantages
Caffe	2013	C++	✔	CNN, LSTM	✔	1. Specializes in computer vision 2. Offers pre-trained models 3. Its scalable, fast and lightweight 4. Image processing and learning speed is great 5. Fast and modular	1. Lacks documentation 2. Difficult to compile 3. No support for distributed training 4. Support for recurrent networks not adequate 5. Cumbersome for huge networks
CNTK	2016	C++	✔	CNN, RNN, FFNN	✔	1. Good performance and scalability 2. Efficient in resource usage	1. Community support is limited 2. Its capabilities on mobile is limited 3. Difficult to install
MXNet	2015	C++, Python, R, Julia, JavaScript, Scala, Go, Perl	✔	LTSM, RNN, CNN	✔	1. Highly scalable tool that can run on variety of devices 2. Supports multiple GPUs 3. Code is easy to maintain and debug 4. Ability to code in a variety of programming language 5. Offers cloud services	1. Smaller community compared to other frameworks 2. Learning to use it is difficult 3. API is not very user-friendly

(continued)

Table 1 (continued)

Framework	Release year	Core language	CUDA support	Support models	Multi GPU	Advantages	Disadvantages
PyTorch	2001	Python, C++, CUDA	✓	CNN, RNN	✓	1. Has many pre-trained models 2. Supports data parallelism and distributed learning model 3. Very flexible and easy to use 4. It allows complex architecture to be built easily 5. Supports dynamic computation graph and efficient memory usage 6. supported by major cloud platforms	1. Third party required for visualization 2. For production API server is required
Theano	2007	Python, CUDA	✓	CNN, RNN	✓	1. Offers decent computational stability 2. Good flexibility 3. Great performance	1. Difficult to learn using it and understanding the syntax 2. Inconvenient to deploy the model 3. Official support ceased in 2017

(continued)

Table 1 (continued)

Framework	Release year	Core language	CUDA support	Support models	Multi GPU	Advantages	Disadvantages
Deeplearning4J	2014	Java, Scala, CUDA, C, C++, Python	✓	RBM, DBN, CNNs, RNNs, RNTNs, LTSM	✓	1. It is supported with distributed frameworks such as Hadoop/Spark 2. Can process huge amount of data without compromising on speed 3. More efficient as it is implemented in Java	1. As java is not immensely popular, so cannot rely on growing codebases
Chainer	2015	Python	✓	RNN, CNN, RL	✓	1. Provides good documentation with examples 2. Faster than another python-oriented framework 3. Supports CUDA computation 4. Requires little effort to run on multiple GPUs 5. Possible to modify existing network on the go	1. Difficult to debug 2. Community is relatively small

4.2   Convolutional Neural Network (CNN/ConvNets)

One of the most popular type of deep neural network architecture is convolutional neural network (CNNs) which has multiple applications such as speech recognition, image/object recognition, image processing, image segmentation, video analysis and natural language processing [15]. Some CNNs are almost bettering the accuracy of diagnosticians when detecting important features in diagnostic imaging investigations. It eliminates the need for manual feature extraction as in the case of ML algorithms. The CNN works by extracting relevant features without being pretrained, they are learned while the network trains on images making it more accurate for computer vision assignments. A CNN breaks down an image into several attributes like edges, contours, strokes, textures, gradients, orientation, color and learn them as representations in different layers. It achieves this by using tens or hundreds of hidden layers. Every hidden layer increases the complexity of the learned image features.

A CNN is composed of an input layer, hidden layer which consists of "convolutional layer" and "pooling layer", fully connected layer and output layer. In hidden layer, each "convolutional layer" comes after "pooling layer" [17, 18]. The convolution phase focuses on extracting high-level features from the input layer. ConvNets are generally not limited to only one convolutional layer. Typically, the first convolutional layer is responsible for capturing the low-level features such as edges, strokes, color, gradient, orientation, etc. Each the other layers in the architecture takes care of the other high-level features, to infer what the image is [7].

The pooling layers are placed between successive convolutional layer [19]. It is responsible for reducing the spatial size of the convolved Feature. Through dimensionality reduction the computational power and the parameters needed to process the data is decreased. The convolutional layer and the pooling layer together form the several layers of a convolutional neural network. Depending on the complexities of the images, the number of such layers may be increased for capturing low levels details.

The full connection layer takes the result of convolution/pool layers and use them to classify image. The classified image is then forwarded to the output layer, in which every neuron represents a classification label [17].

In short CNN can be thought as a model performing two basic tasks on the input namely: extraction and classification. The convolution and pooling layers perform feature extraction while fully connected layers acts as a classifier classifying the images.

4.3   Deep Neural Network (DNN)

DNN deploys a layered architecture with complex function. For this reason, it requires much more processing power and powerful hardware for such complex processing. This model is used for classification or regression. It is more specifically known for its classification performance in complex classifications. Hardware limitations have made DNNs impractical for several years due to high computational demands for both training and processing, especially for real-time applications processing. Recent advances in hardware and parallelization through GPU and cloud computing, these limitations have been overcome to a great extent and have enabled DNNs to be recognized as a major development [20, 21].

4.4   Recurrent Neural Networks (RNN)

RNN is neural network that can analyze sequence of data with variable length input. RNNs uses knowledge of its previous state as an input for its current prediction, and this process is repeated an arbitrary number of steps allowing the network to propagate information by means of its hidden state through time. Doing this, RNNs exploit two sources of input, the present and the recent past, to provide the output of the new data. This feature gives RNNs something like a short-term memory. As RNN is highly effective working with sequence of data that occur in time, it is widely used in language translation, language generation, time series analysis, analysis of text, sound, DNA sequence etc. Although the RNN is a straightforward and robust model, it suffers from the problem of vanishing and exploding gradient [13, 21].

4.5   Long Short-Term Memory (LSTM)/Gated Recurrent Unit (GRU)

The two variants of gated RNN architecture that help in solving the gradient problems of RNNs are Long Short-Term Memory (LSTM) and Gated Recurrent Unit (GRU). Both variants are efficient in applications that requires processing of long sequences [13, 21]. However, GRU is a simplified version of LSTM which came later in 2014. This model gets rid of output gate present in LSTM model. For several applications, the GRU has performance like the LSTM, but as GRU is simpler, it can be trained quickly, and execution is efficient and faster. It can be used in applications of text compression, handwriting recognition, gesture/speech recognition, image captioning [22].

4.6   Restricted Boltzmann Machine (RBM)

RBM is a generative non-deterministic unsupervised deep learning model. It is a variant of Boltzmann machines, with the restriction in connection of pairs of nodes between visible and hidden units. The two layers are connected by a fully bipartite graph, where every node in the visible unit is connected to every node in the hidden unit but no two nodes in the same unit are linked to each other. This restriction allows for more efficient training algorithms and easy to implement than the Boltzmann machines. RBMs can be used for dimensionality reduction, classification, regression, feature learning [13].

4.7 Deep Boltzmann Machine (DBM)

DBM is another variant based on the Boltzmann machines proposed by Salakhut-dinov and Hugo [23]. It is an unsupervised probabilistic and generative model. Like RBM, no two nodes in the same units are connected to each other, connections exist only between units of the neighboring layers. The time complexity required for the inference is considerably higher, making the optimization of the parameters imprac-tical for huge training set. To speed up learning the weights, it uses greedy layer by layer pre training. As DBM is considered as extractor that reduce the dimensionality, Deep-FS a feature selection method uses DBM classifier to reduce the number of features and learning errors without impacting the performance. DBMs have been used in a variety of applications such as image-text recognition, facial expression recognition, audio-visual person identification etc. [24].

4.8 Deep Belief Networks (DBN)

Deep belief network (DBN) is a generative graphical model composed of multiple layers of hidden units. It consists of several layers; the input layer represents the raw sensory inputs middle layers are RBM which learns abstract representation of input and the last layer implements the classification. Training occurs in two steps: unsupervised pretraining and supervised fine-tuning [14].

4.9 Generative Adversarial Networks (GAN)

GAN are generative model for unsupervised and semi-supervised learning models. It has two neural networks that compete against themselves to create the most conceiv-able realistic data. In GAN one of the networks is a generator while the other network is a discriminator. The generator will generate data which is evaluated by the discrim-inator. Discriminators training data comes from two instances, the real pictures and the other fake ones produced by the generator. Generator produces better images maximizing the probability of the discriminator making a mistake, while the discrim-inator tries to achieve higher accuracy in classifying the synthetic image and the real image. Some of the applications include image style transfer, high resolution image synthesis, text-to-image synthesis, image super-resolution, anomaly detection, 3D object generation, music generation, scientific simulations acceleration and many more [10].

4.10 Deep Autoencoders (DAE)

An AE is an unsupervised neural network model used to recreate an output. It is known for dimensionality reduction. AEs are composed of two main parts encoder and decoder. Encoder will convert the input data into an encoded representation by reducing the dimension and compressing the input data while the decoder reconstructs the data as close as possible to the original data from the encoded representation. The design and structure of AE normally have the lesser number of neurons in the hidden layer than the input and output layer. The objective having such a structure is to encode data in low dimensionality space and to achieve extraction of features. However, many AE can be stacked together to create a deep AE architecture stacked auto encoder (SAE) when the dimensionality of the data is high. One of the most essential features of the SAE is to learn highly non-linear and complicated patterns. There are many deviations of AE presented to handle different data patterns for performing specific functions. For example, denoising AE was proposed to increase the robustness of the regular AE model. Another variation is the sparse AE which is used to make the data more separable. Convolutional AE was proposed to preserve spatial locality and process 2-dimensional (2D) patterns. Contractive AE is like denoising AE, but instead of injecting noise to corrupt the training set, it modifies the error function by adding analytic contractive cost [21]. Variational autoencoder (VAEs) is defined as an autoencoder which ensures latent space has good properties enabling generative process and avoids overfitting using regularized training. Simidjievski et al. proposed and implemented different integrative VAE architectures for modeling and analyzing heterogeneous breast cancer data leading to more accurate and stable diagnoses. Typical applications of AEs are dimensionality reduction, image denoising and compression. It can also be used for feature extraction or information retrieval used in data visualization applications [25]. Though each of these architectures have their own advantages and disadvantages the most used DL architectures are CNN (e.g. ResNet, AlexNet, and GoogleNet), RNNs (LSTM, GRU) and stacked autoencoders [10] (Table 2).

5 Comparing the Deep Learning Models

From the deep learning models discussed, it can be noticed that based on the type and need of the application one can opt DNN, CNN, and RNN for deep supervised learning, and DAE, RBM and GAN for unsupervised learning. However sometimes

Table 2 Deep learning architectures used in healthcare applications

DL architecture	Network type	Training type	Author	Applications in healthcare	Reference no.
CNN	FFNN	Supervised	Pereira et al. [26]	Brain tumor segmentation	[26]
			Prasoon et al. [27]	Predicting the risk of Osteoarthritis (OA)	[27]
			Gulshan et al. [28]	Detect diabetic retinopathy	[28]
			Esteva et al. [29]	Classify skin cancer	[29]
			Sathyanarayana et al. [30]	Predicting the quality of sleep from physical activity using wearable data during awake time	[30]
			Alipanahi et al. [31]	DeepBind: predicting the specificities of DNA- and RNA-binding proteins	[31]
LSTM RNN	RNN	Supervised	Lipton et al. [32]	Diagnose patients in Pediatric Intensive Unit Care	[32]
			Wang et al. [33]	Predicting Alzheimer's Disease stage/progression	[33]
			Beeksma et al. [34]	Predicting life expectancy with electronic medical records	[34]
GRU RNN	RNN	Supervised	Choi et al. [35]	Use patient history to predict diagnoses and medications	[35]
			Choi et al. [36]	Early detection of heart failure onset	[36]

(continued)

Table 2 (continued)

DL architecture	Network type	Training type	Author	Applications in healthcare	Reference no.
RBM	FFNN	Unsupervised	van Tulder and de Bruijne [37]	Lung CT analysis	[37]
			Brosch et al. [38]	Detect modes of variations in Alzheimer's Disease	[38]
			Yoo et al. [39]	Segment Multiple Sclerosis (MS) lesions	[39]
GAN	FFNN	Unsupervised/semi-supervised	Phúc [40]	Unsupervised anomaly detection on X-ray images	[40]
				Learning implicit brain MRI manifolds	[41]
DAE	FFNN	Unsupervised	Suk et al. [42]	Alzheimer's Disease/Mild Cognitive Impairment diagnosis	[42]
			Cheng et al. [43]	Diagnose breast nodules and lesions from ultrasound imaging	[43]
			Fakoor et al. [44]	Predicting protein backbones from protein sequences	[44]
			Che et al. [45]	Discovering and detection of characteristic patterns of physiology in clinical time series	[45]
			Miotto et al. [46]	Predicting future clinical events	[46]
DNN	FFNN	Supervised/unsupervised/semi-supervised	Alexander et al. [47]	Predicting pharmacological properties of drugs and drug repurposing	[47]

GAN can also be used for semi-supervised learning tasks. Four deep generative networks are as DBN, Deep Boltzmann Machine (DBM), Generative Adversarial Network (GAN), and Variational Autoencoder (VAE) [48].

For image classification DNN performs better than ANN because of its multiple hidden layer features basically which increases accuracy factors, but the problem of vanishing gradient in DNNs, which requires one to consider Convolution Neural Networks (CNNs). CNNs are preferred as they provide better visual processing models. Suvajit Dutta et al., in their work compared the performance of different NN models for medical image processing without GPU support. The computational time when compared, FFNN took the highest for training followed by DNN and CNN. As per their study with the CPU training DNN performed better than FFNN and CNNs [49].

Tobore et al., in their study points out that RNN and CNN have been increasingly used in applications over the years, with enormous growth rate in CNN. This can be attributed to the success recorded in image data and the many available variants of the model. The CNN and RNN techniques are commonly used DL techniques as they are often used for solving data problems in the form and shape that are either visual or time-dependent, which they are known to handle effectively. PET and CT scan image processing are of many dominant health care applications where CNN has proved to be processing techniques that provides required performance. Although CNN is a feedforward NN the flow of information is in forward direction only, in RNN, the information flows in both direction as it operates on the principle of saving the output of the previous layer and feeding it back to the input to predict the output. The principle of operation of CNN is influenced by the consecutive layer organization so, it is designed to recognize patterns across space. For this reason, CNN is ideal for images magnetic resonance images [MRI], videos (e.g. moving pattern in organs), and graphics (e.g. tumor representation) to recognize features. CNN has been used for several medical applications such as for automatic feature extraction from endoscopy images, as a classifier for detection of lesion in thoraco-abdominal lymph node and interstitial lung disease, Alzheimer's disease detection, Polyp recognition, Diabetic retinopathy, lung cancer, cardiac imaging especially for Calcium score quantification and many more. CNN architectures GoogLeNet, LeNet-5 and AlexNet are used for automatic features extraction and classification achieved 95% and above accuracy [50]. CNNs can be invariant to transformations such as translation, scale, and rotation which allows abstracting an object's identity or category from the specifics of the visual input enabling the network to recognize a given object even when the actual pixel values on the image vary significantly [51].

In contrast, RNN technique is suited to recognize patterns across time, making it a perfect candidate for time series analysis such as sound (e.g. heartbeat, speech), text (e.g. medical records, gene sequence), and signals (e.g. electrocardiogram (ECG)).

The deep generative models DBN and DBM have been least utilized techniques. Major problems are it being computationally expensive and low rate of application. DBN and DBM architectures are derived from RBM and are initialized by layer wise greedy training of RBM. However, they are qualitatively different. The connections between the layers are mixed directed/undirected in case of DBN while fully undirected in DBM. The first 2 layers in a DBN is undirected connection of RBM; the subsequent layers are directed generative connections. As far as DBM is concerned, all the connections between the layers are undirected RBM. Another difference between these 2 techniques are that the connected layers in DBN function as sigmoid belief network but in DBM they are Markov random fields. DBN and DBM can be used to extract features from unprocessed physiological signals and image data (MRI) to reduce the size of features required for classification. These models can be applied for human motions for fall detection, brain disease diagnosis, lesion segmentation, cell segmentation, image parsing and tissue classification. DBN/DBM and SAE can work in unsupervised manner with unlabeled data [51].

Another promising DL method DAE has also shown encouraging performance. Considering the growing size of medical data, an efficient data coding with AE makes it possible to minimize memory requirements and reduce processing costs. Stacked AE (SAE) can be applied in unsupervised feature learning for detection of tumors, cancers, or inflamed organs [21, 52]. However, the disadvantage with autoencoders is that it becomes ineffective if there are errors in the first layers. Training for SAEs is done in real-time while for CNNs and DBNs/DBMs training processes are time-consuming [51].

6 Deep Learning Applications in Healthcare

Applications of deep learning in healthcare provides solutions to medical professionals and researchers for a variety of problems ranging from disease diagnostics to suggestions for personalized treatment for serving the healthcare industry better. It can help physicians analyze electronic health records and detect multiple conditions like analyzing blood samples, detecting underlying health problems, using image analysis to detect tumors, cancerous cells and other health conditions. There are several Deep learning techniques that use EHR records to tackle healthcare issues like reducing the rate of false diagnosis and predicting the outcome of the treatment. Some of the applications of deep learning in healthcare are as stated below [50].

6.1 Drug Discovery

Deep learning is a great tool that can be used in discovery and development of drugs. Alexander et al., in their work proposed a convolution neural network which used transcriptomic data to recognize pharmacological properties of multiple drugs across different biological systems and conditions crucial for Predicting Pharmacological Properties of Drugs and Drug Repurposing, DNN achieved high classification accuracy and it outpaced the support vector machine (SVM) model [47]. Google's AlphaFold [4] was able to predict with excellent precision and speed the 3D structure of proteins an important assessment in drug discovery. Similarly, A model ChemGAN

[53] that uses GANs and DeepChem [54] an open source library has been used for drug discovery.

6.2 Biomedical Image Analysis

Medical imaging techniques such as MRI scans, CT scans, ECG, X-ray are used to diagnose a variety of diseases. These medical images analysis is able help to detect different kinds of cancer at earlier stages with high accuracy. It facilitates doctors to analyze and predict the disease and provide better treatment. The most common CNNs like AlexNet and GoogleNet, designed for the analysis of natural images, have proved their potential in analyzing medical images [55]. Some of the areas where medical images have been used are: According to missingklink.ai, Haenslle et al. trained a CNN model that could as accurately as an expert dermatologist diagnose skin cancer by examining digital image of a skin lesion. Researchers at enclitic created a device to detect lung cancer nodule in CT images [56]. Google AI algorithm LYmph Node Assistant (LYNA) can quickly and accurately detect metastasized breast cancer from pathology images when compared to human doctors achieving a success rate of 99% [57].

6.3 Electronic Health Records Analysis

EHR does not just store basic information of the patient and administrative tasks, but they include a range of data, including the patient's medical history, laboratory reports, demographics, prescription and allergies, sensor measurements, immunization status, radiology images, vital signs etc. A medical concept created by researchers uses deep learning to analyze data stored in EHR and predict heart failures up to nine months prior to doctors [56]. Till recent past EHR analysis was done using ML techniques to convert the data available into knowledge, but now it has been replaced by deep learning techniques.

Leveraging on the EHR, Choi et al.'s Doctor AI framework, was constructed to predict future disease diagnosis along with subsequent medical interventions [35]. They trained an RNN model on patients observed clinical event and time, to predict patients diagnoses, prescription of the medicine and future diagnoses. They found that their system performed differential diagnosis with similar accuracy to physicians, achieving up to 79% recall@30 and 64% recall@10. They then expanded their work [36] by training a GRU network on sequences of clinical event derived from the same skip-gram procedure, and found better performance for predicting the onset of heart disease.

DeepCare framework takes clinical concept vectors via a skip-gram embedding approach creating two separate vectors per patient admission: one for diagnosis codes, and another for intervention codes. After combining these vectors, it is passed into an LSTM network for predicting the next diagnosis, next intervention and future readmission for both diabetes and mental health cohorts. Another system Deepr based on CNN operates with discrete clinical event codes for predicting unplanned readmission following discharge operates. It examines patients' multiple visits each having several associated procedures and diagnosis. It demonstrated a superior accuracy and the capability to learn predictive clinical motifs. Deep Patient framework essentially uses stacked denoising autoencoders (SDA) to learn the patient data representations from multi-domain clinical data. Med2Vec is another algorithm that can efficiently learn code and visit level details by using EHRs datasets improving the prediction accuracy [58, 59].

6.4 Medical Insurance Fraud

Medical insurance fraud claims can be analyzed using Deep learning techniques. With predictive analytics, it is possible to predict fraud claims that are expected to happen in the future. Deep learning also helps insurance industry identify their target patients to send out discounts. Insurance fraud usually occurs in the form of claims. A claimant can fake the identity, duplicate claims, overstate repair costs, and submit false medical receipts and bills. According to Daniel and Prakash, RBMs have proven to have efficiency in identifying cases of treatment overutilization. They recommend RBMs not just for detecting insurance fraud, but in any situation where there is high dimensional categorical outlier detection [60].

6.5 Disease Detection

Alzheimer detection is one of the challenges that medical industry face. Deep learning technique is used to detect Alzheimer's disease (AD) at an early stage. Wang et al., proposed a predictive modelling with LSTM RNN that can effectively predict AD progression stage by using patients' historical visits and medical patterns [33]. Suk et al., proposed a method for Latent feature representation with stacked auto-encoder (SAE) for Alzheimer's disease/Mild Cognitive Impairment diagnosis. It uses two-step learning scheme of greedy layer-wise pre-training and the fine-tuning, to reduce the risk of falling into a poor local optimum, which is the major limitation of the CNN [42].

Nvidia published a study on deep learning model based project, that used NVIDIA Tesla K80 GPUs with the cuDNN-accelerated Caffe framework to drop the breast cancer diagnostic error rates to 85% [61]. Rajaraman et al., proposed a customized and per-trained CNN model that is scalable with end-to-end feature extraction and classification to aid in improved malaria parasite detection in thin blood smear images [62].

6.6 Tumor Detection

Abnormal growth in body tissue is generally referred to as a tumor. Tumors can be either cancerous (malignant), pre-malignant or non-cancerous (benign). So early detection could be of great help for the right treatment.

Ragab et al. [63] proposed a computer aided detection (CAD) system is for classifying benign and malignant tumors in breast mammography images. In the system, AlexNet DCNN is used for feature extraction and the last fully connected (fc) layer of the DCNN is connected to SVM to obtain better classification results. This approach for classifying breast cancer tumors has been compared with results of other researchers, either when using the AlexNet architecture with or other DCNN architectures, where the results of the new proposed methods achieved the highest results. They scored the highest Area Under Curve of 94% for the CBIS-DDSM dataset. Another method that Classified nuclei from breast histopathological images used a stacked sparse autoencoder (SSAE) based algorithm [64].

Several other methods have been discussed and reviewed in [64] for Lung and prostate cancer detection, some of them are: A high-level features of skin samples using pre-trained CNN and AlexNet which used 399 images captured from the standard camera to classify benign nevi from melanoma. For lesion classification, K-nearest neighbor was used, and they were able to achieve the accuracy of 93.62% with a specificity of 95.18% and sensitivity rate of 92.1%. A SkinNet CNN was used in for the segmentation and detection of skin cancer. For prostate segmentation, a combination of sparse patch matching and deep feature learning was proposed to extract the feature representation from the MR images, they used the Stack of Sparse Auto Encoders (SSAE) technique. Another method proposed to detect prostate cancer was using patch-based CNN.

6.7 Disease Prediction and Treatment

The Diabetic Retinopathy (DR) is an eye disease and a diabetes complication which cause damage to retina resulting in eye blindness. If detected on time, at an early stage by retinal screening test, it can be controlled and cured without difficulty. However, Manual screening to detect DR is often time consuming and difficult as during the early stages patients rarely show any symptoms. Deep learning has proven to provide useful and accurate solution that can help prevent this condition. A CNN model can work with data taken from retinal imaging and detect hemorrhages, the early symptoms and indicators of DR for detecting the condition [50]. Gulshan et al. [28] applied Deep Convolutional Neural Network (DCNN) on Eye Picture Archive Communication System (EyePACS-1) dataset and Messidor-2 dataset for automatic classification and detection of diabetic retinopathy in retinal fundus images. The authors claimed 97.5% sensitivity and 93.4% specificity on EyePACS-1: and 96.1% sensitivity and 93.9% specificity on Messidor-1, respectively.

Another method used NVIDIA CUDA DCNN library on Kaggle dataset consisting of above 80,000 digital fundus images. The features vector was fed to Cu-DCNN which classified the images into 5 classes using features like exudates, hemorrhages and micro-aneurysms and achieved up to 95% specificity, 30% sensitivity and 75% accuracy [50]. Chandrakumar and Kathirvel [65] trained a DCNN with publicly available Kaggle, DRIVE and STARE dataset to classify affected and healthy retinal fundus reporting an accuracy of 94%. Kaggle dataset include clinician labelled image across 5 classes: No DR, Mild, Moderate, Severe and Proliferative DR.

Infectious disease occurs when a person is infected by a pathogen transmitted through another person or an animal. Forecasting an infectious disease can be quite a daunting task. Methods of deep learning used for predicting infectious disease can be useful for designing effective models. The aim of the study proposed by Chae et al. [66] was to design an infectious disease (scarlet fever, malaria, chicken pox) prediction model that is more suitable than existing models. The performance of the ordinary least squares (OLS) and autoregressive integrated moving average (ARIMA) analysis was used to assess the deep learning models. From the results for DNN and LSTM, both made much better predictions than the OLS and ARIMA models for all infectious diseases. The best performance was from the DNN models although the LSTM models made more accurate predictions.

Kollias et al. [67] designed novel deep neural architectures, composed of CNN and RNN components, trained with medical imaging data, to obtain excellent performances in diagnosis and prediction of Parkinson's disease.

6.8 Genomic Prediction

Deep learning techniques have a promising future in genomics. It is used to understand a genome and help patients get an idea about disease that might affect them. Kopp et al. [68] created Janggu, a python library that facilitates deep learning for genomics applications. Its key features include special dataset objects, that are flexible data acquisition and pre-processing framework for genomics data, a NumPy-like interface, that is directly compatible with popular deep learning libraries, including Keras. It allows visualization of predictions as genomic tracks or by exporting them to the BIGWIG format.

Several people worldwide suffer from Human Immunodeficiency Virus (HIV). The individuals with HIV require regular dose of medicines to treat their condition. As HIV can mutate very quickly, the drugs that are administered to the patients as a part of their treatment regime needs to be altered accordingly. Reinforcement Learning (RL) a deep learning model can be used to stay ahead of the virus as it can track many biomarkers with every drug administration and provide best solution to alter drug sequence for the dynamic treatment regime [69].

Knowledge about genomes can help researchers discover the underlying mechanisms of diseases and develop cures. Researchers at the University of Toronto created a tool called DeepBind, a CNN model which takes genomic data and predicts the sequence of DNA and RNA binding proteins. DeepBind can be used to create computer models that can reveal the effects of changes in the DNA sequence. The information so obtained can be used to develop more advanced diagnostic tools and medications. DeeperBind, a more recent model predicts the protein-binding specificity of DNA sequences using a LSTM recurrent convolutional network. It employs more complex and deeper layers and showed a better performance than DeepBind for some proteins. Both cannot be used to construct nucleic acid sequences as DeepBind and DeeperBind are classification models rather than generative models [56, 70].

6.9 Sensing

In recent times remote patient monitoring system, wearable devices and telemonitoring systems are frequently used to provide real-time patient care. For such systems biosensors are an integral part used to capture and transmit vital signs to the healthcare providers. It is used to constantly monitor the patient's vitals to detect any abnormalities, to keep track of health status before hospital admission etc. *Iqbal* et al. *used* deep deterministic learning (DDL) to classify cardiac diseases, such as myocardial infarction (MI) and atrial fibrillation (Af) which require special attention. First, they detected an R peak based on fixed threshold values and extracted time-domain features. The extracted features were then used to recognize patterns and divided into three classes with ANN and finally executed to detect MI and Af. For cardiac arrhythmia detection *Yıldırım* et al. proposed a CNN based approach based on long term ECG signal analysis with long duration raw ECG signals. They used 10 s segments and trained the classifier for 13, 15 and 17 cardiac arrhythmia diagnostic classes. *Munir* et al. introduced Fuse AD an anomaly detection technique, with streaming data. The initial step was forecasting models for the next timestamp with Autoregressive integrated moving average (ARIMA) and CNN in each time-series. Next it checked whether each timestamp was normal or abnormal by feeding the forecasted results into the anomaly detector module [71].

7 Challenges of Deep Learning in Healthcare

Deep learning techniques provide range of support in healthcare applications but there are some challenges that needs to be addressed in order to get full benefit of DL techniques. Some of the challenges typically faced by healthcare industry are addressed here.

7.1 Domain Complexity and Limited Datasets

One of the major preconditions to use deep learning is the need for massive amount of training dataset as the quality of deep learning-based classifier greatly relies on amount of the data. Availability of medical imaging data is often limited making it a hinderance in the success of deep learning in medical imaging [72]. Another problem that one would face is that biomedicine and health care domain is generally extremely complicated. The diseases are highly unpredictable with little knowledge on what causes or triggers them and how they progress in individual patients. Besides, the number of patients for certain conditions can be limited which can impact the learning and prediction accuracy of the model [18].

7.2 Preserving Privacy of Deep Learning Models

Despite the insights one could get from the huge amount of data stored in EHR system, there are still risks involved. As sensitive information about patient's personal information such as address, claims details, treatment related matter can be a soft target for attacks. HIPAA (Health Insurance Portability and Accountability Act) provides legal rights to patients to protect their health-related data. So proper protocols must be in place to ensure that EHR data is not misused. For this purpose, automatic deep patient de-identification and deployment of intelligent tools can be considered as another area of future research [58]. Although some work in this area is already available. Dernoncourt et al. built RNN based de-identification system and evaluated their system with i2b2 2014 data and MIMIC de-identification data and showed better performance than the existing systems. Later RNN hybrid model was developed for clinical notes de-identification with bidirectional LSTM model deployed for character-level representation [73].

7.3 Temporality and Irregularity of EHR Data

The longitudinal EHR data describes patient's health condition over a period giving a global context and the short-term dependencies among medical events provide local context to the patient's history. Contexts impact the hidden relations among the clinical variables and future patient health outcomes. As the association between clinical events are complex, it is difficult to identify the true signals from the long-term context. For such irregularities, LTSM and GRUs can be used to solve the challenge of extracting true signals from the long-term context due to their abilities to handle long-term dependencies. Using gated structures [35] is one such example where LSTMs or GRUs were applied to model long-term dependencies between clinical events and making predictions. The existing deep learning models for the healthcare domain cannot handle the time factor in an input. Majority of the time, the clinical data is not a static input as disease activity in a patient keeps changing. Hence, designing models that can handle such temporal healthcare data can solve the issue. There is a wealth of EHR data being collected from continuous time series data available in the form of vital signs and other timestamped data like laboratory test results. Some of such applications that make use of vital signs are used for predicting diagnoses, in-hospital mortality, for distinguishing between gout and leukemia etc. The major concern with this type of EHR data is the irregularity of scale as signals are measured at different time scales like hourly, monthly or yearly time scale. Such irregularities can bring down the performance of the model [58].

7.4   Quality and Heterogeneity of the EHR Data

The nature of the EHR data is heterogenous, as it comes from multiple modalities such as clinical notes, medical reports, Medical images, billing data, patient demographic information, continuous time-series data of vital signs laboratory measurements, prescription, patient history and more. Research suggests that finding patterns between multimodal data can enhance the accuracy of diagnosis, prediction, and overall performance of the learning system. However, multimodal learning is still a challenge due to the heterogeneity of the EHR data [73]. Besides disease-specific multimodal data, some studies used multivariate time series data. For example, CNN was applied on multivariate encephalogram (EEG) signals for automated classification of normal, preictal, and seizure subjects. LSTM based model was developed using vital sign series from the Medical Information Mart for Intensive Care III (MIMIC III) for detection of sepsis. Hierarchical attention bidirectional gated recurrent unit (GRU) model was used in a study where clinical documents from the MIMIC III dataset were automatically tagged with associated diagnosis codes.

To provide an explanation to the classification from clinical notes to diagnosis codes, an interpretable model based on convolution plus attention model architecture was introduced. In another study to automatically extract the primary cancer sites and their laterality, DFFNNs and CNNs were applied, to free-text pathology reports [73].

Several studies focused on vector-based representation of clinical concepts to reduce the dimensionality of the code space and reveal latent relationships between similar types of discrete codes. Deep EHR methods for code representation include the NLP-inspired skip-gram technique for predicting heart failure, a CNN model to predict unplanned readmissions, an RBM based framework for stratifying suicide risk, DBMs for diagnosis clustering and a technique based on LSTMs for modelling disease progression and predicting future risk for diabetes and mental health. While code-based representations of clinical concepts and patient encounters is the method to deal with the problem of heterogeneity, they ignore many important real-valued measurements associated with items such as laboratory tests, intravenous medication infusions, vital signs, and more. In the future, more research should focus on processing diverse sets of data directly, rather than relying on codes from vocabularies that are designed for billing purposes [58]. Multi-task learning is also known to solve the problem of heterogeneity of EHR data. It allows models to jointly learn data across multiple modalities. Some neurons in the neural network model are shared among all tasks, while other neurons are specialized for specific tasks. Si and Roberts proposed MTL framework, based on multi-level CNN that demonstrates the feasibility of using MTL to efficiently handle patient clinical notes and to obtain a vector-based patient representation across multiple predictions [74].

7.5 Lack of Labels

Medical images are often overly complex. There are several kinds of medical imaging devices and each would produce images with thousands of image features. Labels indicates the target of interest such as true clinical outcomes or true disease phenotypes. Labels are not always captured well in EHR data and thus are not available for large number of training models. Label acquisition on the other hand requires domain knowledge or highly trained domain experts. Lack of meaningful labels [75] on EHR records is considered as a major barrier to deep learning models [73]. For supervised learning approaches, labels are manually crafted from the occurrences of codes, such as diagnosis, procedure, and medication codes. Transfer learning could offer alternative approaches. For example, [35] used LSTM to model sequences of diagnostic codes and demonstrated that for the same task the learned knowledge could be transferred to new datasets. An autoencoder variant architecture [71] was applied to perform transfer learning from source tasks in which training labels are ample but of limited clinical value to more specific target tasks such as inferring prescription from diagnostic codes.

7.6 Lack of Transparency and Interpretability

Although deep learning models can produce accurate predictions, they are often termed as black-box models that lack interpretability and transparency. This is a serious concern for both clinicians and patients as they are unwilling to accept machine recommendations without proper reasoning to support it. To tackle this issue, researchers have used a knowledge distillation approach, which compresses the knowledge learned from a complex model into a simpler model that is easier to deploy and interpret. Other mechanisms such as attention mechanism and knowledge injection via attention have been used to add interpretability and to derive a latent representation of medical codes [73].

8 Conclusion

Over the past few years, deep learning for healthcare has grown tremendously. From the several healthcare papers reviewed, deep learning models yield better performance in many healthcare applications when compared to traditional machine learning methods.

Deep learning models require less effort in feature extraction compared to their ML counterparts. Deep learning in healthcare can help physician's diagnose disease, detect and classify tumors, and predict infectious disease outbreaks with higher accuracy. Deep learning models have demonstrated success in several areas of healthcare especially image classification and disease prediction. It can be seen from the review that CNN is very well suited for medical image analysis other frameworks like LSTM, GRU and Autoencoders have been effectively used in several other healthcare tasks.

Deep learning can pave way for next generation of healthcare systems that are more robust, accurate in their predictions, scalable to include huge EHR data, provides better healthcare outcomes and support clinicians in their activities. Although, deep learning techniques are performing well in many analytics tasks, there many inaccuracies and issues in the system such as data heterogeneity, quality, irregularities, security, lack of labels and temporal modelling that still needs to be addressed in order to clearly take advantage of Deep Learning systems further.

References

1. Ed, C.: The Real-World Benefits of Machine Learning in Healthcare. Available from: https://www.healthcatalyst.com/clinical-applications-of-machine-learning-in-healthcare (2017)
2. Anirudh, V.K.: What Are the Types of Artificial Intelligence: Narrow, General, and Super AI Explained. Available from:https://it.toolbox.com/tech-101/what-are-the-types-of-artificial-int elligence-narrow-general-and-super-ai-explained (2019)
3. Michael, C., Kamalnath, V., McCarthy, B.: An executive's guide to AI. Available from: https://www.mckinsey.com/business-functions/mckinsey-analytics/our-insights/an-exe cutives-guide-to-ai
4. Najafabadi, M.M., Villanustre, F., Khoshgoftaar, T.M., et al.: Deep learning applications and challenges in big data analytics. J. Big Data **2**, 1 (2015). https://doi.org/10.1186/s40537-014-0007-7
5. Richa, B.: Understanding the difference between deep learning & machine learning. Available from: https://analyticsindiamag.com/understanding-difference-deep-learning-machine-learning/ (2017)
6. Sambit, M.: Why Deep Learning over Traditional Machine Learning? Available from: https://towardsdatascience.com/why-deep-learning-is-needed-over-traditional-machine-learning-1b6a99177063 (2018)
7. Dino, Q., He, B., Faria, B.C., Jara, A., Parsons, C., Tsukamoto, S., Wale, R.: IBM Redbooks. International technical support organization. IBM PowerAI: Deep Learning Unleashed on IBM Power Systems Serve (2018)
8. Mateusz, O.: Deep Learning Frameworks Comparison—Tensorflow, PyTorch, Keras, MXNet, The Microsoft Cognitive Toolkit, Caffe, Deeplearning4j, Chainer. https://www.net guru.com/blog/deep-learning-frameworks-comparison (2019)

9. Mitul, M.: Top 8 Deep Learning Frameworks. Available from: https://dzone.com/articles/8-best-deep-learning-frameworks (2018)
10. Nguyen, G., Dlugolinsky, S., Bobák, M., et al.: Machine learning and deep learning frameworks and libraries for large-scale data mining: a survey. Artif. Intell. Rev. **52**, 77–124 (2019). https://doi.org/10.1007/s10462-018-09679-z
11. Oleksii, K.: Top 10 Best Deep Learning Frameworks in 2019. Available from: https://toward sdatascience.com/top-10-best-deep-learning-frameworks-in-2019-5ccb90ea6de (2019)
12. Pulkit, S.: 5 Amazing Deep Learning Frameworks Every Data Scientist Must Know! Available from: https://www.analyticsvidhya.com/blog/2019/03/deep-learning-frameworks-comparison/ (2019)
13. Saptarshi, S., Basak, S., Saikia, P., Paul, S., Tsalavoutis, V., Atiah, F.D., Ravi, V., Peters II, R.A.: A review of deep learning with special emphasis on architectures, applications and recent trends. Math. Comput. Sci. Published in ArXiv 2019 (2019). https://doi.org/10.20944/prepri nts201902.0233.v1
14. Simplilearn: Top Deep Learning Frameworks. Available from: https://www.simplilearn.com/deep-learning-frameworks-article (2020)
15. Zhang, Z., Zhao, Y., Liao, X., Shi, W., Li, K., Zou, Q., Peng, S.: Deep learning in omics: a survey and guideline. Deep learning in omics: a survey and guideline. Brief Funct. Genomics **18**(1), 41–57 (2019). https://doi.org/10.1093/bfgp/ely030 (PMID: 30265280)
16. Antonio, H.-B., Herrera-Flores, B., Tomás, D., Navarro-Colorado, B.: A systematic review of deep learning approaches to educational data mining. Complexity **2019**, 22 pp., Article ID 1306039 (2019). https://doi.org/10.1155/2019/1306039
17. Iman, R.V., Majidian, S.: Literature Review on Big Data Analytics Methods (2019). https://doi.org/10.5772/intechopen.86843
18. Riccardo, M., Wang, F., Wang, S., Jiang, X., Dudley, J.T.: Deep learning for healthcare: review, opportunities and challenges. Brief. Bioinform. **19**(6), 1236–1246. PMID: 28481991 (2018) https://doi.org/10.1093/bib/bbx044
19. Athanasios, V., Doulamis, N., Doulamis, A., Protopapadakis, E.: Deep learning for computer vision: a brief review. Recent Dev. Deep Learn. Eng. Appl. **2018**, 13 pp. (2018). https://doi.org/10.1155/2018/7068349 (Article ID 7068349)
20. Daniele, R., Wong, C., Deligianni, F., Berthelot, M., Andreu-Perez, J., Lo, B., Yang, G.-Z., IEEE.: Deep learning for health informatics. IEEE J. Biomed. Health Inform. (2016). https://doi.org/10.1109/jbhi.2016.2636665
21. Tobore, I., Li, J., Yuhang, L., Al-Handarish, Y., Kandwal, A., Nie, Z., Wang, L.: Deep learning intervention for health care challenges: some biomedical domain considerations. JMIR Mhealth Uhealth **7**(8), e11966. https://mhealth.jmir.org/2019/8/e11966. https://doi.org/10.2196/11966 (PMID: 31376272. PMCID: 6696854) (2019)
22. Tim Jones, M.: Deep learning architectures: the rise of artificial intelligence. Available from: https://developer.ibm.com/articles/cc-machine-learning-deep-learning-architectures/ (2017)
23. Salakhutdinov, R., Hugo, L.: Efficient learning of deep Boltzmann Machines. J. Mach. Learn. Res. Proc. Track. **9**, 693–700 (2010)
24. Aboozar, T., Cosma, G., MMcGinnity, T.: Deep-FS: a feature selection algorithm for Deep Boltzmann Machines. Neurocomputing **322**, 22–37 (2018)
25. Nikola, S., Bodnar, C., Tariq, I., Scherer, P., Terre, H.A., Shams, Z., Jamnik, M., Liò, P.: Variational autoencoders for cancer data integration: design principles and computational practice. Front. Genet. (2019). https://doi.org/10.3389/fgene.2019.01205
26. Pereira, S., Pinto, A., Alves, V., Silva, C.A.: Brain tumor segmentation using convolutional neural networks in MRI images. IEEE Trans. Med. Imaging. **35**(5), 1240–1251 (2016) (2016 May). https://doi.org/10.1109/tmi.2016.2538465 (Epub 2016 Mar 4)
27. Prasoon, A., Petersen, K., Igel, C., et al.: Deep feature learning for knee cartilage segmentation using a triplanar convolutional neural network. Med. Image Comput. Comput. Assist. Interv. **16**, 246–253 (2013)
28. Gulshan, V., Peng, L., Coram, M., et al.: Development and validation of a deep learning algorithm for detection of diabetic retinopathy in retinal fundus photographs. JAMA **316**, 2402–2410 (2016)

29. Esteva, A., Kuprel, B., Novoa, R., et al.: Dermatologist-level classification of skin cancer with deep neural networks. Nature **542**, 115–118 (2017). https://doi.org/10.1038/nature21056
30. Sathyanarayana, A., Joty, S., Fernandez-Luque, L., et al.: Sleep quality prediction from wearable data using deep learning. JMIR M health U health **4**, e130 (2016)
31. Alipanahi, B., Delong, A., Weirauch, M.T., et al.: Predicting the sequence specificities of DNA- and RNA-binding proteins by deep learning. NatBiotechnol. **33**, 831–838 (2015)
32. Lipton, Z.C., Kale, D.C., Elkan, C., et al.: Learning to diagnose with LSTM recurrent neural networks. In: International Conference on Learning Representations, SanDiego, CA, USA, pp. 1–18 (2015)
33. Wang, T., Qiu, R.G., Yu, M.: Predictive modeling of the progression of Alzheimer's Disease with recurrent neural networks. Sci. Rep. **8**, 9161 (2018). https://doi.org/10.1038/s41598-018-27337-w
34. Beeksma, M., Verberne, S., van den Bosch, A., et al.: Predicting life expectancy with a long short-term memory recurrent neural network using electronic medical records. BMC Med. Inform. Decis. Mak. **19**, 36 (2019). https://doi.org/10.1186/s12911-019-0775-2
35. Choi, E., Bahadori, M.T., Schuetz, A., et al.: Doctor AI: predicting clinical events via recurrent neural networks. arXiv 2015. http://arxiv.org/abs/1511.05942v11 (2015)
36. Edward, C., Schuetz, A., Stewart, W.F., Sun, J.: Using recurrent neural network models for early detection of heart failure onset. J. Am. Med. Inform. Assoc. **24**(2), 361–370 (2017) (Published online 2016 Aug 13). https://doi.org/10.1093/jamia/ocw112 (PMCID: PMC5391725)
37. van Tulder, G., de Bruijne, M.: Combining generative and discriminative representation learning for lung CT analysis with convolutional restricted Boltzmann machines. IEEE Trans. Med. Imaging **35**(5), 1262–1272 (2016)
38. Brosch, T., Tam, R.: Manifold learning of brain MRIs by deep learning. Med. Image Comput. Comput. Assist. Interv. **16**(Pt 2), 633–640 (2013)
39. Yoo, Y., Brosch, T., Traboulsee, A., et al.: Deep learning of image features from unlabeled data for multiple sclerosis lesion segmentation. In: International Workshop on Machine Learning in Medical Imaging, Boston, MA, USA, pp. 117–124 (2014)
40. Phúc, L.: https://medium.com/vitalify-asia/gan-for-unsupervised-anomaly-detection-on-x-ray-images-6b9f678ca57d (2018)
41. Bermudez, C., Plassard, A.J., Davis, T.L., Newton, A.T., Resnick, S.M., Landman, B.A.: Learning implicit brain MRI manifolds with deep learning. Proc. SPIE Int. Soc. Opt. Eng. 10574–105741L (2018). https://doi.org/10.1117/12.2293515
42. Suk, H., Lee, S., Shen, D.: Latent feature representation with stacked auto-encoder for AD/MCI diagnosis. Brain Struct. Funct. **220**, 841–859 (2015). https://doi.org/10.1007/s00429-13-0687-3
43. Cheng, J.-Z., Ni, D., Chou, Y.-H., et al.: Computer-aided diagnosis with deep learning architecture: applications to breast lesions in US images and pulmonary nodules in CT scans. Sci. Rep. **6**, 24454 (2016)
44. Fakoor, R., Ladhak, F., Nazi, A., et al.: Using deep learning to enhance cancer diagnosis and classification. In: International Conference on Machine Learning, Atlanta, GA, USA (2013)
45. Che, Z., Kale, D., Li, W., et al.: Deep computational phenotyping. In: ACM International Conference on Knowledge Discovery and Data Mining, Sydney, SW, Australia, pp. 507–516 (2015)
46. Miotto, R., Li, L., Kidd, B.A., et al.: Deep patient: an unsupervised representation to predict the future of patients from the electronic health records. Sci. Rep. **6**, 26094 (2016)
47. Alexander, A., Plis, S., Artemov, A., Ulloa, A., Mamoshina, P., Zhavoronkov, A.: Deep learning applications for predicting pharmacological properties of drugs and drug repurposing using transcriptomic data. Mol. Pharm. **13**, 2524–2530. © 2016 American Chemical Society (2016) https://doi.org/10.1021/acs.molpharmaceut.6b00248
48. Md Zahangir, A., Taha, T.M., Yakopcic, C., Westberg, S., Sidike, P., Nasrin, M.S., Hasan, M., Van Essen, B.C., Awwal, A.A.S., Asari, V.K.: A state-of-the-art survey on deep learning theory and architectures. Electronics **8**, 292. https://doi.org/10.3390/electronics8030292 https://www.mdpi.com/journal/electronics (2019)

49. Suvajit, D., Manideep, B.C.S., Rai, S., Vijayarajan, V.: A comparative study of deep learning models for medical image classification. In: 2017 IOP Conference Series: Materials Science and Engineering, vol. 263, pp. 042097 (2017). https://doi.org/10.1088/1757-899x/263/4/042097
50. Muhammad Imran, R., Naz, S., Zaib, A.: Deep Learning for Medical Image Processing: Overview, Challenges and Future. https://arxiv.org/ftp/arxiv/papers/1704/1704.06825.pdf
51. Dinggang, S., Wu, G., Suk, H.-I.: Deep learning in medical image analysis. Annu. Rev. Biomed. Eng. **19**, 221–248 (2017). https://doi.org/10.1146/annurev-bioeng-071516-044442
52. Slava, K.: Deep Learning (DL) in Healthcare. https://blog.produvia.com/deep-learning-dl-in-healthcare-4d24d102d317 (2018)
53. Mostapha, B.: ChemGAN challenge for drug discovery: can AI reproduce natural chemical diversity?. Archiv preprint: 1708.08227v3
54. Ramsundar, B.: deepchem.io. https://github.com/deepchem/deepchem (2016)
55. Fourcade, A., Khonsari, R.H.: Deep learning in medical image analysis: a third eye for doctors. J. Stomatology Oral Maxillofac. Surg. **120**(4), 279–288 (2019)
56. Missinglink.ai. Available From: https://missinglink.ai/guides/deep-learning-healthcare/deep-learning-healthcare/
57. Yun, L., Kohlberger, T., Norouzi, M., Dahl, G.E., Smith, J.L.: Artificial intelligence-based breast cancer nodal metastasis detection insights into the black box for pathologists. Arch. Pathol. Lab. Med. **143**(7), 859–868 (2019). https://doi.org/10.5858/arpa.2018-0147-oa (Epub Oct 8)
58. Benjamin, S., Tighe, P.J., Bihorac, A., Rashidi, P.: Deep EHR: A Survey of Recent Advances in Deep Learning Techniques for Electronic Health Record (EHR) Analysis. arXiv 1706.03446v2 [cs.LG] 24 Feb 2018. https://arxiv.org/pdf/1706.03446.pdf (2018)
59. Luciano, C., Veltri, P., Vocaturo, E., Zumpano, E.: Deep learning techniques for electronic health record analysis. In: 2018 9th International Conference on Information, Intelligence, Systems and Applications (IISA). 978-1-5386-8161-9/18/$31.00 c 2018 IEEE (2018). https://doi.org/10.1109/iisa.2018.8633647
60. Daniel, L., Santhana, P.: Deep Learning to Detect Medical Treatment Fraud Proceedings of Machine Learning Research. KDD 2017: Workshop on Anomaly Detection in Finance, vol. 71, pp. 114–120 (2017)
61. Tony, K.: Deep Learning Drops Error Rate for Breast Cancer Diagnoses by 85%. Nvidia. Available From: https://blogs.nvidia.com/blog/2016/09/19/deep-learning-breast-cancer-diagnosis/ (2016)
62. Rajaraman, S., et al.: Pre-trained convolutional neural networks as feature extractors toward improved malaria parasite detection in thin blood smear images. PeerJ (2018). https://doi.org/10.7717/peerj.4568
63. Ragab, D.A., Sharkas, M., Marshall, S., Ren, J.: (2019) Breast cancer detection using deep convolutional neural networks and support vector machines. PeerJ **7**, e 6201 (2019). https://doi.org/10.7717/peerj.6201
64. Khushboo, M., Elahi, H., Ayub, A., Frezza, F., Rizzi, A.: Cancer diagnosis using deep learning: a bibliographic review. Cancers **11**(9), 1235 (2019) https://doi.org/10.3390/cancers11091235
65. Chandrakumar, T., Kathirvel, R.: Classifying diabetic retinopathy using deep learning architecture. Int. J. Eng. Res. Technol. (IJERT) **05**(06) (2016). http://dx.doi.org/10.17577/IJERTV5IS060055
66. Chae, S., Kwon, S., Lee, D.: Predicting infectious disease using deep learning and big data. Int. J. Environ. Res. Public Health **15**(8), 1596 (2018). https://doi.org/10.3390/ijerph15081596
67. Kollias, D., Tagaris, A., Stafylopatis, A., et al.: Deep neural architectures for prediction in healthcare. Complex Intell. Syst. **4**, 119 (2018). https://doi.org/10.1007/s40747-017-0064-6
68. Wolfgang, K., Monti, R., Tamburrini, A., Ohler, U., Akalin, A.: Janggu—Deep Learning for Genomics. bioRxiv preprint (2019) https://doi.org/10.1101/700450
69. Naveen, J.: Top 5 applications of deep learning in healthcare. Allerin. Available From: https://www.allerin.com/blog/top-5-applications-of-deep-learning-in-healthcare (2018)
70. Im, J., Park, B., Han, K.: A generative model for constructing nucleic acid sequences binding to a protein. BMC Genom. **20**, 967 (2019). https://doi.org/10.1186/s12864-019-6299-4

71. Dubois, S., Romano, N., Jung, K., Shah, N., Kale, D.C.: The Effectiveness of Transfer Learning in Electronic Health Records Data. Available From: https://openreview.net/forum?id=B1_E8xrKe (2017)
72. Sayon, D.: A 2020 Guide to Deep Learning for Medical Imaging and the Healthcare Industry. Nanonets. Available From: https://nanonets.com/blog/deep-learning-for-medical-imaging/ (2020)
73. Xiao, C., Choi, E., Sun, J.: Opportunities and challenges in developing deep learning models using electronic health records data: a systematic review. J. Am. Med. Inform. Assoc. **25**(10), 1419–1428 (2018). https://doi.org/10.1093/jamia/ocy068
74. Si, Y., Roberts, K.: Deep patient representation of clinical notes via multi-task learning for mortality prediction. In: AMIA Joint Summits on Translational Science Proceedings. AMIA Joint Summits Transl. Sci. **2019**, 779–788 (2019)
75. Gloria, H.-J.K., Hui, P.: DeepHealth: Deep Learning for Health Informatics. arXiv:1909.00384 [cs.LG] (2019)

Factors Influencing Electronic Service Quality on Electronic Loyalty in Online Shopping Context: Data Analysis Approach

Ahlam Al-Khayyal⊕, Muhammad Alshurideh⊕, Barween Al Kurdi⊕, and Said A. Salloum⊕

1 Introduction

According to News24 [1], online learning tools for instance Cisco System's WebEx Zoom, Instructure's Canvas, and many other educational tech companies across the U.S. reported instant increase by 700% in sales. While, the online booking and travel sites drops by 20.8%, online grocery sector increases by 19.9%, and there is a fluctuation increase/decrease in the other sectors (Impact of the Coronavirus on Digital Retail: An Analysis of 1.8 Billion Site Visits—Content square, 2020). To maintain this growing shopper's interest in online shopping, retailers should take advantage and invest more in the adoption of e-commerce [2–4]. The world fair of coronavirus encourage them to offer their products and services through e-channels, and to enhance their customer's e-loyalty [5–7]. This challenge for retailers to step forward and attract the attention of the safety of e-channels transactions [8–10]. The rapid growth of online shopping has brought competitive attention to provide higher service quality to ensure retaining customers [11–13]. The improvement and usage of internet e-commerce has become the most popular types of doing business. The importance of this study is improving the elements of electronic channel for the

A. Al-Khayyal · M. Alshurideh · S. A. Salloum (✉)
University of Sharjah, Sharjah, UAE
e-mail: ssalloum@sharjah.ac.ae

M. Alshurideh
Faculty of Business, University of Jordan, Amman, Jordan

B. A. Kurdi
Amman Arab Universities, Amman, Jordan

S. A. Salloum
Research Institute of Sciences and Engineering, University of Sharjah, Sharjah, UAE

© The Editor(s) (if applicable) and The Author(s), under exclusive license to Springer Nature Switzerland AG 2021
A.-E. Hassanien et al. (eds.), *Enabling AI Applications in Data Science*, Studies in Computational Intelligence 911,
https://doi.org/10.1007/978-3-030-52067-0_16

367

service provider to enhance the customer's e-loyalty. This study raises the following research questions:

- **RQ1**: What are the electronic service quality main factors influences customer's e-satisfaction?
- **RQ2**: What are the electronic service quality main factors influences customer's e-trust?
- **RQ3**: What is the impact of customer's e-satisfaction and customer's e-trust on e-shopping?
- **RQ4**: How customers' e-shopping related to customer's e-loyalty?

Following [14–19], this study discusses the related literature review, methodology, result and discussion, and conclusion. This study concludes that electronic service quality highly influences customer e-satisfaction and e-trust. As a result, customer proceeds online shopping while developing the electronic loyalty.

2 Literature Review

A careful literature review is a crucial phase before conducting any further research study. It shapes the basis for knowledge accumulation, which in turn eases the theories' developments and enhancements, closes the existing gaps in research, and reveals areas where previous research has missed [15].

2.1 Electronic Service Quality

Electronic service quality is defined as 'the extent to which a web site facilitates efficient and effective shopping, purchasing, and delivery of products and services' [20, 21]. The importance of studying the electronic service quality dimensions arisen from the extraordinary growth of online shopping [22–24]. An extraordinary perceived service quality will definitely affect customer satisfaction and trust [2]. Similarly, lower service quality will lead to adverse word-of-mouth and therefore reduce sales and revenues as the customer's transfer to rivals [2, 25, 26]. The literature studied on this topic brings attention to two scopes. First, how to evaluate, measures, produce scales for electronic service quality [21]. Second, how electronic service quality influences other factors such as satisfaction, trust, shopping, and loyalty [2].

2.2 Dimensions of Electronic Service Quality: Independent Factors

Scholars have adopted and adjusted different measurements for electronic service quality. This study has focused on three electronic service quality scales, namely E-S-QUAL, eTransQual, and eTailQ. The E-S-QUAL is a multiple-item scale for assessing Electronic Service Quality [21], this scale has four dimensions, namely efficiency, compliance, system availability, and privacy. Where efficiency describe the easy and fast access using the site, implementation is defined as the extent of the promises the website makes regarding the delivery of orders and availability of goods are fulfilled, while system availability indicates for the correct site technical performance, and privacy scale is the site's degree of security protects customer information (Fig. 1).

The eTransQual by [27] have five dimensions, namely responsiveness, reliability, process, functionality/design, and enjoyment. It's a transaction process-based approach for capturing service quality in online shopping. This model demonstrates a significant positive impact on important outcome variables such as customer

Fig. 1 The main dimensions of the E-S-QUAL model

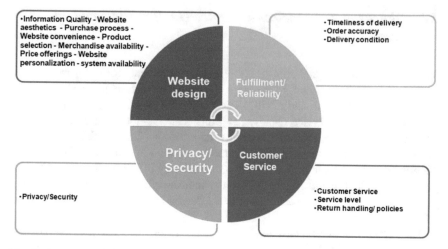

Fig. 2 The eTailQ model

satisfaction and perceived value. Also, it reported that enjoyment is a factor affecting both relationship duration and repurchase intention as major drivers of a customer's lifetime value.

The eTailQ model by [28] has four dimensions including Fulfillment/Reliability, Website design, Privacy/Security, and Customer Service. This scale is described as shown in Fig. 2. This model strongly predicts customer judgments of quality, satisfaction, customer loyalty and attitudes towards the site.

3 Methods

The process of systematic review is to review other researcher findings in particular topic, by identifying the research question, selecting the appropriate databases and publication, assessing the quality of chosen studies, comparing the findings and interpreting the comparison [29]. This systematic review followed [14, 15] systematic review methods. This review is carried out with similar stages: the selection of data sources and research strategies, setting inclusion and exclusion criteria, conducting quality assessment, coding and analyzing data. The details of mentioned stages are described in the following subsections.

3.1 Inclusion/Exclusion Criteria

The inclusion and exclusion criteria described in Table 1 were applied on all researches done on the previous section. Therefore, all included articles in this study should meet these requirements.

Table 1 Inclusion and exclusion criteria

Inclusion criteria	Exclusion criteria
Peer reviewed	Any articles written in other language than English
From 2015 to 2020	Any articles did not include e-service quality in its title
Scholarly journal and articles & conference papers	Any articles did not meet the inclusion criteria

Table 2 Data sources and databases

Database	Result	Final result
ProQuest	28	12
EBSCO Host	7	3
Google scholar	11	5
Emerald	36	2
Science direct	44	4
Total	126	26

3.2 Data Sources and Research Strategies

The articles comprised in this systematic review were collected on February 2020 from five databases: ProQuest One Academic, EBSCO Host, Google scholar, Emerald, Science Direct. The search terminology includes the keywords ("Electronic service quality" AND "Electronic loyalty"), ("E-service quality" AND "E-loyalty"), ("E-service quality" AND "The E-S-QUAL"), ("E-service quality" AND "eTailQ model"), ("Electronic service quality" AND "The E-S-QUAL"), and ("Electronic service quality" AND "eTailQ model"). These searches found 126 articles using the above-mentioned keywords. There were 21 articles found as duplicated and removed from the total result. Therefore, the total number reduced and becomes 106 articles. Table 2 shows the results from the selected databases with numbers of included articles. Figure 3 shows the detailed process of selecting articles for this study.

3.3 Quality Assessment

A quality assessment checklist with eight questions was identified to provide a means for assessing the quality of the articles that were downloaded for additional analysis (N = 26) as seen Table 3. The checklist in Table 3 was adapted from those used by [14, 17, 30–33] who adapted from [34]. Each assessment question was scored with three-point scale, considering "Yes" as 1 point, "No" as 0 point, and "Partially" as 0.5 point. Therefore, each article could score between 0 and 8, with the higher the total score an article achieves, the higher the degree of quality to which this article

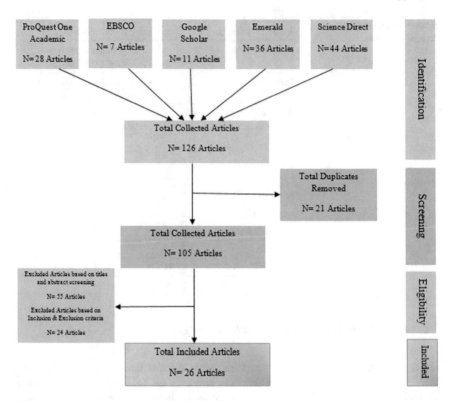

Fig. 3 PRISMA flowchart for the selected studies

Table 3 Quality assessment checklist [15]

#	Questions
1	Are the research objectives clearly defined?
2	Was the study designed to achieve these goals?
3	Are the variables considered by the study clearly defined?
4	Is the context of the study/specialization clearly defined?
5	Are the methods used in collection data sufficiently detailed?
6	Does the study explain the reliability/validity of the measures?
7	Are the statistical techniques used to analyze the data sufficiently described?
8	Do you add results to the literature?
9	Does the study add to your knowledge or understanding?

related to the research questions. Table 4 shows the quality assessment results for all the 26 articles. As, it is clear that all the articles have passed this assessment. This is indicates that all the articles are qualified enough to be used for further analysis and discussion. This assessment done with fully respect to other researcher work and efforts.

Table 4 shows the analysis of the selected articles to assess their quality using the quality assessment questions.

Table 4 Quality assessment results

#	Q1	Q2	Q3	Q4	Q5	Q6	Q7	Q8	Total	Percentage (%)
S1	1	1	1	1	1	1	1	1	8	100
S2	1	1	1	0.5	1	1	1	1	7.5	94
S3	0.5	1	1	1	0.5	0.5	1	1	6.5	81
S4	1	1	1	1	1	0	1	1	7	88
S5	1	1	1	1	1	1	1	1	8	100
S6	1	1	1	0.5	0.5	1	1	1	7	88
S7	0.5	1	0.5	0.5	1	1	1	1	6.5	81
S8	1	1	0.5	1	1	0.5	1	1	7	88
S9	0.5	1	0.5	1	0.5	1	1	1	6.5	81
S10	0.5	1	1	1	1	1	1	1	7.5	94
S11	0.5	1	1	1	1	1	1	1	7.5	94
S12	0.5	1	1	1	1	1	1	1	7.5	94
S13	1	1	1	0.5	1	1	1	1	7.5	94
S14	1	1	1	1	1	1	1	1	8	100
S15	1	1	1	1	1	1	1	1	8	100
S16	0.5	1	1	1	0.5	0.5	1	1	6.5	81
S17	1	1	1	1	0.5	0.5	1	1	7	88
S18	1	1	1	1	0.5	1	1	1	7.5	94
S19	0.5	1	1	1	1	1	1	1	7.5	94
S20	1	1	1	1	1	1	1	1	8	100
S21	0.5	1	1	0.5	0.5	0.5	1	1	6	75
S22	0.5	1	1	1	1	1	1	1	7.5	94
S23	0.5	1	0.5	1	1	1	1	1	7	88
S24	0.5	1	1	1	1	1	1	1	7.5	94
S25	1	1	1	1	0.5	1	1	1	7.5	94
S26	0.5	1	1	1	1	1	1	1	7.5	94

4 Results, Discussions and Recommendations

4.1 Results and Analysis

Based on the 26 research articles from 2015 to 2020, included in this systematic review, this study reported that most of the studies on electronic services quality were conducted on 2016. Figure 4 summarizes and report the numbers of studies conducted between 2015 and 2020. Additionally, Fig. 5 shows the numbers of studies in terms of countries. Figure 6 summarizes the main factors selected frequencies. The results of this systematic review are stated based on the four research questions.

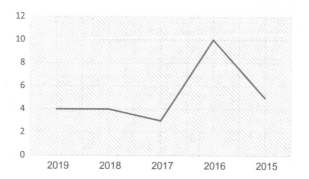

Fig. 4 Number of studies conducted between 2015 and 2020

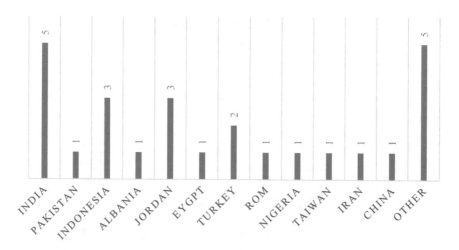

Fig. 5 Number of studies in terms of countries

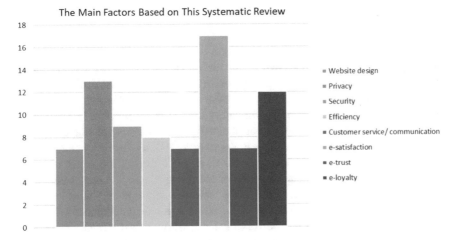

Fig. 6   The main factors with highest frequencies

4.2   Discussion and Recommendations

– What are the e-SQ main factors influences customer e-satisfaction?

Numerous studies examined the relationship between the electronic service quality and the customer electronic satisfaction. This systematic review reported that website design, information quality, security, privacy, fulfilment, customer service/communication, reliability, and efficiency are the most influential electronic service quality factors on customer electronic satisfaction [2, 12, 20, 22, 35–48]. On the other hand, only one study conclude that electronic service quality is partially influences the electronic satisfaction [49].

– What are the e-SQ main factors influences customer e-trust?

Numerous studies examined the relationship between the electronic service quality and the customer electronic satisfaction. This systematic review reported that electronic service quality factors influences customer electronic trust [2, 40, 41, 45, 47]. On the other hand, [37] stated that electronic service quality is partially mediate the electronic service quality and customer electronic satisfaction. However, [50] reported that customer electronic trust not directly influences from electronic service quality. Nevertheless, customers develop electronic trust, when they start have sense that website is trustworthy [51–53].

– What is the impact of customer's e-satisfaction and customer's e-trust on e-shopping?
– How customers' e-shopping related to customer's e-loyalty?

According to [40, 50, 54] the customers should be satisfied and trust the website to proceed shopping online and develop their electronic loyalty.

5 Conclusion and Future Works

This analysis aims to review and synthesize the literature about electronic service quality, illustrates what is known about the topic, and proposed a study model including the hypothesis for future research. After careful review of 26 studies, it was obvious that the main factors influences the electronic service quality are website design, privacy, security, efficiency, and customer service. Limitation for this review includes time allocated for this project, few numbers of studies, and the researcher little experience. The researcher may repeat this review by changing the inclusion/exclusion criteria, expanding the time, and changing the search terms for future research. Online shopping is one of the growing trends among many other online markets. These markets are heading towards a major technological transformation due to new and innovative tools such as Artificial Intelligence (AI), Data Science, Virtual Reality (VR) and Augmented Reality (AR). Customer experience management is greatly influenced by obtaining customer satisfaction through integrated artificial intelligence technology to provide efficient customer service. These technologies will be maintained in future work.

References

1. Al-Maroof, R.S., Salloum, S.A.: An Integrated model of continuous intention to use of google classroom. In: Recent Advances in Intelligent Systems and Smart Applications, pp. 311–335. Springer, Cham
2. Al-dweeri, R., Obeidat, Z., Al-dwiry, M., Alshurideh, M., Alhorani, A.: The impact of e-service quality and e-loyalty on online shopping: moderating effect of e-satisfaction and e-trust. Int. J. Mark. Stud. 9(2), 92–103 (2017)
3. Alshurideh, Masa'deh, R., Al Kurdi, B.: The effect of customer satisfaction upon customer retention in the Jordanian mobile market: an empirical investigation. Eur. J. Econ. Financ. Adm. Sci. 47(12), 69–78 (2012)
4. Alshurideh: Do electronic loyalty programs still drive customer choice and repeat purchase behaviour? Int. J. Electron. Cust. Relatsh. Manag. 12(1), 40–57 (2019)
5. Alshurideh, M., Salloum, S.A., Al Kurdi, B., Al-Emran, M.: Factors affecting the social networks acceptance: an empirical study using PLS-SEM approach. In: 8th International Conference on Software and Computer Applications (2019)
6. Alshurideh, M., Al Kurdi, B., Abumari, A., Salloum, S.: Pharmaceutical promotion tools effect on physician's adoption of medicine prescribing: evidence from Jordan. Mod. Appl. Sci. 12(11), 210–222 (2018)
7. Alshurideh, et al.: The impact of Islamic Bank's service quality perception on Jordanian customer's loyalty. J. Manag. Res. 9 (2017)
8. Obeidat, Z.M., Alshurideh, M.T., Al Dweeri, R., Masa'deh, R.: The influence of online revenge acts on consumers psychological and emotional states: does revenge taste sweet? In: 33 IBIMA Conference Proceedings, 10–11 April 2019, Granada, Spain
9. Kurdi, B.A., Alshurideh, M., Salloum, S.A., Obeidat, Z.M., Al-Dweeri, R.M.: An empirical investigation into examination of factors influencing university students' behavior towards e-learning acceptance using SEM approach. Int. J. Interact. Mob. Technol. 14(2) (2020)
10. Ingham, J., Cadieux, J., Mekki Berrada, A.: E-shopping acceptance: a qualitative and meta-analytic review. Inf. Manag. (2015)

11. Alshurideh: A theoretical perspective of contract and contractual customer-supplier relationship in the mobile phone service sector. Int. J. Bus. Manag. **12**(7), 201–210 (2017)
12. Ammari, G., Al Kurdi, B., Alshurideh, M., Alrowwad, A.: Investigating the impact of communication satisfaction on organizational commitment: a practical approach to increase employees' loyalty. Int. J. Mark. Stud. **9**(2), 113–133 (2017)
13. Alshurideh, M.: Scope of customer retention problem in the mobile phone sector: a theoretical perspective. J. Mark. Consum. Res. **20**, 64–69 (2016)
14. Al-Emran, M., Mezhuyev, V., Kamaludin, A., Shaalan, K.: The impact of knowledge management processes on information systems: a systematic review. Int. J. Inf. Manage. **43**, 173–187 (2018)
15. Al-Emran, M., Mezhuyev, V., Kamaludin, A.: Technology acceptance model in m-learning context: a systematic review (2018)
16. Nedal Fawzi Assad, M.T.A.: Investment in context of financial reporting quality: a systematic review. WAFFEN-UND Kostumkd. J. **11**(3), 255–286 (2020)
17. Alhashmi, S.F.S., Alshurideh, M., Al Kurdi, B., Salloum, S.A.: A systematic review of the factors affecting the artificial intelligence implementation in the health care sector. In: Joint European–US Workshop on Applications of Invariance in Computer Vision, pp. 37–49 (2020)
18. Salloum, S.A., Alshurideh, M., Elnagar, A., Shaalan, K.: Mining in educational data: review and future directions. In: Joint European–US Workshop on Applications of Invariance in Computer Vision, pp. 92–102 (2020)
19. Salloum, S.A., Alshurideh, M., Elnagar, A., Shaalan, K.: Machine learning and deep learning techniques for cybersecurity: a review. In: Joint European–US Workshop on Applications of Invariance in Computer Vision, pp. 50–57 (2020)
20. Zhang, M., Huang, L., He, Z., Wang, A.G.: E-service quality perceptions: an empirical analysis of the Chinese e-retailing industry. Total Qual. Manag. Bus. Excell. **26**(11–12), 1357–1372 (2015)
21. Parasuraman, A., Zeithaml, V.A., Malhotra, A.: E-S-QUAL a multiple-item scale for assessing electronic service quality. J. Serv. Res. **7**(3), 213–233 (2005)
22. Kaya, B., Behravesh, E., Abubakar, A.M., Kaya, O.S., Orús, C.: The moderating role of website familiarity in the relationships between e-service quality, e-satisfaction and e-loyalty. J. Internet Commer. **18**(4), 369–394 (2019)
23. Alshurideh, M.: A qualitative analysis of customer repeat purchase behaviour in the UK mobile phone market. J. Manag. Res. **6**(1), 109 (2014)
24. Alshurideh, M.: The factors predicting students' satisfaction with universities' healthcare clinics' services: a case-study from the Jordanian higher education sector. Dirasat Adm. Sci. **161**(1524), 1–36 (2014)
25. Alshurideh, M.: A behavior perspective of mobile customer retention: an exploratory study in the UK market. The end of the pier? Competing perspectives on the challenges facing business and management British Academy of Management Brighton–UK. Br. Acad. Manag. 1–19 (2010)
26. Al-Dmour, H., Al-Shraideh, M.T.: The influence of the promotional mix elements on Jordanian consumer's decisions in cell phone service usage: an analytical study. Jordan J. Bus. Adm. **4**(4), 375–392 (2008)
27. Bauer, H.H., Falk, T., Hammerschmidt, M.: eTransQual: a transaction process-based approach for capturing service quality in online shopping. J. Bus. Res. **59**(7), 866–875 (2006)
28. Wolfinbarger, M., Gilly, M.C.: eTailQ: dimensionalizing, measuring and predicting etail quality. J. Retail. **79**(3), 183–198 (2003)
29. Khan, K.S., Kunz, R., Kleijnen, J., Antes, G.: Five steps to conducting a systematic review. J. R. Soc. Med. **96**(3), 118–121 (2003)
30. Salloum, S.A., Al-Emran, M., Shaalan, K.: The impact of knowledge sharing on information systems: a review. In: 13th International Conference, KMO 2018 (2018)
31. Salloum, S.A., Alhamad, A.Q.M., Al-Emran, M., Monem, A.A., Shaalan, K.: Exploring students' acceptance of e-learning through the development of a comprehensive technology acceptance model. IEEE Access **7**, 128445–128462 (2019)

32. Salloum, S.A.S., Shaalan, K.: Investigating students' acceptance of e-learning system in higher educational environments in the UAE: applying the extended technology acceptance model (TAM). The British University in Dubai (2018)
33. Alhashmi, S.F.S., Salloum, S.A., Abdallah, S.: Critical success factors for implementing artificial intelligence (AI) projects in Dubai Government United Arab Emirates (UAE) health sector: applying the extended technology acceptance model (TAM). In: International Conference on Advanced Intelligent Systems and Informatics, pp. 393–405 (2019)
34. Kitchenham, S., Charters, B: Guidelines for performing systematic literature reviews in software engineering. Softw. Eng. Group, Sch. Comput. Sci. Math. Keele Univ. 1–57 (2007)
35. Stamenkov, G., Dika, Z.: Bank employees' internal and external perspectives on e-service quality, satisfaction and loyalty. Electron. Mark. 26(3), 291–309 (2016)
36. Ayo, C.K., Oni, A.A., Adewoye, O.J., Eweoya, I.O.: E-banking users' behaviour: e-service quality, attitude, and customer satisfaction. Int. J. Bank Mark. 34(3), 347–367 (2016)
37. Kundu, S., Datta, S.K.: Impact of trust on the relationship of e-service quality and customer satisfaction. EuroMed J. Bus. 10(1), 21–46 (2015)
38. Tiwari, P., Tiwari, S.K., Singh, T.P.: Measuring the effect of e-service quality In online banking. Prestige Int J Manage IT-Sanchayan, 6(1), 43–52 (2017)
39. Jauw, A.L.J., Purwanto, E.: Moderation effects of cultural dimensions on the relationship between e-service quality and satisfaction with online purchase. Qual. Access to Success 18(157), 55–60 (2017)
40. Rita, P., Oliveira, T., Farisa, A.: The impact of e-service quality and customer satisfaction on customer behavior in online shopping. Heliyon 5(10), e02690 (2019)
41. Kao, T.W., Lin, W.T.: The relationship between perceived e-service quality and brand equity: a simultaneous equations system approach. Comput. Human Behav. 57, 208–218 (2016)
42. İdil, K.A.Y.A., Akbulut, D.H.: Big data analytics in financial reporting and accounting. PressAcademia Procedia, 7(1), 256–259 (2018)
43. Khan, M.A., Zubair, S.S., Malik, M.: An assessment of e-service quality, e-satisfaction and e-loyalty: case of online shopping in Pakistan. South Asian J. Bus. Stud. 8(3), 283–302 (2019)
44. Jeon, M.M., Jeong, M.: Customers' perceived website service quality and its effects on e-loyalty. Int. J. Contemp. Hosp. Manag. 29(1), 438–457 (2017)
45. Goutam, D., Gopalakrishna, B.V.: Customer loyalty development in online shopping: an integration of e-service quality model and commitment-trust theory. Manag. Sci. Lett. 8(11), 1149–1158 (2018)
46. Manaf, P.A., Rachmawati, I., Witanto, M., Nugroho, A.: E-satisfaction as a reflection of e-marketing and e-sequal in influencing e-loyalty on e-commerce. Int. J. Eng. Technol. 7(4.44), 94 (2018)
47. Li, H., Aham-Anyanwu, N., Tevrizci, C., Luo, X.: The interplay between value and service quality experience: e-loyalty development process through the eTailQ scale and value perception. Electron. Commer. Res. 15(4), 585–615 (2015)
48. Zeglat, D., Shrafat, F., Al-Smadi, Z.: The impact of the E-service quality of online databases on users' behavioral intentions: a perspective of postgraduate students. Int. Rev. Manag. Mark. 6(1), 1–10 (2016)
49. Mohammed, E.M., Wafik, G.M., Abdel Jalil, S.G., El Hasan, Y.A.: The effects of e-service quality dimensions on tourist' s e-satisfaction. Int. J. Hosp. Tour. Syst. 9, 12–21 (2017)
50. Chek, Y.L., Ho, J.S.Y.: Consumer Electronics e-retailing: why the alliance of vendors' e-service quality, trust and trustworthiness matters. Procedia Soc. Behav. Sci. 219, 804–811 (2016)
51. Al Dmour, H., Alshurideh, M., Shishan, F.: The influence of mobile application quality and attributes on the continuance intention of mobile shopping. Life Sci. J. 11(10), 172–181 (2014)
52. Al-Dmour, H., Alshuraideh, M., Salehih, S.: A study of Jordanians' television viewers habits. Life Sci. J. 11(6), 161–171 (2014)
53. Kurdi: Healthy-food choice and purchasing behaviour analysis: an exploratory study of families in the UK. Durham University (2016)
54. Kalia, P., Arora, R., Kumalo, S.: E-service quality, consumer satisfaction and future purchase intentions in e-retail. e-Service J. 10(1), 24 (2016)

IOT within Artificial Intelligence and Data Science

IoT Sensor Data Analysis and Fusion

Mohamed Sohail

1 Introduction

The Internet of Things (IoT) and Artificial intelligence (AI) are considered the new decade transformative technologies that hold great promises for a tremendous societal, technological, and economic benefits [1]. AI has a great potential to revolutionize how we live, work, learn, discover, and communicate. AI modern research can further our national priorities and day to day habits, including increased economic prosperity, improved educational opportunities, enhancing our quality of life, and boosting our national and homeland security [2]. Because of these potential benefits, many governments and corporations have invested heavily in AI research and initiatives for many years [3]. Yet, as with any significant technology in which all the industries have interest, there are not only tremendous opportunities but also several challenges that should be considered in guiding the overall directions R&D in the next decade [4].

One of the industries that is resistant to change is the Data Center industry, but it cannot afford to resist too much. In large data centers, it is complicated to monitor everything, and it takes too much effort to maintain the environment in a steady state to support the production, self-healing, and self-optimizing data centers became a necessity in this era of the data explosion [5]. A self-healing data center is a data center that is capable of autonomously detecting potential component/module/hardware/software failures before they occur and adjust its configuration in order to achieve optimal function at all times [6]. The concept of detecting malfunctions and data center failures before they happen is called condition-based maintenance (CBM) or predictive maintenance (PdM) [7]. The approach requires a combination of historical, simulation, real-time information, expert heuristics,

M. Sohail (✉)
Senior Engineer, Solutions Architecture, Dell Technologies, Cairo, Egypt
e-mail: Mohamed.Sohail@dell.com

© The Editor(s) (if applicable) and The Author(s), under exclusive license to Springer Nature Switzerland AG 2021
A.-E. Hassanien et al. (eds.), *Enabling AI Applications in Data Science*,
Studies in Computational Intelligence 911,
https://doi.org/10.1007/978-3-030-52067-0_17

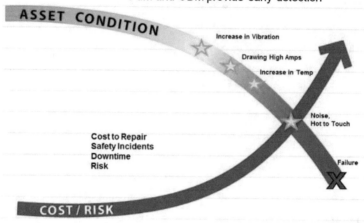

Fig. 1 PDM & CMB

and model design data to enable such capabilities, as we can show in Fig. 1, the diagram below show the correlation between the PdM and CBM that can provide early detection of an event or failure.

The idea of using sensors and actuators to analyze and capture data, then act on all critical events and relevant data issues like (temperature, pressure, humidity, vibration, proximity, etc.) and align it with AI and machine learning algorithms accompanied with smart big data analytics is the basis for any approach that is unique in addressing the concept of self-healing mechanisms specially in the data center industry [8, 9].

One of this decade's facts is that IoT grows at a very fast pace across all industries [10]. The number of connected "things" (from the simple to the most complex) will continue to expand over the near future.

All organizations now recognize the power of their data's value and its impact on their business either to extract value from it or even use it to optimize their operations [11]. The ability to maximize their competitiveness will require deep analytics of all data from all available sources to them. IoT takes that ability to the next level with the very huge number of connected objects as predicted and estimated by many experts and industry main players [12].

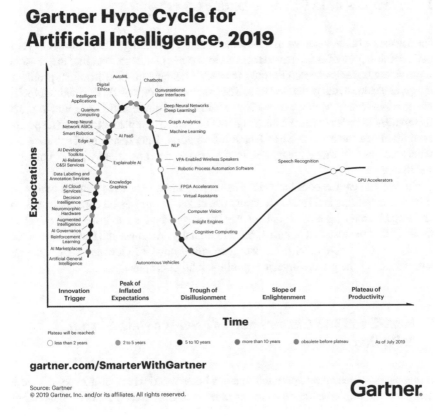

Fig. 2 Gartner hyper cycle for AI

According to Gartner [13], between 2018 and 2019, organizations that have deployed artificial intelligence (AI) grew from 4 to 14%, as shown in Fig. 2 of Gartner hyperscale for AI. According to Gartner's 2019 CIO Agenda survey. So, we are in front of an inevitable opportunity in almost all the industries.

In this chapter, we will show a breakdown of how we can combine AI techniques with IoT sensor data analysis to create a robust model of condition based & predictive maintenance in a self-healing IoT-enabled Data Center. The self-healing model will leverage data analytics techniques, trend analysis and anomaly detection algorithms that aim at achieving a simple and easy approach to provide high accuracy by leveraging the sensors in the form of IoT devices themselves via crowdsourcing fashion to provide high quality results and insights.

2 Principals of IoT Sensor Data Fusion

The sensor Data Fusion is defined as a mechanism of dealing with a subset of synergistic combination of data derived from various types of data sources like IoT sensors, to provide a better understanding of the needed insights to solve stringent problems and gives a value to the generated data. The usage of sensor/data fusion methodologies has provided many advantages such as enabling the combination of multi-sourced information in order to form a unified data lake, and a complete picture of a specific issue thanks to the amount of collected information. Data fusion concepts now are widely adopted in many areas, due to the arising need to know what is behind a specific issue.

No one can ever ignore hat Data Fusion became a very wide subject with a lot of implementations and is used in many use cases. In our treated case study, we are stressing on applying it in the data center industry, where it can bring a lot of values provided that the resistant nature of this old industry. We chose the Data Center field, as it has a lot of potentials from data generation point of view and represents the corner stone of the private–hydride–public–cloud concepts.

3 Next-Gen Data Centers and IoT Sensor Data-Fusion Adoption

In this chapter present an approach that has been tested with a prototype. A goal is to expand the implementation to cover other heterogeneous components including a wide array of different technologies, including servers, coolers, and third-party storage arrays, as shown in Fig. 3 on the high-level structure for aggregating sensors' data; starting from the sensor nodes, transport nodes, edge, and data storage. This can include also the fusion of generated logs from tested components plus externally implemented sensors in order to have multiple sources of data which can be analyzed to produce insightful results.

The need to achieve a self-healing and self-optimized data center is long-standing [14]. This new era of data explosion and specifically the rising adoption of AI and IoT data analysis, has made that far dream a reality [15, 16]. Not like in the past, novel techniques are very welcomed nowadays for a variety of reasons including cost savings, optimal functioning, and dynamic configurations of data center different resources [17]. It is also crucial for the data center decision makers to have meaningful, precise and real-time information about the current conditions of their data centers state, in order to make the meaningful decisions about a technology renovation, or an administrative strategy for both the time and efforts needed for the business continuity without costly disruptions and recovery time objectives [18].

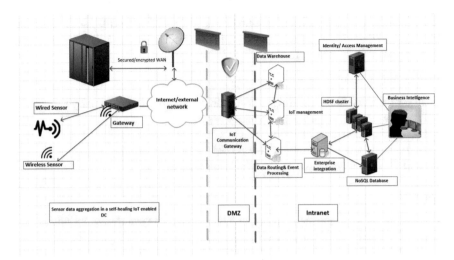

Fig. 3 Sensor data aggregation

To achieve this goal, collect and process the data, including historical, streaming/real-time and simulation data, we used an analytics engine by integrating various big data tools. The analytics engine output is a prediction of the future state of targeted modules, putting into consideration future releases of firmware, hardware lifecycle, and simulation of failures analysis.

The prototype implementation included the following sensors: temperature, current, humidity, and vibration to measure the current working conditions in data center. The framework is elastic enough to include other types of sensors that can be added for enrich the results.

4 Challenges

IoT data can be generated quite rapidly, the volume of data can be huge, and the types of data can be various. We are facing series of challenges. Especially for the type of generated data with the need to an efficient way of sensors' data fusion. The ubiquitous sensors, RFID readers, and other devices involved in the IoT systems can generate data rapidly, so that the data must be stored with a high throughput. Furthermore, because the volume of the data is very large, and can increase rapidly, a data storage solution for the IoT data must not only be able to store massive data efficiently, but also support horizontal scaling, and adoption of various data format due to the implementation of the data fusion. Moreover, the IoT data can be collected from many different sources and consisted of various structured and unstructured data. Data storage components are expected to have the ability to deal with heterogeneous data resources.

Today's emerging problems.

- 59% of fortune 500 companies loose nearly 46$ million dollars per year.
- Reactive rather than pro-active troubleshooting.
- Lack of heterogenous data fusion capabilities.
- Alerts are too late for faster actions.
- Root Cause analysis is done after the issue is raised.

Why now?

- Pressure for more business agility, and abstraction of Data Center hardware.
- Rapid adoption of Software-Defined Data centers.
- Big Data analytics as highest priority for all level customers.
- Infrastructure complexity requires AI based automation to deliver self-healing capabilities.

4.1 Data Centers Industry Constraints

- The data center industry is resistant to change due to complexity and sophisticated nature of the data centers.
- Data centers are often composed of systems silos, similar to the Tower of Babel, wherein every system is communicating in a different language. Which makes it difficult to get a unified language to communicate in.
- Data center operators will not have enough trust in these systems to rely on them. A primary function of the operators is to ensure the availability of data center services. Operators are rightfully hesitant to turn that responsibility over to machines until they are confident that those machines can be trusted.
- The proposed solution will rely on CBM and PdM.
- The solution will rely on some of the innovative ideas and a spatial prediction model for system administrators to get produce proactive notifications about Data Center assets such as drives, arrays, performance, and hosts failures through Software defined Storage (SDS), by integrating multiple data sources.
- The solution is also based on a micro services architecture. It includes independently deployable services that can deliver business value.

In some cases, some storage systems continue to use legacy code and algorithms to deliver features like volume management, protection from media failure and snapshot management. The use of the legacy code and algorithms means that the implementation of each feature adds latency.

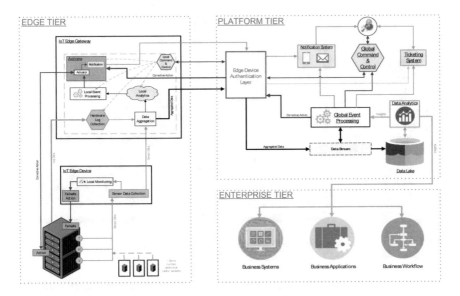

Fig. 4 Different tiers in the solution

In this scenario we use our backend data center management software to determine the areas that consume more power and apply our best practices to reduce this power or seek the peak power saving from the device depending on these operational statistics, this would be illustrated in the Enterprise tier as shown in Fig. 4.

Analysis of the historical data of disk failures will increase productivity, accuracy of results, and can identify the faults in categories. It will help the expected data center facilities impact and identify underlying issues such as:

- RACK location problem (Overheat, High Humidity, Power Interruption, Dusts) which can indicate where most failures/downtimes come from.
- Manufacturer defects affecting disk lifetime, including random failures from different locations.

We employed many algorithms to help us in this journey. The algorithms analyzed the data extracted from the sensors of the data center equipment—servers, arrays, switches, etc.—with the data of some extra sensors within the center like racks, cooling systems, and any possible factors that can affect the data center operations.

This will give us a broad understanding of the environment and will enrich the analytics engine with more data that will help in getting more accurate results.

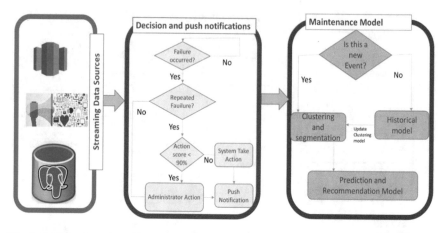

Fig. 5 Decision & push notification—maintenance model

The extracted data will be processed into a private cloud for more proactive analysis for possible failures. This will include an automatic intervention with many systems, for example, the cooling systems and the power plants. In this test phase we used temperature, humidity, and current sensors.

The self-healing algorithm will be responsible for adjusting the cooling system based on the data read from the equipment and the implemented sensors.

The framework was developed through three phases. Figure 5 illustrates the correlation between the decision and push notification phases and the maintenance phase. Here we will depend on three phases including

- **"Simulation analysis"**: In this phase, the framework simulates failures based on data generated from the data center to learn how to detect failures and take appropriate action based on the failure type. Also, in this phase, the framework can use simple data mining and machine learning algorithms to do classification and clustering processes on the data received from the data center components and present all the results in a smart visualization form. Indeed, the importance of this step is to evaluate performance and accuracy for all the algorithms used in this phase before applying it on the real environment.
- **"Historical analysis"**: This phase applied all algorithms used in the previous phase—simulation analysis—plus the mining of the equipment logs and historical data. The historical data means all the data saved on the log files for the tested racks. In this phase, the framework can be used to analyze historical data, detect any type of previous failures, apply machine-learning algorithms, and visualize results, in order to build up insights about the reasons of failures that happened in the past. However, we cannot use this framework in the stream processing or real-time processing to handle any failure that might occur in real time. We tackled this challenge in the next phase.

- **"Streaming analysis"**: This framework supports stream or real time processing, which can react dynamically for any failure occurred and can take the best decision based on the results generated from the machine-learning modules.

As shown in Fig. 5 the machine-learning module is designed to learn how to detect failures and apply the best solutions to solve such events. First, the data wrangling is the most important process in this framework due to the heterogeneous nature of the data center components that generate and store data in different formats.

This presented several challenges for each stage:

- Collecting data from different data format—DB, CSV files, etc.
- Application of the processing algorithm to the data as seen in Fig. 5 explains the steps and the similarity percentage of the events to read data.
- Data collection from different database management systems and different types of files.
- Aggregation and storage of disparate types of data in a standard format before applying any algorithm.
- Figure 6 illustrates how to use Sqoop in our framework to store different data types that come from different data sources in a standard format.
- Applying the algorithms to the standardized data.

However, there are types of failure that should be processed in "hard real time" such as increasingly high temperature in minor time frames. In this case, we

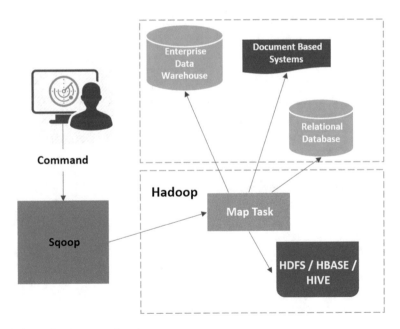

Fig. 6 Sqoop import-export function

might face hardware failure due to the high temperature before getting a fire event notification.

To address this problem, we enhanced our framework by using multi-agent technology. The multi-agent technology allowed for the agent to move through different machines in the cluster with the required data, code, and status. This helps HDFS to support hard real time processing because the agent can run the code on the nearest machine without any permission from the name node on HDFS. Furthermore, there is another advantage from using an agent system a type of agent, called an adaptive agent, which works as a learning agent and helps the system take the best solution from history stored in agent. The Also, there is another type of agent called an interactive agent, which can detect any failure in the real time and move to adaptive agent through to a fire events detected by an adaptive agent to make decisions in real time. All machine-learning algorithms used by the multi-agent system are based on Mahout.

Example of a part of the used code

```
#include <SPI.h>
#include <Ethernet.h>
#include "DHT.h"

#define DHTPIN 2
#define DHTTYPE DHT22

DHT dht(DHTPIN, DHTTYPE);

byte mac[] = {0xDE, 0xAD, 0xBE, 0xEF, 0xFE, 0xED};
IPAddress ip(192,168,1,70);
EthernetClient client;
char server[] = "xx-server.com";
unsigned long lastConnectionTime = 0;
boolean lastConnected = false;
const unsigned long postingInterval = 600*1000;

void setup() {

   Serial.begin(9600);
   Serial.println("DHTxx test!");
   dht.begin();
   //delay(1000);
   Ethernet.begin(mac, ip);
   Serial.print("My IP address: ");
   Serial.println(Ethernet.localIP());
}



void loop() {

  float h = dht.readHumidity();
  float t = dht.readTemperature();
  float f = dht.readTemperature(true);

  if (isnan(h) || isnan(t) || isnan(f)) {
    Serial.println("Failed to read from DHT sensor!");
    return;
  }
```

```
      float hi = dht.computeHeatIndex(f, h);
        if (client.available()) {
          char c = client.read();
          Serial.print(c);
        }

        if (!client.connected() && lastConnected) {
          Serial.println();
          Serial.println("disconnecting.");
          client.stop();
        }

        if(!client.connected() && (millis() -
      lastConnectionTime > postingInterval)) {
          httpRequest();
        }

        lastConnected = client.connected();
      }


    void httpRequest() {

      if (client.connect(server, 80)) {
        Serial.println("connecting...");
        client.println("GET /envparams?temp=t&humidity=h
    HTTP/1.1");
        client.println("Host: xx-server.com");
        client.println("User-Agent: arduino-ethernet");
        client.println("Connection: close");
        client.println();
        lastConnectionTime = millis();
      }
      else {
        Serial.println("connection failed");
        Serial.println("disconnecting.");
        client.stop();
      }
    }
```

5 Visualization of the Idea

In this section we are going to add more illustration how the idea looks like. We preferred to give a good imagination of the data different stages "from creation till execution". This includes detailed information about the workflow and how the sensor data aggregation and fusion should act.

In the Fig. 7 shows the data journey from the perspective of the different available tiers. Starting from the data generation from multiple sensor data, the edge/gateway, till the data center/cloud storage.

In Fig. 8, we illustrate in detail the various steps in the solution, the process starts with the data generation till reaching the data visualization phase.

In Fig. 8 step number 1, we are providing the workflow of the analytics engine to show up how it runs.

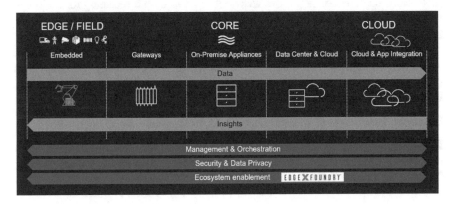

Fig. 7 Different tiers in the eco-system

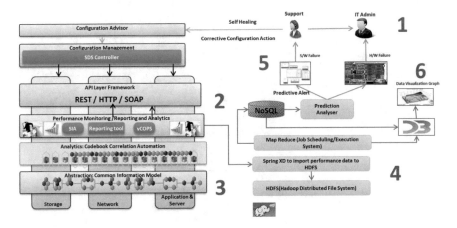

Fig. 8 Holistic view of the sensor data fusion in a self-healed data center

Events and performance metrics are collected from network/storage/virtualized infrastructure.

In Fig. 8 step number 2, Historical performance trending data and object model relationships will be fed into HDFS system. A Correlation engine which uses R Model will obtain the inputs from the HDFS to analyze the current retrieved metrics against historical metric records and events.

In Fig. 8 step number 3, modeling of IT infrastructure can be done using infrastructure management software. "The software can be used to explain the relationship and dependency between the network/storage/virtualized infrastructures. The modeling will consider the relationship with other objects, event conditions, associated problems, symptoms, and thresholds".

In Fig. 8 step number 4, the model will self-learn based on trending data and apply analytics to predict behavior of performance metrics in real-time.

In Fig. 8 step number 5, the Prediction Engine in the statistical analyzer will make predictions on behavior of performance metrics, based on forecasting formulas and probability statistics. A predicted event will be raised based on statistical analysis and will provide detail about impacted objects in the data center network.

In Fig. 8 step number 6, once the prediction alert is raised in the system, corrective action can be applied on the device through REST API commands executed with prior approval from the administrator in case it is new, and will be taken automatically if there is more than 90% similarity to the previous triggered errors.

Example: Prediction of Deterioration in LUN Response Time

In the Fig. 9, we show deterioration of LUN response time, it can be identified by comparison with the normal behavior of the historical response time, and any changes that may occur over time.

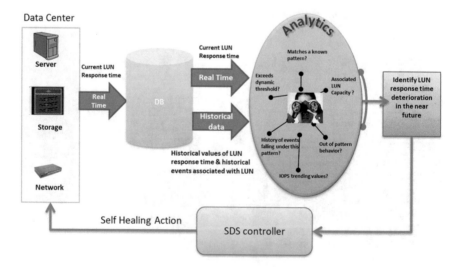

Fig. 9 LUN response time deterioration

6 Benefits

- Prediction of problem ahead of time and providing proactive solutions rather than reactive fixes.
- Quick corrective actions can be taken based on the prediction to avoid issues in the data center.
- Historical trending data is utilized to predict patterns.
- Helps troubleshooting engineer to visualize the statistical data in a datacenter for their troubleshooting—proactive than reactive trouble shooting by Customer Support before customer notice.
- Reduced downtime cost per year and datacenter maintenance cost.
- Increased reliability of the data center components due to early prediction of issues and correction.
- Improved visibility into Key Performance Indicators "KPIs" for customers.
- Networks, storage and virtual infrastructure inside a data center will be better prepared to handle a flood of events, latency issues, and breakdowns.
- The prediction will be used as a self-healing technique using data analytics.

7 Conclusion

The idea of achieving self-optimized and self-healed data center presents its own flavor of technical challenges. With the novel techniques of the IoT sensor data fusion and analysis, we gained some lands towards our goal of making our dream of a self-optimized data center comes true. Self-optimization as concept requires a high degree of intelligence built-into data center systems by leveraging the Fusion analysis and sophisticated AI techniques, because we are never optimizing against just one variable, we optimize for a whole set of a changing land scape of issues every day. For this reason, employing the AI should be seriously taken into consideration. Optimization requires balancing and prioritizing sometimes-competing variables, and that introduces a high degree of complexity we are not close to address today, but we are on the right direction due to the novel and advanced techniques we have now.

In this chapter, we wanted to fill the reader's imagination gap towards the idea of self-healing IoT enabled data center and emphasize the importance of shifting some operations from being manual tasks to AI automated based ones using the IoT sensor data analysis and fusion technology.

References

1. Allam, Z., Dhunny, Z.A.: On big data, artificial intelligence and smart cities. Cities **89**, 80–91 (2019). https://doi.org/10.1016/j.cities.2019.01.032

2. Kirchner, F.: A survey of challenges and potentials for AI technologies. In: Kirchner, F., Straube, S., Kühn, D., Hoyer, N. (eds.) AI Technology for Underwater Robots, pp. 3–17. Springer International Publishing, Cham (2020)

3. Benedikt, L., Joshi, C., Nolan, L., et al.: Human-in-the-loop AI in government. In: Proceedings of the 25th International Conference on Intelligent User Interfaces. ACM, New York, NY, USA, pp. 488–497 (2020)

4. Laï, M.-C., Brian, M., Mamzer, M.-F.: Perceptions of artificial intelligence in healthcare: findings from a qualitative survey study among actors in France. J. Transl. Med. **18**, 14 (2020). https://doi.org/10.1186/s12967-019-02204-y

5. Hosseini Shirvani, M., Rahmani, A.M., Sahafi, A.: A survey study on virtual machine migration and server consolidation techniques in DVFS-enabled cloud datacenter: Taxonomy and challenges. J. King Saud Univ. Comput. Inform. Sci. **32**, 267–286 (2020). https://doi.org/10.1016/j.jksuci.2018.07.001

6. Munn, L.: Injecting failure: data center infrastructures and the imaginaries of resilience. Inform. Soc. 1–10 (2020). https://doi.org/10.1080/01972243.2020.1737607

7. Fadaeefath Abadi, M., Haghighat, F., Nasiri, F.: Data center maintenance: applications and future research directions (2020). Facilities ahead-of-p: https://doi.org/10.1108/F-09-2019-0104

8. El-Din, D.M., Hassanien, A.E., Hassanien, E.E.: Information integrity for multi-sensors data fusion in smart mobility. In: Hassanien, A.E., Bhatnagar, R., Khalifa, N.E.M., Taha, M.H.N. (eds.) Toward Social Internet of Things (SIoT): Enabling Technologies, Architectures and Applications: Emerging Technologies for Connected and Smart Social Objects, pp. 99–121. Springer International Publishing, Cham (2020)

9. Jindal, R., Kumar, N., Nirwan, H.: MTFCT: a task offloading approach for fog computing and cloud computing. In: 2020 10th International Conference on Cloud Computing, Data Science & Engineering (Confluence), pp. 145–149. IEEE (2020)

10. Poongodi, T., Rathee, A., Indrakumari, R., Suresh, P.: IoT sensing capabilities: sensor deployment and node discovery, wearable sensors, wireless body area network (WBAN), data acquisition. In: Peng, S.-L., Pal, S., Huang, L. (eds.) Principles of Internet of Things (IoT) Ecosystem: Insight Paradigm, pp. 127–151. Springer International Publishing, Cham (2020)

11. Bonesso, S., Bruni, E., Gerli, F.: The organizational challenges of big data. In: Behavioral Competencies of Digital Professionals: Understanding the Role of Emotional Intelligence, pp. 1–19. Springer International Publishing, Cham (2020)

12. Hussein, D.M.E.-D.M., Hamed, M., Eldeen, N.: A blockchain technology evolution between business process management (BPM) and Internet-of-Things (IoT). Int. J. Adv. Comput. Sci. Appl. **9**, 442–450 (2018). https://doi.org/10.14569/IJACSA.2018.090856

13. Gartner. https://www.gartner.com/en. Accessed 18 Apr 2020

14. Gupta, P., Gupta, P.K.: Tools for fault and reliability in multilayered cloud. In: Trust & Fault in Multi Layered Cloud Computing Architecture, pp. 181–194. Springer International Publishing, Cham (2020)

15. Singla, S.: AI and IoT in healthcare. In: Raj, P., Chatterjee, J.M., Kumar, A., Balamurugan, B. (eds.) Internet of Things Use Cases for the Healthcare Industry, pp. 1–23. Springer International Publishing, Cham (2020)

16. Amanullah, M.A., Habeeb, R.A.A., Nasaruddin, F.H., et al.: Deep learning and big data technologies for IoT security. Comput. Commun. **151**, 495–517 (2020). https://doi.org/10.1016/j.comcom.2020.01.016

17. Singh, A., Singh, R., Bhattacharya, P., et al.: Modern optical data centers: design challenges and issues. In: Giri, V.K., Verma, N.K., Patel, R.K., Singh, V.P. (eds.) Computing Algorithms with Applications in Engineering, pp. 37–50. Springer Singapore, Singapore (2020)

18. Hauksson, E., Yoon, C., Yu, E., et al.: Caltech/USGS Southern California Seismic Network (SCSN) and Southern California Earthquake Data Center (SCEDC): data availability for the 2019 Ridgecrest sequence. Seismol. Res. Lett. (2020). https://doi.org/10.1785/0220190290

Internet of Things Sensor Data Analysis for Enhanced Living Environments: A Literature Review and a Case Study Results on Air Quality Sensing

Gonçalo Marques ⓘ

1 Introduction

The Internet of Things (IoT) is a ubiquitous behaviour of a variety of entities with intelligence and sensing abilities which can collaborate to achieve shared objectives [1, 2]. IoT technologies offer numerous advantages to several domains, such as smart cities, and smart homes [3, 4].

The smart city can be assumed as an approach to embrace modern urban productivity requirements in a general framework and to focus on the role of information and communication technologies (ICTs) on the creation of enhanced living environments [5, 6]. Cities face relevant open issues regarding the socio-economic development requirements to promote people health and well-being [7, 8]. Moreover, the smart city aims to moderate the obstacles caused by the urban population increase and accelerated urbanisation [9]. One of the most consistent open issues regarding smart cities is the no operability between different systems. IoT can present the interoperability to develop a centralised urban-scale ICT framework for enhanced living environments [8, 10]. Is relevant to mention that smart cities will cause several meaningful effects at distinct layers on science, productivity and technology [11]. The smart city will lead to complex challenges in society regarding the ethical and moral context. The smart city must provide correct information access to the right individuals [12]. The security and privacy of that information must be established because that data is accessible at a distinct spatial scale where people can be distinguished [13, 14].

Furthermore, IoT brings numerous opportunities regarding the development of modern daily routine applications and services for smart cities [15]. Several technologies closely related to the smart city context connected with IoT architecture will

G. Marques (✉)
Instituto de Telecomunicações, Universidade da Beira Interior, Covilhã 6201-001, Portugal
e-mail: goncalosantosmarques@gmail.com

A.-E. Hassanien et al. (eds.), *Enabling AI Applications in Data Science*,
Studies in Computational Intelligence 911,
https://doi.org/10.1007/978-3-030-52067-0_18

397

improve our daily routine and promote health and well-being [16]. IoT architectures can provide ubiquitous and pervasive methods for environmental data acquisition and use wireless communication technologies for enhanced connectivity [17–19].

Air quality has a material adverse impact on occupational health and well-being. Therefore, the air quality supervision is determinant for enhanced living environments and should be a requirement of all buildings and consequently, an integral part of the smart city context [20, 21]. Numerous research studies state the relevant adverse health effects associated with reduced air quality levels, such as premature death, respiratory, and cardiovascular disease [22].

On the one hand, air quality not only has a material responsibility in human exposure to pollutants but is also crucial for specific groups such as older adults and people with disabilities [23, 24]. On the other hand, numerous scientific evidence associated with the negative impacts on health and well-being, particularly on children and older adults related to reduced air quality levels is available on the literature [25]. Every year, air quality concentration levels are responsible for 3.2 million deaths and a relevant increase in heart and asthma attacks, dementia, as well as cancer [26, 27].

The consequences of poor air quality are most severe in developing countries where there is no regulation to control pollutants emissions. However, air quality levels are also a problem in developed countries. Every year in the USA, approximately 60,000 premature deaths are reported and linked to reduced air quality levels, and the healthcare costs related to air quality diseases in healthcare costs reach $150 billion [28]. According to the European Environment Agency, in 2016, air pollution was responsible for 400,000 premature deaths in the European Union (EU). The particulate matter caused 412,000 premature deaths in 41 European countries, and 374,000 occurred in the EU [29]. Moreover, the cost related to the air pollutant emissions effect caused by industrial facilities in 2012 has been estimated as at least 59–189 billion euros in the EU [30]. Even in locations with good air quality, levels are reported in situations of short-term exposure which conduct relevant health symptoms related to sensitive groups such as elderly and children with asthma and cardiovascular problems [31, 32].

On the one hand, outdoor air quality is a significant public health challenge taking into account all the before mentioned facts [33]. On the other hand, indoor air quality (IAQ) is also a critical problem for occupational health. The Environmental Protection Agency (EPA) recognises that indoor air quality values can be one hundred times higher than outdoor pollutant levels and established air quality on the top five environmental hazards to well-being [34].

Moreover, IAQ affects the most underprivileged people worldwide that remain exposed and therefore, can be compared with other public health problems such as sexually transmitted infections [25]. IAQ supervision must be seen as an effective method for the study and assessment of occupational health and well-being. The IAQ data can be used to detect patterns on the indoor quality and to design interventions plans to promote health [35]. The proliferation of IoT devices and systems makes it possible to create automatic methods with sensing, connection and processing capabilities for enhanced living environments [36].

Air quality sensing must be an essential element of smart cities and smart homes. On the one hand, the IAQ can be estimated by providing CO_2 supervision system for enhanced public health and well-being [37]. The CO_2 indoor levels can not only be used to evaluate the necessity of fresh air in a room but also can be used as a secondary sign of unusual concentrations of other indoor air pollutants [38]. On the other hand, cities are responsible for a relevant portion of greenhouse emissions. Therefore, CO_2 monitoring must be provided on both indoors and outdoors environments. The concentrations of CO_2 are growing to 400 ppm, achieving further records every time after they started to be analyzed in 1984 [39]. The outdoor CO_2 monitoring can be relevant to plan interventions on traffic and to detect the emission of abnormal amounts of CO_2 in real-time [40]. Moreover, real-time monitoring of CO_2 levels in smart cities at different locations allows the identification of points of interest to plan interventions to decrease the greenhouse gases.

This paper presents not only present a literature review on IoT sensor data analytics for enhanced living environments but also describes the design development a CO_2 monitoring system based in IoT architecture for enhanced living environments and public health. The main contribution of this paper is to present a comparison summary regarding state of the art in IoT systems for air quality monitoring and present the results of the design and implementation of a cost-effective system for air quality analytics using IoT sensor data.

The rest of the document is written as follows: Sect. 2 presents a literature review on air quality monitoring systems; Sect. 3 presents a case study results of the design and implementation of a CO_2 monitoring system based on IoT, and the conclusion is presented in Sect. 4.

2 Literature Review on IoT Air Quality Sensor Data Analytics

Several IAQ management systems are presented in state of the art; a review summary is presented in this section. Numerous low-cost IoT methods for air quality management that supports open-source and wireless communication for data collection and transmission but also allow different places supervision at the same time through mobile computing technologies are available in the literature.

From the past few years, numerous researchers have contributed to this field; however, it is not possible to include all those studies in this paper. This section provides highlights of 12 studies conducted on air quality monitoring in recent years. The research studies are selected by following these six criteria: (1) present a real case study of design or implementation of IoT architecture for air quality monitoring; (2) use low-cost sensors; (3) incorporate various open-source technologies; (4) based on IoT architecture; (5) indexed in Science Citation Index Expanded and (6) published on or after 2019.

A CO_2 monitoring system based on mobile computing technologies is presented in [41]. This system incorporates a cost-effective CO_2 sensor and uses an ESP8266 microcontroller which supports built-in Wi-Fi connection. The proposed method incorporates a web framework and a mobile application for data consulting. Moreover, it includes a web API for remote notifications to promote intervention planning in real-time.

The authors of [42] propose a multigas monitoring system based on IoT. The proposed method incorporates a MICS-6814 sensor that is capable of detecting several gases such as ammonia (NH_3), carbon monoxide (CO), nitrogen dioxide (NO_2), ethanol (C_2H_5OH), methane (CH_4), and propane (C_3H_8). This system uses a cloud platform for data storage and processing. Furthermore, this system incorporates data consulting methods and real-time notifications using mobile software for iOS. The processing unit is based on the ESP8266.

An air quality monitoring system is presented in [43]. This system uses an Arduino UNO as a microcontroller as a processing unit and integrates several sensors for data collection. The proposed method incorporates sensors for CO and CH_4. The data is stored in an IoT Cloud platform and an analytical module for data prediction. In addition, this service also incorporates an Android mobile application for data consulting.

An assistive robot for IAQ monitoring based on IoT which can with occupants and triggers alerts using social networks to allow the consulting by the caregiver to plan interventions for enhanced living environments on time is proposed by [44]. The system uses wireless sensor networks to collect data from multiple sensor nodes which transmit the data to a gateway connected to the Internet. The sensor used is an LPG (Liquefied petroleum gas). The system is developed using a Notebook running Windows as a processing unit.

An IAQ monitoring system for real-time monitoring which incorporates CO_2, CO, particulate matter (PM), NO_2, temperature and humidity sensor features is presented in [45]. This system is based on an ESP32 microcontroller which provides built-in Wi-Fi and Bluetooth support. The authors have developed an iOS mobile application for data consulting and notifications.

The design and implementation of an IAQ supervisor system with several communication interfaces for data transmission is proposed by [46]. The proposed system supports Modbus, LoRa, W-Fi, general packet radio service, and NB-IoT technologies for long-range and short-range communication. The microcontroller used is an STM32. This system incorporates CO_2, formaldehyde, temperature, humidity and PM sensors. Moreover, this system provides mobile and web compatibility for data consulting.

A real-time air quality monitoring system for public transport vehicles is proposed by [47]. This study has been conducted in Sweden and presents an IoT implementation using low-cost sensors. The system uses a waspmote microcontroller and includes NO_2, CO, temperature and humidity sensors. The cyber-physical system sends data to a backend application using 4G communication. A GPS module is used for localization of the area where the data is collected using geographic coordinates. The proposed method provides web compatibility for data consulting.

The authors of [48], propose an air quality monitoring system based on IoT, which incorporates a low-cost sensor for sulfur dioxide (SO_2), NO_2, CO, ozone (O_3) and PM in real-time. The processing unit is based on a Raspberry Pi microcontroller, and an ESP8266 is used for data transmission using Wi-Fi communication technologies. The proposed system provides web compatibility for data consulting. The Kalman Filter (KF) algorithm is used to improve sensors accuracy by 27%.

An air quality monitoring systems to supervise the indoor living environments of vulnerable groups such as older adults, children, pregnant women and patients is presented in [49]. The proposed model includes an artificial-intelligent-based approach for multiple hazard gas detector (MHGD) system that is installed on a motor vehicle-based robot. This motor vehicle can be controlled remotely and is developed with an Arduino microcontroller. The system includes iOS, Android and desktop software for data consulting. The users used are capable of detecting C_2H_5OH, trimethylamine (CH_3) and hydrogen (H_2). The proposed method is optimized to identify hazardous gases such as ethanol, tobacco smoke and off-flavour from spoiled food.

The authors of [50] propose an IoT architecture for vehicle pollution monitoring. This system uses a Raspberry Pi as a processing unit and an MQ-2 gas sensor for LPG and CO sensing features. Wi-Fi communication technologies are used for data transmission. The data collected is stored in a ThingSpeak cloud service. The proposed method provides a web software for data consulting and incorporated e-mail notifications.

An IoT monitoring system which incorporates PM, CO_2, formaldehyde, Volatile Organic Compounds (VOCs), Benzene (C_6H_6), NO_2, O_3 air quality sensors and humidity, temperature, light and noise sensing capabilities is presented in [51]. A Raspberry Pi microcontroller is used as a processing unit and uses Ethernet technology for the Internet connection.

A study presented by the authors of [52] propose the use of a LoRa communication technology for PM supervision in Southampton, a city of the United Kingdom. The proposed IoT cyber-physical systems are based on the Raspberry Pi microcontroller and incorporate four different PM sensors.

Air quality sensing is an essential requirement for smart cities and further for global public health. Consequently, the design of cost-effective supervision solutions is a trending and relevant research field. Table 1 present a comparison summary of the studies analysed in the literature review.

The distribution of the number of studies according to the processor unit used is presented in Fig. 1. The most used microcontroller in the analyzed studies is the Raspberry Pi (N = 4) which corresponds to 30%, followed by ESP8266 (N = 2) and Arduino (N = 2) responsible for 17% each.

Figure 2 presents the number of studies distributed according to the gas sensor used. Considering the studies analyzed in this literature review, the CO is the most used sensor (N = 6) followed by NO_2 (N = 5) and PM (N = 5). The least used sensors are NH_3, SO_2, CH_3 and VOC, which are used in only one study.

Table 1 Summarised comparison review of air quality sensing solutions

Processing unit	Sensors	Connectivity	Data consulting	Notifications	Reference
ESP8266	CO_2	Wi-Fi	Mobile and Web	Yes	[41]
ESP8266	NH_3, CO, NO_2, C_3H_8, C_4H_{10}, CH_4, H_2, and C_2H_5OH	Wi-Fi	Mobile	Yes	[42]
Arduino	CO and CH_4	Wi-Fi	Mobile	Yes	[43]
Notebook	C_3H_8 and C_4H_{10}	Wi-Fi	Web	Yes	[44]
ESP32	CO_2, CO, PM, NO_2 temperature and humidity	Wi-Fi	Mobile	Yes	[45]
STM32	CO_2, formaldehyde, temperature, humidity and PM	GPRS, Wi-Fi, LoRa and NB-IoT	Mobile and Web	No	[46]
Waspmote	CO, NO_2, temperature and humidity	4G	Web	No	[47]
Raspberry Pi	SO_2, NO_2, CO, O_3, and PM	Wi-Fi	Web	No	[48]
Arduino	C_2H_5OH, CH_3 and H_2.	Ethernet	Web and mobile	No	[49]
Raspberry Pi	C_3H_8, C_4H_{10}, and CO	Wi-Fi	Web	Yes	[50]
Raspberry Pi	PM, CO_2, formaldehyde, VOCs, C_6H_6, NO_2, O_3, humidity, temperature, light and sound	Ethernet	–	–	[51]
Raspberry Pi	PM	LoRa and Wi-Fi	–	–	[52]

According to the communication technology used in the reviewed papers, the most used connectivity methods is Wi-Fi (N = 9) and the least used communication methods are GPRS, 4G and NB-IoT adopted by only one study (Fig. 3).

Furthermore, most of the studies incorporate data consulting methods based on mobile or web technologies. In total, 83% of the studies incorporate consulting methods (N = 10). Four studies incorporate only web software for data consulting (30%), and three studies incorporate only mobile software (25%). Moreover, three studies incorporate both methods for data consulting (30%). Only two studies do not provide any methods for data consulting.

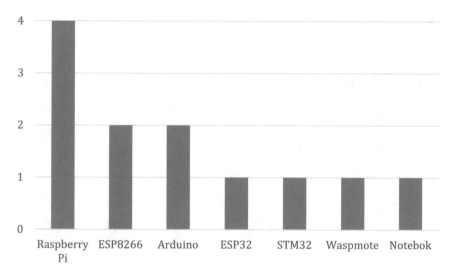

Fig. 1 Number of studies distributed according to the processing unit

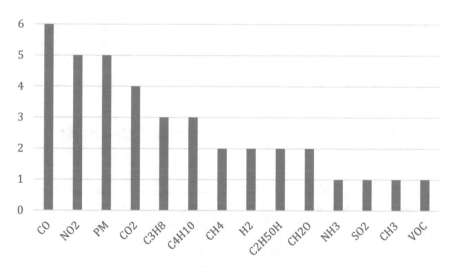

Fig. 2 Number of studies distributed according to the gas sensor used

Notifications features are incorporated in six studies (50%). Remote notifications are relevant since it allows the users to act accordingly in a useful time to promote air quality.

Some studies implement mathematical and artificial methods to promote sensor accuracy and to identify hadrosaurs scenarios. The data collected by the IoT solutions can be enhanced by the adoption of artificial intelligence methods to enhanced living environments [48, 49].

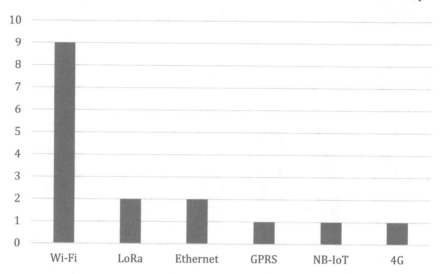

Fig. 3 Number of studies distributed according to the communication technology used

In sum, the literature review presents several automatic systems and projects for air quality sensing, which support the relevance of the IoT architecture and mobile computing technologies as having a significant role in promoting health and well-being.

3 Internet of Things Sensor Data Analysis: A Case Study Results

An air quality monitoring system has been developed to provide real-time CO_2 supervision. The proposed system has been designed using an ESP8266 microcontroller which provides built-in Wi-Fi connectivity capabilities and includes an SGP30 Multi-Pixel Gas sensor as a sensing unit for CO_2 monitoring. Moreover, the proposed architecture incorporates an AM2315 temperature and humidity sensor. The system architecture is based on IoT. The sensors are connected to the ESP8266 via I2C serial communication protocol. A 32-bit MCU and a 10-bit analogue-to-digital converter are incorporated in the ESP8266 microcontroller which also supports built-in Wi-Fi connection features. For continuous humidity compensation of the CO_2 sensor, an AM2315 temperature and humidity sensor have been used to promote data accuracy [53]. The SGP30 is a calibrated sensor developed by Sensirion. The technical data of this sensor is available in [54, 55]. The SGP30 Multi-Pixel Gas sensor is a metal-oxide gas sensor developed which provides a calibrated output with 15% of accuracy [54]. The sensor range for eCO_2 and TVOC concentration is 400–60,000 ppm and 0–60,000 ppb, respectively. The eCO_2 output is based on a hydrogen measurement. The sensor sampling rate is 1 second, and the average current consumption is 50 mA.

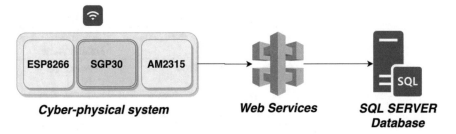

Fig. 4 *MoniCO₂* system architecture

The main objective of the proposed architecture is to promote health and well-being for enhanced living environments. The proposed method provides a continuous stream of IoT data than can be accessed using mobile computing technologies. Furthermore, this data for air quality management to improve public health and for enhanced safety (Fig. 4). The proposed cyber-physical system based on the ESP8266 module that supports the IEEE 802.11 b/g/n protocol provides data acquisition which is stored in a Microsoft SQL Server database using ASP.NET web services.

The acquisition module firmware incorporates the Arduino Core, which is an open-source framework that intends to extend the Arduino libraries support to the ESP8266 microcontrollers.

Furthermore, the proposed data acquisition module has been developed using on open-source frameworks, and is a cost-effective system, with numerous benefits when associated with actual methods. The ES8266 module used is based on FireBeetle ESP8266 (*DFRobot*) board which is used as a processing unit. The CO_2 sensor is a the SGP30 gas sensor module (*Adafruit*) and the temperature and humidity sensor is incorporated in an encased I2C module (*Adafruit*) which provides protection from wind and rain for outdoor use. Table 2 provides a technical description of the sensors used.

The system operation is presented in Fig. 5. The data collected is saved in real-time every 30 seconds using web services.

The proposed method offers a history of the variations of the monitored environment to provide IoT sensor data in a continuous mode. The collected data can support the home or city administrator to deliver an accurate examination of the environment. The data collected can also be applied to assist decision-making on potential arbitrations for enhanced public health and well-being.

The proposed provides an easy configuration of the wireless network. The system is configured as a Wi-Fi client. However, if it is incapable of connecting to the wireless network or if no Wi-Fi networks are ready, the proposed system turn to hotspot mode and start a wireless network. At this time, this hotspot can be used to configure the Wi-Fi network to which the system will be connected through the introduction of the network access credentials.

A mobile application has been developed to provide real-time data to the collected data. This application is compatible with iOS 12 or newer and is represented in Fig. 6.

Table 2 Sensors' technical specification

	SGP30	AM2315
Accuracy	400–60,000 ppm	Temperature: 0.1 °C Humidity: 2%
Range	15%	Temperature: −20 to 80 °C Humidity: 0−+ to 100%
Sensitivity	1 ppm	Temperature: 0.1 °C Humidity: 0.1%
Voltage	3.3–5 V	3.5–5 V
Power consumption	50 mA	10 mA
Operating temperature range	−40 to 85 °C	–
Operating humidity range	10–95% (non-condensing)	–

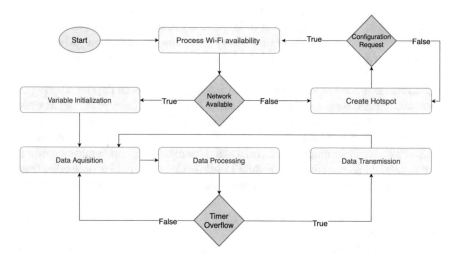

Fig. 5 Diagram of the system operation

The mobile application is used for two primary purposes. This software provides data consulting features in real-time but also allows the user to receive notifications.

As people typically keep mobile phones with them for everyday use, the proposed applications is a powerful tool to access air quality levels. Consequently, people can hold the parameters of chronicle records for additional examination [56]. The introduced method is a decision-making tool to project arbitrations sustained by the data obtained for enhanced public health.

The mobile software provides fast and intuitive access to the collected data as mobile phones have today high performance and storage capacities. In this way, the

Fig. 6 Last collected data represented in the mobile application

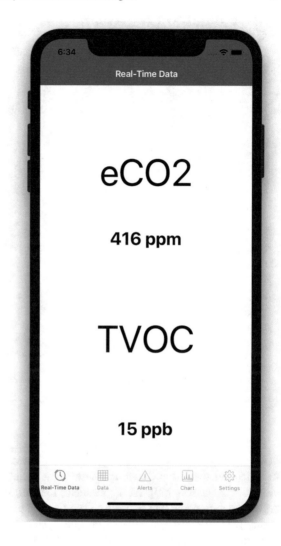

building or city manager can transport the air quality data of their environment for further analysis.

The mobile application provides data consulting as graphical or statistical modes. The graphical representation of the data collected is presented in Fig. 7.

On the one hand, outdoor air quality is conditioned by the levels of numerous substances such as PM, nitrogen dioxide, hydrocarbons, carbon monoxide and ozone which result from combustion sources [57]. On the other hand, air quality is also influenced by meteorology and ventilation conditions such as low wind and convention that promote pollutant concentration. This phenomenon is linked to the increase in carbon dioxide [58].

Fig. 7 Mobile application notification configuration feature and chart view

Furthermore, monitor these numerous substances requires a high number of different sensors and consequently, the overall cost [41, 59]. Several sensors also need maintenance procedures and permanent calibration, such as optical sensors and nitrogen sensors [60]. At the moment, the carbon dioxide sensors are reliable, maintenance-free and can be used to provide a practical assessment of outdoor air quality [61]. Therefore, the cyber-physical system has been installed in an outdoor environment for data collection. The system was tested continually for two months, and several experiments had been conducted with induced simulations. The data has been collected during January of 2020 in continuous mode. These data has been exported using a CSV format and was further analyzed using Python programming language. The module has been powered using a 20,000 mA power bank.

The average values collected according to the hour of the day are represented in Fig. 8 (ppm values). The collected values range from 406 to 546 ppm. The high value is recorded at 13 hours, and the lower value is registered at 8 hours.

The analysis of the CO_2 concentration according to the hours of the day can provide an in-depth analysis of the air pollution of the local since it is possible to correlate events with pollution sources. Moreover, it provides a clear picture of the air quality scenario faced according to the hours at typically people are outside.

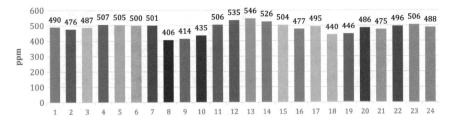

Fig. 8 Carbon dioxide mean levels collected during the research conducted grouped by hour

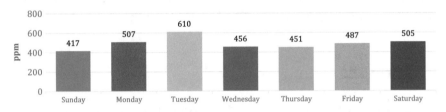

Fig. 9 Carbon dioxide mean levels collected during the research conducted grouped by weekday

According to the collected data, the higher levels of CO_2 are collected between 12 and 14 hours, which can be associated with the lunch break. At this time, people typically use vehicles to leave their jobs and went away for lunch or are going back to work.

Figure 9 (ppm values) represent mean levels collected during the research conducted grouped by weekday. The high concentration values are collected on Tuesday, and the lower values are registered on Sunday. The concentration of CO_2 can be associated with the activities carried at the location of the data collection. Moreover, this data can be used to study the behaviour of people, and activities carried. Sundays are typically a quiet day when people stay in their homes since they do not work. Moreover, industry and companies are typically closed.

The results state that the proposed air quality monitoring system can be used to detect unhealthy scenarios at low-cost. This IoT sensor data can be accessed using a mobile application which provides an improved inspection of observed parameters behaviours when compared with the statistical presentation. This IoT architecture offers chronological data evolution for improved air quality assessment, which is particularly appropriate to identify unhealthy scenarios and project interferences to improve health and well-being.

Numerous air quality monitoring systems are high-priced and are established only provide a random sampling. Therefore, IoT sensor-based architectures can provide cost-effective solutions to provide continuous data collection in real-time and significantly promotes enhanced living environments.

The proposed architecture is scalable since new modules can be added according to the requirements of the scenario. The results are promising as the proposed system can be used to provide a correct air quality assessment at low-cost.

IoT sensor data must be considered as a cost-effective approach to improve people routine and will lead to systematic improvements in the design of automated and intelligent systems to promote health and well-being [62]. Typically the IoT architectures are designed using open-source technologies and incorporate sensors at low-cost that can be used in numerous research domains associated with civil engineering, mechanics, environmental science, healthcare and smart cities [63].

The number of IoT solutions for multiple domains is increasing daily, and residencies incorporate numerous devices for different activities such as energy monitoring and control, environment supervision and security systems. Therefore, the security methods implemented in IoT systems must be a focus of research concerning the impact and acceptability of this kind of methods used by people in their daily routine.

4 Conclusion

IoT sensor data analysis in smart cities will powerfully improve people daily routine. Cyber-physical systems can be used to address multiple complex challenges in multiple fields as IoT can provide the operability between different systems to develop a centralised urban-scale framework for enhanced living environments.

Furthermore, IoT architectures can provide ubiquitous and pervasive methods for environmental data acquisition and provide connectivity for data transmission. These systems incorporate multiple sensors which provide a continuous stream of relevant data. The data collected from IoT cyber-physical systems by their own will not have a high impact on public health and well-being, but the analysis and the adoption of data science methods supported by artificial intelligence methods will significantly contribute for enhanced living environments. The data collection conducted by sensors connected to IoT systems provide an active and continuous stream of data to support multiple activities and significantly improve people daily routine. The adverse effects of poor air quality on occupational health and well-being IoT sensor data analytics must be seen as an integral part of society's everyday activities and must be incorporated in smart cities for enhanced living environments.

Nevertheless, IoT systems have some limitations regarding the quality of the collected data from low-cost sensors. However, several authors are adopting artificial intelligence methods to improve data accuracy that is relevant for specific domains.

This paper has presented a literature review on IoT architectures for air quality monitoring and presents a case study of an IoT sensor data analytics. The results are promising and state CO_2 sensor data from IoT architectures as an effective and efficient method to promote public health.

References

1. Giusto, D. (ed.): The Internet of Things: 20th Tyrrhenian Workshop on Digital Communications. Springer, New York (2010)
2. Marques, G., Pitarma, R., Garcia, N.M., Pombo, N.: Internet of things architectures, technologies, applications, challenges, and future directions for enhanced living environments and healthcare systems: a review. Electronics **8**, 1081 (2019). https://doi.org/10.3390/electronics8 101081
3. Atzori, L., Iera, A., Morabito, G.: The internet of things: a survey. Comput. Netw. **54**, 2787–2805 (2010). https://doi.org/10.1016/j.comnet.2010.05.010
4. Marques, G.: Ambient assisted living and internet of things. In: Cardoso, P.J.S., Monteiro, J., Semião, J., Rodrigues, J.M.F. (eds.) Harnessing the Internet of Everything (IoE) for Accelerated Innovation Opportunities, pp. 100–115. IGI Global, Hershey, PA (2019). https://doi.org/10. 4018/978-1-5225-7332-6.ch005
5. Caragliu, A., Del Bo, C., Nijkamp, P.: Smart cities in Europe. J. Urban Technol. **18**, 65–82 (2011). https://doi.org/10.1080/10630732.2011.601117
6. Abdelaziz, A., Salama, A.S., Riad, A.M., Mahmoud, A.N.: A machine learning model for predicting of chronic kidney disease based internet of things and cloud computing in smart cities. In: Hassanien, A.E., Elhoseny, M., Ahmed, S.H., Singh, A.K. (eds.) Security in Smart Cities: Models, Applications, and Challenges, pp. 93–114. Springer, Cham (2019). https://doi. org/10.1007/978-3-030-01560-2_5
7. Schaffers, H., Komninos, N., Pallot, M., Trousse, B., Nilsson, M., Oliveira, A.: Smart cities and the future internet: towards cooperation frameworks for open innovation. In: Domingue, J., Galis, A., Gavras, A., Zahariadis, T., Lambert, D., Cleary, F., Daras, P., Krco, S., Müller, H., Li, M.-S., Schaffers, H., Lotz, V., Alvarez, F., Stiller, B., Karnouskos, S., Avessta, S., Nilsson, M. (eds.) The Future Internet, pp. 431–446. Springer, Berlin (2011). https://doi.org/10.1007/ 978-3-642-20898-0_31
8. Ahlgren, B., Hidell, M., Ngai, E.C.-H.: Internet of things for smart cities: interoperability and open data. IEEE Internet Comput. **20**, 52–56 (2016). https://doi.org/10.1109/MIC.2016.124
9. Chourabi, H., Nam, T., Walker, S., Gil-Garcia, J.R., Mellouli, S., Nahon, K., Pardo, T.A., Scholl, H.J.: Understanding Smart Cities: An Integrative Framework (2012). https://doi.org/ 10.1109/HICSS.2012.615 (Presented at the January)
10. Allam, Z., Dhunny, Z.A.: On big data, artificial intelligence and smart cities. Cities **89**, 80–91 (2019). https://doi.org/10.1016/j.cities.2019.01.032
11. Caravaggio, N., Caravella, S., Ishizaka, A., Resce, G.: Beyond CO_2: a multi-criteria analysis of air pollution in Europe. J. Clean. Prod. **219**, 576–586 (2019). https://doi.org/10.1016/j.jcl epro.2019.02.115
12. Talari, S., Shafie-khah, M., Siano, P., Loia, V., Tommasetti, A., Catalão, J.: A review of smart cities based on the internet of things concept. Energies **10**, 421 (2017). https://doi.org/10.3390/ en10040421
13. Batty, M., Axhausen, K.W., Giannotti, F., Pozdnoukhov, A., Bazzani, A., Wachowicz, M., Ouzounis, G., Portugali, Y.: Smart cities of the future. Eur. Phys. J. Spec. Top. **214**, 481–518 (2012). https://doi.org/10.1140/epjst/e2012-01703-3
14. Ning, H., Liu, H., Yang, L.T.: Cyberentity security in the internet of things. Computer **46**, 46–53 (2013). https://doi.org/10.1109/MC.2013.74
15. Hernández-Muñoz, J.M., Vercher, J.B., Muñoz, L., Galache, J.A., Presser, M., Hernández Gómez, L.A., Pettersson, J.: Smart cities at the forefront of the future internet. In: Domingue, J., Galis, A., Gavras, A., Zahariadis, T., Lambert, D., Cleary, F., Daras, P., Krco, S., Müller, H., Li, M.-S., Schaffers, H., Lotz, V., Alvarez, F., Stiller, B., Karnouskos, S., Avessta, S., Nilsson, M. (eds.) The Future Internet, pp. 447–462. Springer, Berlin (2011). https://doi.org/10.1007/ 978-3-642-20898-0_32
16. Rashidi, P., Mihailidis, A.: A survey on ambient-assisted living tools for older adults. IEEE J. Biomed. Health Inform. **17**, 579–590 (2013). https://doi.org/10.1109/JBHI.2012.2234129

17. Adams, M.D., Kanaroglou, P.S.: Mapping real-time air pollution health risk for environmental management: combining mobile and stationary air pollution monitoring with neural network models. J. Environ. Manage. **168**, 133–141 (2016). https://doi.org/10.1016/j.jenvman.2015.12.012

18. Cetin, M., Sevik, H.: Measuring the impact of selected plants on indoor CO_2 concentrations. Pol. J. Environ. Stud. **25**, 973–979 (2016). https://doi.org/10.15244/pjoes/61744

19. Shah, J., Mishra, B.: IoT enabled environmental monitoring system for smart cities. In: 2016 International Conference on Internet of Things and Applications (IOTA), pp. 383–388. IEEE, Pune, India (2016). https://doi.org/10.1109/IOTA.2016.7562757

20. Marques, G., Saini, J., Pires, I.M., Miranda, N., Pitarma, R.: Internet of things for enhanced living environments, health and well-being: technologies, architectures and systems. In: Singh, P.K., Bhargava, B.K., Paprzycki, M., Kaushal, N.C., Hong, W.-C. (eds.) Handbook of Wireless Sensor Networks: Issues and Challenges in Current Scenario's, pp. 616–631. Springer, Cham (2020). https://doi.org/10.1007/978-3-030-40305-8_29

21. Saini, J., Dutta, M., Marques, G.: A comprehensive review on indoor air quality monitoring systems for enhanced public health. Sustain. Environ. Res. **30**, 6 (2020). https://doi.org/10.1186/s42834-020-0047-y

22. Stewart, D.R., Saunders, E., Perea, R.A., Fitzgerald, R., Campbell, D.E., Stockwell, W.R.: Linking air quality and human health effects models: an application to the Los Angeles air basin. Environ. Health Insights **11**, 117863021773755 (2017). https://doi.org/10.1177/117863 0217737551

23. Walsh, P.J., Dudney, C.S., Copenhaver, E.D.: Indoor Air Quality. CRC Press, New York (1983)

24. Marques, G., Pitarma, R.: mHealth: indoor environmental quality measuring system for enhanced health and well-being based on internet of things. J. Sens. Actuator Netw. **8**, 43 (2019). https://doi.org/10.3390/jsan8030043

25. Bruce, N., Perez-Padilla, R., Albalak, R.: Indoor air pollution in developing countries: a major environmental and public health challenge. Bull. World Health Organ. **78**, 1078–1092 (2000)

26. World Health Organization: Ambient (outdoor) air quality and health. 2014. Retrieved World Health Organ. Available https://www.who.int/en/news-room/fact-sheets/detail/ambient-(outdoor)-air-quality-and-health. Last Accessed 26 Nov 2015 (2016)

27. Wild, C.P.: Complementing the genome with an "Exposome": the outstanding challenge of environmental exposure measurement in molecular epidemiology. Cancer Epidemiol. Biomarkers Prev. **14**, 1847–1850 (2005). https://doi.org/10.1158/1055-9965.EPI-05-0456

28. National Weather Service: Why air quality is important. https://www.weather.gov/safety/air quality. Accessed 2019 July 21

29. European Environment Agency: Air quality in Europe: 2019 report (2019)

30. Holland, M., Spadaro, J., Misra, A., Pearson, B.: Costs of air pollution from european industrial facilities 2008–2012—an updated assessment. EEA Technical report (2014)

31. Kaiser, J.: Epidemiology: how dirty air hurts the heart. Science **307**, 1858b–1859b (2005). https://doi.org/10.1126/science.307.5717.1858b

32. Weuve, J.: Exposure to particulate air pollution and cognitive decline in older women. Arch. Intern. Med. **172**, 219 (2012). https://doi.org/10.1001/archinternmed.2011.683

33. Liu, W., Shen, G., Chen, Y., Shen, H., Huang, Y., Li, T., Wang, Y., Fu, X., Tao, S., Liu, W., Huang-Fu, Y., Zhang, W., Xue, C., Liu, G., Wu, F., Wong, M.: Air pollution and inhalation exposure to particulate matter of different sizes in rural households using improved stoves in central China. J. Environ. Sci. **63**, 87–95 (2018). https://doi.org/10.1016/j.jes.2017.06.019

34. Seguel, J.M., Merrill, R., Seguel, D., Campagna, A.C.: Indoor air quality. Am. J. Lifestyle Med. 1559827616653343 (2016)

35. Bonino, S.: Carbon dioxide detection and indoor air quality control. Occup. Health Saf. Waco Tex. **85**, 46–48 (2016)

36. Marques, G., Miranda, N., Kumar Bhoi, A., Garcia-Zapirain, B., Hamrioui, S., de la Torre Díez, I.: Internet of things and enhanced living environments: measuring and mapping air quality using cyber-physical systems and mobile computing technologies. Sensors **20**, 720 (2020). https://doi.org/10.3390/s20030720

37. Lu, C.-Y., Lin, J.-M., Chen, Y.-Y., Chen, Y.-C.: Building-related symptoms among office employees associated with indoor carbon dioxide and total volatile organic compounds. Int. J. Environ. Res. Public Health 12, 5833–5845 (2015). https://doi.org/10.3390/ijerph120605833

38. Zhu, C., Kobayashi, K., Loladze, I., Zhu, J., Jiang, Q., Xu, X., Liu, G., Seneweera, S., Ebi, K.L., Drewnowski, A., Fukagawa, N.K., Ziska, L.H.: Carbon dioxide (CO_2) levels this century will alter the protein, micronutrients, and vitamin content of rice grains with potential health consequences for the poorest rice-dependent countries. Sci. Adv. 4, eaaq1012 (2018). https://doi.org/10.1126/sciadv.aaq1012

39. Myers, S.S., Zanobetti, A., Kloog, I., Huybers, P., Leakey, A.D.B., Bloom, A.J., Carlisle, E., Dietterich, L.H., Fitzgerald, G., Hasegawa, T., Holbrook, N.M., Nelson, R.L., Ottman, M.J., Raboy, V., Sakai, H., Sartor, K.A., Schwartz, J., Seneweera, S., Tausz, M., Usui, Y.: Increasing CO_2 threatens human nutrition. Nature 510, 139–142 (2014)

40. Afshar-Mohajer, N., Zuidema, C., Sousan, S., Hallett, L., Tatum, M., Rule, A.M., Thomas, G., Peters, T.M., Koehler, K.: Evaluation of low-cost electro-chemical sensors for environmental monitoring of ozone, nitrogen dioxide, and carbon monoxide. J. Occup. Environ. Hyg. 15, 87–98 (2018). https://doi.org/10.1080/15459624.2017.1388918

41. Marques, G., Ferreira, C.R., Pitarma, R.: Indoor air quality assessment using a CO_2 monitoring system based on internet of things. J. Med. Syst. 43 (2019). https://doi.org/10.1007/s10916-019-1184-x

42. Marques, G., Pitarma, R.: A cost-effective air quality supervision solution for enhanced living environments through the internet of things. Electronics 8, 170 (2019). https://doi.org/10.3390/electronics8020170

43. Dhingra, S., Madda, R.B., Gandomi, A.H., Patan, R., Daneshmand, M.: Internet of things mobile-air pollution monitoring system (IoT-Mobair). IEEE Internet Things J. 6, 5577–5584 (2019). https://doi.org/10.1109/JIOT.2019.2903821

44. Marques, G., Pires, I., Miranda, N., Pitarma, R.: Air quality monitoring using assistive robots for ambient assisted living and enhanced living environments through internet of things. Electronics 8, 1375 (2019). https://doi.org/10.3390/electronics8121375

45. Taştan, M., Gökozan, H.: Real-time monitoring of indoor air quality with internet of things-based E-nose. Appl. Sci. 9, 3435 (2019). https://doi.org/10.3390/app9163435

46. Zhao, L., Wu, W., Li, S.: Design and implementation of an IoT-based indoor air quality detector with multiple communication interfaces. IEEE Internet Things J. 6, 9621–9632 (2019). https://doi.org/10.1109/JIOT.2019.2930191

47. Kaivonen, S., Ngai, E.C.-H.: Real-time air pollution monitoring with sensors on city bus. Digit. Commun. Netw. 6, 23–30 (2020). https://doi.org/10.1016/j.dcan.2019.03.003

48. Lai, X., Yang, T., Wang, Z., Chen, P.: IoT implementation of Kalman Filter to improve accuracy of air quality monitoring and prediction. Appl. Sci. 9, 1831 (2019). https://doi.org/10.3390/app9091831

49. Wu, Y., Liu, T., Ling, S., Szymanski, J., Zhang, W., Su, S.: Air quality monitoring for vulnerable groups in residential environments using a multiple hazard gas detector. Sensors 19, 362 (2019). https://doi.org/10.3390/s19020362

50. Gautam, A., Verma, G., Qamar, S., Shekhar, S.: Vehicle pollution monitoring, control and Challan system using MQ2 sensor based on internet of things. Wirel. Pers. Commun. (2019). https://doi.org/10.1007/s11277-019-06936-4

51. Sun, S., Zheng, X., Villalba-Díez, J., Ordieres-Meré, J.: Indoor air-quality data-monitoring system: long-term monitoring benefits. Sensors 19, 4157 (2019). https://doi.org/10.3390/s19194157

52. Johnston, S.J., Basford, P.J., Bulot, F.M.J., Apetroaie-Cristea, M., Easton, N.H.C., Davenport, C., Foster, G.L., Loxham, M., Morris, A.K.R., Cox, S.J.: City scale particulate matter monitoring using LoRaWAN based air quality IoT devices. Sensors 19, 209 (2019). https://doi.org/10.3390/s19010209

53. Pedersen, T.H., Nielsen, K.U., Petersen, S.: Method for room occupancy detection based on trajectory of indoor climate sensor data. Build. Environ. 115, 147–156 (2017). https://doi.org/10.1016/j.buildenv.2017.01.023

54. Rüffer, Daniel, Hoehne, Felix, Bühler, Johannes: New digital metal-oxide (MOx) sensor platform. Sensors **18**, 1052 (2018). https://doi.org/10.3390/s18041052
55. Sensirion: Datasheet SGP30 Sensirion gas platform. https://www.sensirion.com/fileadmin/user_upload/customers/sensirion/Dokumente/0_Datasheets/Gas/Sensirion_Gas_Sensors_S GP30_Datasheet.pdf. Accessed 08 Feb 2020
56. Marques, G., Pitarma, R.: Smartwatch-based application for enhanced healthy lifestyle in indoor environments. In: Omar, S., Haji Suhaili, W.S., Phon-Amnuaisuk, S. (eds.) Computational Intelligence in Information Systems, pp. 168–177. Springer, Cham (2019). https://doi.org/10.1007/978-3-030-03302-6_15
57. Gurney, K.R., Mendoza, D.L., Zhou, Y., Fischer, M.L., Miller, C.C., Geethakumar, S., de la Rue du Can, S.: High resolution fossil fuel combustion CO_2 emission fluxes for the United States. Environ. Sci. Technol. **43**, 5535–5541 (2009). https://doi.org/10.1021/es900806c
58. Abas, N., Khan, N.: Carbon conundrum, climate change, CO_2 capture and consumptions. J. CO_2 Util. **8**, 39–48 (2014). https://doi.org/10.1016/j.jcou.2014.06.005
59. Marques, G., Pitarma, R.: IAQ evaluation using an IoT CO_2 monitoring system for enhanced living environments. In: Rocha, Á., Adeli, H., Reis, L.P., Costanzo, S. (eds.) Trends and Advances in Information Systems and Technologies, pp. 1169–1177. Springer, Cham (2018). https://doi.org/10.1007/978-3-319-77712-2_112
60. Karagulian, F., Barbiere, M., Kotsev, A., Spinelle, L., Gerboles, M., Lagler, F., Redon, N., Crunaire, S., Borowiak, A.: Review of the performance of low-cost sensors for air quality monitoring. Atmosphere **10**, 506 (2019). https://doi.org/10.3390/atmos10090506
61. Honeycutt, W.T., Ley, M.T., Materer, N.F.: Precision and limits of detection for selected commercially available, low-cost carbon dioxide and methane gas sensors. Sensors **19**, 3157 (2019). https://doi.org/10.3390/s19143157
62. Marques, G., Pitarma, R.: Using IoT and social networks for enhanced healthy practices in buildings. In: Rocha, Á., Serrhini, M. (eds.) Information Systems and Technologies to Support Learning, pp. 424–432. Springer, Cham (2019). https://doi.org/10.1007/978-3-030-03577-8_47
63. Marques, G., Pitarma, R.: An internet of things-based environmental quality management system to supervise the indoor laboratory conditions. Appl. Sci. **9**, 438 (2019). https://doi.org/10.3390/app9030438

Data Science and AI in IoT Based Smart Healthcare: Issues, Challenges and Case Study

Sohail Saif, Debabrata Datta, Anindita Saha, Suparna Biswas, and Chandreyee Chowdhury

1 Introduction to IoT Based Smart Applications

The idea of Internet of Things (IoT) comes from connecting various physical objects and devices to world using internet. A basic example of such object and device can be HVAC (Heating, Ventilation, and Air Conditioning) monitoring and controlling which enables a smart home. There are plenty other domains where IoT plays an important role that improve our quality of living. In recent years this technology has been vastly adopted by industries, healthcare, transportation, agriculture etc. IoT enables the physical objects to see, hear perform specific jobs, which makes the objects smart. Over the time, A lot of home and business applications will be based on IoT which not only improve quality of living but also grow the world's economy. For instance, smart home can help it residents to automatically turn on the lights, fans

S. Saif · S. Biswas (✉)
Department of Computer Science and Engineering, Maulana Abul Kalam Azad
University of Technology, Kolkata, India
e-mail: mailtosuparna@gmail.com

S. Saif
e-mail: sohailsaif7@gmail.com

D. Datta
Department of Computer Science, St. Xavier's College (Autonomous), Kolkata, India
e-mail: debabrata.datta@sxccal.com

A. Saha
Department of Information Technology, Techno Main Salt Lake, Kolkata, West Bengal, India
e-mail: anindita_saha03@yahoo.co.in

S. Saif · C. Chowdhury
Department of Computer Science and Engineering, Jadavpur University, Kolkata,
West Bengal, India
e-mail: chandreyee.chowdhury@gmail.com

© The Editor(s) (if applicable) and The Author(s), under exclusive license
to Springer Nature Switzerland AG 2021
A.-E. Hassanien et al. (eds.), *Enabling AI Applications in Data Science*,
Studies in Computational Intelligence 911,
https://doi.org/10.1007/978-3-030-52067-0_19

Fig. 1 Overall application domain of IoT

when they reach home, control the air conditioning system etc. Smart healthcare can reduce the need of visit to physicians for regular health checkups. Smart Agriculture can help the farmers to sprinkle water to the plants automatically. Figure 1 shows the various trending application domain of IoT.

1.1 Need of IoT Based Smart Applications

IoT is one of the essential piece of technology that is set to improve significantly throughout the time. There are several advantages of having things connected with each other. With the help of sensors, it can collect a huge amount of data which are beneficial for analysis. For example analyzing the data of a smart refrigerator, information like power consumption, temperature etc. can be extracted so the power efficiency can be increased. So the gathered data can be helpful to make decisions another benefit can be ability to track and monitor things in a real time manner. A patient suffering from critical diseases can be remotely monitored and with the help of Artificial Intelligence (AI), Machine Learning (ML), Deep Learning (DL) it is possible to design personalized drugs. IoT Application can lighten the workload with the help of automation. For Example in a smart home environment, doors of the garage can be opened automatically when the owner reaches the home. So this can greatly reduce the human effort. IoT applications can increase efficiency in terms of saving money and other resources. For instance, Smart Lights can turn themselves off if no one present in the room, this can cut the electricity bills. IoT applications can improve the quality of lifestyle, health, wellness. For example, Light weight wearable health bands can help a person to track his body vitals, such as temperature, heart rate, oxygen levels.

1.2   Architectures and Building Components of Such Systems

Basic architecture of any IoT application consists of four layers such as, Sensing, Network, Processing and Application. Figure 2 depicts the traditional architecture of an IoT Application. Sensing layer contains various sensors, such as biomedical, environmental, electrical etc. Network layer is consists of heterogeneous networking de-vices such as access point, router, switch etc. Components of processing layers are CPU, GPU, cloud servers etc. Application layer is the output of feedback layer.

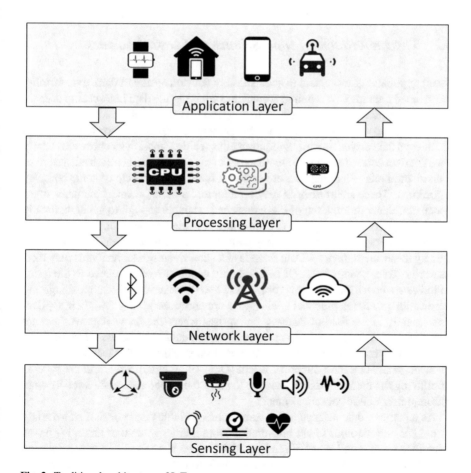

Fig. 2 Traditional architecture of IoT

2 Need of Data Analytics/Science in Smart Applications

Data science is a relatively new term introduced to identify the technique to find the hidden patterns that may be present in a collection of data. Essentially, data science works with a substantially large collection of data, often termed as big data. So, handling big data becomes an integral part of data science. Many algorithms have been incorporated to deal with a massive collection of data of importance. On the other hand, smart applications need to primarily deal with Internet of Things (IoT), which may be a source of terabytes of data, if not more. Now-a-days, smart cities, smart healthcare systems etc. have become commonplace specifically in the developed nations. So handling that huge amount of data generated by IoT or smart devices as a whole has become a topic of concern and hence a buzzing topic of research as well. This chapter discusses about the utility of data science and different algorithms that can analyze data through data science in the field of smart computing.

2.1 The Importance of Data Science and Data Analytics

Smart applications are related to acquisition of data, cleansing of data, transforming the cleaned data into the required format and applying the data in a specific domain to make a better user experience. The applications should have insights into the features related to the desired domain of user applications. Hence, smart applications are always data driven as well as requirement specific. Smart devices are essentially used for data collection and communication of data to either a centralized system or a distributed one where the data are to be analyzed with the help of some complex algorithms. These smart devices may be a laptop, a tablet, a smart phone, a smart watch and many more. Each of these devices is capable enough to generate data in its own capacity and if required to send data to the required system.

The challenge remains in dealing with this massive collection of data generated by these smart devices. Data science and subsequently data analytics play their respective roles exactly here. Of late, industry people are often related to the terms called data science and data analytics. These two terms have different meanings and implications as far as practical applications are concerned. They find their applications mainly in the field of business enterprises where the knowledge workers are constantly occupied in analyzing past or historical data to make the future business better, more profitable and perhaps more customer friendly. The technical persons involved in the Decision Support System (DSS) of an enterprise are in the process of utilizing the theory of data science to identify the importance of data fed through different user queries or requirements.

As a subject, data science has a number of specialties and consists of a variety of models and methods to get information from a large volume of data. It is a part of applied science and hence different models, methods, tools specifically belonging to mathematics and statistics come under the purview of data science. These tools consist of the processes to analyze or extract meaningful information out of the gathered data. So working with data science is nothing but connecting information and data points to find relations among the data that can be made useful for the overall growth of the business. Data science explores a seemingly unknown world trying to find some new patterns and insights. Hence, data science tries to build connections and plans for the future. Data science often provides new perspective of understanding and projection of data for the benefits of the organization as a whole.

Data Analytics is essentially a part of data science. It is more specific and target oriented in the sense that instead of just looking for the connections among data, data analysts are more engaged in satisfying the business requirements of the organization for which the data analysis is done. At the same time, the process of data analytics is often automated to provide insights in certain goal oriented areas. So the technique of data analytics tries to find the hidden patterns that may be present in the gathered data. These patterns are generated only to comply with the company's true goals and tends to be slightly more business and strategy focused.

The relationship between data science and data analytics can actually be very fruitful and can have a huge impact on a business enterprise. Though data scientists and data analysts perform apparently different duties, if they can be utilized in a cohesive manner, at the end of the day, the organization will be benefitted. Data analytics and data science can be used to find different things, and while both are useful to companies, they both won't be used in every situation. Generally, data science with its methodologies plays a significant role in the development of artificial intelligence and machine learning. On the other hand, data analytics is often used in industries like healthcare, gaming, and travel.

Though there are few subtle differences between the approaches followed by data science and data analytics, both of them can be utilized together to handle, analyze, understand the enormous collection of data coming from smart devices in the name of smart applications.

2.2 Issues and Challenges in Handling Big Data

Now-a-days, the use of big data has become quite commonplace because most of the business organizations, especially the e-commerce sites, social networking sites need to deal with huge collection of data in order to understand their business requirements. Ideally, big data means the volume of data that cannot be processed by traditional database methodologies and algorithms. Big data should be able to deal with data in heterogeneous formats. So data can be structured, semi-structured and unstructured. In fact, most of the gathered data come in a massive format where unstructured information. So there may be another way of defining big data as the amount of data that cannot be stored and managed efficiently with the traditional storage medium. A big data set is traditionally identified by three V's—Volume, Velocity and Variety. Volume identifies the amount of data stored in the form of big data. Velocity identifies the speed of data acquisition and aggregation. Finally, Variety identifies the heterogeneity in the nature of acquired data. The major challenge with the present scenario with big data is that the current growth rate with which the data are acquired is astounding. So, the IT researchers and practitioners are facing the problem with this exceedingly fast growth rate. The related challenges are to design an appropriate system to handle the data effectively as well as efficiently. At the same time, another major issue comes into the picture regarding the analysis of big data in connection with the extraction of relevant meaning for the decision making.

A number of issues come up to understand and discuss the challenges while working with big data. An extensive research work is required to exactly point out the areas to be highlighted. A significant point has been put forward in [1] where the researchers mentioned that the usefulness of the big data depends on the areas where the big data are to be used. It means that the utility of the big data depends on the requirements of the concerned enterprise. Not only this, the availability of the technologies tackling the big data is also a point of concern [1]. In this context, it is quite possible that the newly envisioned data are neither known to the knowledge workers nor to the designers. It leads to an additional challenging situation where the technical persons need to improvise on the interfaces, application organizations, metaphors, conceptual models and metadata in such a way that it can be presented in a comprehensive manner to the end users of the system. There may be a tricky problem arising out of unknown challenges that will arise with the increase in the scale and development of new analytics [2]. Some of these challenges may be intractable with the tools and techniques presently dealing with big data.

A major demanding issue raised in big data design is the output process as pointed out in [3] where it has been briefly stated that it has been always easier to get data into a system than to get the required output from the system. The same research work has also shown that the traditional transaction processing system is not at all suitable to deal with the big data analytics as the latter demands OnLine Analytical Processing (OLAP) instead of OnLine Transaction Processing (OLTP). So new tools and techniques are required to process and analyze big data. This leads to another challenging factor to work with big data.

The very basic nature of big data is its heterogeneity. This creates another problem in context with big data as for the analysis purpose, all the gathered data are to be converted into a suitable format. Converting heterogeneous data sets in the form of structured, semi-structured and unstructured formats into a common format needs an extensive and complex methodology.

Periodically updating the content of a big data set leads to another issue concerning the handling of big data. As the outcome of the analysis has to be effectively and periodically updated, the corresponding data set needs to be modified as and when the case arises. Now the most challenging issue comes regarding this is to find the exact amount of time after which the modification has to be reflected in the data set. Different researchers have raised different factors concerning this, but the major factor lies in the fact that the decision support system that is running in the background of the system should be responsible to identify the duration of this period and fundamentally this duration can be both static as well as dynamic as the case may be.

Another point to be considered as a challenge for big data users is a trade-off be-tween quantity against quality of big data. For a set of users and the related applications, the quantity, i.e., the volume of big data matters as larger amount of data means more flexibility in terms of training the data and subsequently testing the data. On the other hand, there are another set of big data users who mainly focus on the quality of the acquired data as noisy data may actually ruin the purpose of the applications. Effectively, a high quality data set can lead to a more precise analytical conclusion. Hence, depending on the emphasis given either on quantity or on quality, the algorithms, methodologies, tools are to be selected and this may lead to a really challenging job.

Another major challenge issue is in the field of data dissemination. The bottleneck can be the communications middleware which may happen when the communication hardware speeds tend to increase with the new technologies and at the same time, message handling speeds decrease. The heterogeneity in the nature of big data may also lead to a hindrance to the data dissemination process.

Another factor that is of key concern is the privacy and the security of the data [4]. As big data sometimes deal with storing sensitive as well as personal information, ensuring security of this data is to be seriously dealt with. But this may lead to a severe security concern, especially in the scenario where many individuals provide very personal information. Many countries have already incorporated laws to tackle this sensitive issue. It is the information about you that is much greater than the information you create and/or release about yourself.

Processing big data is also a big challenge as there are many types of big data analytics, viz., descriptive, estimative, predictive etc.; leading to various types of decision and optimization models. A research work has been depicted in [5] where different models have been discussed based on the type of problems to be addressed and subsequently the nature of the data analytics to be incorporated. These models include econometric models, game theory, control theory, evolutionary computation, and simulation models. A particular model may not be sufficient to solve a particular problem and hence there may be a requirement of the collaboration of multiple models to attack the issue. This leads to an open research issue – whether for any given type of algorithm, there is a fundamental limit to its scalability.

2.3 Data Science and Data Analytics Techniques in Smart Applications

Data science and data analytics are together being utilized in different applications related to smart computing. Any smart computing means adding a computing power to all the connected devices like mobile phone, television, watch etc. These devices become smart devices when they are provided with computing power and internet connectivity and put together they are often termed as Internet of Things (IoT). The main focus area of smart computing is to provide a real time solution to some real

Fig. 3 Data feed from an IoT infrastructure for the analytics through a big data storage medium

world problems with a seamless integration of hardware, software and network platforms through which the devices are connected. This section discusses about the
usefulness of the methodologies provided by both data science and data analytics in
the domain of smart computing.

Figure 3 depicts the association between an IoT platform and the data analysis
part. The massive amount of data generated in an IoT infrastructure are fed and
stored into a big data storage platform. The rightmost component of Fig. 3 collects
the data from the data storage platform and uses the data for the analysis purpose
and for successive report generation. This component utilizes the methodologies of
data science and data analytics. The usefulness of this diagram is further explained
next.

The smart devices are capable enough to produce an enormous volume data. The
problem lies in a rational, useful as well efficient analysis of these data. If they can
be properly analyzed, a huge benefit may usher in the daily life. From the business
organization's point of view, the effectiveness may bring a more profitable scenario.
One of the application areas that is much talked about of late is in the field of power
grid. Since the traditional fossil fuels are facing the problem of depletion and the
de-carbonization demands the power system to reduce the carbon emission, smart
grid system has been identified to be an effective solution to provide electrification in
the society [6]. Different methodologies provided by data science and data analytics
are being utilized for a better regulation and dispatch of renewable energy and this
has become a major area of research topic among the researchers engaged in the
development of smart applications. Smart applications can further be extended in
the electric billing process [7]. Whereas the traditional billing system depends on
the manual collection of data, a smart grid system can be used to analyze the usage
pattern and the status of the electricity networks. This analysis can later be utilized
for demand forecasting and energy generation optimization as well.

Another significant area of application of smart computing using data science and
data analytics is in the field of smart healthcare system. With a tremendous growth of
IoT devices, tracking data from different smart devices has become quite a commonplace affair especially in the areas considered to be smart cities [8]. Data science and
data analytics are together utilized to provide innovative ideas and solutions to the
healthcare problems after analyzing the data received through smart devices present

at the patients' ends. The effective union of data science, data analytics and IoT with telemedicine, e-health, m-Health for patient monitoring has been significantly improved in the last decade and thus reformed the personalized healthcare system by constantly monitor-ing the patients' conditions irrespective of his/her locations [9]. At the same time, the system is cost effective as well. So the usefulness of data science and data analytics has been tremendous and the subsequent innovations are considered to resolve significant issues of remote patient monitoring. The real challenge lies in constantly acquiring data related to patients' health and thereafter analyzing the same. For this, different algorithms of data science and data analytics are serving the purpose. Various biomedical sensors as IoT devices have also been used in the patient monitoring system which provides vital biological and genetic information. Different health-related parameters like body temperature, blood pressure, pulse rate, blood glucose level, ECG, EEG etc. can be tracked using these smart devices associated with the smart healthcare system. The data received from the smart healthcare devices through a network are then analyzed in a centralized system through numerous data analytic techniques in order to realize the status of the concerned patients and accordingly either a diagnosis or an advice is prescribed. This application is specifically useful in the areas where there is a scarcity of medical staff or the number of ageing population is substantially more. According to an estimation of The World Health Organization (WHO) the requirement and actual number of the medical staff throughout the world will be about 82 million and 67 million, respectively, by 2030 [10]. This huge amount of shortage in the medical staff can be fruitfully tackled with the help of smart healthcare systems in collaboration with data science and data analytics as mentioned above. The table shown in the next page mentions some emergent areas of research in the field of smart healthcare system and the corresponding data science or data analytics algorithms used. The table also shows the citations of the work. The issues and their respective remedial methodologies as depicted in Table 1 are discussed next.

Table 1 Summary of few medical issues and the proposed algorithms used to solve the issues

Nature of the disease	Proposed algorithm(s)
Prediction of type-2 diabetes	Density-based spatial clustering
Prediction of hypertension	Synthetic minority over-sampling
Prediction of tuberculosis	Regression model and Cuckoo search optimization algorithm
Optic disk localization	Statistical edge detection
Lung cancer classification	Deep convolution neural network and support vector machine
Estimation of skin aging	Polynomial regression and color histogram intersection
Classification of organ inflammation	Genetic algorithm and support vector machine

A random forest-based algorithm for the prediction of the type-2 diabetes has been proposed in [11] where the authors have highlighted the use of a technique called density-based spatial clustering. In the same research paper, the researchers have proposed another technique called synthetic minority over-sampling to predict the hypertension issue prevalent among the patients. The techniques mentioned had shown a significant improvement in precision and accuracy using three benchmark datasets.

For the prediction of tuberculosis, a research work has shown the use of both the regression model and Cuckoo search optimization algorithm [12]. This novel work has shown that the goodness-of-fit with adjusted R-squares was more than 0.96.

Another application of data analytics was seen in the detection of an optic disk to be applied in the retinal images [13]. In this research work, the authors have used the statistical edge detection algorithm where more than 250 samples were tested and the accuracy rate was more than 95%. According to the authors, the proposed methodology could be utilized in the detection of various retinal diseases like glaucoma, diabetic retinopathy etc.

A combined approach using deep convolution neural network and support vector machine was taken up for the classification of lung cancers in [14]. This research work has used more than 2000 pulmonary computed tomography images to use in the algorithm and for the analysis of the obtained result. In this work also, the authors have claimed an accuracy of over 90%.

A research article titled "Skin aging estimation scheme based on lifestyle and ceroscopy image analysis" has shown the use of two data science algorithms, viz., polynomial regression and color histogram intersection [15] to attack the problems related to the topics of skin condition tracing and skin texture aging estimation based on lifestyle. The authors have involved more than 350 volunteers in the performance evaluation and the results indicated an accuracy of more than 90%. Another research work [16] that has been shown in Table 1 had talked about a multi-objective optimization problem that was formulated for the support vector machine classification of organ inflammations and solved by a genetic algorithm. In this research work, to achieve a better accuracy, the authors had used a combination of kernels of support vector machine. Effectively, the accuracy obtained through this approach was shown to be more than 90%.

An emerging area of application of data science and data analytics is in the development of a smart city. A smart city is a relatively new term to define a city having some smart solutions in its infrastructure including electric supply, water supply, solid waste management, traffic mobility, health, education, e-governance etc. According to [17], a smart city should consist of the following dimensions:

- Smart people
- Smart mobility
- Smart environment
- Smart living
- Smart governance
- Smart economy

To implement each of these infrastructural issues efficiently, a constant monitoring of data is required. For that, different smart devices should work in tandem to gather data and essentially analyze the gathered data through some high end algorithms. The major components of a smart city [18] and the related roles of data science and data analytics are discussed next:

- A major component of a smart city should be a smart healthcare system. Applications of data science and data analytics in smart healthcare system have already been discussed.
- Another component is smart energy system which is related to smart grid sys-tem. This has already been discussed.
- Smart transportation should be an inherent component of a smart city. For this, it is required to recognize traffic patterns by analyzing real time data. It is very useful as it reduces the congestion in the traffic system by predicting traffic conditions and adjusting traffic controls. With the help of a predictive data analysis model, the smart city will be able to deal with the traffic congestion issue. It may also provide information related to available car parking and alternative roads. A very significant component of a smart transportation system is the smart traffic lights which would auto-mate the traffic flow avoiding traffic jam like situation.
- Smart governance should also be an integral part of a smart city. It should work with different types of data collected from the citizens. The next work is to analyze these data and accordingly suggest an improvement of its citizens. This will enhance the quality of the government agencies as well.
- Planning a city in an effective way is also a part of smart city development. This will include the distribution of electricity, proper sewerage system and some relevant factors.

3 Need of AI Techniques in Smart Applications

Smart Applications refer to those applications that assimilate data-driven and action oriented insights with the experience of the user. These insights are delivered as features in smart applications that empower the users to complete a task more efficiently and effectively. For these smart applications IoT is the key enabler and the confluence of AI with IoT is believed to bring myriad changes in technological space around mankind. IoT sensors generate and transmit a huge amount of data (Big Data), and the conventional processes of collection, analysis and processing this huge volume of data are falling short day by day. This urges the need for AI techniques to be introduced along with IoT Smart Applications. With the help of AI, machines acquire the ability to simulate the basic and complex cognitive function of the human being, through hardware implementations, software programs, and applications. Combining AI with IoT can improve the operational efficiency of the smart applications, through simple steps like tracking (collection of data), monitoring (analyzing the data), controlling, optimizing (training the model), and automating (testing and predicting). When programmed with ML (its subset) AI can be used widely in IoT for analyzing huge data generated by it and make appropriate decisions.

As IoT generates a huge amount of data, it is essential to ensure that only adequate data is collected and processed, as otherwise it becomes time-consuming and affects operational efficiency. AI with several steps of data mining like data integration, data selection, data cleaning, data transformations, and pattern evaluation can accomplish the task. A significant number of applications are powered by AI in IoT like the smart healthcare industry where it can be useful for analyzing the symptoms, prediction, and detection of severe diseases. Smart agricultural industries can use AI techniques to perform several tasks like identifying and predict crop rotation, assess risk management, evaluate soil properties, predict climate change, and many more. Smart homes are provided with many features that might be enabled with AI like enhanced security, automation of household equipments, and recognition of activities of the dwellers especially the elders and the disabled.

3.1 Why Artificial Intelligence?

By definition, AI would be read as "a field of science concerned with the computational understanding of what is commonly called intelligent behavior, and with the creation of intelligent agents that exhibit such behavior" [19]. Simply put, AI is the science of making machines more intelligent, and make them think or act or take decisions rationally like human beings. These machines would be termed as "intelligent agents" that can utilize the operating environment around it to take in-puts from it and also take actions by it without any human interventions. The environment is perceived through sensors and acted upon through effectors. Hence in AI, reasoning and knowledge are essential components through which the Intelligent Agents can manipulate input data to produce actions. Earlier AI involved algorithms that were based on a step by step reasoning for predicted problems however, this approach was not suitable for uncertain situations or in an environment with incomplete information. Hence current AI models have been developed to respond to such unfavorable situations through concepts of probability and statistics. So AI models can gain widespread knowledge about the environment of the problem (certain or uncertain) and solve them with computational efficiency. Another reason for AI to gain popularity in smart applications is its capability to represent knowledge through computational resources and programming languages like Python that enables deployment of sophisticated algorithms to resolve complex problems [20]. An essential component of Intelligence is the learning and modeling different aspects of learning process of Intelligent Agents is done by Machine Learning which is a subset of AI. ML aims towards self-learning of machines without any human interventions and improving its present outputs through knowledge from past experiences. Machine Learning has given AI an edge to gain rapid technological advances in the fields of wireless technology and computing power. Figure 4 illustrates the efficient and intelligent handling of big data using IoT and AI-enabled technologies for improved outcomes.

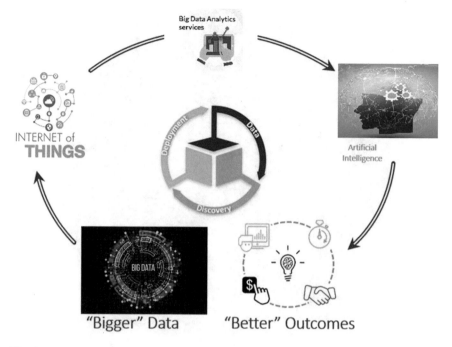

Fig. 4 IoT and AI

3.2 *AI, Machine Learning and Deep Learning Techniques Applied*

When it comes to smart healthcare applications, conventional diagnostic methods are often found to be dependent on the depth of knowledge and experience of the medical practitioner's which might lead to misdiagnosis in the majority of cases. This results in wastage of huge amounts of medical data. Hence a profound integration of Artificial Intelligence, Machine Learning, and Deep Learning can revolutionize the conventional disease diagnostic systems and improve the potential of smart healthcare applications. Artificial Intelligence may be considered as the umbrella that encompasses the Machine Learning as well as Deep Learning as shown in Fig. 5, apart from others like conventional machine learning, natural language processing, robotics, computer visions, expert systems, general intelligence, super intelligence, narrow intelligence, automated learning and many more.

Machine Learning is a subset of Artificial Intelligence that intends to train machines with the help of algorithms and data, without being explicitly programmed to make decisions without human intervention. Machine Learning also attempts to reduce errors, and maximize the chances of the predictions being accurate. It is considered to be dynamic which refers to its capability to modify itself when the size

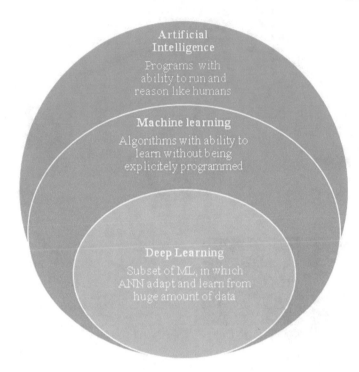

Fig. 5 AI with ML and DL

of data increases, which is an essential criterion as in IoT-Big Data. Popular applications of Machine Learning can be speech recognition, face recognition, handwriting recognition, medical practice, and machine translation. In short, Machine Learning is a simple technique to realize Artificial Intelligence in real life.

Deep Learning is a special subset of Machine Learning incorporating algorithms and computational models to mimic the architecture of the biological neural network of the human brain. It tries to impose an artificial replication of the structure as well as functionality of the brain. In a simple ANN there are three major layers to perform a complex task namely input layers to receive the data, the output layer to produce the result of data processing, and the hidden layer that extracts the pattern within the input data by learning underlying features of the data. As the input data travels from one hidden layer to the other, simple features recompose and recombine to produce complex features. By "Deep", the system refers to those extra numbers of hidden layers for processing that adds on to a simple ANN to produce DNN or Deep Neural Network. These additional numbers of hidden layers empower DNN to solve more complex tasks. Deep Learning mainly focuses on the representation of data and not on task-specific algorithms like Machine Learning. It is in-fact perfect for unstructured data, with better efficiency than of Machine Learning. The only drawback is that due to its huge number of hidden layers, Deep Learning needs a huge volume of training data as well as expensive software and hardware combinations to give accurate results. Therefore it is suitable for Big Data analytics because as datasets grow, Deep Learning can give better outcomes and exhibit improved efficiency. It is used in various tasks like natural language processing, image recognition, language translations, etc. [21].

3.3 Influence of AI and ML and DL in Smart Applications

The wide acceptance of Artificial Intelligence in Smart Applications like Healthcare is due to its advanced computing ability to perform three major medical tasks namely Diagnosis, Prognosis, and Therapy. However, the combination of Machine Learning and Deep Learning has enhanced the power of Artificial Intelligence to handle this diagnostic-therapeutic cycle far better. For example Deep Learning is useful for image analysis that includes classification, segmentation as well as restoration of biological images. Hence they are applied in dermatology where skin lesion images can be classified into different types of skin cancers through Convolutional Neural Network [22]. In recent researches it has been shown by Google researchers, that Convolutional Neural Network, which is a class of Deep Neural Networks, is capable of detecting lymph node metastasis, thus outperforming certified pathologists, whose processes were found labor-intensive and prone to error, on large datasets like Camelyon16. [23]. Experimental sensitivity recorded was more than 90% in comparison to a human pathologist which was close to 75%. The false-negative rate was reduced drastically comparatively while detecting very small tumors in gigapixel pathology slides.

   Pneumonia is one of the major causes of mortality and morbidity in humans. Early diagnosis and treatment of studying chest X-rays can save lives. Researchers have developed algorithms that are 121 layers of Convolutional Neural Network trained on the ChestX-Ray14 dataset that can detect Pneumonia from frontal view X-Ray images of the chest [24]. It claims to exceed the performance of radiologists and can also detect the other 14 diseases present in the dataset. In literature there exist several types of researches that have established the contribution of Artificial Intelligence along with Machine Learning and Deep Learning, in detecting and curing severe diseases like a Brain tumor and Stroke. Biologically, brain tumors can be of different shape and texture and can have different locations and if not detected early, can harm the patient. Residual Networks (RN), which is a specialized Neural Network, for handling more sophisticated deep learning models and perform complex tasks, may be used to classify three types of brain tumors from MRI images [25]. Several data augmentation techniques were adopted to improve accuracy like flips, shifting, rotation, zooming, shearing, brightness manipulation, etc. More than 3000 MRI images were evaluated under performance metrics like f1-score, precision, recall, and balanced accuracy. Stroke is another medical emergency that needs immediate and timely treatment to save lives. A scaled PCA Deep Neural Networks is proposed in [26], to automatically detect strokes, from health behavior data and medical utilization history of more than 15000 patients. With simple inputs, the Deep Neural Network can study the variables of interest and the scaled PCA could generate improved inputs for better accuracy.

Thus patients detected with a high risk of stroke can be diagnosed and sent for additional checkups.

Apart from disease detection, motion analysis is widely used in computer vision to pick behavior of moving objects in image sequences. In cardiology, images of heart, taken using cardiac magnetic resonance imaging or CMRI, can be used to detect cardiac patterns using Deep Learning that reveals risks of heart failure for the patient [27]. Patients with ACHD (Adult Congenital Heart Disease) can also be identified, based on the severity and complexity of the disease, using machine learning algorithms, trained on huge datasets, for prognosis assessments and guide therapy [28]. Artificial Intelligence can be useful in Ambient Assisted Living, especially for elder and disabled people as HAR (Human Activity Recognition) has emerged substantially in modern times. HAR can combine features from the Smartphone accelerometer, as well as heart-rate measuring sensors worn by users, and finer variations of heart-rate can be estimated in comparison to the resting heart-rate of the same users using an ensemble model based on several classifiers [19, 29].

When it comes to severe neurodegenerative disorders like Parkinson's, there are quite a few Artificial Intelligent applications that can diagnose, monitor, and manage the disease using data that describes the motion of lower and upper extremities. Based on this body kinematics, therapy assessment and progress prediction of Parkinson's Disease can be done, especially in the early stages [30]. Another disease that is common in occurrence nowadays is Alzheimer's that involves high dimensional data in computer vision and Deep Learning, with its advanced neuro-imaging techniques, can detect and automate the classification of Alzheimer's Disease. The combination of SAE (Stacked Auto Encoder) along with Machine Learning, for feature selection, can give accuracy more than 98% for Alzheimer's Disease and 83% for MCI (Mild cognitive impairment), which is just the prodromal stage to the disease. Deep learning approaches with fluid biomarkers and multimodal neuroimaging can give results with better accuracy. [31]. Diabetes is another severe ailment that causes other disorders in the human body. DKD or Diabetic Kidney Disease is one of them that can be detected using natural language as well as longitudinal Big-data with Machine Learning. Artificial Intelligence when combined with the Deep learning concept of Convolutional Neural Network, could extract features for more than 60,000 diabetic patients, based on their medical records, using convolutional auto-encoder and logical regression analysis which resulted in more than 70% of accuracy [32]. Nevertheless, one of the biggest threats to mankind is Cancer that causes death at any stage of its severity. Researches reveal that Deep Learning technologies like Convolutional Neural Network can be applied in cancer imaging that assists pathologist to detect and classify the disease at earlier stages, thus improving chances of the patient to survive, especially for lung, breast and thyroid cancer [33]. Contributions of AI in the healthcare paradigm as discussed above in detail are presented in Fig. 6.

Fig. 6 Contribution of AI in healthcare

4 Case Studies on Application of Data Science

In last few years, there is an exponential growth of fields like ecommerce, Internet of Things, healthcare, social media etc. which generates more than 2.5 quintillion bytes of data every day. Possible sources of data can be the purchase made though online shopping, various sensors installed in the smart cities, health information of human body etc. These huge data are termed as Big Data and processing those data to extract meaningful information is known as Data Science. Some of the potential applications has been shown in Fig. 7. In this section we have done few case studies on the application of Data Science in smart applications.

In [34], Authors has used clickstream data to predict the shopping behavior of ecommerce users. This can reduce the cost of digital marketing and increase the revenue. They have observed that supervised machine learning (SML) technique is not suitable for analysis of these kind clickstream data due its sequential structure. They have also applied a model based on Recurrent Neural Network (RNN) to the real world e-commerce data for campaign targeting. Experimental results show the comparison between the RNN based method and traditional SML based techniques. Based on the results they have developed an ensemble of sequence and conventional classifiers which outperforms in terms of accuracy.

One of the potential application of IoT is Smart home where energy consumption are monitored through some sensors. Authors [35] has proposed an energy management scheme which can decrease the cost of residential, commercial and industrial sectors still after meeting the energy demand. Some sensors have been in home appliances used to gather the energy consumption data and those data are sent centralized

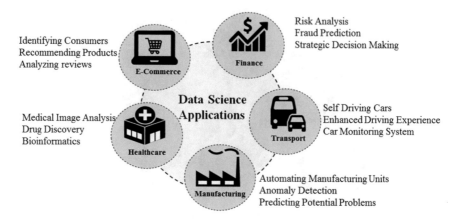

Fig. 7 Potential applications of data science

server. At the server end further processing and analysis has been done thorough Big Data analytics package. Since 60% of the energy demand is generated from Air conditioning in Arab Countries, author has considered managing the devices related with air conditioning. They has made a prototype and also deployed that in a small area for testing.

In Smart healthcare, wearable sensors generate enormous data, which need to be stored securely as well processed for knowledge extraction. An architecture has been proposed by authors [36] to store and process scalable health data for health care applications. Proposed architecture consists of two sub architecture namely Meta Fog-Redirection (MF-R) and Grouping and Choosing (GC) architecture. Where Apache Pig and Apache HBase has been used in MF-R architecture to efficiently collect and store health data. On the other hand GC is responsible for to secure the integration of Fog with cloud computing. Key management and categorization of data such as Normal, Critical and Sensitive is also ensured by GC architecture. Map Reduce based prediction model has been applied on the health data for heart disease prediction.

In Recent years, Diabetes diagnosis and personalized treatment has been proposed by Author [37] where vital data of patients suffering from Diabetes has been used. Body vitals such as blood glucose, temperature, ecg and blood oxygen has been collected using wearable sensors. 12,366 persons were involved in the experiment and 757,732 number health vitals were collected. By removing the irrelevant data they found 716,173 records of 9594 persons. Now they have identified that only 469 persons are suffering from Diabetes and rest of them are normal. Based on these dataset three machine learning algorithms has been used to establish different models for diagnosis of diabetes and to provide personalized treatment.

In [38], authors have proposed a method to detect urban emergency through crowdsourcing technique. Crowdsourcing is a technique of collection, integration and analysis of enormous data generated by various sources such as devices, vehicles, sensors, human etc. Due to growth in use of social media, it has become a big data store, authors has used geographic data and events if users to identify urban emergency event in a real time manner, their proposed model is called 5 W (What, Where, When, Who and Why).They have used a social media called "Weibo", which is similar to "Twitter" and popular in China. They have gathered the data based on 5 W for emergency event detection. Case studies on real dataset has been also conducted by them and results show that their proposed model obtained good performance and effective ness in terms of analysis and detection of such emergency events.

Fraud detection in finance sector is an important application of data science. Authors have introduced a Fraud Risk Management model for well-known organization Alibaba by analyzing Big Data [39]. They have prepared a fraud risk monitoring system based on real-time big data and processing models. It can identify frauds by analyzing huge volume of user behavior and network data. They have termed the system as AntBuckler which is responsible to identify and prevent all kind of malicious activity of users related to payment transactions. It uses RAIN score engine to determine the risk factor of a user and shows it through user friendly visualization.

When it comes Smart manufacturing, Data science has a big contribution in it. In [40], author has presented a system architecture for analyzing manufacturing process based on event logs. Well known big data toolbox Hadoop has been used for process mining. They have also prepared a prototype where an event log is need to supplied, and it can discover the manufacturing process and prepare animation depending on those discovered process model. Based on the activities it shows the working time of each activity, total execution time for each case.

5 Case Studies on Application of ML and DL

Machine Learning (ML) is a one of the technique of data analysis which is used for analytical model building. It is an application of Artificial Intelligence (AI) that makes a system learn automatically and improve without being programmed every time. Similarly Deep Learning (DL) is a branch of Machine Learning based on Artificial Neural Network. It is more capable of unsupervised learning where data is mostly unlabeled or unstructured. For ML based technique it requires structured data, so the main difference between ML and DL is the way to represent data in the system. Table 2 shows a small comparison between ML and DL, basic architecture has been shown in Fig. 8. In this section we have made few case studies on application of ML and DL.

Table 2 Machine learning versus deep learning

Characteristics	ML	DL
Data preprocessing	Required	Not required
Dataset size	Small	Large
Training time	Less	More
Hardware requirement	Simple (CPU)	High-end (GPU)

Machine Learning

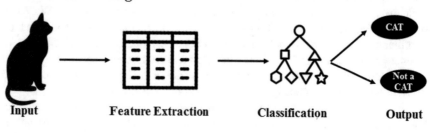

| Input | Feature Extraction | Classification | Output |

Deep Learning

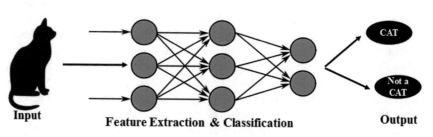

Input — Feature Extraction & Classification — Output

Fig. 8 Basic architecture of machine learning and deep learning

Machine Learning and Deep Learning has been widely adopted by various domains such as, Healthcare, Banking, Natural Language Processing, Information Retrieval etc.

One of the popular application of Machine Learning is Drug Discovery. In [41] authors has discussed about biological-structure-activity-relation (QSAR) modeling, which is a ML based drug discovery method. It is very much effective in identifying potential biological active modules from a lot of candidate compounds. Commercial Drugs such as CCT244747, PTC725, RG7800, and GDC-0941 has been discovered by ML technique in the year 2012,2014,2016,2015 respectively. In recent years drug discovery has been replaced by Deep learning methods. Since DL has ability of powerful and parallel computing using GPU, so it helps in accumulation of massive amount of biomedical data.

Another Drug Discovery based on chemo informatics has been proposed in the year 2018. Here authors in [42] has shown the use of compound database to extract the chemical features by characterizing them in chemical substructure fragments. Based on the absense or presense of substructure fragments. Chemical fingerprint has been created and finally QSAR approach has been used to train the machine learning model so the compound property can be predicted.

A framework for prediction of chronic disease such as heart failure, kidney failure and stroke has been proposed by Authors in [43] using Deep Learning Techniques. Electronic Health Record (EHR) information has been used in their work which contains free-text medical notes and structured information. This framework can accept negations and numerical values present in the text and it does not require any feature extraction based on disease. Authors has compared the performance of various deep learning architecture such as CNN, LSTM etc. Experimental results on cohort of 1 million patient shows that the models using text performs better than model using structured data.

Smart Grid is one of the potential application of Machine Learning. Authors [44] a hybrid technique to predict the electricity consumption of air conditioner for the next day. This forecast helps the power grids to efficiently balance the power demand. Their technique is a combination of linear autoregressive integrated moving average model and a nonlinear nature-inspired meta-heuristic model. They have collected Training and Testing data using various environmental sensors, infrared sensor and smart meters from an office. Experimental results shows that the hybrid model performs better than traditional conventional linear and nonlinear models in terms of prediction accuracy. Proposed model has achieved correlation factor R = 0.71 and Error rate = 4.8%.

Deep Learning algorithms has been used by authors [45] propose a Distributed Intelligent Video Surveillance system (DIVS). Proposed system uses a three layer architecture, first consists of monitoring cameras connected with edge servers, in second layer another edge servers are present, in the third layer cloud server are present which are connected the edge servers present in second layer. Proposed system uses parallel training, model synchronization and workload balancing. This parallel processing technique accelerates the video analysis process. CNN model has been used for vehicle classification and for traffic flow prediction LSTM has been applied. Both of this models are working on edge nodes which results parallel training.

Anomaly detection in network traffic is another important application of Machine Learning. In [46], authors proposed anomaly detection in wide area network meshes. Two machine learning algorithm, Boosted Decision Tree (BDT) and a simple feed forward neural network has been applied on network data. Authors has used PerfSO-NAR servers data which are collected using Open Science Grid. Objective of this work is increase the network performance by reducing anomalies, such as packet loss, throughput etc. Performance evaluation depicts that Boosted Decision Tree works better in terms of speed and accuracy in detection of anomalies.

Deep Learning approaches has been widely adopted in IoT, One of the recent work [47] shows the use Deep Learning technique for traffic load prediction and efficient channel allocation for IoT Application. Since IoT generates a huge volume of data, hence more transmission speed as an necessity. In conventional method fixed channels are used for data transmission which is ineffective now days. Authors has proposed an intelligent technique based on deep learning for Traffic Load (TL) Prediction system to forecast the future traffic load and congestion in the network. Also a Channel Allocation algorithm terms as DLPOCA which is based on Partially Overlapping Channel Assignment (POCA) has been described by them. They have three types of proposed traffic load prediction such as Central controlbased Traffic load Prediction (CTP), Semi-Central control Traffic load Prediction(S-CTP), and

Distributed control Traffic load Prediction (DTP). Deep Belief Architecture (DBA) and Deep Convolutional Neural Network (CNN) has be used for this purpose.

In [48], Authors has applied machine learning algorithms such as Decision Tree, naive Bayes and maximum entropy level for detection of radiology reports where follow-up is mentioned. This is one of the application of Natural Language Processing using feature engineering and machine learning technique. This application is very much effective for the patients who are from possible cancer and require follow-up. Most of the radiology reports contain follow-up information. A dataset containing 6000 free text reports has been used for training and testing, 1500 features has been determined using NLP. Experimental results show that decision tree achieves F1 score of 0.458 and accuracy of 0.862 which is better than other two algorithms.

6   Conclusion

In this literature a review on Internet of things, ML, DL and AI based smart healthcare ensuring improved human living specially for elderly and diseased people with support has been presented. For continuous monitoring of health vitals for regular health checkup purposes of elderly suffering from age related ailments without physically seeing a makes this IoT enabled smart applications usable and convenient for them. Health vitals such as blood pressure, blood sugar, heart rate, pulse, body temperature, ECG signal etc. can easily be transmitted to remote medical cloud server where from doctor can access the analysis report of health vitals processed using AI, ML and DL techniques or directly health vitals can be seen by the doctor if it is need-ed. Accordingly if any medicinal advice is sufficient then that easily be given to the patient or the relative through SMS or smartphone applications (APP). Otherwise if hospitalization is required that also can immediately be intimated from remote end. Health data analysis using ML, DL and AI has open up several revolutionary research wings such as disease, predictions, disease detection, drug discovery etc. Also smart health sensor based motion monitoring helps to identify activity or posture and fall event from remote end maintaining privacy, accordingly alert can be generated for ensuring timely support. While realizing all these features to live improved quality of life comes with a challenge to handle big health data efficiently taking care of issues like data redundancy, erroneous or insignificant data, accuracy of diagnosis or analysis by executing necessary processing like data integration, data selection, data cleaning, data transformations, and pattern evaluation applying AI with several steps of data mining. Several case studies on application of data science and machine learning helps to get practical systems and applications realized through Internet of things, data science and AI support.

References

1. Stonebraker, M., Hong, J.: Researchers' big data crisis; understanding design and functionality. Commun. ACM **55**(2), 10–11 (2012)
2. Kaisler, S., Armour, F., Espinosa, J.A., Money, W.: Big Data: issues and challenges moving forward. In: Proceedings of 46th International Conference on System Sciences, pp. 995–1004 (2013)
3. Jacobs, A.: Pathologies of Big Data. Commun. ACM **52**(8), 36–44 (2009)
4. Saif, S., Biswas, S., Chattopadhyay, S.: Intelligent, secure big health data management using deep learning and blockchain technology: an overview, in deep learning techniques for biomedical and health informatics. Studies Big Data **68**, 187–209 (2020)
5. Kaisler, S.: Advanced analytics. CATALYST Technical Report, i_SW Corporation, Arlington, VA (2012)
6. Ak, R., Fink, O., Zio, E.: Two machine learning approaches for short-term wind speed time-series prediction. IEEE Trans Neural Netw. Learn. Syst. **27**(8), 1734–1747
7. Zhang, Y., Huang, T., Bompard, E.F.: Big data analytics in smart grids: a review. Energy Inform. (2018)
8. Lytras, M.D., Chui, K.T., Visvizi, A.: Data analytics in smart healthcare: the recent developments and beyond. Appl. Sci. (2019)
9. Lytras, M.D., Visvizi, A.: Who uses smart city services and what to make of it: toward interdisciplinary smart cities research. Sustainability (2018)
10. Scheffler, R., Cometto, G., Tulenko, K., Bruckner, T., Liu, J., Keuffel, E.L., Preker, B., Stilwell, B., Brasileiro, J., Campbell, J.: Health workforce requirements for universal health coverage and the sustainable development goals. World Health Organization (2016)
11. Ijaz, M.F., Alfian, G., Syafrudin, M., Rhee, J.: Hybrid prediction model for type 2 diabetes and hypertension using DBSCAN-based outlier detection, synthetic minority over sampling technique (SMOTE), and random forest. Appl. Sci. (2018)
12. Wang, J., Wang, C., Zhang, W.: Data analysis and forecasting of tuberculosis prevalence rates for smart healthcare based on a novel combination model. Appl. Sci. (2018)
13. Ünver, H.M., Kökver, Y., Duman, E., Erdem, O.A.: Statistical edge detection and circular hough transform for optic disk localization. Appl. Sci. (2019)
14. Polat, H., Mehr, H.D.: Classification of pulmonary CT Images by using hybrid 3D-deep convolutional neural network architecture. Appl. Sci. (2019)
15. Rew, J., Choi, Y.H., Kim, H., Hwang, E.: Skin aging estimation scheme based on lifestyle and dermoscopy image analysis. Appl. Sci. (2019)
16. Chui, K.T., Lytras, M.D.: A novel MOGA-SVM multinomial classification for organ inflammation detection. Appl. Sci. (2019)
17. Moustaka, V., Vakali, A., Anthopoulos, L.G.: A systematic review for smart city data analytics. ACM Comput. Surv. **51**(5), 103–0143 (2018)
18. Nuaimi, E.A., Neyadi, H.A., Mohamed, N., Al-Jaroodi, J.: Applications of big data to smart cities. J. Internet Serv. Appl. 6–30 (2015)
19. Shapiro, S.C.: Encyclopedia of Artificial Intelligence, 2nd edn. Wiley, New York (1992)
20. Reddy, S.: Use of artificial intelligence in healthcare delivery. In: eHealth-Making Health Care Smarter. IntechOpen (2018)
21. Jakhar, D., Kaur, I.: Artificial intelligence, machine learning and deep learning: definitions and differences. Clin. Exp. Dermatol. **45**(1), 131–132 (2020)
22. Cullell-Dalmau, M., Otero-Viñas, M., Manzo, C.: Research techniques made simple: deep learning for the classification of dermatological images. J. Invest. Dermatol. **140**(3), 507–514 (2020)
23. Liu, Y., Gadepalli, K., Norouzi, M., Dahl, G.E., Kohlberger, T., Boyko, A., Venugopalan, S., et al.: Detecting cancer metastases on gigapixel pathology images. arXiv preprint arXiv:1703. 02442 (2017)

24. Rajpurkar, P., Irvin, J., Zhu, K., Yang, B., Mehta, H., Duan, T., Ding, D., et al.: Chexnet: radiologist-level pneumonia detection on chest x-rays with deep learning. arXiv preprint arXiv: 1711.05225 (2017)

25. Ismael, S.A.A., Mohammed, A., Hefny, H.: An enhanced deep learning approach for brain cancer MRI images classification using residual net-works. Artif. Intell. Med. **102** (2020)

26. Cheon, S., Kim, J., Lim, J.: The use of deep learning to predict stroke patient mortality. Int. J. Environ. Res. Pub. Health **16**(11) (2019)

27. Bello, G.A., Dawes, T.J.W., Duan, J., Biffi, C., De Marvao, A., Howard, L.S.G.E., Simon, J., Gibbs, R., et al.: Deep-learning cardiac motion analysis for human survival prediction. Nat. Mach. Intell. **1**(2), 95–104 (2019)

28. Diller, G.-P., Kempny, A., Babu-Narayan, S.V., Henrichs, M., Brida, M., Uebing, A., Lammers, A.E., et al.: Machine learning algorithms estimating prognosis and guiding therapy in adult congenital heart disease: data from a single tertiary centre including 10 019 patients. Eur. Heart J. **40**(13), 1069–1077 (2019)

29. Saha, J., Chowdhury, C., Roy Chowdhury, I., Biswas, S., Aslam, N.: An ensemble of condition based classifiers for device independent detailed human activity recognition using smartphones. Inf. MDPI **9**(4), 94, 1–22 (2018)

30. Belić, M., Bobić, V., Badža, M., Šolaja, N., Đurić-Jovičić, M., Kostić, V.S.: Artificial intelligence for assisting diagnostics and assessment of Parkinson's disease–a review. Clin. Neurol. Neurosurg. (2019)

31. Jo, T., Nho, K., Saykin, A.J.: Deep learning in Alzheimer's disease: diagnostic classification and prognostic prediction using neuroimaging data. Front. Aging Neurosci. **11** (2019)

32. Makino, M., Yoshimoto, R., Ono, M., Itoko, T., Katsuki, T., Koseki, A., Kudo, M., et al.: Artificial intelligence predicts the progression of diabetic kidney disease using big data machine learning. Sci. Rep. **9**(1), 1–9 (2019)

33. Coccia, M.: Deep learning technology for improving cancer care in society: new directions in cancer imaging driven by artificial intelligence. Technol. Soc. **60** (2020)

34. Koehn, D., Lessmann, S., Schaal, M.: Predicting online shopping behaviour from clickstream data using deep learning. Expert Syst. Appl. **150** (2020)

35. ALi, A.R.Al., Zualkernan, I.A., Rashid, M., Gupta, R., Alikarar, M.: A smart home energy management system using IoT and big data analytics approach. IEEE Trans. Consumer Electron. **63**(4), 426–434 (2017)

36. Manogaran, G., Varatharajan, R., Lopez, D., Kumar, P.M., Sundarasekar, R., Thota, C.: A new architecture of Internet of Things and big data ecosystem for secured smart healthcare monitoring and alerting. Future Gener. Comput. Syst. 375–387 (2017)

37. Chen, M., Yang, J., Zhou, J., Hao, Y., Zhang, J., Youn, C.: 5G-smart diabetes: toward personalized diabetes diagnosis with healthcare Big Data clouds. IEEE Commun. Mag. **56**(4), 16–23 (2018)

38. Xu, Z., Liu, Y., Yen, N., Mei, L., Luo, X., Wei, X., Hu, C.: Crowdsourcing based description of urban emergency events using social media big data. IEEE Trans. Cloud Comput. (2016)

39. Chen, J., Tao, Y., Wang, H., Chen, T.: Big data based fraud risk management at Alibaba. J. Finance Data Sci. **1**(1), 1–10 (2015)

40. Yang, H., Park, M., Cho, M., Song, M., Kim, S.: A system architecture for manufacturing process analysis based on big data and process mining techniques. In: IEEE International Conference on Big Data (Big Data). Washington, DC, pp. 1024–1029 (2014)

41. Zhang, L., Tan, J., Han, D., Zhu, H.: From machine learning to deep learning: progress in machine intelligence for rational drug discovery. Drug Discov. Today **22**(11), 1680–1685 (2017)

42. Lo, Y.-C., Rensi, S.E., Torng, W., Altman, R.B.: Machine learning in chemoinformatics and drug discovery. Drug Discov. Today **23**(8), 1538–1546 (2018)

43. Liu, J., Zhang, Z., Razavian, N.: Deep EHR: chronic disease prediction using medical notes. arXiv:1808.04928, https://arxiv.org/abs/1808.04928

44. Chou, J., Hsu, S., Ngo, N., Lin, C., Tsui, C.: Hybrid machine learning system to forecast electricity consumption of smart grid-based air conditioners. IEEE Syst. J. **13**(3), 3120–3128 (2019)

45. Chen, J., Li, K., Deng, Q., Li, K., Yu, P.S.: Distributed deep learning model for intelligent video surveillance systems with edge computing. IEEE Trans. Ind. Inform. (2019)
46. Zhang, J., Gardner, R., Vukotic, I.: Anomaly detection in wide area network meshes using two machine learning algorithms. Future Gener. Comput. Syst. **93**, 418–426 (2019)
47. Tang, F., Fadlullah, Z.M., Mao, B., Kato, N.: An intelligent traffic load prediction-based adaptive channel assignment algorithm in SDN-IoT: a deep learning approach. IEEE Internet Things J. **5**(6), 5141–5154 (2018)
48. Lou, R., Lalevic, D., Chambers, C., Zafar, H.M., Cook, T.S.: Automated detection of radiology reports that require follow-up imaging using natural language processing feature engineering and machine learning classification. J. Digit Imaging **33**, 131–136 (2020)

IoT Sensor Data Analysis and Fusion Applying Machine Learning and Meta-Heuristic Approaches

Anindita Saha, Chandreyee Chowdhury, Mayurakshi Jana, and Suparna Biswas

1 Introduction

With the incorporation and advancement of pervasive and ubiquitous computing, a continuous improvement in existing technology is exhibited worldwide. In the last decade, due to rapid progress in computing as well as communication techniques, Machine Learning (ML) has drawn worldwide attention in a widespread variety of applications, all across the globe. ML is a sub-category of Artificial Intelligence (AI) that encompasses several algorithms which learn from different datasets prepared on the basis of past experiments and can make predictions for future benefits. To enhance the power of ML, Meta-heuristics approaches have been quite effective as proved in literature. Meta-heuristics are high level, general purpose, problem independent, and heterogeneous algorithmic framework, that are useful to solve similar or dissimilar but complex optimization problems. They provide a near optimal solution within a reasonable computing timeframe. Recently, the trend of living is changing and Internet of Things (IoT) plays an important role in urbanizing several applications.

A. Saha
Department of Information Technology, Techno Main Salt Lake, Kolkata, West Bengal, India
e-mail: anindita_saha03@yahoo.co.in

C. Chowdhury
Computer Science and Engineering, Jadavpur University, Kolkata, India
e-mail: chandreyee.chowdhury@gmail.com

M. Jana
Department of Computer Science, Bijay Krishna Girls' College, Howrah, India
e-mail: mjcompsc@gmail.com

S. Biswas (✉)
Computer Science and Engineering, Maulana Abul Kalam Azad University of Technology, Kolkata, West Bengal, India
e-mail: mailtosuparna@gmail.com

© The Editor(s) (if applicable) and The Author(s), under exclusive license to Springer Nature Switzerland AG 2021
A.-E. Hassanien et al. (eds.), *Enabling AI Applications in Data Science*, Studies in Computational Intelligence 911, https://doi.org/10.1007/978-3-030-52067-0_20

441

By definition, (IoT) is the network of physical objects that contain embedded technology to communicate and sense or interact with their internal states or the external environment[1]. IoT generates massive amount of heterogeneous data, and the collection, processing, and storage of such big volume of data is cumbersome for humans. Further, tasks like proper classification of data collected by IoT sensors, defining accurate patterns & behaviors, processing them for assessment and prediction, and finally sending it back to the device for decision making-all these need additional mechanisms like ML to provide embedded intelligence to IoT. ML encompasses a wide range of applications like computer vision, bio-informatics, fraud detection, IDS, face recognition, speech recognition and many more. In this chapter, several smart application domains that require IoT data analysis are discussed, followed by influence of ML in such smart applications. Further, it is discussed how data analysis of IoT sensors can be improved by applying meta-heuristics approaches, and subsequently, a detailed analysis of the overall effect of ML and Meta-Heuristics on smart applications is presented for ready reference.

2 The Architecture of IoT

IoT applications mostly follow a 3-tier architecture (as shown in Fig. 1) where the first layer is the end user layer consisting of the sensors carried by people or placed at certain dedicated places. Smartphone is a potential source of IoT data. For smart home applications, consumer electronic devices with embedded sensors are also potential

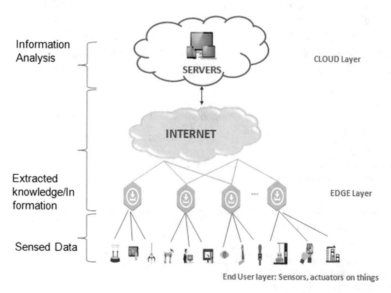

Fig. 1 An architecture of IoT enabling sensing data analysis

data sources along with the Smartphones. The sensed data from the end users are sent to an edge device for transmitting it to cloud servers over the Internet. The edge devices can also perform certain computations, such as, data preprocessing and extracting knowledge from data so that the cloud servers only receive the extracted knowledge and not the actual data. In this way, not only the Internet traffic gets reduced but also data privacy over the Internet could be preserved. Cloud servers actually perform data analysis to find meaningful patterns and accordingly provide services back to the end user layer.

3 IoT Application Domains Requiring Data Analysis

IoT as a technology can be applied to a wide variety of heterogeneous domains, which in future, are going to be the crucial drivers of innovations and investments for the world. The major areas revolutionized by IoT are:

- Smart City that encompasses several components as application domains for IoT like Smart Parking, Smart Traffic Management, Smart Environment, Smart Waste Management
- Smart Home
- Smart Healthcare
- Smart Agriculture
- Smart Energy
- Smart Industry
- Smart Business

This is summarized in Fig. 2. The integration of IoT devices generating a huge amount of real time data and processing it for data analysis is a "Smart" step towards developing a **Smart City**.

A **Smart City** incorporates ICT and IoT to manage resources and assets effectively to enhance the performance and quality of urban services provided to the citizens. IoT can be an enabling technology for smart cities where all its components can be supervised by implanting sensors, and actuators that can collect data in huge

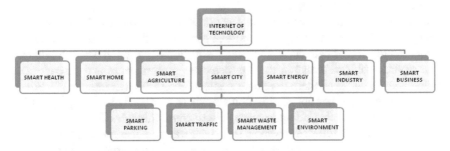

Fig. 2 IoT application domains requiring data analysis

volumes (millions of terabytes), called Big Data, that has varied characteristics and velocities. It can be audio, video, images, logs, posts, emails, health records, social networking interactions and many more. The various steps of data analysis can be to capture, store, manage, analyze and visualize these data in massive volumes, which will efficiently and effectively support decision making process in present and future.

Several components of a smart city need data analysis for separate reasons. For example, in a **Smart Parking** system, which are real time solutions to detect empty parking spaces in a geographical area with the help of sensors and cameras and place vehicles at the right position, data analysis can inform a driver about the nearest and most suitable parking slot for his car. In **Smart Traffic**, proper management of traffic signals on road, with help of IoT sensors, an efficient real time analysis can be beneficial for both the citizens as well as government. Traffic information received from these sensors can provide useful information like number of vehicles on road, current location and the distance between two vehicles at any instance, of time. Further in case of an accident, sensors can send timely alerts to the police as well as the doctors simultaneously. Smart cities must ensure a **Smart Environment** which refers to an ecosystem that has embedded sensors, actuators and other computational devices to monitor and control environmental factors like air pollution, noise, effective water supply, planning green areas, and many more. So environmental IoT sensors should be able to inform about several environmental anomalies like increment of toxic gases in air, excess noise in environment, or even the proper balance of natural gases like O_2, CO, CO_2, SO_2 etc. **Smart Waste** management is another important component of a smart city where sensors are placed inside waste bins all across the city, to measure filling levels, and provide notification when they need to be emptied. IoT can optimize the data driven collection services by analyzing the historically collected data to identify the filling patterns, especially in densely populated area of the city. IoT can also optimize driver routes, garbage pick up schedules, and this significantly reduce operational costs.

Other than Smart city, data analysis is utmost essential for other application domains like Smart Home and Smart Healthcare. **Smart home** refers to a home where appliances (lights, fans, fridge, TV) and devices are controlled automatically and remotely through Internet connections, via smart-phones or other networked devices. The home may be monitored on a continuous basis through several sensors like smoke detectors for file alarms, and temperature detectors for extreme weathers. Sensors can also inform about power electricity consumptions and gas usage, water consumption etc. for a trouble free household. **Smart Healthcare** defines the technology and devices (sensors, actuators) that provides better tools for diagnosis and better treatments for patients as a whole. The technology used in Smart Healthcare can record health information through deployed sensors, store and process them automatically and send these to remote medical servers through Internet [2]. **Smart Agriculture** is another application domain where data analytics can play a huge role. It is a concept that refers to manage agriculture through ICT, sensors, which can collect valuable information about soil, light, water, temperature, humidity of the farm and many more, thus minimizing or eliminating human labor needed to process these.

IoT thus provides the communication infrastructure to interconnect every sensor of all application domains mentioned, and also helps in local or remote data procuring, in cloud storage, data access for analysis and decision making, data visualization, user interfacing and operation automation.

4 Learning in the Cloud Versus Edge Learning

Machine Learning and Deep Learning algorithms have gained immense popularity in various applications such as smart healthcare, computer vision, surveillance natural language processing etc. These techniques are resource hungry specially deep learning algorithms demanding higher computational power, memory and battery power requirements etc., due to algorithmic complexity. Also it demands high level of resources for training deep learning models to get trained using huge dataset created by edge devices such as smartphone sensors or Internet of things enabled sensors. Now for learning at cloud, these huge dataset needs to be transferred to cloud server through internet which is vulnerable to security attacks and threats which may violate security and privacy of sensitive data such as patients' health vitals. Also long distance transmission causes latency between data acquisition to data processing and consumes higher bandwidth, energy etc. Along with that one more vital point gets added in this era of Internet of things is scalability issue which can better be addressed by computing learning algorithms through edge computing [3, 4] by exploiting hierarchical structure formed by edge devices edge compute node such as local server and cloud server at distance. This also helps to avoid single point failure due to central cloud server failure issue. So in short, advantages learning at advanced edge devices equipped with high volume of resources e.g. memory, processing ability, energy source, IoT compatibility etc. are: (i) Localized learning hence low latency, (ii) Data gets processed for training learning models close to point of origin in learning at edge hence security attacks and threats towards confidentiality, integrity, authentication, data privacy while getting transmitted over the open internet gets much reduced. (iii) Correctness/accuracy in learning at edge devices: Learning using acquired dataset at edge devices reduces probability of erroneous data due to illegal modifications by security attacks and threats at open wireless channel during transmission from edge devices to cloud server. Thus learning using correct data ensures proper and correct knowledge building.

In spite of a number of obvious advantages of edge computing, it is only not being used rather learning at cloud server or a combination of edge devices and cloud server (as shown in Fig. 3) are preferred due to several reasons which are being reported here. Edge devices such as sensor nodes with computation ability, memory and battery power, smartphone, tablet computer, laptop have limited resources. Machine and deep learning models need extensive training using huge dataset (70–80% of total dataset) for knowledge building. These training algorithms are resource hungry. Also trained models need to be tested using 20%–30% of total dataset which is also not of very low volume. Sometimes to execute such volume of computations in resource

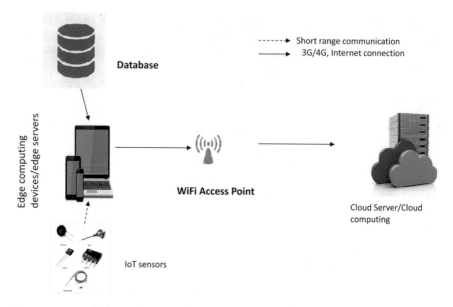

Fig. 3 Computing at edge devices or cloud server

constrained edge devices may not be possible at all or may be possible for few times. Besides this, some applications such as smart healthcare or smart transportation etc. are time critical hence latency tolerance is very low. Such specialized applications may be based on edge computing.

Deep learning neural network models are prevalent so edge computing. Cloud server execution of heavyweight deep learning and machine learning models using high volume of Electronic health record (EHR) has proved to be convenient in terms of resource consumption and rigorous learning supported by as complex algorithms can be repeatedly executed to find optimum result with as many rounds or hidden layers or data set to avoid problems of overfitting or underfitting so that higher accuracy can be achieved. Deep learning models execution in edge computing needs comparative performance evaluation by computing models on heterogenous hardware devices such as including the simple devices (e.g., Raspberry Pi), smartphones, home gateways, and edge servers. Much of the current work has focused on either on powerful servers or on smartphones [5].

5 Review of ML Approaches for IoT Sensor Data Analysis

In past few years Internet of Things (IoT) has rapidly developed for people's comfort. The IoT basically provides the connection between various physical devices, hardware and software through internet and the main purpose is exchange of data among these devices. The IoT enhances the standard of people's living like smart cities have so many opportunities- to predict traffic jams, reduces congestion and pollution, detection of traffic violations, searching of suitable parking space etc.; smart home services improving the safety measures, connection between the home appliances; healthcare- continuous monitoring and reporting, assurance of real time medical data delivery, disease prediction and identification, remote medical assistance. For efficient processing and intelligent analysis of IoT sensor data so that evitable information extraction can be done towards monitoring, support, disease identification, prediction and diagnosis. Besides, smart healthcare, there are several other smart applications are there in which enabling technologies such as sensor, IoT, cloud etc. are there for data acquisition, transmission and storage. But to handle huge data intelligent ML algorithms are being applied. Followings are some of the ML applications in handling IoT sensor data in several smart applications.

Smart Healthcare: Smart Healthcare is one of the most important paradigms under IoT. Now a days, it is not only limited to patients who are getting admitted to hospitals but also for those patients who are taken care at their homes or any care-giving centers. Hence we have to design a framework to give equal importance to all patients. IoT together with cloud computing facility give us an opportunity to provide a patient-centric environment where high quality treatment can be given in low cost budget. In general, the IoT architecture consists of sensor devices to collect real time data of a patient, a suitable transmission medium (e.g. Wi-Fi, Bluetooth etc.) to transmit data in real-time environment, cloud platform which is a large storage area with complex computing capacity, various machine learning approaches are used to predict and detect diseases, the front-end user-friendly accessibility where a doctor can easily monitor a lot of patients at the same time. To get the result with high accuracy IoT-cloud computing integration should be done effectively.

There are several types of sensor devices used in IoT environment for monitoring patients' activities. It may be temperature sensor, pressure sensor, blood-oxygen level measurement sensor, blood-glucose level measurement sensor, accelerometer, Electrocardiogram (ECG), Electroencephalogram (EEG), altimeter, GPS location tracker etc. They can either work individually or as a combination of sensors. In case of heart related problem the ECG signal is used to detect it. In [6], authors represent how ECG signals with SVM and other ML classifiers are used for arrhythmic beat classification, as shown in Fig. 4. Generally two common heart diseases- Myocardial ischaemia and Cardiac arrhythmias are reported which can be identified using electrocardiogram (ECG) based analysis. The collected ECG signals from MIT-BIH databases are preprocessed using ML based filters i.e. LMS algorithm. The delayed LMS algorithm is used to calculate updated weights (Weight update equation: $\mathbf{w(n + 1) = w(n) + \mu x(n\text{-}kD)e(n\text{-}kD)}$ where, n, w(n), w(n + 1) denote time step, old weight, updated weight respectively. M is the step size, x (n) is the filter input and e (n) is the error signal.), error calculation $\mathbf{(e(n\text{-}kD) = d(n\text{-}kD) - y(n\text{-}kD))}$ and output estimation $\mathbf{(y(n\text{-}kD) = w^T(n\text{-}kD)\ x(n\text{-}kD))}$ kD defines number of delays used in the pipelining stage.) in pipelined form. Finally a Normalized LMS (NLMS) algorithm is used for coefficient update equation $\mathbf{(w(n + 1) = w(n) + \mu_n x(n)\ e(n))}$ where, μ_n

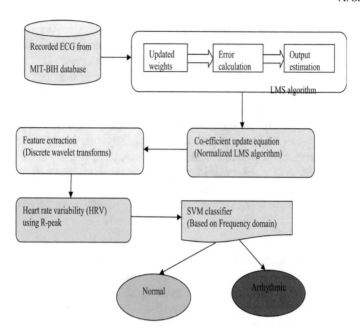

Fig. 4 Architecture of SVM classifier based abnormality detection

is the normalized step size which is defined **as $\mu_n = [\mu/p + x^t(n)\, x(n)]$**. After noise removal ECG signal is used for feature extraction using Discrete Wavelet Transform (DWT) technique. Authors have chosen coiflet wavelet for extraction of R-peak. This R-peak is based heart rate function to calculate RR interval i.e. continuous monitoring heart rate within a minute; practically it is ±1 ms. This R-peak is used to determine heart rate variability (HRV). This extracted HRV features can be represented in time domain or frequency domain. Authors used 14 time domains and frequency domain HRV features. The support vector machine (SVM) is used to classify the extracted HRV feature based on frequency domain (Fig. 4). Three frequency bands (VLF, LF, and HF) and frequency ratio (LF/HF) are used. Total 200 HRV data is classified among them 180 (90%) is training data and 20 (10%) is test data set. The SVM classifier gives 96% accurate results which is higher than any other ML technique. Therefore, on SVM classifier HRV data can be easily classified as normal or arrhythmic.

Generally, machine learning techniques are categorized into two types- 1. Supervised learning, 2. Unsupervised learning. In Supervised learning, a set of person's statistical data that is height, weight, smoking status, lifestyle disease (e.g. diabetes) is successfully trained to a model. The dataset may be discrete or continuous. For discrete category, the model should answer either positive or negative based on probability. Any probability greater than threshold value is classified as a positive. For continuous value, regression algorithm is used to predict a person's life expectancy or dose of chemotherapy. In unsupervised learning, datasets are analyzed and grouped into clusters which represent similar types of data points. Other uncommon data are just excluded. Here authors follow supervised learning and develop a model using R statistical programming environment. They use "Breast cancer Wisconsin Diagnostic Dataset" which is available from University of California Irvine (UCI). It is a clinical dataset of the fine-needle aspiration (FNA) of cell nuclei taken from Breast masses. The features of FNA samples which are available as digitize image are extracted using linear programming, pattern separation or Decision tree construction etc. the collected FNA sample is divided into Malignant and benign depending on FNA. Furthermore, the classification is done with respect to features of the dataset. Now the machine learning approach is used on the dataset. The downloaded dataset from UCI repository are represented as a matrix and missing data are removed. The 67% of whole dataset is used as training and 33% is used for evaluation. Various machine learning algorithms are used for training. The first one generalized linear model (GLM) is used to reduce features because the number of features is greater than number of instances. Generally the features are selected using (LASSO). Now the Support vector machine (SVM) algorithm is used to separate two classes using a hyper plane. Artificial neural network (ANN) takes features in the dataset as input and passes it into hidden layers before presenting the final decision at output layer. For error minimization ANN follows the following equation to reduce error

$$Y = \text{activation} \left(\sum (\text{weight} \times \text{input}) + \text{bias} \right)$$

Also Deep Neural Network is used for high dimensional data like image recognition. The ML algorithms are classified into three performance parameters

Sensitivity = true positives/true negatives
Specificity = true negatives/actual negatives
Accuracy = (true positives + true negatives)/total predictions

A confusion matrix is represented using SVM and is compared to true classifications. The result will generate ether positive or negative based on threshold value. Now, new data is applied to the trained model with various parameters like thickness, cell size etc. the trained model returns predictions nearly accurate.

The increasing number of neurological disorder reaches to a significant number where approx. 700 million people in the world currently suffering from Epilepsy. The Electroencephalograms (EEG) signals are used not only to determine this neurological disorder but also it is reused to study patient's clinical history for disease diagnosis. In [7], authors present "Differentiation between Normal and Epileptic EEG using k-Nearest-Neighbors technique". This paper shows the classification of EEG signals are either normal or epileptic in nature using k-nearest-neighbor (kNN) algorithm. International 10–20 system electrodes are used to collect EEG signals. The EEG database is used from http://www.oracle.com/technetwork/java/index.html. The collected data is sampled into five sets with respect to some physical activities i.e.

A: Healthy person's recordings with eyes open;

B: Healthy person's recordings with eyes closed;

C: Recordings of the hippocampal formation of the opposite hemisphere of the brain from patients with epilepsy;

D: Epileptogenic zone recordings from patients with epilepsy;

E: Seizure activity recordings, which were selected from all recording sites showing ictal activity from patients with epilepsy.

Now, cross-correlation (CC) method is used to extract features from these EEG segments. The features are Peak value (PV $=$ max (CCo)), Root mean squared (RMS $= 0.707*$PV), Centroid (CE $= \frac{\sum_{i=-n}^{n} i*CC0(i)}{\sum_{i=-n}^{n} CC0(i)}$), Equivalent width (EW $= \frac{\sum_{i=-n}^{n} CC0(i)}{PV}$), Mean square abscissa (MSA $= \frac{\sum_{i=-n}^{n} i^2*CC0(i)}{\sum_{i=-n}^{n} CC0(i)}$) analyzed for both healthy and epileptic segments. The extracted features are classified into two classes- normal and abnormal using kNN algorithm. The kNN method usually examines a new example with training set. For this purpose, Euclidean distance is used to measure the similarity. Generally, kNN method classifies a new example with predominant class in its neighborhood. Also this algorithm is easy to implement and understanding of extracted features are less time consuming. The cross-validation (CV) method is used for performance evaluation. The data is divided into k-equal sized samples where kth sample is test set and k-1th samples are training set. After that statistical measures (average error and Standard deviation) are used. Now, contingency table (CT) method is used to verify the examples belonging in the same class or not. The attributes are used for calculation CT are-Negative Predictive Value (NPV), Positive Predictive Value (PPV), Specificity, Sensitivity. The result shows that k $=$ 1 is the highest value for NPV (91.18%) and Sensitivity (90.91%) and k $=$ 7 is the highest value for PPV (98.88%) and Specificity (99.00%) respectively. Therefore, k $=$ 1 is good for epileptic EEG detection and k $=$ 7 is good for healthy EEG detection. The authors also indicate the future work can be done using other ML techniques and more number of classes for epilepsy or other brain diseases diagnosis.

In Parkinson's disease (PD) physical disorders get developed due to the damaging of dopaminergic neurons in Substantia Nigra (mid brain) [8]. They studied the requirements of EUH-2020 project PD manager to develop a patient-centric platform. Applying data mining techniques Parkinson's disease can be diagnosed, analysed, predicted etc. The significant indication is tremor assessment where postural action, handwriting, drawing etc. A recent study presents that six body sensors are used to collect real time raw data. This data is classified into two groups- one set for posture recognition and another one for tremor detection and quantification. Two Hidden Markov Models (HMM) is used for this purpose and high accuracy results are obtained. Another method is Gait Analysis. Various researchers show that different ML techniques are used to distinctly identify normal and PD gait pattern. Freezing of Gait Detection (FoG) is another important indication of PD. There are some computational methods developed for this purpose to analyze foot pressure, motion signal using accelerometers an gyroscope sensors, electromyography signals, 3D motion etc. Different ML techniques such as Naïve Bayes, Random Forest, Decision tree, and Random tree are used. The accuracy for this system is 96.11%. The vocal impairments of PD patient are analyzed using ML technique like Neural Network, DMneural, Regression and Decision Tree where Neural network gives the accuracy of 92.9%. Also the vocal features are extracted to discriminate between healthy people with PD patient. Therefore all methods are combined together to predict PD with accuracy ranging from 57.1 to 77.4%. So there is a chance to improve prediction accuracy. Here also particular dataset is used with respect to time and all ML techniques are used on it. But if there is a chance to collect more datasets coming from different sensor devices and fuse those heterogeneous data sets. Meta machine learning can be used to identify which ML algorithm is appropriate for a particular context. Hyper heuristics method can be applicable to identify a compatible ML algorithm.

Sometimes heterogeneous sensor devices are used to collect complex signals generated from human body. Smartphones have been extensively used for pervasive computing especially for Human Activity Recognition, as discussed in [9], where a two phase HAR framework has been proposed to identify both dynamic and static activities, through feature selection and ensemble classifiers. Alhussein et al. [10] proposes the framework that takes input from the sensors or smart devices and gives output on a patient's state based on Deep learning algorithm. The framework consists of various types of sensors like Accelerometer, ECG, EEG, GPS location, Altimeter, Thermometer etc. The collected data is sent to the smart devices that is Smartphone or laptop via Local Area Network. These smart devices transmit data to the cloud via 4G, 5G, Wi-Fi. Further this data is analyzed for seizure detection and detailed analysis is sent to doctors. The authors used the freely available CHB-MIT dataset foe EEG epileptic seizure detection collected from Children's Hospital, Boston. It contains dataset of 23 epilepsy patients and total 686 EEG recordings are collected, out of which 198 contain one or more seizures. Deep CNN algorithm is used to extract features from raw data and higher-level features are represented as combination of lower level features. In CNN, convolution layer consists of filters or kernels which takes 2D input and gives output as feature maps. Now, a three-layer neural network or auto-encoder is used to train the input which is mapped to the hidden layer as well as the output of hidden layer is mapped to the output layer. This is known as unsupervised learning algorithm. Authors used cropped training technique to train the input data but there are chances to over fit. Therefore the results of this architecture are

Average value of sensitivity = 93%
Average value of recognition accuracy = 99.2%
Overall accuracy of Deep CNN-stacked auto-encoder = 99.5%

Smart Transportation: With increasing number of vehicles in the road, it is very difficult to identify a route which is less time consuming and cost-effective. In [11] the authors show that advancement of IoT, video cameras and radio frequency identification devices are used to collect traffic data. These collected data are transferred through wired media like optical fiber or wireless media like 3G/4G, Wi-Fi etc. AI based techniques are used to analyze these data, for improving individual safety, comfort, convenience as well as emission in connected vehicles. To provide sustainable transportation authors of [12] divides the smart transportation problem into layers- physical layer, communication layer, operation layer, service layer. Physical layer collects the data from roadside sensors, onboard sensors, social media etc. Operation layer is used to collect and translate data into information and knowledge using vision analysis, data fusion and data mining technique. The result of operation layer is presented to the user as the form of services which is presented in the service layer.

Smart Parking: With increase of urban population and traffic congestion it is very difficult to find a suitable place for parking. Many researchers have worked on it to remove unnecessary parking on the roads and to create a better urban planning. The researchers of [13] show that crowdsensing and gaparking is used to detect parking applicants. Then various softwares are used to analyze the collected data. Some smart parking city projects are PARK Smart (New York City), Park Right (London), MOSCOW Fastprk (Moscow), Berlin Pilot (Berlin) etc. After analyzing the collected data parking vacancy prediction is made. Researchers of [14] propose mobile on-street parking space detection. Ultrasonic range finder sensors are implemented on the vehicles to measure the distance between vehicles and roadside. A supervised learning algorithm is designed to map the parked vehicle's position together with GPS to the central server. This is how the occupied parking is created. Similarly it is used for unoccupied parking and space availability information can be accessed by the users through web portal or mobile App. The backbone data transmission can be 3G/4G and the mobile App or web-portal consists of Linux, Apache, MySQL and PHP.

6 Review of Meta-Heuristic Approaches to IoT Sensor Data Analysis

6.1 Introduction to Meta-Heuristics

At present, adapting a meta-heuristic approach for IoT sensor data analysis can be a smart move for future. To solve optimization problems like Travelling Salesman problem, where the best solution needs to be extracted from all feasible solutions, heuristic or meta-heuristic approaches are more suitable than linear or non-linear programming. Heuristic algorithms look for acceptable and reasonable solutions but it is quite subjective and depends on the type of optimization problem [15].

The term "meta" in meta-heuristics refers to a higher level heuristic approach that explores and exploits a larger search space, performing multiple iterations on the subordinate heuristic and finding a near optimal solution in lesser time. These iterations evolve multiple solutions and continue until solutions close to the predefined converging criteria is found. This newly found near optimal solution is accepted and system is considered to have reached to a converged state.

In most cases, heuristic approaches don't guarantee the best or global optima of a problem. For example in hill climbing method, while going uphill, the best functional value often gets stuck in the local optima, as the search space is limited and movement is influenced by the initial point. Meta-heuristics techniques like simulated annealing solve this problem by expanding the search space, with possible movement in both uphill and downhill from any random starting point. Thus getting trapped in local optima can be restricted in meta-heuristics which widens the horizon of feasible solution to an optimization problem.

There are several meta-heuristic algorithms like genetic algorithms, evolutional algorithms, random optimizations, local search, reactive search, greedy algorithm, hill-climbing, best-first search, simulated annealing, ant colony optimization, stochastic diffusion search, harmony search, variable neighborhood search and many more, which are widely used by researchers to solve optimization problems. Out of these, the most popular ones are as follows, which are mainly inspired by nature or social behavior:

Genetic Algorithm: This algorithm is a classic choice for problems which are complex, non-linear, stochastic, and non-differentiable. Genetic algorithm proves to be a good optimization technique which iterates across different stages as selection of individuals, and applies genetic operators like crossover and mutation to obtain an optimal solution. Based on the age old theory of the survival of the fittest, it simply uses natural selection as its base. A set of individuals known as initial population is chosen where each such individual is characterized by a set of parameters or variables called genes. These genes are joined to form a string or chromosomes which are a potential solution to the optimization problem and are binary encoded. A fitness function is applied on a set of chromosomes within the population and thus the fitness of each solution is calculated. The survival or selection of an individual to the next level of iteration is purely determined by the fitness value of the same and will be known as parents. The individuals that possess the least fitness value will be eliminated from the population. Genetic operators will work on these parents and thus good characteristics of the parents will be propagated to the next generation (off-springs).

Particle Swarm Optimization: Similar to Genetic Algorithm, Particle Swarm Optimization is another popular technique to solve complex optimization problems that considers a population of individuals or swarm, rather than a single individual. Every member in the swarm is a particle, with a predefined position and velocity that modifies at each step of the algorithm, in the search space, based on its own best previous location (p-best) as well as best previous location of the entire swarm (g-best).

Ant Colony Optimization: Ant Colony Optimization is another population based meta-heuristic approach where ants (software agents) find the best solution to an optimization problem which is analogous to finding the best path on a weighted graph. The agents move on that graph, and are influenced by pheromone model (set of parameters associated to edges/nodes of the graph), and final solution is reached by continuously modifying these set of parameters during runtime.

6.2   Influence of Meta-Heuristics on IoT Data Analysis

The application meta-heuristics can give substantial advantages to Big data analysis in IoT. The power to solve complex optimization problem gives meta-heuristics the capability to ameliorate the data intensive processes of IoT. In addition, these approaches are suitable to provide security and, robustness to IoT data apart from making them process faster and more efficiently than heuristic approaches.

Smart City: There is an enormous contribution of IoT technologies and application for the urbanization of cities, to make them "Smart" (as shown in Fig. 5). However, meta-heuristic approaches can help IoT framework to get one step ahead with its ability to solve complex optimization problems faced during analysis of data collected by IoT devices and in decision making processes. The various aspects of **Smart city** where Meta-heuristics can be widely accepted are as follows:

Smart Parking: Smart parking, a component of a smart city, can use IoT and meta-heuristic algorithm like Genetic Algorithm, to solve problems of congestion in shopping malls. Genetic Algorithm can provide an optimal solution to find out the best parking space as well as the parking slot for any vehicle entering the mall [16]. The

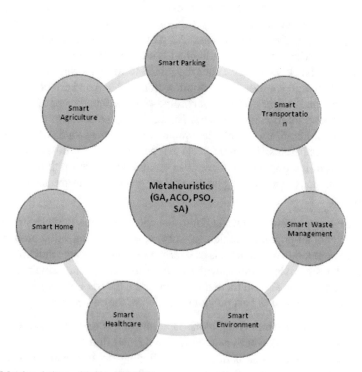

Fig. 5   Metaheuristics and IoT application

parking space is considered to be a chromosome and the parking slots as genes. With the use of Roulette wheel selection method, 3 parent crossover operator and bit string mutation operator, the optimum parking space and slot can be found out which will reduce the waiting time of the mall visitors. Other meta-heuristic approaches like Ant Colony Optimization [17] can also be useful in finding free parking slots where the current location of the car and the desired location of parking determines the right location.

Smart Traffic Management: Meta-Heuristic algorithms can treat on- road traffic management as a complex optimization problem and find optimal solutions for reducing congestions and prioritize traffic as per real time demands. Genetic Algorithm is particularly useful to develop intelligent traffic management at intersections. In [18], from the large number of vehicles on road at intersections (initial population), the fittest can be selected by human community Genetic Algorithm, from the data collected by cameras, fitted at the intersections. The genes are represented by various traffic parameters like vehicle mean speed, traffic flow and traffic density.

Smart Environment: In highly populated cities, air pollution is a major problem. For that, proper air monitoring system can be designed through energy efficient and low cost sensors, as proposed in [19] and Hierarchical based Genetic Algorithm can be used to optimize energy dissipation for extended network lifetime. It can be used for real time air quality monitoring as well for IoT devices

Smart Waste Management: The smart city concept allows municipal waste management more effective and efficient through IOT infrastructure [20]. Here, a meta-heuristic approach like Genetic Algorithm can be used as an effective tool for garbage collection optimization, by tracking garbage trucks especially at overloaded places in the city. Further it is important to identify locations to dump as well as process solid waste, violation of which may cause unwanted hazards to health, environment and ecosystem. In [21], this issue is solved through Genetic Algorithm that can predict locations through Geo-points in the state of Coimbatore, from where the data is collected. The fitness criterion is formulated by Ministry of Urban Development, Government of India.

Apart from Smart City, other IoT application domains have profound influence of meta-heuristic approaches while trying to improve the functionality and utility of the framework. They are as follows:

Smart Home: Meta-heuristic approaches like Genetic Algorithm can be used to design an automated and adaptive lighting system in home that can turn on or off lights automatically, provide user's desired brightness, and also thus save energy. A control algorithm is designed based on Genetic Algorithm that considers a light turning pattern as a chromosome containing genes, with fitness function that decides which light turning pattern will be eliminated after crossover and mutation [22]. Other meta-heuristic approaches like Particle Swarm Optimization can also contribute much towards betterment of a Smart Home. In [23], Particle

Swarm Optimization is applied with ZigBee Tree network in smart homes to ensure minimum energy consumption of the deployed sensor nodes and minimum cost path of the same. The application of meta-heuristics in Zigbee reduces its router dependency, increases energy efficiency as well as network lifetime. Interestingly Genetic Algorithm can also be applied with Zigbee along with Particle Swarm Optimization, for routing so that the optimal route can be meta-heuristically searched when the complexity of the network increases [24].

Smart Healthcare: Meta-heuristic approaches like Particle Swarm Optimization and Genetic Algorithm can play a significant role in Smart Healthcare. Genetic Algorithm can be extensively used for ambulance response time for emergency cases in a particular geographical area [25]. The algorithm can compare the number of ambulances with their current and future optimal location and reduce response time for patients by conducting an adequate spatial analysis. Genetic Algorithm can also effectively handle home caregiver scheduling and routing in a smart city [26]. Here the chromosomes (individuals) are represented by multiple vectors. In IoT based smart healthcare it is also important to store the medical data of the patient for further analysis and future monitoring. Hence a virtual machine Selection and migration algorithm, along with Particle Swarm Optimization is proposed in [27] that effectively stores data in minimal virtual machines, with lesser energy consumptions and response time thus serving more users at a time. At this point, security is a major concern, especially in IoT framework, while storing medical images of a patient, which can be provided by Particle Swarm Optimization, by optimizing standard cryptographic solutions [28]. Capturing data from mobile devices (as users requiring smart healthcare are moving) to cloud resources is a challenging task in IoT framework. Meta-heuristics can play an important role here to migrate data to a virtual machine for mobile patients, with Ant Colony Optimization based joint VM migration model [29].

Smart Agriculture: Researches have revealed a significant contribution of meta-heuristic approaches in Smart agriculture. Genetic Algorithm can be typically applied for tasks like weather prediction before harvesting crops. Farmers can plan before-after agricultural strategies based on the predicted weather conditions, and can also get crop damage alerts as well as water management [30]. Swarm Intelligence is also a major contributor in smart farming, like in [31], the authors have show how sensor node deployment in agriculture faces path loss as a major threat due to presence of tall trees and dense grass, in a farm. Particle Swarm Optimization can be used to combine with two path loss models in order to find out the optimal co-efficient of functions, for improving the models accuracy. This in turn improves the RSSI of the propagated signal, for proper data communication among the deployed sensor nodes in the farming land. In developing countries like China, agricultural afforestation is essential for carbon reduction in the environment, but that would affect food productivity.

6.3   Influence of Meta-Heuristics in Other IoT Application Domains

There are several other IoT application domains where meta-heuristic approaches can help them to solve optimization problem in a better manner. They are as follows:

Smart Energy: Energy management is vital for both facilities and utilities as the demand for energy supply is increasing day by day. IoT is capable of revolutionizing energy management by making it "Smart". In smart cities, energy management in buildings (smart buildings) can be controlled by an efficient control scheme. Multi-agent control systems can be developed with stochastic intelligent optimization, and Genetic Algorithm can be the best choice for this [32].

Smart Industry: Smart Industry (Industry 4.0) may be defined as an industrial framework with radical digitalization, and connecting people, products, machines through upcoming technology like IoT. Supply chain management is an integral part of any industry and to optimize the inventory cost of these supply chain, an improved Genetic Algorithm with a better search efficiency is proposed that can deal with IoT based supply chains [33]. Other meta-heuristics approach like Simulated Annealing can be used in combination with Genetic Algorithm, to obtain better path of AGV transportation system, through optimizations [34]. This can be beneficial for a manufacturing industry for product automation, reduce unnecessary labor cost, and get promoted to Industry 4.0.

Smart Business: Meta-heuristic approaches like Genetic Algorithm can be implemented to maximize economic profit, in a business against traditional methods like Lewis Model. The real parameter objective function can be optimized by using Genetic Algorithm to trade of between Total revenue and Total cost [35]. Genetic Algorithm can be applicable in hospitality industry as well to understand customer preferences and satisfaction, with Online Travel Agencies websites. As mentioned in application of Genetic Algorithm can help the business to understand customer preferences based on different criteria for different segments [36].

This is summarized in Fig. 6. It is apparent from the figure that GA is applied in almost every aspect of smart applications considered in this chapter.

7   Review of Machine Learning and Meta-Heuristics Based Approaches to IoT Sensor Data

7.1   Influence of Machine Learning on Meta-Heuristics

Meta-heuristics encompasses several algorithms which are complex and need to perform well to solve complex optimization problems. It is observed that if the parameters of these complex algorithms are fine tuned, the performance can be enhanced to

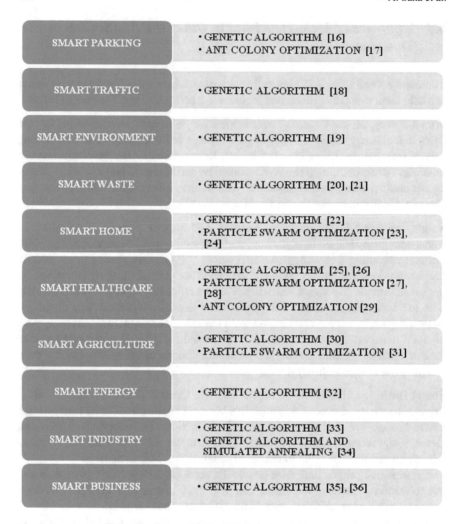

Fig. 6 Taxonomy of Meta-heuristics on IoT application domains

considerable extents. The parameters can be controlled either by some deterministic rules or through feedback from search and at this point, machine learning approaches like SVM and Fuzzy Logic can be quite effective. The regression approach of supervised machine learning can also be used for fine tuning the parameters of meta-heuristic algorithms. Researchers have also highlighted the usage of popular techniques like Bayesian Network, neural networks as well as case base reasoning or CBR to be effective in fine tuning instance specific parameters of Meta-heuristic algorithms. Usually initial population of such algorithms are generated randomly but at some instances, popular ML techniques like CBR OR hop-field neural network can be used as well, to initialize Genetic Algorithm. Q-learning can be implemented in the constructive phase of GRASP [37]. While evaluating the initial population, machine learning techniques like Markov Fitness Model, neural networks, Polynomial regression can be utilized to find out computationally inexpensive objective functions for a problem. For example in [38], Extreme Learning Machine, is used as a proposed surrogate model in order to approximate the fitness values of majority of individuals in the population in Genetic Algorithm and in [39], a fuzzy logic based multi-objective fitness function, is proposed for Genetic Algorithm to perform better. For population management in evolutionary algorithms, clustering analysis is one of the most common techniques used, in practice. Meta-Heuristic algorithms like Genetic Algorithm can use machine learning techniques like clustering analysis, in order to manage the population and also select parents for producing off-springs [40].

7.2 Influence of Meta-Heuristics on Machine Learning

Lately, meta-heuristics have been widely used in order to improve machine learn-ing tasks in various aspects of its application. For example, in *Classification* problems of supervised Learning, several tasks like feature selection & extraction, and parameter fine tuning can be done using meta-heuristics. In computer vision problems like Bag of Visual Words [41], an evolutionary algorithm has been proposed that can automatically learn and detect supervised as well as unsupervised weighting schemes. Meta-heuristic approaches like Genetic Algorithm can be employed to select important features in Intrusion Detection System for Machine Learning classifiers like decision tree [42]. In ANN, the difficulty to choose optimal network parameters is considered to be a major challenge, which can be solved by adaptive multi-objective genetic evolutionary algorithms for optimizing such parameters and escape local optimum [43]. Meta-heuristic approaches like Genetic Algorithm are powerful enough for feature selection as well as parameter optimization simultaneously for Support Tucker Machine. As proposed in [44], the algorithm can delete irrelevant information from tensor data and provide better accuracy in general. The influence of meta-heuristics on machine learning can be useful in health care as well. For example in [45], a multi-objective Genetic Algorithm is proposed to select appropriate features for microarray cancer data classification using SVM classifier. Interestingly other meta-heuristic approaches like Particle Swarm Optimization can also be used for feature selection and maximizing performance in multi-label classification problems [46].

In order to provide a better solution in *Regression* problems, meta-heuristics approaches can be reasonably viable. To begin with, neural networks are evolved using Genetic Algorithm, a process known as Neuro-evolution, with an objective to recreate the CNS of a natural organism, for an artificial creature in a virtual environment. Genetic Algorithm can create the architecture of the artificial creator based on fitness function and also train it to behave like living organisms [47]. Deep Learning is an emerging field in machine learning and meta-heuristic approaches like Genetic Algorithm can competitively train deep neural networks so that challenging reinforcement Learning Tasks can be performed successfully. This can be termed as Deep Genetic Algorithm [48]. Further in Feed-Forward Neural Network or FNN, which is a special type of ANN, meta-heuristic approaches can be extensively used to find out various optimization methods [49]. The simple reason is that, the conventional gradient based algorithms that were employed earlier, to train FNN, could only exploit current solutions in order to generate new solutions but could not explore the search space and hence used to get stuck in the local optima. Meta-heuristic algorithms could do both, (exploit and explore) in order to reach to a near optimal solution, of a complex optimization problem.

It is observed, that meta-heuristics have similar influence to improve unsupervised machine learning algorithms, like *Clustering* problems. Recently social network has become an essential component of our lives and detecting a community structure in these complex social networks, can also be considered as an optimization problem. Hence the concept of clustering can be used and evolutionary algorithms like Genetic Algorithm can be applied to identify these communities from dense networks and also can figure out cohesive node groups in a sparse network as well [50]. Other than Genetic Algorithm, meta-heuristics algorithms like Ant Colony Optimization and Particle Swarm Optimization can contribute much in solving clustering problems in machine learning. In VANET, efficient clustering can improve communication and routing of vehicles. Here Ant Colony Optimization based approach can generate near optimal solutions, because vehicles are always in motion. The entire approach minimizes the clusters and eventually the routing costs of the VANET [51]. Efficient routing protocols in VANET can also be developed using Clustering via meta-heuristic approaches like Simulated Annealing where cluster heads can be chosen using RBF neural network algorithms [52]. Another application of Clustering problem in machine learning can be Virtual Learning, which is gaining worldwide popularity day by day. Since the pedagogical approach of a learner differs from the other, meta-heuristic approaches like Particle Swarm Optimization can be utilized to satisfy the need of the learners and simultaneously improve quality of learning. The Particle Swarm Optimization based approach can capture data from learners, analyze them, and cluster the data based on accuracy, quality and efficiency of each learner [53]. In computational bio-informatics, clustering can group the genes and thus distinguish the disease prone malicious ones from others. Popularly known as Gene clustering this can be achieved by applying meta-heuristics approaches like Genetic Algorithm and fuzzy c- means clustering techniques [54].

7.3   *Enhancing IoT with Machine Learning and Meta-Heuristics*

The classic combination of machine learning with meta-heuristics has the power to make IoT framework even more effective and efficient for the society (as shown in Fig. 7). For example in **Smart Traffic management** it is possible to reduce traffic

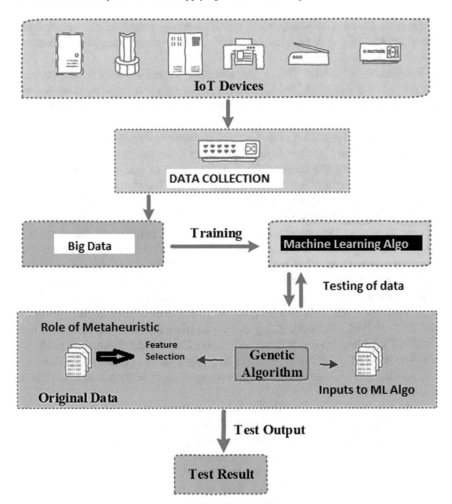

Fig. 7 Influence of Metaheuristics and Machine Learning on IoT

congestion through a co-operative light controlling algorithm using machine learning algorithms like linear regression and meta-heuristic approaches like Genetic Algorithm. The former is used for trend prediction and the latter is used to calculate optimum plan of light timings [55]. In [56], an intelligent transportation system is proposed by performing forecasting of short term traffic flow. This method is a combination of Random forest and Support Vector Regression with Genetic algorithm for feature selection, prediction as well as parameter optimization respectively. Ant Colony Optimization is an also implemented to precisely predict, real time traffic with Machine Learning algorithms like ANN, SVM that can successfully discover the nonlinear information hidden in real time traffic data as in [57]. The combination of this meta-heuristic approach with machine learning technique also exhibits anti-jamming capability with fault tolerance and can forecast accuracy with a shortened operation time.

Researchers have also found out extensive contributions of meta-heuristic approaches combined with machine learning algorithms in ensuring **Smart Environment**, around. A real time fine grained IoT air quality sensing system is proposed in [58] that minimizes average joint error of real time air quality map. Here deep Q-Learning solution is used for power control problem while sensing the tasks online, whereas the genetic algorithm is used for proper location selection problems while efficiently deploying the sensors. The location selection problem is done using k-means clustering for initialization and the genetic algorithm is used for improvement. Air pollutants like benzene is hazardous to human health, leading to blood cancer. Swarm Intelligence meta-heuristic approaches like Particle Swarm Optimization can be used to predict the concentration of this harmful pollutant in air, along with machine learning techniques like Fuzzy logic, considering triangular mutation operator. The proposed model successfully improves searching capabilities of Particle Swarm Optimization and increases the speed of convergence [59]. In Smart Cities, airborne pollution can be dangerous for human health and the non linear behavior of such pollution paves the way for meta-heuristics to be applied in order to combat with them. In [60], machine learning techniques like neuro-fuzzy systems have been combined with improved Ant Colony Optimization, in order to improve prediction of airborne pollution. It is observed that this approach can predict contaminants in air through shortest search and uses pheromone to approximate real time data for accurate results.

Meta-heuristics and machine learning can also make a **Smart Home** even smarter. The most important criterion of a smart home is ensuring security and that can be accomplished through IDS (Intrusion Detection System) in Smart homes. This setup is beneficial for elderly people for detecting their daily activities to monitor and intervene when required. In [61], the authors have employed a Genetic K-means algorithm to detect unknown patterns (human behaviors) and thus help in accurate detection of attacks, if any, through pattern analysis techniques. The combination of Genetic Algorithm with K-means helps in eliminating redundant input features for accurate prediction of malicious patterns, and faster response. In [61], authors have proposed a recurrent output neural network that predicts any unusual behavior of the elderly person under observation. The Genetic Algorithm is integrated in the learning step in order to get a better accuracy of the results so that the actual situation can be appropriately monitored remotely. The algorithm works on the basis of real time data fetched from IoT sensors. In a smart home, it is possible to detect locating users inside the premises (residents or criminals) through the data collected by routers based on Wi-Fi signal strength, from personal devices worn by people. Machine learning techniques like ANN can be trained by using weights obtained from Particle Swarm Optimization and an accurate optimization strategy can be developed [62]. Ant Colony Optimization can be implemented in Smart Homes along with machine learning technique like Decision Tree, as seen in [63]. This combination can be useful in home automation system, speech recognition, and device control with embedded system.

A substantial amount of work is done in the field of **Smart healthcare** which takes into account both meta-heuristics and machine learning concepts. It is observed that in smart cities, a global vision can merge several aspects together like AI, ICT, machine learning, Big data and IoT and this confluence can successfully predict and detect certain deadly diseases like Alzheimer's, Tuberculosis and Dementia. Genetic Algorithms can be extensively used along with SVM, KNN, RF in order to accomplish the task [64]. Common health issues like diabetes are faced by almost millions of people all across the world. Hence an ensemble (group) learning approach is proposed through reinforcement learning that accurately predicts diabetes mellitus and meta-heuristic approach like Genetic algorithm is used for selecting hybrid features for the task [65]. Other researches like [66], applies multi-objective evolutionary Fuzzy Classifier to the Pima Indians Diabetes Dataset and uses Goldberg's Genetic Algorithm to predict type 2 diabetes. The feature selection method using Genetic Algorithm helps in reducing number of features in the dataset and hence prediction can be done more accurately in lesser time. It is observed that meta-heuristics and machine learning can be useful for predicting cardiovascular diseases as well. In [67], a hybrid method to enhance the neural network is proposed where meta-heuristic approaches like Genetic Algorithm can enhance the initial weight of the NN and improve accuracy and performance in detecting severe heart diseases in patients. Unlike Angiography, this method is cost effective and has almost no side effects. Other meta-heuristic approaches like Particle Swarm Optimization can successfully predict coronary heart disease using ensemble classifiers like AdaBoost, Random Forest and bagged tree [68]. Here Particle Swarm Optimization is used as a method for feature selection to reduce least ranked features and ensemble methods improve classification performance. This accurately predicts and detects heart diseases earlier than others. Ant Colony Optimization is another meta-heuristic approach that can be useful in predicting the risk of a disease. In Big data, it is difficult to extract features from huge datasets, especially if the data is unstructured, as shown in [69]. So an improved Ant Colony Optimization is presented along with neural network machine learning technique, so that the best features can be selected for better prediction accuracy.

Smart Agriculture: It is observed that the confluence machine learning with meta-heuristics can also have significant contributions in Smart Agriculture as well. Genetic Algorithm can be used successfully in precision agriculture as shown in [70], where translation of satellite imagery using convolutional neural network can be done for accurate decision making. This can also be used to reduce carbon dioxide emission, and minimize land degradation. Other approaches which are effective in Smart Farming can be the integration of Particle Swarm Optimization with hybrid machine learning model based on ANN as in [71] which results in improvement in performance of common combine harvesters which are essential for the agricultural industry. The combination of meta-heuristic approach like Particle Swarm Optimization with ANN provides higher accuracy than simple ANN. Ant colony optimization also has significant contribution in Smart Agriculture when combined with ELM (Extreme Learning machine) for wind power production forecasting. Here the space of possible input features is represented as nodes of a solution graph, which is explored by using Ant Colony Optimization Technique, based on a certain fitness calculated by the cross validated prediction of ELM model. ELM also ensures light optimization due to low complexity in training process [72]. This regression model is in practice in two farms in Spain with distinct patterns of wind. This is summarized as Table 1. It is interesting to note that combination of meta-heuristics techniques and ML techniques have found wide applications in IoT eneabled applications ranging from environment monitoring to healthcare to traffic monitoring applications.

8   Conclusion

This chapter presents a detailed review of data intensive IoT based smart applications especially, smart city and its associated domains. Learning the heterogeneous data from various sources and prediction from it are crucial to build a robust system. Different machine learning techniques, along with their utilities in IoT based smart applications are discussed in detail. Interestingly, meta-heuristics techniques are found to have been applied in many such applications either as a standalone method for optimization or in combination with machine learning techniques to create smarter applications for a smarter world of internet of things.

Table 1 Comparison of different works on IoT based smart applications that incorporates both machine learning and meta-heuristics techniques for data analysis

References	IoT area of application	Meta-heuristic approach	Machine learning approach	Remarks
Wang et al. [55]	Smart traffic	Genetic algorithm	Linear-regression	Reduces waiting time, balances traffic pressure at intersections.
Zhang et al. [56]	Smart traffic	Genetic algorithm	SVR & rRandom forest	Accurate forecasting of short term traffic flow
Song [57]	Smart traffic	Ant colony optimization	ANN	Fault Tolerance, anti jamming, shorten operation time
Hu et al. [58]	Smart environment	Genetic algorithm	Q-learning, K-means	Reduces average joint error, finds suitable locations for limited sensors
Kaur et al. [59]	Smart environment	Particle swarm optimization	Fuzzy logic	Improves searching capability of PSO and speeds up convergence

(continued)

Table 1 (continued)

References	IoT area of application	Meta-heuristic approach	Machine learning approach	Remarks
Martinez-Zeron et al. [60]	Smart environment	Ant colony optimization	Neuro-Fuzzy	Improves prediction accuracy, reduces prediction errors
Sandhya and Julian [61]	Smart home	Genetic algorithm	K-means	Identifying patterns and attacks, in IDS, reduces redundant features
Narayanan et al. [62]	Smart home	Genetic algorithm	RO-neural network	Detects Unusual behavior of elderly patients
Soni and Dubey [63]	Smart home	Particle swarm optimization	ANN	Locates residents, criminals through wearable devices nearby homes
Chui et al. [64]	Smart home	Ant colony optimization	Decision tree	Home automation, speech recognition, device control
Abdollahi et al. [65]	Smart healthcare	Genetic algorithm	Ensemble learning	Diagnose,predict diabetes accurately, through hybrid feature selection
Vaishali et al. [66]	Smart healthcare	Genetic algorithm	MO volutionary fuzzy	Reduced features, accurate prediction in lesser time
Arabasadi et al. [67]	Smart healthcare	Genetic algorithm	Neural network	Enhances initial weight of NN, cost effective and no sideeffects
Yekkala et al. [68]	Smart heathcare	Particle swarm optimization	Ensemble methods	Improved feature selection, improved classification, early diagnosis
Joel et al. [69]	Smart healthcare	Ant colony optimization	Neural networks	Effective feature selection from Big Data
Arkeman et al. [70]	Smart agriculture	Genetic algorithm	Convolutional NN	Translation of satellite imagery, reduce CO_2 minimize land degradation
Nádai et al. [71]	Smart agriculture	Particle swarm optimization	ANN	Performance enhancement of combine harvesters, better accuracy
Carrillo et al. [72]	Smart agriculture	Ant colony optimization	ELM	Features as nodes of graph, light optimization for low complexity

References

1. URL: https://www.gartner.com/en/information-technology/glossary/internet-of-things
2. Mallick, A., Saha, A., Chowdhury, C., Chattopadhyay, S.: Energy efficient routing protocol for ambient assisted living environment. Wirel. Pers. Commun. **109**(2), 1333–1355 (2019)
3. Shi, W., Cao, J., Zhang, Q., Li, Y., Xu, L.: Edge computing: Vision and challenges. IEEE Internet Things J. **3**(5), 637–646 (2016)
4. Li, H., Ota, K., Dong, M.: Learning IoT in edge: Deep learning for the Internet of Things with edge computing. IEEE Netw. **32**(1), 96–101 (2018)
5. Chen, J., Ran, X.: Deep learning with edge computing: a review. Proc. IEEE **107**(8) (2019)
6. Venkatesan, C., Karthigaikumar, P., Paul, A., Satheeskumaran, S., Kumar, R.J.I.A.: ECG signal preprocessing and SVM classifier-based abnormality detection in remote healthcare applications. IEEE Access 6, 9767–9773
7. Oliva, J.T., Garcia Rosa, J.L.: Differentiation between normal and epileptic eeg using k-nearest-neighbors technique. In: Machine Learning for Health Informatics, pp. 149–160. Springer, Cham (2016)
8. Miljkovic, D., Aleksovski, D., Podpečan, V., Lavrač, N., Malle, B., Holzinger, A.: Machine learning and data mining methods for managing Parkinson's disease. In: Machine Learning for Health Informatics, pp. 209–220. Springer, Cham (2016)
9. Saha, J., Chowdhury, C., Biswas, S.: Two phase ensemble classifier for smartphone based human activity recognition independent of hardware configuration and usage behaviour. Microsyst. Technol. **24**(6), 2737–2752 (2018)
10. Alhussein, M., Muhammad, G., Shamim Hossain, M., Umar Amin, S.: Cognitive IoT-cloud integration for smart healthcare: case study for epileptic seizure detection and monitoring. Mobile Netw. Appl. **23**(6), 1624–1635
11. Sumalee, A., Wai Ho, H.: Smarter and more connected: future intelligent transportation system. IATSS Res. **42**(2), 67–71
12. Lin, Y., Wang, P., Ma, M.: Intelligent transportation system (its): concept, challenge and opportunity. In: 2017 IEEE 3rd International Conference on Big Data Security on Cloud (bigdatasecurity), IEEE International Conference on High Performance and Smart Computing (hpsc), and IEEE International Conference on Intelligent Data and Security (ids), pp. 167–172. IEEE (2017)
13. Lin, T., Rivano, H., Le Mouël, F.: A survey of smart parking solutions. IEEE Trans. Intell. Transp. Syst. **18**(12), 3229–3253
14. Roman, C., Liao, R., Ball, P., Ou, S., de Heaver, M.: Detecting on-street parking spaces in smart cities: performance evaluation of fixed and mobile sensing systems. IEEE Trans. Intell. Transp. Syst. **19**(7), 2234–2245
15. Khosravanian, R., Mansouri, V., Wood, D.A., Reza Alipour, M.: A comparative study of several metaheuristic algorithms for optimizing complex 3-D well-path designs. J. Petrol. Expl. Product. Technol. **8**(4), 1487–1503
16. Thomas, D., Kovoor, B.C.: A genetic algorithm approach to autonomous smart vehicle parking system. Procedia Comput. Sci. **125**, 68–76 (2018)
17. Balzano, W., Stranieri, S.: ACOp: an algorithm based on ant colony optimization for parking slot detection. In: Workshops of the International Conference on Advanced Information Networking and Applications, pp. 833–840. Springer, Cham (2019)
18. Hnaif, A.A., Nagham, A.-M., Abduljawad, M., Ahmad, A.: An intelligent road traffic management system based on a human community genetic algorithm. In: 2019 IEEE Jordan International Joint Conference on Electrical Engineering and Information Technology (JEEIT), pp. 554–559. IEEE (2019)
19. Khedo, K.K., Chikhooreeah, V.: Low-cost energy-efficient air quality monitoring system using wireless sensor network. In: Wireless Sensor Networks-Insights and Innovations. IntechOpen (2017)

20. Paulchamy, B., Babu Thirumangai Alwar, E., Anbarasu, K., Hemalatha, R., Lavanya, R., Manasa, K.M.: IOT based waste management in smart city. Asian J. Appl. Sci. Technol. 2(2), 387–394
21. Ramasami, K., Velumani, B.: Location prediction for solid waste management—a Genetic algorithmic approach. In: 2016 IEEE International Conference on Computational Intelligence and Computing Research (ICCIC), pp. 1–5. IEEE (2016)
22. Ngo, M.H., Viet Cuong Nguyen, X., Duong, Q.K., Son Nguyen, H.: Adaptive Smart Lighting Control based on Genetic Algorithm, pp. 320–325
23. Fernando, S.L., Sebastian, A.: IoT: Smart Homeusing Zigbee clustering minimum spanning tree and particle swarm optimization (MST-PSO). Int. J. Inf. Technol. (IJIT) 3(3) (2017)
24. Jiang, D., Yu, L., Wang, Xiaoxia Xie, F., Yu, Y.: Design of the smart home system based on the optimal routing algorithm and ZigBee network. PloS One 12(11)
25. Sasaki, S., Comber, A.J., Suzuki, H., Brunsdon, C.: Using genetic algorithms to optimise current and future health planning-the example of ambulance locations. Int. J. Health Geograph. 9(1), 4 (2010)
26. Borchani, R., Masmoudi, M., Jarboui, B.: Hybrid genetic algorithm for home healthcare routing and scheduling problem. In: 2019 6th International Conference on Control, Decision and Information Technologies (CoDIT), pp. 1900–1904. IEEE (2019)
27. Ambigai, S.D., Manivannan, K., Shanthi, D.: An efficient virtual machine migration for smart healthcare using particle swarm optimization algorithm. Int. J. Pure Appl. Math. 118(20), 3715–3722
28. Elhoseny, M., Shankar, K., Lakshmanaprabu, S.K., Maseleno, A., Arunkumar, N.: Hybrid optimization with cryptography encryption for medical image security in internet of things. Neural Comput. Appl., 1–15
29. Islam, Md.M., Abdur Razzaque, Md., Mehedi Hassan, M., Ismail, W.N., Song, B.: Mobile cloud-based big healthcare data processing in smart cities. IEEE Access 5, 11887–11899 (2017)
30. Gumaste, S.S., Kadam, A.J.: Future weather prediction using genetic algorithm and FFT for smart farming. In: 2016 International Conference on Computing Communication Control and automation (ICCUBEA), pp. 1–6. IEEE (2016)
31. Jawad, H.M., Jawad, A.M., Nordin, R., Kamel Gharghan, S., Abdullah, N.F., Ismail, M., Jawad Abu-Al Shaeer, M.: Accurate empirical path-loss model based on particle swarm optimization for wireless sensor networks in smart agriculture. IEEE Sens. J. (2019)
32. Shaikh, P.H., Mohd Nor, N.B., Nallagownden, P., Elamvazuthi, I., Ibrahim, T.: Intelligent multi-objective control and management for smart energy efficient buildings. Int. J. Electr. Power Energy Syst. 74, 403–409 (2016)
33. Wang, Y., Geng, X., Zhang, F., Ruan, J.: An immune genetic algorithm for multi-echelon inventory cost control of IOT based supply chains. IEEE Access 6, 8547–8555 (2018)
34. Fan, C., Li, S., Guo, R., Wu, Y.: Analysis of AGV optimal path problem in smart factory based on genetic simulated annealing algorithm. In: 4th Workshop on Advanced Research and Technology in Industry (WARTIA 2018). Atlantis Press (2018)
35. Chatterjee, S., Nag, R., Dey, N., Ashour, A.S.: Efficient economic profit maximization: genetic algorithm based approach. In: Smart Trends in Systems, Security and Sustainability, pp. 307–318. Springer, Singapore (2018)
36. Hao, J.-X., Yan, Yu., Law, R., Fong, D.K.C.: A genetic algorithm-based learning approach to understand customer satisfaction with OTA websites. Tour. Manag. 48, 231–241 (2015)
37. De Lima, F.C., De Melo, J.D., Doria Neto, A.D.: Using the Q-learning algorithm in the construc-tive phase of the GRASP and reactive GRASP metaheuristics. In: 2008 IEEE International Joint Conference on Neural Networks (IEEE World Congress on Computational Intelligence), pp. 4169–4176. IEEE (2008)
38. Guo, P., Cheng, W., Wang, Y.: Hybrid evolutionary algorithm with extreme machine learning fitness function evaluation for two-stage capacitated facility location problems. Expert Syst. Appl. 71, 57–68 (2017)
39. Téllez-Velázquez, A., Molina-Lozano, H., Villa-Vargas, L.A., Cruz-Barbosa, R., Lugo-González, E., Batyrshin, I.Z., Rudas, I.J.: A feasible genetic optimization strategy for parametric interval type-2 fuzzy logic systems. Inte. J. Fuzzy Syst. 20(1), 318–338 (2018)

40. Chehouri, A., Younes, R., Khoder, J., Perron, J., Ilinca, A.: A selection process for genetic algorithm using clustering analysis. Algorithms 10(4), 123 (2017)
41. Escalante, H.J., Ponce-López, V., Escalera, S., Baró, X., Morales-Reyes, A., Martínez-Carranza, J.: Evolving weighting schemes for the bag of visual words. Neural Comput. Appl. 28(5), 925–939
42. Azad, C., Mehta, A.K., Jha, V.K.: Evolutionary decision tree-based intrusion detection system. In: Proceedings of the Third International Conference on Microelectronics, Computing and Communication Systems, pp. 271–282. Springer, Singapore (2019)
43. Ibrahim, A.O., Mariyam Shamsuddin, S., Abraham, A., Noman Qasem, S.: Adaptive memetic method of multi-objective genetic evolutionary algorithm for backpropagation neural network. Neural Comput. Appl. 31(9), 4945–4962
44. Zeng, D., Wang, S., Shen, Y., Shi, C.: A GA-based feature selection and parameter optimization for support tucker machine. Procedia Comput. Sci. 111, 17–23 (2017)
45. Rani, M.J., Devaraj, D.: Microarray data classification using multi objective genetic algorithm and SVM. In: 2019 IEEE International Conference on Intelligent Techniques in Control, Optimization and Signal Processing (INCOS), pp. 1–3. IEEE (2019)
46. Zhang, Y., Gong, D.-w., Sun, X.-y., Guo, Y.-n.: A PSO-based multi-objective multi-label feature selection method in classification. Scientific Reports 7(1), 1–12 (2017)
47. Jha, S.K., Josheski, F.: Artificial evolution using neuroevolution of augmenting topologies (NEAT) for kinetics study in diverse viscous mediums. Neural Comput. Appl. 29(12), 1337–1347
48. Such, F.P., Madhavan, V., Conti, E., Lehman, J., Stanley, K.O., Clune, J.: Deep neuroevolution: Genetic algorithms are a competitive alternative for training deep neural networks for reinforcement learning. arXiv preprint arXiv:1712.06567 (2017)
49. Ojha, V.K., Abraham, A. Snášel, V.: Metaheuristic design of feedforward neural networks: A review of two decades of research. Eng. Appl. Artif. Intell. 60, 97–116
50. Said, A., Ayaz Abbasi, R., Maqbool, O., Daud, A., Aljohani, N.R.: CC-GA: A clustering coefficient based genetic algorithm for detecting communities in social networks. Appl. Soft Comput. 63, 59–70 (2018)
51. Aadil, F., Bashir Bajwa, K., Khan, S., Majeed Chaudary, N., Akram, A.: CACONET: Ant colony optimization (ACO) based clustering algorithm for VANET. PloS One 11(5), e0154080 (2016)
52. Bagherlou, H., Ghaffari, A.: A routing protocol for vehicular ad hoc networks using simulated annealing algorithm and neural networks. J. Supercomput. 74(6), 2528–2552 (2018)
53. Govindarajan, K., Selvi Somasundaram, T., uresh Kumar, V.: Particle swarm optimization (PSO)-based clustering for improving the quality of learning using cloud computing. In: 2013 IEEE 13th International Conference on Advanced Learning Technologies, pp. 495–497. IEEE (2013)
54. Dutta, P., Saha, S., Pai, S., Kumar, A.: A protein interaction information-based Generative Model for enhancing Gene clustering. Scientific Reports 10(1), 1–12 (2020)
55. Wang, H., Liu, J., Pan, Z., Takashi, K., Shimamoto, S.: Cooperative traffic light controlling based on machine learning and a genetic algorithm. In: 2017 23rd Asia-Pacific Conference on Communications (APCC), pp. 1–6. IEEE (2017)
56. Zhang, L., Alharbe, N.R., Luo, G., Yao, Z., Li, Y.: A hybrid forecasting framework based on support vector regression with a modified genetic algorithm and a random forest for traffic flow prediction. Tsinghua Sci. Technol. 23(4), 479–492 (2018)
57. Song, L.: Improved intelligent method for traffic flow prediction based on artificial neural networks and ant colony optimization. J. Convergence Inf. Technol. 7(8), 272–280 (2012)
58. Hu, Z., Bai, Z., Bian, K., Wang, T., Song, L.: Implementation and optimization of real-time fine-grained air quality sensing networks in smart city. In: ICC 2019–2019 IEEE International Conference on Communications (ICC), pp. 1–6. IEEE (2019)
59. Kaur, P., Singh, P., Singh, K.: Air pollution detection using modified Traingular mutation based particle swarm optimization (2019)

60. Martinez-Zeron, E., Aceves-Fernandez, M.A., Gorrostieta-Hurtado, E., Sotomayor-Olmedo, A., Ramos-Arreguín, J.M.: Method to improve airborne pollution forecasting by using ant colony optimization and neuro-fuzzy algorithms. Int. J. Intell. Sci. **4**(4), 81 (2014)

61. Sandhya, G., Julian, A.: Intrusion detection in wireless sensor network using genetic K-means algorithm. In: 2014 IEEE International Conference on Advanced Communications, Control and Computing Technologies, pp. 1791–1794. IEEE (2014)

62. Narayanan, S.J., Perumal, B., Joe Baby, C., Bhatt, R.B.: Fuzzy decision tree with fuzzy particle swarm optimization clustering for locating users in an indoor environment using wireless signal strength. In: Harmony Search and Nature Inspired Optimization Algorithms, pp. 217–225. Springer, Singapore (2019)

63. Soni, N., Dubey, M.: A review on home automation system with speech recognition and machine learning. IJARCSMS **5**(4) (2017)

64. Chui, K.T., Alhalabi, W., Han Pang, S.S., de Pablos, P.O., Liu, R.W., Zhao, M.: Disease diagnosis in smart healthcare: Innovation, technologies and applications. Sustainability **9**(12), 2309 (2017)

65. Abdollahi, J., Moghaddam, B.N., Effat Parvar, M.: Improving diabetes diagnosis in smart health using genetic-based ensemble learning algorithm. Approach to IoT Infrastructure. Future Gen Distrib Systems J **1**, 23–30 (2019)

66. Vaishali, R., Sasikala, R., Ramasubbareddy, S., Remya, S., Nalluri, S.: Genetic algorithm based feature selection and MOE Fuzzy classification algorithm on Pima Indians Diabetes dataset. In: 2017 International Conference on Computing Networking and Informatics (ICCNI), pp. 1–5. IEEE (2017)

67. Arabasadi, Z., Alizadehsani, R., Roshanzamir, M., Moosaei, H., Asghar Yarifard, A.: Computer aided decision making for heart disease detection using hybrid neural network-Genetic algorithm. Comput. Methods Programs Biomed. **141**, 19–26 (2017)

68. Yekkala, I., Dixit, S., Jabbar, M.A.: Prediction of heart disease using ensemble learning and particle swarm optimization. In: 2017 International Conference On Smart Technologies For Smart Nation (SmartTechCon), pp. 691–698. IEEE (2017)

69. Joel, G.N., Manju Priya, S.: Improved ant colony on feature selection and weighted ensemble to neural network based multimodal disease risk prediction (WENN-MDRP) classifier for disease prediction over big data. Int. J. Eng. Technol. **7**(3.27), 56–61 (2018)

70. Arkeman, Y., Buono, A., Hermadi, I.: Satellite image processing for precision agriculture and agroindustry using convolutional neural network and genetic algorithm. In: IOP Conference Series: Earth and Environmental Science, vol. 54, no. 1, p. 012102. IOP Publishing (2017)

71. Nádai, L., Imre, F., Ardabili, S., Mesri Gundoshmian, T., Gergo, P., Mosavi, A.: Performance analysis of combine harvester using hybrid model of artificial neural networks particle swarm optimization. arXiv preprint arXiv:2002.11041 (2020)

72. Carrillo, M., Del Ser, J., Nekane Bilbao, Perfecto, M.C., Camacho, D.: Wind power production forecasting using ant colony optimization and extreme learning machines. In: International Symposium on Intelligent and Distributed Computing, pp. 175–184. Springer, Cham (2017)

Geo-Spatiotemporal Intelligence for Smart Agricultural and Environmental Eco-Cyber-Physical Systems

Babak Majidi, Omid Hemmati, Faezeh Baniardalan, Hamid Farahmand, Alireza Hajitabar, Shahab Sharafi, Khadije Aghajani, Amir Esmaeili, and Mohammad Taghi Manzuri

1 Introduction

In the recent years, climate change created adverse effects on various sectors of the societies around the world. Significant increase of the drought cycles created severe problems for the agricultural sector. Deforestation due to the natural or man-made fires, illegal over-logging and illegal construction in the forests created severe problems for the natural environment. The central plateau of Iran is one of the areas in which the effect of the climate change is increasingly creates significant problems. The deforestation in Zagros area, the drought in the agricultural centers of southern provinces of Iran, drying of the major lakes in central Iran and air pollution resulting from dust particles are only some of these problems which are increasing in scale by the year.

B. Majidi (✉) · S. Sharafi
Department of Computer Engineering, Khatam University, Tehran, Iran
e-mail: b.majidi@khatam.ac.ir

B. Majidi
Emergency and Rapid Response Simulation (ADERSIM) Artificial Intelligence Group, Faculty of Liberal Arts & Professional Studies, York University, Toronto, Canada

O. Hemmati · H. Farahmand
School of Civil Engineering, Sharif University of Technology, Tehran, Iran

F. Baniardalan · A. Esmaeili
Department of Computer Engineering, Iran University of Science and Technology, Tehran, Iran

A. Hajitabar · M. T. Manzuri
Department of Computer Engineering, Sharif University of Technology, Tehran, Iran

K. Aghajani
Department of Computer Engineering, University of Mazandaran, Babolsar, Iran

© The Editor(s) (if applicable) and The Author(s), under exclusive license   471
to Springer Nature Switzerland AG 2021
A.-E. Hassanien et al. (eds.), *Enabling AI Applications in Data Science*,
Studies in Computational Intelligence 911,
https://doi.org/10.1007/978-3-030-52067-0_21

Therefore, for decision makers as well as farmers and natural environment management organization it is very important to investigate the scale of drought and deforestation caused by the climate change. The access to real time information allows the decision makers to have an encompassing view of the environment and to take precautionary actions rapidly. The changes in the environment requires smart solutions for energy and water management by deployment of optimized automated systems in agriculture. Rural communities should adapt climate smart strategies in order to be able to mitigate the effects of climate change in the next decades. Agricultural production is highly dependent on weather and climate. Without adequate rainfall and appropriate temperatures, crops fail and pastures become barren. Interestingly, the opposite is also true and weather and climate are influenced by agricultural practices. The changes in the environment requires smart solutions for energy and water management by deployment of optimized automated systems in agriculture, forestry, livestock, poultry and fishery industries. The smart villages and rural communities provide the means for this essential transformation by optimizing the energy consumption and intelligent management of human impact on the environment as well as rapid response to the climate related disasters and business disruptions.

Intelligent decision support and robotic solutions provide rural communities and the related industries with a full stack artificial intelligence based software and hardware platforms for climate smart autonomy. These solutions can deploy and manage swarms of unmanned ground vehicles, aerial vehicles and other robotics systems through a single unified management interface whether on-premises or over the cloud. The investment required for providing smart solutions for the Agri-food industry and rural communities is relatively small compared to the profits and importance of this sector. A growing world population and diversification in diet requires rapid increase in the production of the global food and agriculture industry. Another important issue in management of rural communities is disaster management. In 2019 many governments around the world declared national climate emergency. Wildfires rage stronger and harsher than ever before. The cities and homes are burning and each fire season, more communities displaced by the flames. Many places around the globe has seen an increase in hot extremes.

Currently, the main information technology solution for agricultural management is static sensor networks combined with traditional smart phone applications. However, end-to-end descriptive, predictive and prescriptive cloud-based solutions which use a combination of mobile sensor networks and machine learning based predictive models can provide solutions which help rural communities and Agri-food industry to remain sustainable and continue their business in face of disasters caused by climate change. In recent years, the smart cities analyze various forms of data from the city environment and the behavior of the people in that environment and recommend policies and products to them based on the collected personalized information.

The Eco-Cyber-Physical-System (ecoCystem) is a combination of the living entities of the ecosystem in conjunction with the Cyber-Physical System (CPS) based components of the smart rural environments, interacting as a system. The ecoCystem monitors the rural environment and gradually learn to optimize the behavior of the people and businesses to adhere with climate smart and sustainable policies. The goal of the ecoCystem is to use the power of artificial intelligence combines with robotics and sensor networks in order to provide solutions for rural, agricultural and natural ecosystems. The main modules of the proposed ecoCystem are the Geolyzer and the Modular Rapidly Deployable Internet of Robotic Things (MORAD IoRT). The Geolyzer platform is the focus of this paper. The Geolyzer platform automatically indexes and processes the satellite and drone data in order to provide descriptive, prescriptive and predictive solutions for rural communities. Until recently the processing of visual information gathered by the drones are performed after transmission of visual data to a ground station either by a computer or manually by a human expert. The advances in onboard processing systems in recent years give the drones the ability to process the visual data onboard and perform intelligent decisions on the flight. In this paper, a computationally light deep neural network is used for fast and accurate interpretation of rural environment for use in decision management of an autonomous UAV. Several training scenarios are investigated and various applications for such system in agriculture, forestry and climate change management are proposed. The experimental results show that the proposed system have high accuracy for various agricultural and forestry applications. This drones can act as agents in MORAD IoRT [1–4]. MORAD IoRT is a collection of intelligent unmanned vehicles which can monitor the environment around them using a collection of sensors including visual sensors. These robots use localized and distributed intelligence to model the environment, to make decisions based on their goals, to act on these decisions and to communicate with each other over internet and wireless networks. This mobile sensor network is used for localized model generation and validation of Geolyzer models.

The rest of this paper is organized as follows. The related works are discussed in Sect. 2. The Geolyzer is detailed in Sect. 3. The experimental design and the simulation scenarios are discussed in Sect. 4. Finally Sect. 5 concludes the paper.

2 Related Works

In the past few years, deep neural networks provided a good solution for many complex pattern recognition and visual interpretation problems. The application of deep neural networks in satellite and high altitude remote sensing is investigated significantly. Ammour et al. [5] proposed an asymmetric adaption neural network (AANN) for cross-domain classification in remote sensing images. They used three remote sensing data sets (i.e. Merced, KSA, ASD) with six scenarios. Chen et al. [6] proposed two semantic segmentation frameworks (SNFCN & SDFCN) which use deep Convolutional Neural Network (CNN). The proposed framework increases the accuracy compared to the FCN-8 and Segnet model and post processing raises the accuracy by about 1–2%. Chen et al. [7] presented a method for feature fusion framework based on deep neural networks. The proposed method uses a CNN to extract features and a fully connected DNN which fuses the heterogeneous features that generated by the previous CNN. They showed that the deep fusion model has better results in classification accuracy.

Cheng et al. [8] proposed a method using Discriminative CNNs to improve their performance for remote sensing image scene classification. In contrast to traditional CNN that minimized only the cross entropy loss, D-CNN are trained by optimizing a new discriminative objective function. Experimental result show that this method perform better than existing methods. Gavai et al. [9] uses MobileNets model to categorize flower datasets, which can minimize the time for flower classification. Gong et al. [10] investigates the Deep Structural ML (DSML) for remote sensing classification and uses structural information during training of remote sensing images. Experiments over the six remote sensing datasets show better performance than other algorithms.

Ben Hamida et al. [11] investigated the application of deep learning for classification of remote sensing hyper spectral dataset and proposed the new 3D DL approach for join spectral and spatial information processing. Kussul et al. [12] compared CNN with traditional fully connected Multi-Layer Perceptron (MLP). Li et al. [13] presented a new approach to retrieve large-scale remote sensing images based on deep hashing neural networks and showed that this approach performs better than state-of-the-art methods on two selected public dataset. Scot et al. [14] investigated application of deep CNN in conjunction with two techniques (Transfer learning and data augmentation) for remote sensing imagery. They achieved land-cover classification accuracies of $97.8 \pm 2.3\%$, $97.6 \pm 2.6\%$, and $98.5 \pm 1.4\%$ with CaffeNet, GoogLeNet, and ResNet, respectively on UC Merced dataset.

Shao et al. [15] proposed a remote sensing fusion method based on deep convolutional network that can extract spectral and spatial features from source images. This method perform better than traditional methods. Shi et al. [16] proposed a framework for remote sensing image captioning by leveraging the fully convolutional networks. The experimental results show that this method generate robust and comprehensive sentence description with desirable speed. Xiao et al. [17] presented a new method which uses a multi-scale feature fusion to represent the information of each region using GoogleNet model. In the next level this feature is the input to a support vector machine. Finally a simplified localization method is used. The experimental results demonstrates that the feature fusion outperforms other methods. Xie et al. [18] proposed a new method based on deep learning for detecting clouds in remote sensing images by Simple Linear Iterative Clustering (SLIC) method for segmentation and a deep CNN with two branches to extract multi-scale features. This method can detect clouds with high accuracy.

Xu et al. [19] investigated two-branch CNN and showed that this method can achieve better classification performance than previously proposed methods. Yang et al. [20] proposed a new detection algorithm named Region-based Deep Forest (RDF) which consists of a simple region proposal network and a deep forest ensemble and showed that this method performs better than state-of-the-art methods on numerous thermal satellite image datasets. Yu et al. [21] proposed an unsupervised convolutional feature fusion network in order to formulate an easy-to-train and effective CNN representation of remote sensing images. The experimental results demonstrate the efficiency and effectiveness of this method for unsupervised deep representations and the classification of remote sensing images. Yuan et al. [22] introduced the multi-scale and multi-depth CNN for the pan-sharpening of remote sensing imagery.

3   End-to-End Intelligent Geo-Spatiotemporal Analyzer

In the past few years there was a significant increase in the amount of low-cost geospatial data in forms of satellite and drone data. The resolution and coverage of this data is increasing rapidly. The intelligent analytics of this massive data can help increasing the productivity across a large group of industries such as agriculture, livestock and environment industries. In this paper, two project related to processing of remote sensing images for agricultural and environment management are discussed. The first project is a computationally light deep neural network which is used for fast and accurate interpretation of rural environment using an autonomous unmanned aerial vehicle. The second project is a platform for large scale interpretation of satellite images for agricultural and environmental applications which is called SatPlat. In order to effectively use the massively available geospatial data, the SatPlat has custom-made big spatial data processing stack to provide descriptive, prescriptive and predictive solutions for agriculture and environment management.

3.1   Deep Learning for Aerial MORAD IoRT

Despite achieving good results compared to traditional remote sensing methods, a main drawback of deep neural networks is that they need a significant amount of processing power and time for the training phase. A method for reducing this processing requirement is application of transfer learning. Transfer learning is reusing the lower layers of a deep neural network and retraining only the top layers for new instances of data. In this paper, a combination of transfer learning and a computationally light deep neural is used for fast and accurate interpretation of rural environment for use in an autonomous UAV. The reduced computational complexity resulting from combined transfer learning and a light weight deep neural network make the proposed model applicable for several practical scenarios in agriculture, forestry and climate change management. In this paper, four scenarios are investigated for using a UAV for natural resource management. Instead of transferring the aerial video to the ground station it is assumed that the interpretation is performed on-board the UAV. This process will help rapid decision making. The four scenarios investigated in this paper are: 1. Detection of construction of the houses in the forest territory, 2. Illegal widening of dirt roads and jungle roads, 3. Illegal farming in the forest territory and 4. Erosion of the jungle due to the drought.

There are various architecture proposed for deep neural networks. MobileNets [9] is a light model that can perform the pattern recognition tasks efficiently on a device with limited resources. Therefore, it can be used in a portable solution like recognition of the objects on-board a drone. MobileNets is a model based on depthwise separable convolutions. As it is demonstrated in Fig. 1, in this architecture, a standard convolution factorizes into a depth wise convolution and a 1×1 point-wise convolution. It uses two step filters to reduce the computation with a small reduction in accuracy.

If the kernel size is $D_k \times D_k$, there are M input channels and N output channels and the feature map size is $D_F \times D_F$. In a standard CNN each stride is calculated as:

$$G_{k,l,n} = \sum_{i,j,m} \mathbf{K}_{i,j,m,n} . \mathbf{F}_{k+i-1,l+j-1,m} \tag{1}$$

Fig. 1 MobileNets architecture

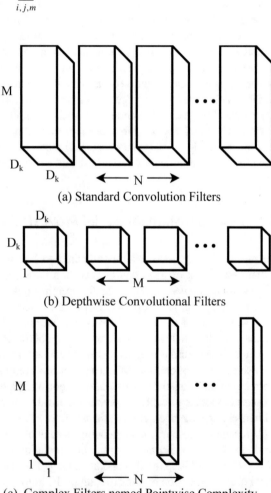

(a) Standard Convolution Filters

(b) Depthwise Convolutional Filters

(c) Complex Filters named Pointwise Complexity

Fig. 2 Depth wise convolution versus the standard convolution

(a) Standard convolutional layer

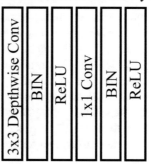

(b) Depthwise Separable convolutions

and the computational cost is:

$$D_K \times D_K \times M \times N \times D_F \times D_F \tag{2}$$

whereas in the MobileNets architecture the depth-wise convolution can be written as:

$$G_{k,l,n} = \sum_{i,j,m} \hat{\mathbf{K}}_{i,j,m} . \mathbf{F}_{k+i-1,l+j-1,m} \tag{3}$$

And the computational cost is reduced to:

$$D_K \times D_K \times M \times D_F \times D_F \tag{4}$$

Figure 2 shows this comparison and in the final architecture each layer is followed by a batch-normalization layer and the final fully connected layer feeds into a Softmax for classification.

3.2 SatPlat

In the past few years there has been a significant increase in the amount of available low-cost satellite data. The intelligent analytics of this massive data can help increasing the productivity across a large group of industries such as agriculture industry. Using the digital technologies is usually very expensive for smallholder farmers. This is due to the high cost of buying computers and hardware required for process and storage of the data needed for digital agriculture. The SatPlat is an end-to-end big data platform for decision agriculture based on satellite remote sensing. SatPlat is currently working as a startup company in Iran which is founded by a multidisciplinary group of environment enthusiast graduates. The team realized that in order to provide scalable cost effective solutions, a proprietary geospatial big data analytics solution should be developed. This realization led to obtaining enterprise level data and processing servers and starting the development of the backend data Extraction, Transformation and Load (ETL) process for the geospatial data of Iran. This process led to an automatic end-to-end solution which can collect the Sentinel satellite data automatically from European Space Agency servers in China and indexes and calculates various agricultural indices and stores this data in a proprietary database suitable for rapid access and processing. Currently the 150 TB server stored more than 100 TB of processed and raw data which has the historical agricultural remote sensing data for the entire country of Iran from October 2018 until March 2020.

The big data analytics and cloud computing platforms provided by SatPlat helps the farmers to have insights about their farms and to optimize their farming process without having to buy expensive infrastructure. The farmers can select their farms using either SatPlat smartphone application or the website and the data for crop health, irrigation and fertilization is provided for them on weekly basis. Currently there are seven measures which are calculated automatically in weekly intervals for the entire country of Iran from September 2018 until March 2020. The currently available measures are: 1. NDVI: which is used for crop health and density detection. It can be used for weed detection or water and fertilizer stress management or detecting presence of crop diseases, 2. CHL: which depends on adequate irrigation and fertilizer and measures the healthiness of the crops, 3. NDRE: which shows the amount of Nitrogen in the crops and is used for smart fertilization, 4. NDWI: which shows the amount of moisture in crops and is used for smart irrigation management, 5. LAI: which is the index for the amount of tree leaves and can be used for detection of damages from hailstorms to the trees, 6. VHL: which is used for measurement of the effect of drought on the crop health and 7. Evapotranspiration: which is also used for smart irrigation and calculation of the climate related stress on crops.

SatPlat also calculates the Surface Energy Balance Algorithm for Land (SEBAL) to estimate aspects of the hydrological cycle. The clustering and classification of various land covers are performed automatically by machine learning and remote sensing algorithms on time series of indices. The satellite radar images are used for calculation of subsidence which has application in earthquake damage assessment. As the amount of data for these measures for the entire country of Iran is significantly large, proprietary software solutions are used by SatPlat for storage and processing of the remote sensing data. Currently two enterprise web servers host the SatPlat platform. The amount of data is currently exceeded 100 TB and the processing of this data considered an enterprise level professional big data solution which is developed for the first time in Iran by SatPlat team.

4 Experimental Results and Discussion

4.1 Deep Learning for Aerial MORAD IoRT

For the first section of the experimental results the study area is in Mazandaran and Kerman regions of Iran. The data from these regions are classified into 4 classes: Agricultural fields (C1), Drought stricken areas (C2), Construction in the forest territory (C3) and Dirt roads in the forest territory (C4). The training data set consists of 5000 images from Google earth. Google Earth has been chosen for training due to availability of the visual information across a large range of training classes. A sample of each class is presented in Fig. 3. In this paper, the implementation of MobileNets deep neural network is performed using Tensorflow machine learning environment. Figure 4 shows the Tensor Board architecture of the used deep neural network.

The results are tested on a dataset of aerial images captured from a UAV flying over Melbourne, Australia [23–26] (Fig. 5).

In order to investigate the effect of the altitude of UAV on the accuracy of interpretation, two deep neural networks are trained. The first network is trained using 2800 images from 100 m altitude from the Google Earth. Then the results are tested on the actual UAV images and the accuracy and cross entropy are presented in Fig. 6.

The results for each class using the 100 m training dataset is presented in Table 1. The second network is trained using 2000 images with 50 m altitude from the Google Earth. Then the results are tested and the accuracy and cross entropy are presented in Fig. 7. The experimental results show that the proposed model has high accuracy for the intended application. Furthermore, the experimental results show that decreasing the altitude of the training set will result in increasing the accuracy of the interpreted images.

The results for each class using the 50 m training dataset is presented in Table 2.

4.2 SatPlat

In this section various projects which are performed by SatPlat team are discussed and the viability of the proposed solution for commercial applications is investigated. The SatPlat projects can be categorized in three levels: The government related projects (B2G), the projects in which SatPlat collaborated with other businesses (B2B) and finally the direct to customer (B2C) initiatives in which SatPlat engaged the farmers directly and provided solutions for smallholders.

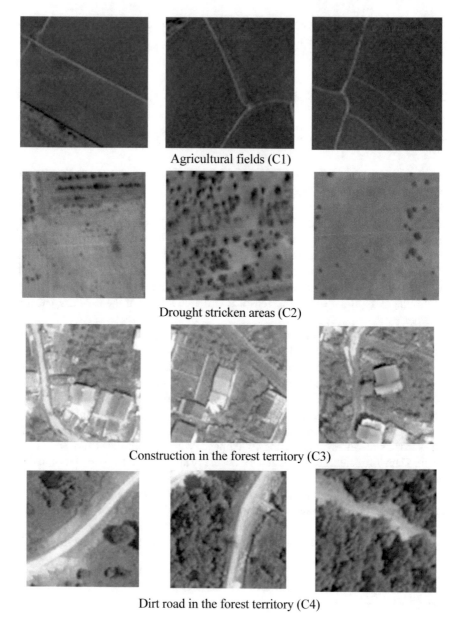

Fig. 3 Sample of training dataset classes

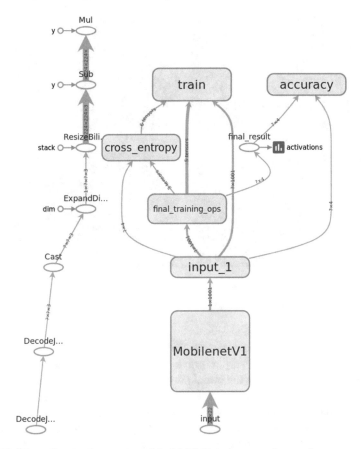

Fig. 4 The Tensorflow implementation of the MobileNets deep neural network

The first main B2G project of SatPlat which lead to the development and testing of the platform was with the Agricultural Insurance Fund (AIF) of Iran. The AIF has many human agents and the traditional method for verification of the insurance claims is manually by these agents. This project was started in August 2018 and the goal of the project was development of autonomous end-to-end agricultural insurance fraud detection and other insurance related sub projects. The first phase of the project was a pilot for investigation of 17 land parcels with total area of 200 ha in the Sabzevar-Bijar area in Kurdistan province of Iran. The goal of this phase was detection of whether these parcels are under cultivation of wheat or left as fallow automatically and using satellite imagery. Then the temperature stress are calculated for these land parcels and the required recommendation for irrigation deficiency in drought conditions are presented to the farmers.

In this project, Sentinel 2 and Landsat 8 satellite images are used for generation of recommendations. For validation of the recommendations, the SatPlat team interviewed the farmers during and after the project and a set of questionnaires are

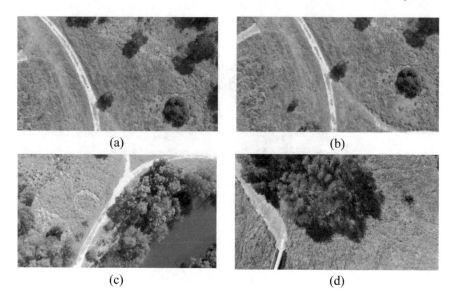

Fig. 5 The test dataset captured using a UAV

used for data collection and improvement of the SatPlat platform. After the project, the recommendations of SatPlat platform is validated by the AIF and the results are considered satisfactory. After the initial pilot, AIF requested the SatPlat services for other projects. The insurance claim verification extended to land use detection in Kerman, Ardabil and Mahabad regions with total land size of 200 ha. The outcome of project was development of autonomous remote sensing models and solutions for automatic crop type and crop health investigation using satellite imagery. Figure 8 shows some of the results of this projects.

Another SatPlat project for AIF was automatic monitoring of change detection during the massive floods of March 2019 in Khuzestan province of Iran using satellite imagery. Figure 9 shows the flooded area calculated in this project.

After this pilot, SatPlat had the initial models for land use and crop mask detection which expanded to other geographical regions in 2020 projects. The 2018–2019 pilots resulted in an end-to-end platform for providing autonomous services to the large scale agricultural organizations. This project lead to an investment from Iranian NS-Fund which is awarded to AIF for this new initiative on November 2019.

For B2B projects, SatPlat collaborated with several agricultural Internet of Things (IOT) companies and provided these companies with the insights for optimization of their smart irrigation solutions. SatPlat team modelled the requirements for all the stakeholders of smart rural communities from farmers to other Agritech companies. In another service, SatPlat provided artificial intelligence and digital image processing solutions for agricultural drone service providers. SatPlat provided image processing services for an agricultural drone company for management and tree

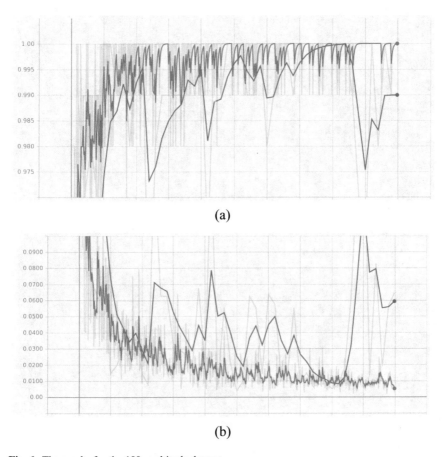

(a)

(b)

Fig. 6 The results for the 100 m altitude dataset

Table 1 The results for each class using the 100 m training dataset

Class	Accuracy (%)
C1	90
C2	100
C3	100
C4	100
UAV	100
Overall	97.8

modelling in a Pistachio farm in Yazd during September 2019. The pilot was implemented on 1 ha of a 500 ha pistachio farm. SatPlat is also provided remote sensing insights for smart irrigation platform by Rayan Arvin Algorithm Company for 40 ha of apple orchards in Mashhad on July 2019. The insight from this project lead to improved smart irrigation solutions for orchards. Smart irrigation solutions are also

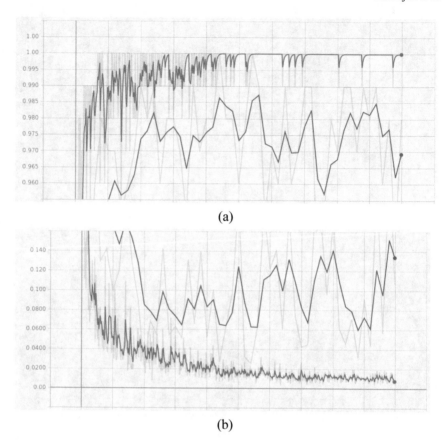

(a)

(b)

Fig. 7 The results for the 50 m altitude dataset

Table 2 The results for each class using the 50 m dataset

Class	Accuracy (%)
C1	100
C2	100
C3	100
C4	90
UAV	100
Overall	98

used for various other crops. Figure 10 shows the results of pistachio tree counting and corn field irrigation decision support.

For B2C projects, in order to design a Platform usable by different groups of famers in Iran (from smallholder farmers to larger agricultural companies) and across various cultural and geographical domains (From Chabahar area at the shores of Persian Gulf

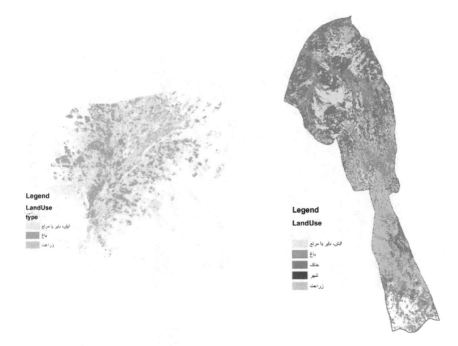

Fig. 8 VCI index and land use calculation using satellite imagery for AIF

to Ardabil at north of Iran), SatPlat interviewed various farming groups. For example, SatPlat provided solutions for a private sector farmer with an 80 ha farm at Shahin Shahr in Isfahan region of Iran in July 2019. The project lead to new insights to the SatPlat team about thr arid regions on the farm (known as Oily regions locally). In another project SatPlat estimated the crops of a pistachio farm using remote sensing in Rafsanjan region on September 2019. Some of the indices provided by SatPlat smartphone application is presented in Fig. 11.

SatPlat website currently has 237 active farmers as its users. Most of these farmers required remote help for registration of their lands on the website and SatPlat team gathered information and insights about improvements required for connecting with smallholder farmer. An extremely valuable insight was that after receiving help for initial registration of their land the farmers managed to registered and model the other farms by themselves. During the last two months only 5% of the farmers manage to perform the entire registration process by themselves. From the remaining farmers 15% required help for registering their farm and close to 40% of the farmers required help for registering themselves. This indicates that the user experience of the platform should be improved and also educational material and courses should be provided for the farmers. SatPlat also provided initiative solution for various companies and governmental organization in order to educate these organizations about the possible applications of remote sensing and artificial intelligence. For

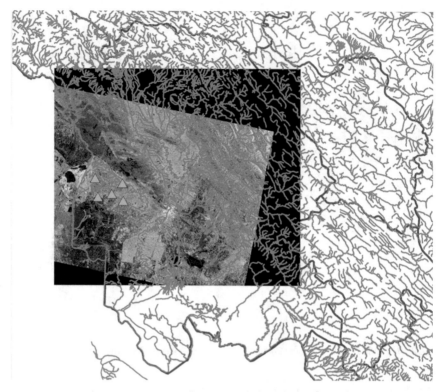

Fig. 9 The flooded area calculations in Khuzestan province

example as demonstrated in Fig. 12 SatPlat voluntarily introduced change detection calculation to the municipal governments.

SatPlat solution received great interest in Bakutel 2019 expo in Baku, Azerbaijan which shows the great potential for the proposed solution in various markets around the world (Fig. 13).

The main challenges about the implementation of autonomous remote sensing based agricultural management system is not the technical side but from cultural and low readiness of the users. Although the provided solutions are scalable, reliable and cost effective as an alternative solution for traditional agricultural tasks, we encountered resistance from the traditional agricultural management ecosystem despite encountering great enthusiasm from the smallholder farmers. We solved this problem by creating the SatPlat smartphone application and providing the direct interface for the smallholder farmers to directly select their farms using the SatPlat website.

(a)

(b)

Fig. 10 **a** The results of Pistachio tree counting and **b** the remote sensing insight provided by SatPlat for smart irrigation

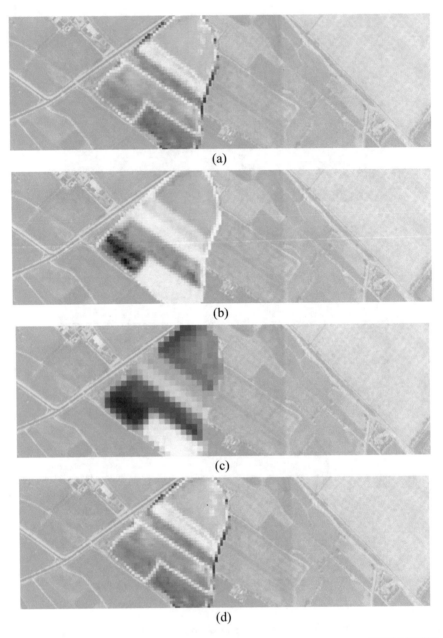

(a)

(b)

(c)

(d)

Fig. 11 Some of indices provided by SatPlat for a farm during the farming year: **a** NDVI, **b** NDRE, **c** NDWI and (d) LAI

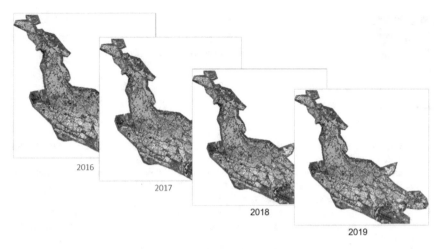

Fig. 12 Change detection for city of Shiraz from 2016 to 2019

Fig. 13 SatPlat team at Bakutel Expo 2019, Azerbaijan with president Ilham Aliyev

5 Conclusion

In this paper, two artificial intelligence based solutions for the applications of aerial and satellite imagery in forestry and agriculture are presented. The presented machine learning based remote sensing solutions give the local governments and various businesses the ability to have an encompassing view of large agricultural, horticultural and grazing fields and the forests. This ability enables them to make optimal decision in adverse scenarios caused by climate change and drought. A series of projects and solutions for validation of the proposed framework are discussed in this paper.

References

1. Abbasi, M.H., et al.: deep visual privacy preserving for internet of robotic things. In: 2019 5th Conference on Knowledge Based Engineering and Innovation (KBEI) (2019)
2. Nazerdeylami, A., Majidi, B., Movaghar, A.: Smart coastline environment management using deep detection of manmade pollution and hazards. In: 2019 5th Conference on Knowledge Based Engineering and Innovation (KBEI) (2019)
3. Sanaei, S., Majidi, B., Akhtarkavan, E.: Deep Multisensor dashboard for composition layer of web of things in the smart city. In: 2018 9th International Symposium on Telecommunications (IST) (2018)
4. Norouzi, A., Majidi, B., Movaghar, A.: Reliable and energy-efficient routing for green software defined networking. In: 2018 9th International Symposium on Telecommunications (IST) (2018)
5. Ammour, N., et al.: Asymmetric adaptation of deep features for cross-domain classification in remote sensing imagery. IEEE Geosci. Remote Sens. Lett. **15**(4), 597–601 (2018)
6. Chen, G., et al.: Symmetrical dense-shortcut deep fully convolutional networks for semantic segmentation of very-high-resolution remote sensing images. IEEE J. Sel. Topics Appl. Earth Observ. Remote Sens. **11**(5), 1633–1644 (2018)
7. Chen, Y., et al.: Deep fusion of remote sensing data for accurate classification. IEEE Geosci. Remote Sens. Lett. **14**(8), 1253–1257 (2017)
8. Cheng, G., et al.: When deep learning meets metric learning: remote sensing image scene classification via learning discriminative CNNs. IEEE Trans. Geosci. Remote Sens. **56**(5), 2811–2821 (2018)
9. Gavai, N.R., et al.: MobileNets for flower classification using TensorFlow. In: 2017 International Conference on Big Data, IoT and Data Science (BID) (2017)
10. Gong, Z., et al.: Diversity-promoting deep structural metric learning for remote sensing scene classification. IEEE Trans. Geosci. Remote Sens. **56**(1), 371–390 (2018)
11. Hamida, A.B., et al.: 3-D deep learning approach for remote sensing image classification. IEEE Trans. Geosci. Remote Sens. **56**(8), 4420–4434 (2018)
12. Kussul, N., et al.: Deep learning classification of land cover and crop types using remote sensing data. IEEE Geosci. Remote Sens. Lett. **14**(5), 778–782 (2017)
13. Li, Y., et al.: Large-scale remote sensing image retrieval by deep hashing neural networks. IEEE Trans. Geosci. Remote Sens. **56**(2), 950–965 (2018)
14. Scott, G.J., et al.: Training deep convolutional neural networks for land-cover classification of high-resolution imagery. IEEE Geosci. Remote Sens. Lett. **14**(4), 549–553 (2017)
15. Shao, Z., Cai, J.: Remote sensing image fusion with deep convolutional neural network. IEEE J. Sel. Topics Appl. Earth Observ. Remote Sens. **11**(5), 1656–1669 (2018)
16. Shi, Z., Zou, Z.: Can a machine generate humanlike language descriptions for a remote sensing image? IEEE Trans. Geosci. Remote Sens. **55**(6), 3623–3634 (2017)
17. Xiao, Z., et al.: Airport detection based on a multiscale fusion feature for optical remote sensing images. IEEE Geosci. Remote Sens. Lett. **14**(9), 1469–1473 (2017)
18. Xie, F., et al.: Multilevel cloud detection in remote sensing images based on deep learning. IEEE J. Sel. Topics Appl. Earth Observ. Remote Sens. **10**(8), 3631–3640 (2017)
19. Xu, P., et al.: Full-Wave simulation and analysis of bistatic scattering and polarimetric emissions from double-layered sastrugi surfaces. IEEE Trans. Geosci. Remote Sens. **55**(1), 292–307 (2017)

20. Yang, F., et al.: Ship detection from thermal remote sensing imagery through region-based deep forest. IEEE Geosci. Remote Sens. Lett. **15**(3), 449–453 (2018)
21. Yu, Y., et al.: An unsupervised convolutional feature fusion network for deep representation of remote sensing images. IEEE Geosci. Remote Sens. Lett. **15**(1), 23–27 (2018)
22. Yuan, Q., et al.: A multiscale and multidepth convolutional neural network for remote sensing imagery pan-sharpening. IEEE J. Sel. Topics Appl. Earth Observ. Remote Sens. **11**(3), 978–989 (2018)
23. Majidi, B., Bab-Hadiashar, A.: Land cover boundary extraction in rural aerial videos. In: MVA (2007)
24. Majidi, B., Bab-Hadiashar, A.: Aerial tracking of elongated objects in rural environments. Mach. Vis. Appl. **20**(1), 23–34 (2009)
25. Sharafi, S., Majidi, B., Movaghar, A.: low altitude aerial scene synthesis using generative adversarial networks for autonomous natural resource management. In: 2019 5th Conference on Knowledge Based Engineering and Innovation (KBEI) (2019)
26. Majidi, B., Patra, J.C., Zheng, J.: Modular interpretation of low altitude aerial images of non-urban environment. Digit. Signal Proc. **26**, 127–141 (2014)

SaveMeNow.AI: A Machine Learning Based Wearable Device for Fall Detection in a Workplace

Emiliano Anceschi, Gianluca Bonifazi, Massimo Callisto De Donato, Enrico Corradini, Domenico Ursino, and Luca Virgili

1 Introduction

Slips, trips and falls are among the main causes of accidents in a workplace in all the countries of the world. For this reason, many fall detection approaches have been proposed in the past literature. A possible taxonomy for them can be based on the environment surrounding the user and the employed sensors. According to this taxonomy, we can distinguish between ambient sensor based approaches, vision based approaches and wearable based approaches [24].

Ambient sensor based approaches analyze the recordings of audio and video streams from the work environment [33, 37] and/or track vibrational data derived from the usage of pressure sensors [2, 29]. They are little intrusive for the final user; however, they have high costs and could generate many false alarms.

Vision based approaches [23, 25] exploit image processing techniques, which use cameras to record workers and detect their falls. They are not intrusive and can achieve a great accuracy. However, they require to install cameras in each room to monitor and can return many false alarms.

E. Anceschi · M. C. De Donato
Filippetti S.p.A., Falconara Marittima (AN), Italy
e-mail: emiliano.anceschi@gruppofilippetti.it

M. C. De Donato
e-mail: massimo.callistodedonato@gruppofilippetti.it

G. Bonifazi · E. Corradini · D. Ursino (✉) · L. Virgili
DII, Polytechnic University of Marche, Ancona, Italy
e-mail: d.ursino@univpm.it

G. Bonifazi
e-mail: g.bonifazi@univpm.it

E. Corradini
e-mail: e.corradini@pm.univpm.it

L. Virgili
e-mail: l.virgili@pm.univpm.it

© The Editor(s) (if applicable) and The Author(s), under exclusive license
to Springer Nature Switzerland AG 2021
A.-E. Hassanien et al. (eds.), *Enabling AI Applications in Data Science*,
Studies in Computational Intelligence 911,
https://doi.org/10.1007/978-3-030-52067-0_22

Wearable based approaches make use of wearable devices [17, 22, 34] which workers are provided with, and, in some cases, they are combined with Machine Learning algorithms to process data provided by these devices [26, 30]. They are cost-effective and easy to install and setup. Moreover, they are strictly related to people and can detect falls regardless the environment where workers are operating. However, they can be bulky and intrusive, and their energy power and computation capabilities are limited. Finally, analogously to what happens for the approaches belonging to the other two categories, they could generate many false alarms [19] because realizing a model that accurately represents the daily activities of workers is difficult.

Nevertheless, we think that the advantages provided by this category of approaches are extremely relevant and the current problems affecting them are, actually, challenging open issues that, if successfully faced, can open many opportunities in preventing, or at least quickly facing, accidents in a workplace. For this reason, in this paper, we aim at proposing a contribution in this context presenting a wearable device called SaveMeNow.AI. This device aims at maintaining all the benefits of the previous wearable devices proposed for the same purposes and, simultaneously, at avoiding most, or all, of the problems characterizing them.

The hardware at the core of SaveMeNow.AI is SensorTile.box.[1] This is a device containing the ultra-low-power microcontroller STM32L4R9[2] and several sensors. Among them the one of interest for us is LSM6DSOX,[3] which is a six-axis Inertial Measurement Unit (hereafter, IMU) and Machine Learning Core.

The fall detection approach we propose in this paper, which defines the behavior of the SaveMeNow.AI device of a given worker, receives the data continuously provided by its six-axis IMU and processes it by means of a customized Machine Learning algorithm, conceived to determine if the corresponding worker has fallen or not. In the affirmative case, it immediately sends an alarm to all the near workers, who receive it through the SaveMeNow.AI device worn by them. This approach, once defined, trained and tested, can be natively implemented in the Machine Learning Core of LSM6DSOX.

As we will see in the following, SensorTile.box is very small and not bulky and, as we said above, it is provided with an ultra-low-power microcontroller. We implemented in it the Machine Learning approach presented in this paper and, therefore, we optimized the exploitation of the limited computation power characterizing this device. Finally, as we will show below, the accuracy of the defined fall detection approach is very satisfying, and false alarms are very few. As a consequence, Save-MeNow.AI is capable of addressing all the four open issues for wearable devices that we mentioned above.

This paper is organized as follows: In Sect. 2, we define and, then, illustrate the implementation and testing of the approach underlying SaveMeNow.AI. In Sect. 3, we illustrate all the features characterizing both the hardware and the software of

---

[1] https://www.st.com/en/evaluation-tools/steval-mksbox1v1.html.

[2] https://www.st.com/en/microcontrollers-microprocessors/stm32l4r9-s9.html.

[3] https://www.st.com/en/mems-and-sensors/lsm6dsox.html.

SaveMeNow.AI. In Sect. 4, we describe related literature and evidence the differences between several past approaches and ours. Finally, in Sect. 5, we draw our conclusion and have a look to some future developments of our approach.

2 Applying Machine Learning to Evaluate Fall Detection in a Workplace

In this section, we illustrate the customized Machine Learning approach for fall detection we have defined for SaveMeNow.AI. As a preliminary activity, we consider relevant describing the data sources we have used for both its training and its testing.

2.1 Construction of the Support Dataset

In recent years, thanks to the pervasive diffusion of portable devices (e.g., smartphones, smartwatches, etc.), wearable based fall detection approaches have been increasingly investigated by scientific community [6]. Thanks to this great interest, it is possible to find online many public datasets to perform analyses on slips, trips and falls or to find new approaches for their detection and management. After having analyzed many of these datasets, we decided to select four of them for our training and testing activities. In particular, we chose some datasets that would help us to define a generalized model, able to adapt to the activities carried out by workers and operators from various sectors, and performing very different movements during their tasks.

The first dataset used is "SisFall: a Fall and Movement Dataset" (hereafter, Sis-Fall), created by SISTEMIC. This is the Integrated Systems and Computational Intelligence research group of the University of Antioquia [32]. This dataset consists of 4505 files, each referring to a single activity. All the activities are grouped in 49 categories: 19 refer to ADLs (Activities of Daily Living) performed by 23 adults, 15 concern falls (Fall Activities) that the same adults had, and 15 regard ADLs performed by 14 participants over 62 years of age. Data was collected by means of a device placed at the hips of the volunteers. This device consists of different types of accelerometers (ADXL345 and MMA8451Q) and of a gyroscope (ITG3200).

The second dataset used is "Simulated Falls and Daily Living Activities" (hereafter, SFDLAs), created by Ahmet Turan Özdemir of the Erciyes University and by Billur Barshan of the Bilkent University [26]. It consists of 3060 files that refer to 17 volunteers who carried out 36 different kinds of activity. Each of them was repeated by each volunteer about 5 times. The 36 categories of activities are, in turn, partitioned in 20 Fall Activities and 16 ADLs. The dataset was recorded using 6 positional devices placed on the volunteers' head, chest, waist, right wrist, right thigh and right ankle. Each device consists of an accelerometer, a gyroscope and a magnetometer.

Table 1 Structure of the new dataset

Measure	Acceleration			Rotation		
Axis	X	Y	Z	X	Y	Z

The third dataset used is "CGU-BES Dataset for Fall and Activity of Daily Life" (hereafter, CGU-BES) created by the Laboratory of Biomedical Electronics and Systems of the Chang Gung University [8]. This dataset contains 195 files that refer to 15 volunteers who performed 4 Fall Activities and 9 ADLs. Data was collected by a system of sensors consisting of an accelerometer and a gyroscope.

The fourth, and last, dataset used is the "Daily and Sports Activities Dataset" (hereafter, DSADS) of the Department of Electrical and Electronic Engineering of the Bilkent University [1]. This dataset comprises 9120 files obtained by sampling 152 activities carried out by 8 volunteers. Each activity had a duration of about 5 min, split into 5-s recordings. This dataset does not contain fall activities, but sport activities. We chose it in order to make our model generalizable and, therefore, more adaptable to most of the various situations that may occur in the working environment. Data was collected through 5 sensors containing an accelerometer, a gyroscope and a magnetometer, positioned on different parts of the volunteer's body.

From these four datasets, we decided to extrapolate only the accelerometric and gyroscopic data. This choice was motivated by two main reasons. The first concerns data availability; in fact, the only measurements common to all dataset are acceleration and rotation. The second regards the ability of Machine Learning models to obtain better performance than thresholding-based models when using accelerometric data, as described in [13]. By merging the acceleration and rotation data extrapolated from the four datasets we obtained a new dataset whose structure is shown in Table 1. It stores data from 8579 activities. 4965 of them do not represent falls, while the remaining 3614 denote falls. Each activity has associated a file that stores the values of the 6 parameters of interest for a certain number of samples. Since data comes from different datasets, the number of samples associated with the various activities is not homogeneous; in fact, it depends on the length of the activity and the sampling frequency used in the dataset where it was originally registered. With regard to this aspect, it should be noted that having datasets characterized by different activity lengths and sampling frequencies does not significantly affect the final result, as long as the sampling frequency is very high compared to the activity length, as it is the case for all our datasets. This is because our features are little influenced by the number of samples available. This happens not only for the maximum and the minimum values, which is intuitive, but also for the mean value and the variance, because, in this case, as the number of samples increases, both the numerator and the denominator of the corresponding formulas grow in the same way.

After building the new dataset, we applied a Butterworth Infinite Impulse Response (i.e., IIR) second order low-pass filter with a cut-off frequency of 4 Hz to the data stored therein. The purpose of this task was keeping the frequency response module

Table 2 Feature definition

Feature	Definition
Maximum value	$\max\limits_{k=1..n} (\zeta[k])$
Minimum value	$\min\limits_{k=1..n} (\zeta[k])$
Mean value	$\mu = \frac{1}{n} \sum_{k=1}^{n} \zeta[k]$
Variance	$\sigma^2 = \frac{1}{n} \sum_{k=1}^{n} (\zeta[k] - \mu)^2$

as flat as possible in the pass band in such a way as to remove noise. Instead, the choice of the Butterworth filter was motivated by its simplicity and low computational cost [15]. These features make it perfect for a possible future hardware implementation.

After performing data cleaning, through which we eliminated excess data, and data pre-processing, through which we reduced the noise as much as possible, we proceeded to the feature engineering phase. In particular, given a parameter ζ, whose sampled data was present in our dataset, we considered 4 features, that is the maximum value, the minimum value, the mean value and the variance of ζ. If n is the number of samples of ζ present in our dataset and $\zeta[k]$ denotes the value of the kth sample of ζ, $1 \leq k \leq n$, the definition of the 4 features is the one shown in Table 2.

As shown in Table 1, the parameters present in our dataset are 6, corresponding to the values of the X, Y and Z axes returned by the accelerometer and the gyroscope. As a consequence, having 4 features for 6 parameters at disposal, each activity can have associated 24 features.

Finally, in a very straightforward way, each activity can have associated a two-class label, whose possible values are *Fall Activity* and *Not Fall Activity*.

The result of all these operations is a 8579×25 matrix that represents the training set used to perform the next classification activity.

2.2 Descriptive Analytics on the Support Dataset

In this section, we illustrate some of the analyses that we conducted on the support dataset and that allowed us to better understand the reference scenario and, then, to better face the next challenges.

The first activity we performed was the creation of the correlation matrix between features, which is reported in Fig. 1.

What clearly emerged when looking at this matrix was the presence of some evident negative correlations between the maximum and minimum values of some parameters. Moreover, a positive correlation between the maximum values (resp., minimum values, variances) calculated on the various axes and on the two sensors could be noticed. Finally, there were some parameters that had no significant correlation, either positive or negative. This is particularly evident for all the cases in which

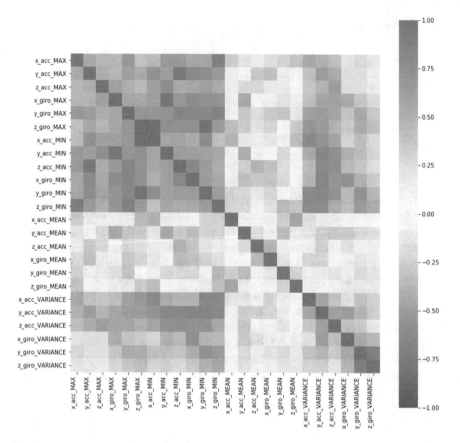

Fig. 1 Correlation matrix between the features

the feature "mean value" is involved. From this analysis, we intuitively deduced that exactly these last parameters would have played a fundamental role in the next classification activity.

To verify if this last intuition was right, we ran a Random Forests algorithm [5] with a 10-Fold Cross Validation [14] that allowed us to generate the list of features sorted according to their relevance in identifying the correct class of activities.

In particular, in order to compute the relevance of features, this algorithm operates as follows. Given a decision tree $\mathcal{D}$ having N nodes, the relevance ρ_i of a feature f_i is computed as the decrease of the impurity of the nodes splitting on f_i weighted by the probability of reaching them [12]. The probability of reaching a node n_j can be computed as the ratio of the number of samples reaching n_j to the total number of samples. The higher ρ_i, the more relevant f_i will be. Formally speaking, ρ_i can be computed as:

$$\rho_i = \frac{\sum_{n_j \in N_{f_i}} \vartheta_j}{\sum_{n_j \in N} \vartheta_j}$$

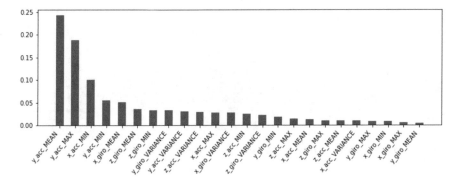

Fig. 2 Feature relevance in identifying the correct class of activities

Here, N_{f_i} is the set of the nodes of N splitting on f_i. ϑ_j is the relevance of the node n_j. If we assume that n_j has only two child nodes n_l and n_r, then:

$$\vartheta_j = w_j C_j - w_l C_l - w_r C_r$$

Here:

- w_j (resp., w_l, w_r) is the fraction of samples reaching the node n_j (resp., n_l, n_r);
- C_j is the impurity value of n_j;
- n_l (resp., n_r) is the child node derived from the left (resp., right) split of n_j.

The value of ρ_i can be normalized to the range $[0, 1]$. For this purpose, it must be divided by the sum of the relevances of all features.

$$\overline{\rho_i} = \frac{\rho_i}{\sum_{f_k \in F} \rho_k}$$

where F denotes the set of all the available features.

The final relevance of a feature f_i returned by Random Forests is obtained by averaging the values of the normalized relevances $\overline{\rho_i}$ computed on all the available trees:

$$\widehat{\rho_i} = \frac{\sum_{t_q \in T} \overline{\rho_i}}{|T|}$$

Here, T is the set of all the trees returned by Random Forests.

The result obtained by applying the approach illustrated above to the features of our interest is shown in Fig. 2.

To check if what suggested by Random Forests made sense, we considered the two features with the highest relevance returned by this algorithm, i.e., the mean and the maximum accelerations computed on the Y axis. Starting from these two features, we built the scatter diagram shown in Fig. 3. Here, an orange dot is visualized for

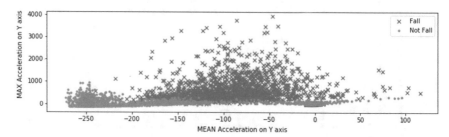

Fig. 3 Activities labeled as *Not Fall* and *Fall* against the mean and the maximum accelerations on the *Y* axis

each activity labeled as *Not Fall*, while a blue cross is visualized for each activity labeled as *Fall*. Looking at this diagram, we can observe that the activities labeled as *Not Fall* have a very negative mean acceleration and a much lower maximum acceleration than the ones labeled as *Fall*. This allows us to conclude that Random Forests actually returned a correct result when it rated these two features as the most relevant ones. In fact, their combination makes it particularly easy to distinguish falls from not falls.

2.3 Applying Machine Learning Techniques on the Available Dataset

After having constructed a dataset capable of supporting the training task of our Machine Learning campaign, the next activity of our research was the definition of the classification approach to be natively implemented in the Machine Learning Core of LSM6DSOX, i.e., the sensor at the base of SaveMeNow.AI. The first step of this activity was to verify if one (or more) of the existing classification algorithms, already proposed, tested, verified and accepted by the scientific community, obtained satisfactory results in our specific scenario. Indeed, in that case, it appeared us natural to adopt a well known and already accepted approach, instead of defining a new one, whose complete evaluation in real scenarios would have required an ad-hoc experimental campaign in our context, the publication in a journal and the consequent evaluation and possible adoption by research groups all over the world, in order to find possible weaknesses that could have been overlooked during our campaign.

In order to evaluate the existing classification algorithms, we decided to apply the classical measures adopted in the literature, i.e., Accuracy, Sensitivity and Specificity. If we indicate by: *(i)* TP the number of true positives, *(ii)* TN the number of true negatives, *(iii)* FP the number of false positives, and *(iv)* FN the number of false negatives, these three measures can be defined as:

$$Accuracy = \frac{TP + TN}{TP + TN + FP + FN}$$

$$Sensitivity = \frac{TP}{TP + FN}$$

$$Specificity = \frac{TN}{TN + FP}$$

Accuracy corresponds to the number of correct forecasts on the total input size, and represents the overall performance of the algorithm. Sensitivity denotes the fraction of positive samples that are correctly identified. In our scenario, it stands for the fraction of *Fall Activities* that are properly identified by the algorithms. Finally, Specificity corresponds to the fraction of negative samples correctly identified, so it represents the fraction of *Not Fall Activities* properly identified by the algorithms.

In Table 3, we report a summary of all the tested classification algorithms; in particular, we show the mean values of Accuracy, Sensitivity and Specificity obtained through a 10-Fold Cross Validation.

Depending on the application scenario, a metric can be more important than another one. In our case, in which we want to detect falls in a work environment, Sensitivity has a higher importance than Specificity. In fact, a missed alarm (corresponding to a *Not Fall* prediction of a *Fall Activity*) leads to a lack of assistance to the worker. Furthermore, a false alarm can be mitigated by providing the worker with the possibility to interact with the device and turn off the alarm.

From the analysis of the Table 3, we can observe that the Machine Learning model that has the highest Accuracy (and, therefore, the best overall performance) is the Decision Tree—C4.5. This model obtains excellent results also in terms of Sensitivity and Specificity. Another interesting results was obtained through the Quadratic Discriminant Analysis, which achieves a Specificity value equal to 0.9680. However, this last algorithm obtains low values for Accuracy and Sensitivity, which led us to discard it.

Table 3 Accuracy, sensitivity and specificity values achieved by several classification algorithms when applied to our dataset

Algorithm	Accuracy	Sensitivity	Specificity
Decision tree—C4.5	0.9487	0.9391	0.9566
Decision tree—CART	0.9128	0.8910	0.9223
Multilayer perceptron	0.9270	0.8829	0.9363
k-Nearest neighbors (k = 3)	0.8790	0.8747	0.9263
Logistic regression	0.7707	0.8599	0.7057
Quadratic discriminant analysis	0.7664	0.4956	0.9680
Linear discriminant analysis	0.7557	0.4956	0.9663
Gaussian naive bayes	0.7175	0.4947	0.8989
Support vector machine	0.7141	0.4103	0.9486

Based on all these considerations, we decided that, among the classification algorithms of Table 3, the best one for our scenario was the Decision Tree—C4.5. Furthermore, we evaluated that the performance it achieved was so good that it could be adopted for our case, without the need to think about a new ad-hoc classification algorithm, which would have hardly achieved better performance than it and would have been exposed to all the problems mentioned at the beginning of this section.

3 Design, Realization and Testing of SaveMeNow.AI

In this section, we explain how we realized SaveMeNow.AI starting from the device SensorTile.box. Specifically, in Sect. 3.1, we describe the main characteristics of the hardware adopted. Then, in Sect. 3.2, we outline how we implemented the logic of our approach in the device. Finally, in Sect. 3.3, we show how we tested it.

3.1 Hardware Characteristics of the IoT Device at the Core of SaveMeNow.AI

The choice of the IoT device for implementing our approach and constructing SaveMeNow.AI was not simple. Indeed, it had to comply with some requirements. First, as outlined previously, the device had to be small and ergonomic in order to be worn by a user. Afterwards, it should have an Inertial Measurement Unit (i.e., IMU), which, in its turn, should have contained an accelerometer and a gyroscope, as well as a Bluetooth module able to manage the Bluetooth Low Energy (i.e., BLE) protocol. One of the possible devices compliant with all these requirements is SensorTile.box provided by STMicroelectronics. In Fig. 4, we report a picture of this device.

SensorTile.box was designed for supporting the development of wearable IoT devices. It contains a BLE v4.2 module and a ultra-low-power microcontroller STM32L4R9 that manages the following sensors:

- STTS751, which is a high precision temperature sensor;
- LSM6DSOX, which is a six-axis IMU and Machine Learning Core (i.e., MLC);
- LIS3DHH and LIS2DW12, which are a three-axis accelerometer;
- LIS2MDL, which is a magnetometer;
- LPS22HH, which is a pressure sensor;
- MP23ABS1, which is an analogic microphone;
- HTS221, which is a humidity sensor.

As we said in the previous sections, in the current version of SaveMeNow.AI, the only sensor we used is LSM6DSOX. However, we do not exclude that we will employ one or more of the other sensors in the future.

Fig. 4 SensorTile.box (STEVAL-MKSBOX1V1)

LSM6DSOX contains everything necessary for our approach. Indeed, it is a system-in-package (i.e., SIP) that contains a three-axis high precision accelerometer and gyroscope. Beside from the advantages of being a low-power sensor and having a small size, the really important feature of LSM6DSOX is the MLC component. In fact, it is able to directly implement Artificial Intelligence algorithms in the sensor, without involving a processor. MLC uses data provided by the accelerometer, the gyroscope and some possible external sensors for computing some statistical parameters (such as mean, variance, maximum and minimum values, etc.) in a specific sliding time window. These parameters can be provided in input to a classification algorithm (in our case, a decision tree) previously loaded by the user. The whole workflow of MLC is reported in Fig. 5.

As reported in Fig. 5, some filters can be applied to provided data. Specifically, the possible filters are a low-pass filter, a bandwidth filter, a First-Order IIR and a Second-Order IIR. This last feature was very important for our approach in that it allowed us to implement the Butterworth filter to be applied on the data provided by the accelerometer and the gyroscope to reduce noise (see Sect. 2.1).

3.2 Embedding the Logics of SaveMeNow.AI in the IoT Device

In order to implement our approach, we had to develop a firmware that can be loaded in SensorTile.box. This device accepts a firmware written in the C language, which must contain all the instructions for the initialization of the

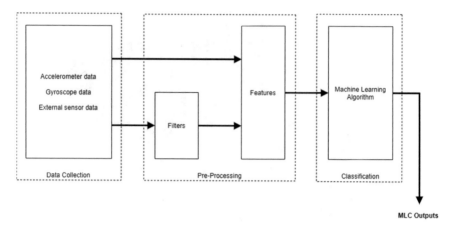

Fig. 5 Workflow of the Machine Learning Core of LSM6DSOX

micro-controller and the configuration of the Machine Learning Core. To support these tasks, STMicroelectronics provides two software tools (i.e., STM32CubeMX and STM32CubeIDE) allowing users to develop C code for the microcontroller STM32L4R9. STM32CubeMX is a graphic tool to initialize the peripherals, such as GPIO and USART, and the middlewares, like USB or TCP/IP protocols, of the microcontroller. The second software is an IDE allowing users to write, debug and compile the firmware of the microcontroller.

The firmware that we developed contains three main functions, namely:

- `HAL_init()`: it initializes the Hardware Abstraction Layer, which represents a set of APIs above the hardware allowing developers to interact with the hardware components in a safe way.
- `Bluetooth_init()`: it initializes the whole Bluetooth stack. Such a task comprises the setting of the MAC address, the configuration of the HCI interface, the GAP and GATT protocols, and so forth.
- `MLC_init()`: it initializes the MLC component of LSM6DOX and enables the interruption of the output of the decision trees. The MLC initialization is performed through the loading of a specific header file that configures all the registers of LSM6DOX. We dive into this file below.

The MLC configuration is not trivial, because it implies to also configure the sensors of LSM6DSOX and to set all its registers. To perform this task, STMicro-electronics provides a software tool called Unico. This is a Graphical User Interface allowing developers to manage and configure sensors, like accelerometers and gyro-scopes, along with the Machine Learning Core of LSM6DSOX. The output of Unico is a header file containing the configurations of all the registers and all the information necessary for the proper functioning of the Machine Learning models. Indeed, thanks to Unico, it is possible to set the configuration's parameters of MLC and the sensors of LSM6DSOX, like the output frequency of MLC, the full scale of the

Table 4 Adopted configuration of the MLC component

Measure	Setting
Input data	Three axis accelerometer and gyroscope
MLC output frequency	12.5 Hz
Accelerometer sampling frequency	12.5 Hz
Gyroscope sampling frequency	12.5 Hz
Full scale accelerometer	± 8 g
Full scale gyroscope	± 2000 dps
Sample window	37 samples
Filtering	Second-Order IIR filter with cutting frequency at 4 Hz

accelerometer and gyroscope, the sample window of reference for the computation of features, and so on. We report our complete configuration in Table 4.

With this configuration, at each clock of MLC, the output of the classification algorithm implemented therein is written to a dedicated memory register. In this way, it is possible to read this value and, in case this last is set to *Fall* (which implies that the worker who is wearing it has presumably fallen), to activate the alarm. At this point, all the problems concerning the communication between SaveMeNow.AI devices in presence of an alarm come into play.

In Fig. 6, we show a possible operation scenario of such an alarm. Each Save-MeNow.AI device continuously checks its status and determines whether or not there is a need to send an alarm. If the MLC component of the SaveMeNow.AI device worn by a worker reports a fall, the device itself sends an alarm in broadcast mode. All the other SaveMeNow.AI devices that are in the signal range receive the alarm and, then, trigger help (for example, workers wearing them go to see what happened). If no SaveMeNow.AI device is in the range of the original alarm signal, the alarm is managed by the Gateway Devices. These must be positioned in such a way as to cover the whole workplace area. A Gateway Device is always in a receiving state and, when it receives an alarm, it sets a 30 s timer. After this time interval, if no SaveMeNow.AI device was active in the reception range of the original alarm, the Gateway Device itself sends an alarm and activates rescue operations.

As mentioned above, communications are managed through the Bluetooth protocol, in its low-energy version, called BLE. Each SaveMeNow.AI device has two roles, i.e., Central and Peripheral. The BLE protocol is ideal for our scenario because it allows SaveMeNow.AI to switch its role at runtime. During its normal use, a Save-MeNow.AI device listens to any other device; therefore, it assumes the role of Central. When the worker who wears it falls, and its MLC component detects and reports this fall, it switches its role from Central to Peripheral and starts sending the advertising data connected to the alarm activation.

Fig. 6 A possible
emergency scenario

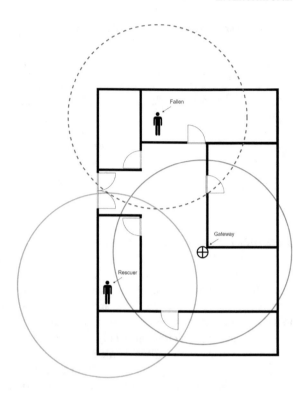

3.3   Testing of SaveMeNow.AI

After having deployed the logic of our approach in the SensorTile.box, we proceeded
with the testing campaign. Specifically, we selected 30 volunteers, 15 males and 15
females, of different age and weight, and asked them to perform different kinds of
activity. In particular, the considered activities include all the ones mentioned in
the past literature. They are reported in Table 5. Some of them could be labeled as
Fall Activity, whereas other ones could be labeled as *Not Fall Activity*. In all these
activities, SaveMeNow.AI was put at the waist of the volunteers.

In Table 6, we report the confusion matrix obtained after all the activities, and the
corresponding output provided by SaveMeNow.AI.

From the analysis of this table, we can observe that the number of real *Fall
Activities* was 1,205; 1,170 of them were correctly recognized by SaveMeNow.AI,
whereas 35 of them were wrongly categorized by our system. On the other hand, the
number of real *Not Fall Activities* was 595; 540 of them were correctly recognized by
SaveMeNow.AI, whereas 55 of them were wrongly labeled by our system. Observe
that the number of real *Fall Activities* is much higher than the one of real *Not Fall
Activities*. This fact is justified because, in our scenario, Sensitivity is much more
important than Specificity. Starting from these results, we have that the Sensitivity of

Table 5 A taxonomy for *Not Fall Activities* (on the left) and *Fall Activities* (on the right)

Not Fall Activity
Walk slow (<6 km/h)
Walk fast (≥6 km/h)
Run slow (<8 km/h)
Run fast (≥8 km/h)
Sit slowly in a chair
Sit slowly on the ground
Sit abruptly in a chair
Jump to reach an object located at the top
Go up and down the stairs slowly (<6 km/h)
Go up and down the stairs quickly (≥6 km/h)
Walk and stumble without falling down
Jump forward from an elevated position
Jump forward from the floor
Fall activity
Walk and fall forward after tripping
Walk and fall sideways (right) after tripping
Walk and fall to the side (left) after tripping
Fake fainting and fall on the right while standing
Fake fainting and fall forward while standing
Fake fainting and fall on the left while standing
Run and fall forward after stumbling

Table 6 Confusion matrix for the output provided by SaveMeNow.AI

	(Real) fall	(Real) not fall
(Evaluated) fall	1170 (TP)	55 (FP)
(Evaluated) not fall	35 (FN)	540 (TN)

SameMeNow.AI is equal to 0.97; its Specificity is equal to 0.91. Finally, its Accuracy is 0.95.

After having tested SaveMeNow.AI, we can conclude that its performance is very satisfying during both the training and the test phases. In our opinion, the dataset used for training played a key role in obtaining this successful results, because it contained very heterogeneous activities, which allowed us to create a generalized model. Indeed, our model is able to distinguish sport activities from fall activities, which is a difficult task to achieve. A careful reader could point out that a generalized model like ours sacrifices performance to generalizability. However, we observe that

Sensitivity (that is the most important parameter to evaluate in our scenario) is very high; only Specificity is not particularly high. This could lead to some false alarms that, in most cases, could be directly stopped by the worker wearing the alarming device. On the other hand, in a work environment, which is the reference application scenario of our approach, it is really common to assist to activities like running or jumping, which could generate many false alarms if the model would not be sufficiently generalized to handle them, at least partially.

4   Related Literature

Before examining related approaches in detail, we believe that a preliminary observation is necessary. In fact, unlike most approaches present in the literature, which focus on elderly fall detection, our approach was specifically conceived to detect falls in work environment. In our setting, during the working hours, an operator can perform running and/or jumping activities that can be easily confused with fall activities. Actually, in most past approaches, a wearable device contains accelerometers and gyroscopes registering the user behavior. From these sensors' perspective, sport activities and fall activities have some common points. This is the reason why we also used the Daily and Sports Activities Dataset (see Sect. 2.1) to train our classification algorithm. This choice allowed us to better train our classification algorithm in order to make it more capable of distinguishing sport activities from fall activities.

After this clarification, we start our detailed analysis of related literature by observing that, as pointed out in the Introduction, the different techniques developed for fall detection can be categorized in three different classes, depending on the environment where the user and the employed sensors operate. Specifically, it is possible to distinguish three categories of fall detection systems, namely ambient sensor based, vision based, and wearable device based [24].

The first category is based on the recording of audios and videos of the environment and/or the monitoring of vibrational data [7, 9, 33, 35, 37]. In the former case, the fall detection techniques exploit audio and video streams for object detection and tracking. For instance, in [33], the authors present an image sensing and vision-based reasoning for analyzing and verifying sensor-transmitted events. Specifically, a wireless badge node is placed between the user and her network; it detects falls through event sensing functions. Furthermore, there is a continuous tracking of the approximate location of the user performed through signal strength measurements provided by the network nodes.

Another interesting approach is proposed in [37], where the authors use the audio signal from a single far-field microphone. In particular, they create a Gaussian Mixture Model (i.e., GMM) super vector to model each fall as a noise segment, and then compute the difference between audio segments by means of the Euclidean distance. The kernel between the GMM super vectors makes up the one of Support Vector Machine employed for the classification of various types of audio and noise segments into falls.

Ambient sensors based approaches exploiting vibrational data are focused on the usage of pressure sensors. For example, in [2], the authors design a floor vibration-based fall detector. It considers the vibrations caused by objects moving on the floor, because the vibrations generated by a human fall are different from the ones related to normal activities. In this perspective, they use a special piezoelectric sensor, coupled with the floor, and generate a binary fall signal in case of a fall event.

Another proposal in this setting can be found in [29]. Here, the authors propose to use a floor sensor based on near-field imaging. This sensor detects the locations and patterns of people by measuring the impedances with respect to a matrix of thin electrodes under the floor. Then, a collection of features is computed starting from the cluster of observations associated with a person. In this way, a Bayesan filter and a Markov chain can be adopted to estimate the posture of the user and, finally, to detect a possible fall.

The approaches based on ambient sensors are not intrusive for the final user. However, they have two main disadvantages. The former regards their cost, while the latter concerns the difficulty of installing them because it is necessary to setup the whole room with sensors.

The second type of fall detection approaches concerns those based on vision [10, 11, 21, 23, 25]. The reasoning underlying this kind of system is that cameras are increasingly present in our daily environment and are less intrusive than other kinds of objects (for instance, the ones that should be worn by the user). In [23], the authors present a fall detector for smart homes based on artificial vision algorithms. The overall system is developed through a single-board computer with an external camera, placed in the room to be monitored. The approach consists of different phases. First, it acquires an image and subtracts the subject from the background. Then, it uses a Kalman filter to reduce noise in the data. Afterwards, it starts to study the changes in the human actions. Finally, it applies a Machine Learning algorithm to the data obtained to classify the current state of the subject.

Another interesting fall detection system is reported in [25], where the authors propose a framework for indoor scenarios using a single-camera system. This approach is based on the analysis of motion orientation, motion magnitude and human shape changes. According to the authors, the duration of a fall is often less than $2s$, starting when the balance is lost until the fallen person completely lies on the floor. Specifically this system works as follows: when it detects an abnormally large motion, whose direction is less than $180°$, it continues to monitor the next 50 frames. Then, if there is a downward movement, followed by the exceeding of AR ratio (which represents a body width-to-height ratio) and the inclination of the angle of the major axis of the person, a fall might have happened. Then, it monitors the next 25 frames and, if no further movement, or just a small movement, occurs, it concludes that the motion is a fall. If none of the above conditions is satisfied, no warning signal is sent out and the monitoring continues.

Vision based approaches are really interesting and can achieve a great accuracy in the fall detection process. Their main drawback concerns the necessity to install cameras in each room to monitor, which, in turn, leads to a high installation cost.

Finally, the last category of fall detection systems is based on wearable devices [3, 4, 16, 18, 20, 27, 31, 36]. These approaches rely on smart garments with embedded sensors capable of detecting the motion and location of the user body. In the literature, there are many interesting proposals, each one with different employed sensors. For instance, in [17], the authors present a posture-based fall detection algorithm that operates starting from the reconstruction of the posture of a user. Several wireless tags are placed on some parts of the body, such as hips, ankles, knees, wrists, shoulders and elbows. The locations of these tags are detected by a motion capture system, so that it can reconstruct the complete posture of a person in a 3D plane. Finally, acceleration thresholds, along with velocity profiles, are applied to detect falls.

A less invasive approach, based on an accelerometer, is presented in [22]. Here, the authors use an integrated approach of waist-mounted accelerometers, so that a fall is detected when a negative acceleration is suddenly increased, due to the change in orientation from an upright to a lying position. A similar proposal can be found in [34], where the authors design a wearable airbag containing an accelerometer and a gyroscope. This airbag is inflated when acceleration and angular velocity thresholds are exceeded.

There are also interesting fall detection proposals using Machine Learning algorithms. An example is reported in [30], where the authors propose a fall detection system consisting of a sensing unit (such as a mobile phone) and a threshold for acceleration along three axes specific to a patient. The overall system is based on monitoring the tri-axial accelerometer data in three different sliding time windows, each one lasting one second. Depending on this information and the threshold related to a patient, the authors exploit a Machine Learning algorithm to predict if she is falling or she is conducting normal daily activity.

A similar proposal is reported in [26]. Here, the authors describe a fall detection system with wearable motion sensor units fitted to the subject's body at six different positions. Each of these units comprises three tri-axial devices, i.e., an accelerometer, a gyroscope, and a magnetometer. Then, six different Machine Learning algorithms are tested to evaluate which one performs better than the others. Finally, the overall system is tested in a real world scenario, obtaining interesting results. Even if this approach achieves a high accuracy with an acceptable computation time, it could be invasive for the final user to be adopted in everyday life.

Analogously to the other categories of fall detection approaches, the wearable based ones have their advantages and disadvantages. The most important advantages are their cost efficiency, their easy installation and setup. Furthermore, these systems are not directly connected with only one place, but with a person; therefore, they can identify falls regardless of the environment. On the other hand, some disadvantages concern the low computation power and the high energy power consumption characterizing wearable devices. Another possible disadvantage could be the intrusiveness of the system in the user's life, even if researchers are constantly offering increasing small and ergonomic wearable devices.

In any case, since SaveMeNow.AI belongs to the category of wearable based fall detection approaches, we consider appropriate to present a further comparison

Table 7 Comparison between SaveMeNow.AI and several wearable based fall detection approaches proposed in past literature

Research	Sensors	Sensors'position	Algorithm	Results
SaveMeNow.AI	Accelerometer, gyroscope	Waist	Decision tree	Sensitivity: 0.97
				Specificity: 0.91
				Accuracy: 0.95
Pannurat et al. [27]	Accelerometer	Waist	Gaussian mixture model	Accuracy: 0.91
Sabatini et al. [31]	Acceleromter, gyroscope, barometric altimeter	Upper right iliac bone	Decision tree	Sensitivity: 0.8
				Specificity: 0.99
Jian and Chen [16]	Accelerometer, gyroscope	Shirt	k-NN	Sensitivity: 0.95
				Specificity: 0.96
Karantonis et al. [18]	Accelerometer	Waist	Binary classifier	Accuracy: 0.95
Zhang et al. [36]	Gyroscope	Waist	Decision tree	Specificity: 1.00
Bourke et al. [4]	Accelerometer	Waist	One-Class SVM	Accuracy: 0.96
Anania et al. [3]	Accelerometer	Jacket collar	Decision tree	Sensitivity: 0.98
Lai et al. [20]	Accelerometer	Waist, neck, right and left hands	Decision tree	Accuracy: 0.92

between it and several other approaches belonging to this category, particularly those ones that, like ours, use accelerometers and gyroscopes.

In Table 7, we report a comparison between SaveMeNow.AI and most wearable based fall detection approaches proposed in the literature. This comparison considers several characteristics, namely the position of sensors, the adopted Machine Learning algorithm and the results obtained. From the analysis of this table, we can see that SaveMeNow.AI returns results equivalent or better than the ones characterizing the other approaches. In particular, the Sensitivity of SaveMeNow.AI (which, we recall, is much more important than Specificity in our application scenario) is higher than the one of all the other approaches, except the approach of [3] that presents a Sensitivity slightly higher (0.98 against 0.97 reached by SaveMeNow), even if no information about Specificity and Accuracy is provided by the authors.

5 Conclusion

In this paper, we have proposed SaveMeNow.AI, a Machine Learning based wearable device for fall detection in a workplace. To realize it, we have preliminarily created a new dataset by merging four datasets available online in order to obtain more data on the classical activities that a worker can perform in the workplace. Then, we tested different classification algorithms and found that at least one of them, i.e. Decision Tree based on C4.5, can reach very satisfactory results when applied on the created dataset.

After this, we selected an IoT device available on the market and we natively implemented the logic of our approach on it. As for this aspect, we observe that the choice of SensorTile.box as the starting IoT device where implementing SaveMeNow.AI helped us very much reaching our goals. Indeed, thanks to SensorTile.box, we were able to implement all the operations for data collection, data pre-processing, feature engineering and classification directly into the STM32L4R9 microcontroller and the LSM6DSOX sensor contained in this device. This fact allowed us to obtain a relevant energy saving and the optimization of the limited computation power characterizing our device, as well as all the other wearable ones available in the market.

Afterwards, we tested SaveMeNow.AI in a real world scenario and found that its performance is very satisfying, especially for Sensitivity. Finally, we proposed a comparison between SaveMeNow.AI and several wearable based fall detection approaches proposed in the literature.

Regarding some possible future developments of our research, we note that, currently, the only sensor of SensorTile.box used in SaveMeNow.AI is LSM6DSOX. However, other sensors in the device may be useful to monitor some parameters to predict and/or report possible emergency situations in a workplace. For example, humidity, pressure and temperature sensors could be used for this purpose. In addition, the set of SaveMeNow.AI devices worn by operators in a delimited place can be seen as a Wireless Sensor Network that could be used, similarly to what proposed in [28], to detect emergency situations, such as fires or harmful gas leaks.

Another interesting development could be the implementation of a routing system that can show a rescuer the shortest route to the fallen worker. Finally, SaveMeNow.AI could be transformed into a non-invasive garment that allows a worker to perform operations and movements in total freedom. The simplest solution would be the insertion of the various sensors on a shirt that, once worn, would allow the evaluation of the accelerometric and gyroscopic data in a solidary way with the body, making data processing even more accurate. Last, but not the least, other sensors could be added to evaluate vital parameters, such as blood pressure and heartbeat. This would open up new frontiers in the use of SaveMeNow.AI which would also (at least partially) become a medical device.

Acknowledgments This work was partially funded by the Department of Information Engineering at the Polytechnic University of Marche under the project "A network-based approach to uniformly extract knowledge and support decision making in heterogeneous application contexts" (RSAB 2018), and by the Marche Region under the project "Human Digital Flexible Factory of the Future Laboratory (HDSFIab)—POR MARCHE FESR 2014-2020—CUP B16H18000050007'.

References

1. Altun, K., Barshan, B., Tunçel, O.: Comparative study on classifying human activities with miniature inertial and magnetic sensors. Pattern Recognit. **43**(10), 3605–3620 (2010)
2. Alwan, M., Rajendran, P.J., Kell, S., Mack, D., Dalal, S., Wolfe, M., Felder, R.: A smart and passive floor-vibration based fall detector for elderly. In: Proceedings of the International Conference on Information & Communication Technologies (ICICT'06), Damascus, Syria, vol. 1, pp. 1003–1007. IEEE (2006)
3. Anania, G., Tognetti, A., Carbonaro, N., Tesconi, M., Cutolo, F., Zupone, G., De Rossi, D.: Development of a novel algorithm for human fall detection using wearable sensors. In: Sensors, pp. 1336–1339. IEEE (2008)
4. Bourke, A.K., Lyons, G.M.: A threshold-based fall-detection algorithm using a bi-axial gyroscope sensor. Med. Eng. Phys. **30**(1), 84–90 (2008)
5. Breiman, L.: Random forests. Mach. Learn. **45**(1), 5–32 (2001)
6. Casilari, E., Santoyo-Ramón, J., Cano-García, J.: Analysis of public datasets for wearable fall detection systems. Sensors **17**(7), 1513 (2017)
7. Chaccour, K., Darazi, R., El Hassans, A.H., Andres, E.: Smart carpet using differential piezoresistive pressure sensors for elderly fall detection. In: Proceedings of the International Conference on Wireless and Mobile Computing, Networking and Communications (WIMOB'15), Abu-Dhabi, United Arab Emirates, pp. 225–229. IEEE (2015)
8. Chan, H.L.: CGU-BES Dataset for Fall and Activity of Daily Life, p. 8 (2018)
9. Chandra, I., Sivakumar, N., Gokulnath, C.B., Parthasarathy, P.: IoT based fall detection and ambient assisted system for the elderly. Clust. Comput. **22**(1), 2517–2525 (2019)
10. Cucchiara, R., Prati, A., Vezzani, R.: A multi-camera vision system for fall detection and alarm generation. Expert Syst. **24**(5), 334–345 (2007)
11. Diraco, G., Leone, A., Siciliano, P.: An active vision system for fall detection and posture recognition in elderly healthcare. In: Proceedings of the Design, Automation & Test in Europe Conference & Exhibition (DATE'10), Dresden, Germany, pp. 1536–1541. IEEE (2010)
12. Genuer, R., Poggi, J.M., Tuleau-Malot, C.: Variable selection using random forests. Pattern Recognit. Lett. **31**(14), 2225–2236 (2010)
13. Gibson, R.M., Amira, A., Ramzan, N., Casaseca de-la Higuera, P., Pervez, Z.: Multiple comparator classifier framework for accelerometer-based fall detection and diagnostic. Appl. Soft Comput. **39**, 94–103 (2016)
14. Han, J., Kamber, M., Pei, J.: Data Mining: Concepts and Techniques, 3rd edn. Morgan Kaufmann notes (2011)
15. Hussain, F., Umair, M.B., Ehatisham ul Haq, M., Pires, I.M., Valente, T., Garcia, N.M., Pombo, N.: An Efficient Machine Learning-based Elderly Fall Detection Algorithm (2019). arXiv preprint 1911.11976
16. Jian, H., Chen, H.: A portable fall detection and alerting system based on k-NN algorithm and remote medicine. China Commun. **12**(4), 23–31 (2015)
17. Kaluža, B., Luštrek, M.: Fall detection and activity recognition methods for the confidence project: a survey, A:22–25 (2009)
18. Karantonis, D.M., Narayanan, M.R., Mathie, M., Lovell, N.H., Celler, B.G.: Implementation of a real-time human movement classifier using a triaxial accelerometer for ambulatory monitoring. IEEE Trans. Inform. Technol. Biomed. **10**(1), 156–167 (2006)
19. Kwolek, B., Kepski, M.: Human fall detection on embedded platform using depth maps and wireless accelerometer. Comput. Methods Program. Biomed. **117**(3), 489–501 (2014)
20. Lai, C.F., Chang, S.Y., Chao, H.C., Huang, Y.M.: Detection of cognitive injured body region using multiple triaxial accelerometers for elderly falling. Sensors **11**(3), 763–770 (2010)
21. Mastorakis, G., Makris, D.: Fall detection system using Kinect's infrared sensor. J. Real-Time Image Process. **9**(4), 635–646 (2014)
22. Mathie, M.J., Coster, A.C.F., Lovell, N.H., Celler, B.G.: Accelerometry: providing an integrated, practical method for long-term, ambulatory monitoring of human movement. Physiol. Meas. **25**(2), R1 (2004)

23. De Miguel, K., Brunete, A., Hernando, M., Gambao, E.: Home camera-based fall detection system for the elderly. Sensors **17**(12), 2864 (2017)
24. Mubashir, M., Shao, L., Seed, L.: A survey on fall detection: principles and approaches. Neurocomputing **100**, 144–152 (2013)
25. Nguyen, V.A., Le, T.H., Nguyen, T.H.: Single camera based fall detection using motion and human shape features. In: Proceedings of the Symposium on Information and Communication Technology (SoICT'16), Ho Chi Minh, Vietnam, pp. 339–344 (2016)
26. Özdemir, A.T., Barshan, B.: Detecting falls with wearable sensors using machine learning techniques. Sensors **14**(6), 10691–10708 (2014)
27. Pannurat, N., Thiemjarus, S., Nantajeewarawat, E.: A hybrid temporal reasoning framework for fall monitoring. IEEE Sens. J. **17**(6), 1749–1759 (2017)
28. Qandour, A., Habibi, D., Ahmad, I.: Wireless sensor networks for fire emergency and gas detection. In: Proceedings of the International Conference on Networking, Sensing and Control (ICNSC'12), Beijing, China, pp. 250–255. IEEE (2012)
29. Rimminen, H., Lindström, J., Linnavuo, M., Sepponen, R.: Detection of falls among the elderly by a floor sensor using the electric near field. IEEE Trans. Inform. Technol. Biomed. **14**(6), 1475–1476 (2010)
30. Saadeh, W., Altaf, M.A.B., Altaf, M.S.B.: A high accuracy and low latency patient-specific wearable fall detection system. In: Proceedings of the International Conference on Biomedical & Health Informatics (BHI'17), Orlando, FL, USA, pp. 441–444. IEEE (2017)
31. Sabatini, A.M., Ligorio, G., Mannini, A., Genovese, V., Pinna, L.: Prior-to-and post-impact fall detection using inertial and barometric altimeter measurements. IEEE Trans. Neural Syst. Rehabil. Eng. **24**(7), 774–783 (2015)
32. Sucerquia, A., López, J.D., Vargas-Bonilla, J.F.: SisFall: a fall and movement dataset. Sensors **17**(1), 198 (2017)
33. Tabar, A.M., Keshavarz, A., Aghajan, H.: Smart home care network using sensor fusion and distributed vision-based reasoning. In: Proceedings of the International Workshop on Video Surveillance & Sensor Networks (VSSN'06), Santa Barbara, CA, USA, pp. 145–154 (2006)
34. Tamura, T., Yoshimura, T., Sekine, M., Uchida, M., Tanaka, O.: A wearable airbag to prevent fall injuries. IEEE Trans. Inform. Technol. Biomed. **13**(6), 910–914 (2009)
35. Wang, F., Wang, Z., Li, Z., Wen, J.R.: Concept-based short text classification and ranking. In: Proceedings of the International Conference on Information and Knowledge Management (CIKM'14), Shangai, China, pp. 1069–1078. ACM (2014)
36. Zhang, T., Wang, J., Xu, L., Liu, P.: Fall detection by wearable sensor and one-class SVM algorithm. In: Intelligent Computing in Signal Processing and Pattern Recognition, pp. 858–863. Springer (2006)
37. Zhuang, X., Huang, J., Potamianos, G., Hasegawa-Johnson, M.: Acoustic fall detection using Gaussian mixture models and GMM supervectors. In: Proceedings of the International Conference on Acoustics, Speech and Signal Processing (ICASSP'09), Taipei, Taiwan, pp. 69–72. IEEE (2009)

Artificial Intelligence and Security

Fraud Detection in Networks

Paul Irofti, Andrei Pătraşcu, and Andra Băltoiu

1 Introduction

The value of fraudulent transactions in the EU was about 1.8 billion euro in 2016 for card fraud [20] only. Many other financial criminal activities, like money laundering (e.g. through multiple or over-invoicing), corruption-related transfers, VAT evasion, identity theft, affect individuals, banks and state activities like tax collection. As the number of transactions increases and criminal behavior becomes more sophisticated, fraud detection requires more attention and time from human experts employed by banks or state authorities. The need of performant automatic tools for at least selecting the most likely fraudulent activities, but aiming also to detect new types of ill-intentioned activities, is imperative. This is by no means valid only in the case of financial crimes. Insurance frauds, e-commerce or social networks misconducts such as fake reviews are other examples of domains where securing activities are critical.

Money transactions (payments, transfers, cash withdrawals, etc.) can be described by a vector of characteristics. The suspect transfers—the anomalies—may be regarded as outliers in a given set of training vectors. However, treating transactions as independent vectors is an over-simplification, due to the intricacies of many types of criminal behavior. It is much more adequate to treat the transactions in their natural form, that of a *graph* whose nodes are the financial entities (individuals, firms, banks) and whose edges are transactions data (amount, time, payment mode, etc.).

P. Irofti (✉) · A. Pătraşcu
Department of Computer Science, University of Bucharest, Bucharest, Romania
e-mail: paul@irofti.net

A. Pătraşcu
e-mail: andrei.patrascu@fmi.unibuc.ro

A. Băltoiu
The Research Institute of the University of Bucharest (ICUB), Bucharest, Romania
e-mail: andra.baltoiu@fmi.unibuc.ro

© The Editor(s) (if applicable) and The Author(s), under exclusive license
to Springer Nature Switzerland AG 2021
A.-E. Hassanien et al. (eds.), *Enabling AI Applications in Data Science*,
Studies in Computational Intelligence 911,
https://doi.org/10.1007/978-3-030-52067-0_23

A directed edge between two nodes illustrates that there is a money transfer in the respective direction, where the weight on the edge is the transferred amount. Graphs allow to model inter-dependencies, capture the relational nature of transactions and are also a more robust tool, as fraudsters usually do not have a global view of the graph. Some frauds, in other words, imply a particular scheme of relationships that can be revealed only by looking at the underlying graph structure. In those cases, taken in isolation, an individual transaction may display little or no indication of being fraudulent but its nature becomes apparent when the larger context is considered.

This review is concerned with such frauds that take place over networks. Most of the times, the data consists of an attributed graph, with attributes either on the nodes, edges or both. Fraud detection methods thus need to integrate both relational knowledge, accessible from the network structure, and the numeric features that describe graph components. Depending on the application, anomalies can be considered at the level of a node, an edge, or a subgraph. Our present review is concerned mostly with finding unusual subgraph patterns, as they frequently appear in the types of application we consider. We survey several types of topological features that can be exploited in order to differentiate legitimate graph entities from fraudulent ones and examine how relational information can be used in tandem with other numerical data sources.

Robustness to camouflage strategies is becoming one of the requirements for fraud detection methods. In some cases disguise is impossible without knowledge of the network structure, which is unavailable to all users including fraudsters. Others, still, are more prone to be affected by deceitful actions, even in the absence of this type of information. Throughout the survey, we signal the solutions that take camouflage into consideration.

Whenever applicable, we also focus on solutions that take into account the evolution of a graph. The need for including the temporal dimension is two-fold. On one hand, in some cases it is the temporal pattern that defines the anomaly. Several types of frauds, including credit card frauds, network attacks and spam, involve some sort of high-frequency activity. It is not the activity per se that gives away its fraudulent nature, but the fact that it occurs often or in a specific pattern. If no temporal information were considered, the behavior could pass as legitimate. On the other hand, there is the pressing need to react as fast as possible to the occurrence of a misconduct. Fraudulent event detection therefore requires temporal representation.

As an overall perspective dictated by the above anomaly environments, fraud detection methods are thus expected to:

- function in semi-supervised or unsupervised scenarios; this requirement is motivated by the need to discover unknown fraud patterns and keep pace with novel fraud schemes.
- be robust to camouflage; as previously mentioned in the Introduction and in part related to the previous requirement, robustness to camouflage is needed since particular forms of frauds, especially in finance, take considerable measures in appearing legitimate.

- function in (quasi) real-time; with the definition of real-time differing slightly from one application to another, what is common to all fraud detection methods is the need that the estimates they provide are actionable in due time.
- result in high specificity; since the discovery of frauds is intended to be coupled with corrective actions that in one form or another restrict access to the network, solutions are preferred that do not hinder the legitimate activity.

1.1   Preliminaries and Chapter Outline

We continue our **Introduction** by referring to existent work in surveying methods for fraud identification. While, to our knowledge, no review addresses the exact problematic we do, some tackle similar topics. Section 1.2 presents related work and states the differences from our approach.

In Sect. 2 (**Locality**) we survey methods that examine the local patterns of the graph, by working with neighborhoods. It is a top-down view of the network, where communities (Sect. 2.1) and other compact regions (Sects. 2.2, 2.3 and 2.4) are analyzed and evaluated in terms of their conformity to normal network structures.

Section 3 (**Hybrid Clustering**) deals with integrating topological information with node or edge numerical attributes in order to go beyond subgraph pattern matching. The majority of the solutions reviewed here create hybrid measures and employ clustering techniques for the task.

Opposite to Sect. 2, in Sect. 4 (**Perspectives**) we present different views on the topology of a network. The top-down approach is replaced by singular viewpoints, as in the case of egonets or when information contained in the attributes of the nodes or edges can also provide new perspectives in representing the graph (Sect. 4.1). The situation where the graph can be decomposed in several overlapping networks is treated in Sect. 4.2.

We end our review (Sect. 5, **Challenges**) with a word on dealing with missing information and present solutions for estimating the graph when it is not available and for managing the lack of positive examples.

1.1.1   Notations

We now briefly define some of the notions most present in our survey.

Definition   The adjacency matrix of graph G is the matrix $A \in \mathbb{R}^{V \times V}$ with elements in $\{0, 1\}$ where V is the number of vertices and $A_{ij} = 1$ if there is an edge between vertices i and j, and $A_{ij} = 0$ otherwise. For multi-graphs A_{ij} is equal to the number of edges from x to y.

Definition   The vertex indexed Laplace matrix L of graph G is $L = D - A$, where D is the diagonal degree matrix associated with each vertex. For undirected graphs L is symmetric with zero row sums.

Definition   The spectrum of a graph G is the spectrum of the adjacency matrix A.

1.2 Related Work

Several surveys exist that tackle similar topics. We briefly state the main differences from our present review. Whenever appropriate, we focus on works newer than those covered by existing surveys.

Dating 2010, [21] focuses on fraud detection techniques from an audit perspective, however it does not cover only machine learning approaches. A classification-oriented survey on financial fraud identification can be found in [47], yet it does not review solutions for graphs. Accuracy and specificity summaries obtained by various machine learning methods on different types of financial frauds can be found in [68]. The survey in [53] presents supervised and unsupervised methods and have a dedicated section to similar applications. Credit card fraud detection methods are reviewed in [13] and more recently in [61].

Fraud detection can be cast as an anomaly detection (AD) task, since frauds are rare events that distinguish themselves from normal behavior. As such, methods for anomaly identification on graphs represent a great pool of solutions to the problem of fraud detection. Nonetheless, it must be noted that in the absence of an application-driven definition of a graph anomaly, some general methods can be ineffective for our task. A comprehensive survey on graph anomaly detection can be found in [2]. The taxonomy is constructed along the following lines: (a) quantitative descriptions of graph anomalies and qualitative explanations/explorations; (b) methods that consider static graphs and those that deal with the evolution of the graphs; (c) graph analysis based on the structure of the graph and on communities [2]. The distinction between structure-based and community-based analysis is relevant to the works we present herein and is also a point where our approach diverges from the one in [2].

In the case of simple, non-attributed graphs, the above cited survey defines structure-based methods as those which seek information in the characteristics of the graph to define (ab)normality: node-level measures such as degree, between nodes measures such as number of common neighbors. Community-based methods, on the other hand, assume in their perspective that anomalous nodes do not belong to a community, and are found to be linking communities together. While the assumption is true when working with a broad definition of anomalies and may apply to some types of network frauds as well (especially those involving fake accounts of different sorts), it does not cover a wide range of misconducts that are performed by otherwise legitimate entities (e.g. financial frauds). Moreover, particularly in large networks it is to be expected that not all legitimate nodes belong to communities [9]. The definition of a community, as well as that of an anomaly is application-dependent.

The two above categories are also used in [2] when considering the case of attributed graphs, albeit with a slightly different meaning. Attributed graphs contain additional information, either on the nodes, thus describing features of the respective entity or on the edges, characterizing the relationship between two entities. Therefore, in this case, structure-based methods are, for the authors of [2], those methods that look for unusual subgraph patterns. The definition is particularly relevant to the fraud detection problem, since frauds are often performed by a group of entities, as a scheme. As mentioned earlier, taken individually, the events making up the scheme may look legitimate, yet they form an anomalous structure. More on several such structures, successfully identified on a dataset of real financial transactions data, however not dealing with attributed graphs, can be found in [17].

Returning to the perspective of [2], community-based methods are now concerned with finding the odd node out in a given community. The authors introduce a third category of relational learning. Unlike the regular learning paradigm, where independence is assumed between entities, relational learning seeks to incorporate connectivity information, for example by looking at one node's neighbours as well.

2 Locality

The large variety of anomalies types arising in networked environments have led to different research directions depending of applications. In this section we survey literature sectors which orbit around the idea of detecting anomalies in attributed networks through promotion of particular subgraphs having unusual structure, such as high density or specific connectivity patterns like rings, cliques or heavy paths.

2.1 Communities

Perhaps the most widespread approach when considering anomalous patterns is that of using community information to train a supervised learning system.

In [52] a normality measure is used to quantify the topological quality of communities as well as the focus attributes of communities in attributed graphs. In other words, normality quantifies the extent to which a neighborhood is internally consistent and externally separated from its boundary. The proposed method discovers a given neighborhood's latent focus through the unsupervised maximization of its normality. The respective communities for which a proper focus cannot be identified receive low score, and are deemed as anomalous.

A modularized anomaly detection hierarchical framework has been developed in [17] to detect static anomalous connected subgraphs, with high average weights. For this purpose, particular community detection strategies are tailored based on 140 features (including Laplacian spectral information) and network comparison tests (such as NetEMD). Then, a classification via random forests or simple sum of individual (feature-based) scores is performed to highlight the anomalous subgraphs.

In directed trading networks, blackhole and volcano patterns represent groups of nodes with inlinks only from the rest of nodes or outlinks only towards the rest nodes, respectively. These kinds of patterns, which often have fraudulent nature, are isolated in [40] through pruning (divide-et-impera) schemes based on structural features of blackholes and volcanoes.

In [56], community detection methods are combined with supervised learning for detecting money laundering groups in financial transactions. Community detection begins with extraction of (possibly overlapping) connected components from the transactions attributed multi-graph. Since typical fraudulent communities contain a small number of vertices (less than 150), the excessively large connected components extracted from the AUSTRAC dataset [37] are further decomposed through a k−step neighborhood strategy. This entire process leads to a collection of small communities which are classified through a supervised learning scheme (Support Vector Machine and Random Forests).

Within communities, uncommon subgraphs can be mined using structural information, as shown in [7]. The authors propose adaptations of the dictionary learning problem to incorporate connectivity patterns. One such adaptation involves imposing that the dictionary atoms express a Laplacian structure, thus creating a dictionary of elementary relational patterns.

Evolutionary networks are considered in [10], where a community detection strategy is used to highlight anomalies based on the temporal quantitative evolution of network communities.

2.2 Dense Subgraphs

The intuition behind searching for dense blocks in graphs as signs of anomalies is that some frauds are performed by repetitive activity bursts. When looking at the graph connectivity, these activities form a dense subgraph that stands out from the sparser normal activity. In yet other cases, the density is a consequence of multiple malicious users acting similarly and synchronously, a behavior known as lockstep.

The dense subgraph detection problem is approached in [75]. The authors consider a hierarchical framework of subgraph detection, where the initial graph is successively filtered in k steps until a dense cluster results. After modeling this problem as the maximization of a nonconvex quadratic finite-sum (with k term) over integers, several relaxations are applied: (i) ordering in node vectors (equivalent with replacing binary constraint $\{0, 1\}$ with convex interval $[0, 1]$); (ii) penalization of hierarchical density order such that the k−th subgraph is more dense than the subgraph $(k − 1)$. Furthermore, as typical for continuous nonconvex problems, a block-coordinate gradient descent algorithm with Armijo stepsize policy is presented and its convergence to a stationary point of QP model is proved. For numerical evaluation, the authors use the AMiner co-authorship citation network and a financial bank accounts network. Large density clusters detection is proved along with superiority over existing 2-hierarchies strategies.

The approach in [5] considers tensors for modelling large scale networked data. Starting from the fact that formation of dense blocks is the result of certain entities sharing between two or more entries in the tensor, they construct an Information Sharing Graph (ISG) illustrating these relations. Since the dense blocks in the tensor model leads to dense subgraphs in the ISG, an efficient D-Spot algorithm that detects these dense subgraphs is proposed. Along with some theoretical guarantees for the subgraphs densities generated by the algorithm, the empirical evaluation shows that D-Spot has better accuracy than other tensor-based schemes on synthetic and real datasets such as Amazon, DARPA, Yelp and AirForce.

The work in [32] sets the goal to detect suspicious nodes in a directed graph that have synchronized and abnormal connectivity patterns. First, a mapping is proposed that embeds the data into a chosen feature space. Then, synchronicity and normality measures are introduced. Similarity is computed between the points resulting from the embedding as well as normality of the given data features relative to the rest of the data. Parabolic lower bounds are shown for the synchronicity-normality function. Superior performance (precision, recall, robustness) over well-known state-of-the-art static graph anomaly detection techniques such as Oddball and OutRank is shown on three real world datasets, namely TwitterSG, WeiboJanSG an WeiboNovSG, all of which are complete graphs with billions of edges.

Coordinated activity can also be detected in edge streams. The problem of near real-time detection of fraudulent edges is addressed in [19]. The authors propose a combined metric for labeling attacks that takes into account activity bursts, as well as path weights between source and destination regions of the graph. The solution keeps a memory-reasonable edge sample for comparison and uses a random walk method with a modified score in order to label fraudulent edges. Furthermore, the work in [72] sets to uncover both structural and weight changes in graphs. A structural anomaly is considered when the process of adding or deleting edges is not smooth, namely when the first and second derivative of the node scores are large. The score of a node is given a PageRank-like interpretation and is updated dynamically as modifications appear on the graph.

In [55] constrained cycles are detected in dynamic graphs and labeled as fraudulent activities in financial payments system (fake transactions). The authors consider a directed attributed graph with varying edge-structure over time. For each incoming edge between vertices (u, v), efficient algorithms are given to generate all fixed $k-$length cycles between u and v. Based on empirical observed issues in the case when high-degree vertices (hot points) are encountered in the generated paths, some indexing procedures are proposed to boost time performance of the brute force depth-first search algorithm. The evaluation data is based on real activity from Alibaba's e-commerce platform, containing both static and dynamic edges, resulting in a graph with $\approx 10^9$ vertices and $\approx 10^9$ edges. Results show a somewhat improved performance, guaranteed by the indexing procedure.

An altogether different way of looking at graphs is through the concept of k-core structures, which represent "the maximal subgraph in which all vertices have degree at least k" [59]. The authors study a number of real world networks (such as social networks and citation networks) and observe several patterns related to the coreness property. One such model, named the "Mirror Pattern", correlates the core number to the degree of a vertex. The authors are thus able to identify anomalies deviating from this pattern, specifically lockstep-type attacks, as vertices with low degree and high coreness.

2.3   Bipartite Graphs

A bipartite graph is a graph that can be split into two distinct subsets such that all edges connect a node from the first subset to one from the second subset. In some fraud detection problems, a bipartite (sub)graph occurs naturally as a result of scams, or is a convenient way of representing the data.

One reason for the evasive nature of network frauds is the difficulty of finding the right focus: looking at individual nodes/edges may not reveal anything suspicious, and likewise, considering a too large set of entities may obscure fraudulent activities occurring within the group [69]. The work in [49] considers two types of identities in auction networks besides honest users: frauds and accomplices. The latter category supports frauds, but also acts as a camouflage, by adding legitimate activities to the frauds' repertory. The networks these two types of users create form a bipartite core within the large graph. The authors then develop a belief propagation algorithm, which infers the identity of a node by evaluating the neighbours. An adaptation is also provided that efficiently solves the identification problem when the graph structure evolves in time.

Cases when the two classes of nodes involve mutual relations are approached in [42], where bipartite graphs are also used to represent data. The fraudulent instances considered here are assumed to satisfy some given empirically observed traits such as: fraudsters engage as much firepower as possible to boost customer objects, suspicious objects seldom attract non-fraudulent users and fraudulent attacks are well represented by bursts of activity. Further, they detect fraudulent blocks corresponding to both vertices sets in the bipartite graph and formulate a metric that measures to what extent a given block obeys the fixed traits. By maximizing this metric over the entire data, suspicious blocks are labeled. The experiments show that the solution achieves significant accuracy improvements on synthetic and real data, compared to other fraud detection methods.

More particular bipartite reviewer-product data is considered in [14] and, using unsupervised algorithmic heuristics, the authors aim to find fraudulent groups of reviewers that typically write fraud reviews to promote/demote certain products. DeFrauder detects suspicious groups by several coherent behavioral signals of reviewers based on particular quantitative measures such as: reviewer tightness, neighbor tightness and product tightness. Also, the ranking of groups over the spamicity degree is realized through a specific ranking strategy. Experiments on four real-world labeled datasets (including Amazon and Playstore) show that the DeFrauder algorithm outperforms certain baselines, having 11.35% higher accuracy in group detection.

Furthermore, deep learning is used in [66] to design novel graph fraud detection methods. The data, representable as a bipartite graph (e.g. nodes are users on one side and products on the other), is embedded into a latent space such that the representations of the suspicious users in the same fraud block sit as close as possible, while the representations of the normal users are distributed uniformly in the remaining latent space. In this way, the additional density-based detection methods might easily detect the fraud blocks. In fact, the deep model from [66] involves an autoencoder used to reconstruct the "user" nodes information from the bipartite graph and, at the same time, to ensure that in the low dimensional latent space the anomalous instances are sufficiently similar with respect to a proposed similarity measure. Thus, the objective function of the minimization problem contains the nonlinear composite terms associated to reconstruction and similarity. Experimental evaluations are given for some synthetic datasets and a real-world network attack dataset. The tests show that the model is able to robustly detect multiple fraud blocks without predefining the number of blocks, in comparison with other state-of-the-art baselines (which do no rely on deep learning) such as HoloScope, D-cube and others.

2.4 Spectral Localization

As stated in [17], *spectral localization is the phenomenon in which a large amount of the mass of an eigenvector is placed on a small number of its entries.* The methods surveyed in this section use spectral information to find particular subgraphs and thus distinguish anomalies. While not all focus on attributed graphs and some works rely only on topological information, the selection is relevant for the descriptive power of eigenvalues and eigenvectors. This potential can be used for a subsequent extension towards nodes or edges attributes.

Related to Laplacian matrices, the dominant components of their eigenvectors correspond to nodes in the network with special properties and thus constitute good candidates for the anomaly detection task (see [12, 50]). In the series of papers [43–46] a set of schemes are developed in order to uncover anomalies using spectral features of the modularity matrix. Furthermore, the authors of [45] extend these methods to use the ℓ_1 norm for eigenvectors of sparse Principal Component Analysis (PCA), which performs well at the cost of being more computationally intensive.

Another spectral framework is presented in [71], that exploits the localization of attackers in a different spectral region compared to regular users. Moreover, their identification process of anomalous patterns in graph spectral spaces supports collaborative attacks detection. The authors start from a clean graph G that has been perturbed by additional anomalous nodes and vertices N resulting in graph $\tilde{G} = G + N$. The nodes N create a clique amongst themselves to mimic authentic communities and avoid detection through standard topological based methods. But these nodes also have to reach-out to the existing nodes in G for the purpose of fraud or other malign intentions. The N nodes randomly attack victim nodes from G in order to further avoid detection. The combination of these techniques is called Random Link Attack (RLA) in [71]. In this setup, anomaly detection is performed by taking into account the k largest eigenvalues and their associated eigenvectors.

The authors provide a result that pinpoints the change of each eigenvector element which is directly tied to an attacker node based on its victims. Following this insight attacking nodes can be separated from victim nodes due to the special distribution of their corresponding eigenvector elements that can be upper bounded, thus unveiling a non-random pattern that is not observed in victim nodes. This spectrum-based detection approach shows superior performance compared with the topology-based detection, especially on mixed RLAs models.

A similar approach is presented in [69] where the attacker nodes are considered subtle subgraphs with minor impact on the overall spectrum of the general graph (termed the background graph). The authors motivate the selection of the first largest k eigenvalues by showing that when adding attacker nodes N to the existing graph G, the added information can only be separated in the first two eigenvectors x_1 and x_2. Afterwards the victim and attacker node information is indistinguishable. Unlike before, the authors of [69] provide exact formulation for the entries in the first two eigenvectors without the use of the power method. With this result they continue with the actual anomaly detection where the two types can be separated in similar fashion: based on the distribution of their associated entries.

Lockstep behavior can also be revealed by spectral properties, as shown in [31]. The authors identify different anomalous connectivity patterns by bridging adjacency information with information from the singular vectors.

A different spectral approach, proposed in [63], is based on matrix factorization in bipartite networks and leverages the intuition that the nodes and edges which are badly represented by the factorization should be considered as anomalies.

The authors of [74] use spectral information to train a deep autoencoder and a convolutional neural network based on the RLA attack model and the spectral strategy from [71]. They start with the k largest eigenvectors of the graph G that they arrange as the columns of matrix $M \in \mathbb{R}^{k \times V}$ and for each vertex v_j they select the corresponding row from M. Then they proceed to compute the mean vector spectral of its n-nearest neighbours, iterating from 1 to n. The input for the neural network is obtained by concatenating the above vectors into a single large feature vector.

3   Hybrid Clustering

This section is concerned with attempts at integrating topological information with the contextual information of the graph attributes. The solutions described here depart from the local view present in the previous section and from the aim of finding particular structures. Instead, the methods surveyed here use hybrid measures that aggregate structure data with node or edge descriptions.

In many financial databases, individual isolated fraudulent transactions appear as normal entries if one simply relies only on local statistical feature analysis. A higher-level perspective based on the nodes and edge attributes and on the dependencies between transactions leads to a better indicator of anomalous payments. Inspired by these observations, in [8] the authors seek fraudulent payments in e-commerce transactions networks, based on inter-transaction graph dependencies. The work models the real payment transaction Electronic Arts database as a Heterogenous Information Networks (HIN), which represents an attributed graph with multiple nodes and edge types. The approach is strongly based on several meta-path concepts that are introduced in order to extract semantic relations among multiple transactions (using the paths connecting them) and aggregation of label information. Over baseline methods such as Support Vector Machine (SVM) and Random Forests, the proposed HitFraud algorithm provides a boost of 7.93% in recall and 4.63% in F-score on the EA data.

An integrated anomaly detection framework for attributed networks is proposed in [41]. A preliminary clustering strategy is presented, which provides the degrees to which attributes are associated to each cluster. Then, a subsequent unsupervised learning procedure is applied based on the representation of the links and data attributes by the set of outcome vectors from the clustering stage. Finally, the abnormal attributes and their corresponding degrees of abnormality are computed on the basis of these representations. In [3], clusters are constructed of nodes that have "similar connectivity" and exhibit "feature coherence", based on the intuition that clusters are a way to compress the graph. As such, the Minimum Description Length (MDL) principle is used to derive a cost function that encodes both the connectivity matrix and the feature matrix.

Edge-attributed networks are considered in [58] from an information-theoretic perspective similarly based on MDL. The algorithm consists of a combination between an aggregation step of neighborhood attributes information and a clustering step used to provide an abnormality score on each node using the aggregated data. MDL is also used in [64], where normative graph substructures are identified by taking into account some coverage rules and their quantitative occurrence is established. Anomalous substructures are selected from those with least occurrences in the graph.

In [65] certain types of anomalies are detected based on scoring each node (or an entire subgraph) using statistical neighborhood information, such as the distance between the attributes of the node and its neighbors. A combination of this scoring procedure with a deep autoencoder is also provided. Several social network statistical metrics and clustering techniques are also used in [11] to detect fraud in a factoring company.

The model from [38] defines a normal instance as one that has a sparse representation on a set of some representative instances. The problem of anomaly detection is thus cast as a minimization problem of the representation residual. The network structure is included in the model through a Laplacian type quadratic penalty. Furthermore, the model developed in [51] selects a subset of representative instances on the space of attributes that are closely hinged with the network topology based on CUR decomposition, and then measures the normality of each instance via residual analysis.

The authors of [39] adopt a finite mixture model to interpret the underlying attribute distributions in a vertex-attributed graph. In order to accommodate graph structure, network entropy regularizers are proposed to control the mixture proportions for each vertex, which facilitates assigning vertices into different mixture components. A deterministic annealing expectation maximization is considered to estimate model parameters.

In [36] a hybrid anomaly detection approach is considered that employs clustering (Euclidean Adaptive Resonance Theory) to establish customers' normal behaviors and then uses statistical anomaly index to determine deviation of a particular transaction from the corresponding group behavior.

Following a different approach, other attributed graph clustering methods seek to jointly optimize the potential of a cluster to incorporate both topological information and node/edge attributes. Such an approach is proposed in [26] and uses nonnegative matrix factorization to cluster the graph based on topology and on features at the same time. The objective function is composed of a term that models structure and a term that captures information on attributes.

Yet another approach uses a Bayesian interpretation and computes the maximum a posteriori estimate of the joint probability distribution over all possible clusters and all attributed graphs [70].

A latent attributed network representation is learned in [73] by using a number of network walks. The representation is obtained through maintaining the pairwise vertex-distance in the local walks and by hybridizing it with the hidden layer of deep auto-encoder, such that the resulting embedding is guaranteed to faithfully reconstruct the original network. Then, a dynamic clustering model is used to flag anomalous vertices or edges based on the learned vertex or edge representations. Moreover, leveraging a reservoir sampling strategy, any dynamic network change induces only modest updates on the learned representations.

The vulnerability of a deep fraud detector is analyzed in [23] using adversarial examples based on small perturbations of input. The purpose is to increase the robustness of the detector by finding representative adversarial instances. Based on the iterative Fast Gradient Sign Method, the authors propose the Iterative Fast Coordinate Method for discrete L_1 and L_2 attacks which is efficient in generating large amounts of instances with satisfactory effectiveness. The other two attack algorithms are based on minimization of the fraudulent score (from the output logit layer) over a positive integers ball. Besides the one which repeatedly searches for perturbations, the Rounded Relaxation scheme approaches minimization of the fraudulent score over a relaxation of the integer ball to the real ball. Using some reparametrizations, the ball-constrained problem is converted into and unconstrained one. Experimental evaluation is performed on data generated from TaoBao platform, containing millions of real-world transactions. Results show that the fraud detector significantly outperforms the adaptions of the state-of-the-art attacks and is significantly robust well on the unperturbed data.

In [6] the dynamic graphs are modeled as sets of triplets. The triplet is composed of source-destination nodes and the timestamp that it arrived at. They aim to find (spatial and temporal) clusters of similar edges that show a large burst of activity in a short period of time within the graph structure. The anomaly detection technique considered in [6] relies on dynamic computation of an anomaly score for each edge, based on chi-squared statistic, which allows to derive false positives probability bounds. Empirical effectiveness is proved on DARPA, TwitterSecurity and TwitterWorldCup datasets.

4 Perspectives

So far, we have surveyed methods that consider a panoptic view of the graph, where topological information is typically conveyed by connectivity matrices such as the adjacency matrix and graph Laplacian or is summarized using different network statistics. We turn now to alternative ways of exploiting connectivity information, as well as to the case where a single graph has limited descriptive power.

4.1 Egonets and Multiview

An egonet is the induced subgraph formed by the neighbors of a single node. The authors of [1] have done an extensive research on several real-world networks and found that using carefully extracted, but otherwise intuitive features for describing an egonet, unforeseen normality power laws appear. These patterns describe dependencies such as between the number of nodes and edges, between the weights and number of edges, the principal eigenvalue and total edge weight. The power laws, which have been tested for both uni- and bipartite, weighted and unweighted graphs, encourage the use of various outlier detection metrics. A mix of distance-based heuristic and a density-based score is used in [1]. The method is used to identify several types of graph anomalies such as (near-)cliques, stars, heavy vicinities and heavy links.

The work in [57] uses the egonet approach and a thresholding technique to detect anomalous cliques, considering statistical measures of the underlying graph model. A similar solution uses egonets for fraud detection in e-banking transaction data [67]. Authors use the Mahalanobis distance metric to label anomalous accounts.

A combination of egonet attributes and node features is used in [25], together with a set of recursive features. The latter consist of aggregated means and sums of different network metrics. The method thus constructs an abstract characterization of a node that serves in classification and de-anonymization (identity resolution) tasks.

A different connectivity formulation can also be obtained by casting the multi-attribute representation of a graph as a multi-view formulation. This implies looking at the relationship between the entities (nodes) from different perspectives, namely one for each attribute. The authors of [48] propose a metric for quantifying how suspicious a group of nodes is, by extending previous single-view metrics such as mass (sum of edge weights) or average degree to the multi-view setting. Aggregating across views led to 89% precision in detecting organizations that violated the Snapchat Terms of Service through different fraud schemes.

4.2 Multiple Graphs

If the nodes (or edges) of a network are highly heterogeneous, working with the entire graph might leave out important relationship attributes. Consider the case where nodes can represent actors that interact differently across domains. Example:

individuals connections with colleagues, friends and family which can be represented as three separate graphs or a single graph with multiple edges between nodes (e.g. colleagues can also be friends). In other words, the underlying graph is multi-relational. In [22] anomalies are viewed as nodes that perform inconsistently across multiple sources (e.g. a close family member that is not friendly). Each source S_i is used to build a simple graph G_i containing the same vertices V but with different edges E_i. Let v_{ij} represent vertex j from graph G_i, the multigraph G is built by connecting the identical vertices from all source nodes, i.e. there will exist an edge between v_{ij} and v_{kj} with $k \neq i, \forall j \in V$. Let m_{ij} be the constraint placed between two sources S_i and S_j with $M_{ij} = m_{ij} I$. The adjacency matrix of the resulting graph is built by placing on the diagonal the adjacency matrix of each graph G_i and M_{ij} on the corresponding off-diagonal entries. With this the authors build the Laplace matrix and use it to select the largest k eigenvectors that they further split into two parts P and Q, thus performing soft spectral clustering. Finally they assign an anomalous score to each vertex v_j through the cosine distance between the corresponding vectors from p_j and q_j, respectively. We note that a similar multi-source approach, called unmasking, has been applied in computer vision for deep anomaly detection in video content [27].

5 Challenges

In this section we briefly mention two general issues that occur in fraud detection applications. The first concerns the case where the real graph is actually not known or is incomplete. In practice, this situation can arise for several reasons, including privacy protocols. The second issue is that of missing positive examples, which often occurs as a consequence of anomalies being extremely scarce.

5.1 Estimating the Graph

While some methods for fraud detection are graph-agnostic and treat transactions as regular signals, as the field matures, it becomes a requirement to include information on the underlying network. In some applications however, the real network is not available. Dictionary learning (DL) [16] offers a solution for estimating the graph, as well as classifying the signals. In DL, one seeks for a sparse representation of the samples, as well as the basis (dictionary) for that representation. When the signals lie on a weighted graph, the dictionary is also required to capture the underlying graph structure. One approach is to include the graph Laplacian, which incorporates graph patterns, into the learning problem.

Often, signals are similar when the nodes they rest on are connected, contributing to the smoothness property of the graph. Integrating smoothness ensures that the structure of the dictionary captures the graph topology and is generally obtained via a regularization term that controls the similarity between dictionary atoms with respect to local patterns. An additional way to address smoothness is by requiring the sparse representations to follow a similar rule. This task can be achieved by taking another Laplacian, called the manifold Laplacian, which considers the feature space defined by the signals.

DL classification has been successfully applied to the task of anomaly detection in water distribution networks [30]. A different formulation, that is suited for online, semi-supervised applications, has been tested on the task of detecting malware files [28]. Subsequent tests on credit card data has shown good results in detecting fraud as well [29]. Moreover, the latter work also addresses the problem of unsupervised anomaly detection for fraud identification.

5.2 Dealing with Label Unbalance

Oftentimes, the datasets for fraud detection are extremely unbalanced, containing only few illegitimate activities. Moreover, new fraudulent schemes may constantly appear, as a response to increasing anti-fraud methods and policies. In those cases, no positive examples exist in the databases.

One-class classification problems refers precisely to those two-class problems where the main class is the well sampled ("normal" samples), and the other one is a severely undersampled because of its extremely diverse nature ("abnormal" samples). The main objective of one-class classification technique is to distinguish between a set of target objects and the rest of existing objects, which are defined as anomalies or outliers [60]. One-class Support Vector Machine (OC-SVM) represents an effective boundary based classification method which provides an optimal hyperplane with maximum margin between the data points and the origin. A new data sample will be classified as normal if it is located within the boundary and conversely, as abnormality when if lies outside the boundary [4, 18, 35, 62]. The method has been applied on money laundering applications and for finding fraudulent credit card transactions [24, 33].

One approach to solving the problem of missing positive examples in training data can be found in [34], that uses the Markov decision framework to estimate the distribution of the missing anomalous samples. Experiments show encouraging results even in the case where the estimated distribution is not the real one. A newer approach uses generative adversarial networks (GAN) to generate positive examples [76]. The method was tested on a Wikipedia database with the aim of identifying editing vandals. Results show improved precision and accuracy over three baseline algorithms: one-class SVM, one-class nearest neighbors and one-class Gaussian process.

Other approaches consider human intervention. In [15], the anomaly detection problem in interactive attributed networks is approached by allowing the system to proactively communicate with the human expert in making a limited number of queries about ground truth anomalies. The problem is formulated in the multi-armed bandit framework and after applying some basic clustering methods, it aims to maximize the true anomalous nodes presented to the human expert in the given number of queries. The results show certain improvements compared to similar approaches.

In [54] it is shown that unsupervised anomaly detection is an undecidable problem, requiring priors to be assumed on the anomaly distribution. In the expert feedback context, a new layer extension is analyzed, that can be applied on top of any unsupervised anomaly detection system based on deep learning to transform it in an active anomaly detection system. In other words, the strategy is to iteratively select a number of the most probable samples to be audited, wait for the expert to select their label, and continue training the system using the new information. Various improvements have been shown over the state-of-the-art deep approaches.

6 Conclusions

From transaction attributes to relational structures and encoding information on the time varying properties of a graph, we searched for solutions that promote the aggregation and integration of multiple sources of data. While anomaly detection on graphs is a well studied field, the particularities of fraud detection, especially in the face of fraudsters camouflaging their activity, calls for specialized methods. We also devoted space to describing the problem of using deep learning for anomaly detection, despite the fact that efforts in applying the framework for fraud detection are still scarce.

Acknowledgements This work was supported by BRD Groupe Societe Generale through Data Science Research Fellowships of 2019.

References

1. Akoglu, L., McGlohon, M., Faloutsos, C.: Oddball: Spotting anomalies in weighted graphs. In: Pacific-Asia Conference on Knowledge Discovery and Data Mining, pp. 410–421 (2010)
2. Akoglu, L., Tong, H., Koutra, D.: Graph based anomaly detection and description: a survey. Data Min. Knowl. Discov. **29**, 626–688 (2014)
3. Akoglu, L., Tong, H., Meeder, B., Faloutsos, C.: PICS: Parameter-free Identification of Cohesive Subgroups in Large Attributed Graphs. In: Proceedings of the 2012 SIAM International Conference on Data Mining, pp 439–450 (2012)
4. Amer, M., Goldstein, M., Abdennadher, S.: Enhancing one-class support vector machines for unsupervised anomaly detection. In: Proceedings of the ACM SIGKDD Workshop on Outlier Detection and Description, pp. 8–15 (2013)
5. Ban, Y., Liu, X., Duan, Y., Liu, X., Xu, W.: No place to hide: catching fraudulent entities in tensors. In: The World Wide Web Conference, pp. 83–93 (2019)
6. Bhatia, S., Hooi, B., Yoon, M., Shin, K., Faloutsos, C.: Midas: microcluster-based detector of anomalies in edge streams. In: Association for the Advancement of Artificial Intelligence (2020)
7. Băltoiu, A., Pătraşcu, A., Irofti, P.: Graph anomaly detection using dictionary learning. In: The 21st World Congress of the International Federation of Automatic Control, pp. 1–8 (2020)

8. Cao, B., Mao, M., Viidu, S., Yu, P.S.: Collective fraud detection capturing inter-transaction dependency. In: Proceedings of Machine Learning Research, KDD 2017, vol. 71, pp. 66–75 (2017)
9. Chen, J., Saad, Y.: Dense subgraph extraction with application to community detection. IEEE Trans. Knowl. Data Eng. **24**(7), 1216–1230 (2012)
10. Chen, Z., Hendrix, W., Samatova, N.F.: Community-based anomaly detection in evolutionary networks. J. Intell. Inf. Syst. **39**(1), 59–85 (2012)
11. Colladon, A.F., Remondi, E.: Using social network analysis to prevent money laundering. Expert Syst. Appl. **67**, 49–58 (2017)
12. Cucuringu, M., Blondel, V.D., Van Dooren, P.: Extracting spatial information from networks with low order eigenvectors. Phys. Rev. E **87**, 032803 (2013)
13. Delamaire, L., Abdou, H., Pointon, J.: Credit card fraud and detection techniques: a review. Banks Bank Syst. **4**, 57–68 (2009)
14. Dhawan, S., Gangireddy, S.C.R, Kumar, S., Chakraborty, T.: Spotting collective behaviour of online frauds in customer reviews. In: Proceedings of the Twenty-Eighth International Joint Conference on Artificial Intelligence (IJCAI-19), pp. 245–251 (2019)
15. Ding, K., Li, J., Liu, H.: Interactive anomaly detection on attributed networks. In: Proceedings of the Twelfth ACM International Conference on Web Search and Data Mining, pp. 357–365 (2019)
16. Dumitrescu, B., Irofti, P.: Dictionary Learning Algorithms and Applications. Springer (2018)
17. Elliott, A., Cucuringu, M.C., Luaces, M.M., Reidy, P., Reinert, G.: Anomaly detection in networks with application to financial transaction networks. Arxiv: arXiv:1901.00402 [stat.AP] (2018)
18. Erfani, S.M., Rajasegarar, S., Karunasekera, S., Leckie, C.: High-dimensional and large-scale anomaly detection using a linear one-class svm with deep learning. Pattern Recognit. **58**(C), 121–134 (2016)
19. Eswaran, D., Faloutsos, C.: Sedanspot: detecting anomalies in edge streams. In: IEEE International Conference on Data Mining (ICDM), pp. 953–958 (2018)
20. European Central Bank. Ecb report shows a fall in card fraud in 2016. https://www.ecb.europa.eu/press/pr/date/2018/html/ecb.pr180926.en.html, 26 September 2018. Accessed 29 Feb 2020
21. Flegel, U., Vayssiere, J., Bitz, G.: A state of the art survey of fraud detection technology. In Probst, C., Hunker, J., Gollmann, D., Bishop, M. (eds.) Insider Threats in Cyber Security, pp. 73–84. Springer (2010)
22. Gao, J., Du, N., Fan, W., Turaga, D., Parthasarathy, S., Han, J.: A multi-graph spectral framework for mining multi-source anomalies. In: Graph Embedding for Pattern Analysis, pp. 205–227. Springer (2013)
23. Guo, Q., Li, Z., An, B., Hui, P., Huang, J., Zhang, L., Zhao, M.: Securing the deep fraud detector in large scale e-commerce platform via adversarial machine learning approach. In: Proceedings of the 2019 World Wide Web Conference (WWW'19), pp. 616–626 (2019)
24. Hejazi, M., Singh, Y.P.: One-class support vector machines approach to anomaly detection. Appl. Artif. Intell. **27**(5), 351–366 (2013)
25. Henderson, K., Gallagher, B., Li, L., Akoglu, L., Eliassi-Rad, T., Tong, H., Faloutsos, C.: It's who you know: graph mining using recursive structural features. In: Proceedings of the 17th ACM SIGKDD International Conference on Knowledge Discovery and Data Mining, KDD '11, pp. 663–671. New York, NY, USA. Association for Computing Machinery (2011)
26. Huang, Z., Ye, Y., Li, X., Liu, F., Chen, H.: Joint weighted nonnegative matrix factorization for mining attributed graphs. In: Pacific-Asia Conference on Knowledge Discovery and Data Mining, pp. 368–380 (2017)
27. Ionescu, R.T., Smeureanu, S., Alexe, B., Popescu, M.: Unmasking the abnormal events in video. In: Proceedings of the IEEE International Conference on Computer Vision, pp. 2895–2903 (2017)
28. Irofti, P., Băltoiu, A.: Malware identification with dictionary learning. In: 27th European Signal Processing Conference, pp. 1–5 (2019)

29. Irofti, P., Băltoiu, A.: Unsupervised dictionary learning for anomaly detection. Arxiv: arXiv:2003.00293 (2019)
30. Irofti, P., Stoican, F.: Dictionary learning strategies for sensor placement and leakage isolation in water networks. In: The 20th World Congress of the International Federation of Automatic Control, pp. 1589–1594 (2017)
31. Jiang, M., Cui, P., Beutel, A., Faloutsos, C., Yang, S.: Inferring strange behavior from connectivity pattern in social networks. In: Pacific-Asia Conference on Knowledge Discovery and Data Mining, pp. 126–138. Springer (2014)
32. Jiang, M., Cui, P., Beutel, A., Faloutsos, C., Yang, S.: Catching synchronized behaviors in large networks: a graph mining approach. ACM Trans. Knowl. Discov. Data 10(4), 1–27 (2016)
33. Jun, T., Jian, Y.: Developing an intelligent data discriminating system of anti-money laundering based on SVM. In: 2005 International Conference on Machine Learning and Cybernetics, vol. 6, pp. 3453–3457 (2005)
34. Kocsis, L., György, A.: Fraud detection by generating positive samples for classification from unlabeled data. In: Proceedings of the 27th International Conference on Machine Learning. Workshop on Machine Learning and Games (2010)
35. Lamrini, B., Gjini, A., Daudin, S., Pratmarty, P., Armando, F., Travé-Massuyès, L.: Anomaly detection using similarity-based one-class svm for network traffic characterization. In: 29th International Workshop on Principles of Diagnosis (2018)
36. Larik, A.S., Haider, S.: Clustering based anomalous transaction reporting. Procedia Comput. Sci. 3, 606–610 (2011)
37. Latimer, P.: Australia: Australian transaction reports and analysis centre (austrac). J. Financ. Crime 3, 306–307 (1996)
38. Li, J., Dani, H., Hu, X., Liu, H.: Radar: residual analysis for anomaly detection in attributed networks. In: Proceedings of the Twenty-Sixth International Joint Conference on Artificial Intelligence, pp. 2152–2158 (2017)
39. Li, N., Sun, H., Chipman, K.C., George, J., Yan, X.: A probabilistic approach to uncovering attributed graph anomalies. In: Proceedings of the 2014 SIAM International Conference on Data Mining, pp. 82–90 (2014)
40. Li, Z., Xiong, H., Liu, Y., Zhou, A.: Detecting blackhole and volcano patterns in directed networks. In: 2010 IEEE International Conference on Data Mining, pp. 294–303 (2010)
41. Liu, N., Huang, X., Hu, X.: Accelerated local anomaly detection via resolving attributed networks. In: Proceedings of the Twenty-Sixth International Joint Conference on Artificial Intelligence, pp. 2337–2343 (2017)
42. Liu, S., Hooi, B., Faloutsos, C.: Holoscope: Topology-and-spike aware fraud detection. In: Proceedings of the 2017 ACM on Conference on Information and Knowledge Management, pp. 1539–1548 (2017)
43. Miller, B.A., Arcolano, N., Bliss, N.T.: Efficient anomaly detection in dynamic, attributed graphs: emerging phenomena and big data. In: IEEE International Conference on Intelligence and Security Informatics, pp. 179–184 (2013)
44. Miller, B.A., Beard, M.S., Bliss, N.T.: Eigenspace analysis for threat detection in social networks. In: Proceedings of the 14th International Conference on Information Fusion (FUSION), pp. 1–7 (2011)
45. Miller, B.A., Beard, M.S., Wolfe, P.J., Bliss, N.T.: A spectral framework for anomalous subgraph detection. IEEE Trans. Signal Process. 63(16), 4191–4206 (2015)
46. Miller, B.A., Bliss, N.T., Wolfe, P.J.: Toward signal processing theory for graphs and non-euclidean data. In: IEEE International Conference on Acoustics Speech and Signal Processing (ICASSP), pp. 5414–5417 (2010)
47. Ngai, EWT., Hu, Y., Wong, Y.H., Chen, Y., Sun, X.: The application of data mining techniques in financial fraud detection: a classification framework and an academic review of literature. Decis. Support Syst. 50(02), 559–569 (2011)
48. Nilforoshan, H., Shah, N.: Slicendice: mining suspicious multi-attribute entity groups with multi-view graphs. In: 2019 IEEE International Conference on Data Science and Advanced Analytics (DSAA), pp. 351–363 (2019)

49. Pandit, S., Chau, D.H., Wang, S., Faloutsos, C. Netprobe: a fast and scalable system for fraud detection in online auction networks. In: Proceedings of the 16th international conference on World Wide Web, pp. 201–210 (2007)
50. Pastor-Satorras, R., Castellano, C.: Distinct types of eigenvector localization in networks. Sci. Rep. **6** (2016)
51. Peng, Z., Luo, M., Li, J., Liu, H., Zheng, Q.: Anomalous: a joint modeling approach for anomaly detection on attributed networks. In: Proceedings of the Twenty-Seventh International Joint Conference on Artificial Intelligence, IJCAI-18, pp. 3513–3519 (2018)
52. Perozzi, B., Akoglu, L.: Scalable anomaly ranking of attributed neighborhoods. In: Proceedings of the 2016 SIAM International Conference on Data Mining, pp. 207–215 (2016)
53. Phua, C., Lee, V., Smith, K., Gayler, R.: A comprehensive survey of data mining-based fraud detection research. Intell. Comput. Technol. Autom. (ICICTA), pp. 50–53 (2010)
54. Pimentel, T., Monteiro, M., Viana, J., Veloso, A., Ziviani, N.: A generalized active learning approach for unsupervised anomaly detection. CoRR, abs/1805.09411 (2018)
55. Qiu, X., Cen, W., Qian, Z., Peng, Y., Zhang, Y., Lin, X., Zhou, J.: Real-time constrained cycle detection in large dynamic graphs. Proc. VLDB Endow. **11**(12), 1876–1888 (2018)
56. Savage, D., Wang, Q., Chou, P., Zhang, X., Yu, X.: Detection of money laundering groups using supervised learning in networks. arXiv preprint arXiv:1608.00708 (2016)
57. Sengupta, S.: Anomaly Detection in Static Networks using Egonets (2018)
58. Shah, N., Beutel, A., Hooi, B., Akoglu, L., Günnemann, S., Makhija, D., Kumar, M., Faloutsos, C.: Edgecentric: anomaly detection in edge-attributed networks. In: 2016 IEEE 16th International Conference on Data Mining Workshops (ICDMW), pp. 327–334 (2016)
59. Shin, K., Eliassi-Rad, T., Faloutsos, C.: Patterns and anomalies in k-cores of real-world graphs with applications. Knowl. Inf. Syst. 677–710 (2017)
60. Skretting, K., Engan, K.: Intrusion detection in computer networks by a modular ensemble of one-class classifiers. Inf. Fus. **9**(1), 69–82 (2008)
61. Sorournejad, S., Zojaji, Z., Atani, R.E., Monadjemi, A.H.: A survey of credit card fraud detection techniques: data and technique oriented perspective. arXiv: abs/1611.06439 (2016)
62. Tian, Y., Mirzabagheri, M., Bamakan, H., Wang, S.M.H., Qu, Q.: Ramp loss one-class support vector machine; a robust and effective approach to anomaly detection problems. Neurocomputing **310**, 223–235 (2018)
63. Tong, H., Lin, C.-Y.: Non-negative residual matrix factorization with application to graph anomaly detection. In: Proceedings of the 2011 SIAM International Conference on Data Mining, pp. 143–153 (2011)
64. Velampalli, S., Eberle, W.: Novel graph based anomaly detection using background knowledge. In: FLAIRS Conference (2017)
65. Vengertsev, D., Thakkar, H.: Anomaly detection in graph: unsupervised learning, graph-based features and deep architecture. Tech. Rep. (2015)
66. Wang, H., Zhou, C., Wu, J., Dang, W., Zhu, X., Wang, J.: Deep structure learning for fraud detection. In: IEEE International Conference on Data Mining, pp. 567–576 (2018)
67. Wang, Y., Wang, L., Yang, J.: Egonet based anomaly detection in e-bank transaction networks. IOP Conf. Ser. Mater. Sci. Eng. **715**, 012038 (2020)
68. West, J., Bhattacharya, M., Islam, R.: Intelligent financial fraud detection practices: an investigation. In: International Conference on Security and Privacy in Communication Networks, pp. 186–203. Springer (2014)
69. Wu, L., Wu, X., Lu, A., Zhou, Z.H.: A spectral approach to detecting subtle anomalies in graphs. J. Intell. Inf. Syst. **41**(2), 313–337 (2013)
70. Xu, Z., Ke, Y., Wang, Y., Cheng, H., Cheng, J.: A model-based approach to attributed graph clustering. In: Proceedings of the 2012 ACM SIGMOD International Conference on Management of Data, SIGMOD '12, pp. 505–516. New York, NY, USA, ACM (2012)
71. Ying, X., Wu, X., Barbará, D.: Spectrum based fraud detection in social networks. In: 2011 IEEE 27th International Conference on Data Engineering, pp. 912–923. IEEE (2011)
72. Yoon, M., Hooi, B., Shin, K., Faloutsos, C.: Fast and accurate anomaly detection in dynamic graphs with a two-pronged approach. In: Proceedings of the 25th ACM SIGKDD International

Conference on Knowledge Discovery & Data Mining, KDD '19, pp. 647–657. Association for Computing Machinery (2019)
73. Yu, W., Cheng, W., Aggarwal, C..C, Zhang, K., Chen, H., Wang, W.: Netwalk: a flexible deep embedding approach for anomaly detection in dynamic networks. In: Proceedings of the 24th ACM SIGKDD International Conference on Knowledge Discovery & Data Mining, pp. 2672–2681 (2018)
74. Yuan, S., Wu, X., Li, J., Lu, A.: Spectrum-based deep neural networks for fraud detection. In: Proceedings of the 2017 ACM on Conference on Information and Knowledge Management, pp. 2419–2422 (2017)
75. Zhang, S., Zhou, D., Yildirim, M.Y., Alcorn, S., He, J., Davulcu, H., Tong, H.: Hidden: hierarchical dense subgraph detection with application to financial fraud detection. In: Proceedings of the 2017 SIAM International Conference on Data Mining, pp. 570–578 (2017)
76. Zheng, P., Yuan, S., Wu, X., Li, J., Lu, A.: One-class adversarial nets for fraud detection. In: Proceedings of the AAAI Conference on Artificial Intelligence, pp. 1286–1293 (2018)

Biomorphic Artificial Intelligence: Achievements and Challenges

D. O. Chergykalo and D. A. Klyushin

1 Introduction

Artificial intelligence technologies are increasingly moving away from where they started—from modeling human behavior. Currently, quite a few people use processes associated with neural networks of the brain to implement software, and all funding depends on the specific tasks performed on time.

If to take into account the generally accepted point of view "Everything is complicated," as well as bitter experience in this area (for example, two AI winters), then we have no choice but to accept status quo, and slowly make small steps improving existing algorithms in order to increase profit for various companies by increasing the accuracy of methods.

But is the brain really so complex to simulate it as a black box? Below we describe what technologies applicable to create humanoid AI have appeared recently. But before that, we emphasize that the practical ideas of AI undergo the same evolution as the behavior and brain of animals.

Scientists are constantly improving old models, combining methods, experimenting with their models, choosing the best ones, and when it comes to practice, those models are selected that are best suited for specific tasks, finding their niche, so AI ideas have even more similarities with biology than the researchers themselves might suppose at first glance. In the next subsection we will discuss just that.

D. O. Chergykalo · D. A. Klyushin (✉)
Faculty of Computer Science and Cybernetics, Taras Shevchenko National University of Kyiv,
Akademika Glushkova Avenu, 4D, Kyiv 03680, Ukraine
e-mail: dokmed5@gmail.com

D. O. Chergykalo
e-mail: denischergicalo@gmail.com

© The Editor(s) (if applicable) and The Author(s), under exclusive license
to Springer Nature Switzerland AG 2021
A.-E. Hassanien et al. (eds.), *Enabling AI Applications in Data Science*,
Studies in Computational Intelligence 911,
https://doi.org/10.1007/978-3-030-52067-0_24

2 The Parallel Path of Smart Technologies and the Simulation of Biological Processes

The term "biomorphic AI" in the title can confuse those who are familiar with modern developments of biomorphic neural networks, for example Blue Brain, but here it's better to clarify. This term is used in art and means something based on natural patterns or forms resembling nature and living organisms. By biomorphism, we mean precisely the similarity of the general form and organization of processes in AI that we offer.

In this subsection, we show the similarity of the processes that were created for AI and the real organization of processes in the brain and psyche. Since we describe humanoid AI, it is quite logical that it will contain neural networks, as already noted, we consider the term "biomorphism" more generally, and we believe that traditional neural networks are efficient in use and, if used correctly, are no worse than natural neural networks. Why we chose them in general terms will be explained in the Sect. 3.

As is well known, neural networks have recently undergone a period of intensive development. There are many intriguing parallels between artificial and natural neural networks. An interesting example is dropout algorithm—random removal of neurons in an artificial neural network that helps to make the neural network more stable and reduces overfitting on the available data [24]. The authors of that papers note that neurons during the dropout process can be compared with genes during sexual reproduction—only a certain part of them is realized in the descendant. Therefore, if an individual possessed good features, but these features were determined by a large set of genes, in other words, a whole large co-adaptation of genes, then this set is very likely to be passed on to the next generation due to the random nature of their transmission, therefore only genes that themselves remain in nature benefit, or small sets of such genes.

However, there is a more direct biological rationale for the dropout algorithm, because in the human brain neurons do experience similar processes. As noted in [2], neurons are trying to effectively adapt to the data, trying not to succumb to large co-adaptations and their random dropout occurs. Oddly enough, few people mention such a direct connection.

The artificial neural networks that we program usually develop normally in isolation from the outside world. They have the opportunity to learn all the variety of information at the same time, without intermediate tests of performance. On the contrary, a person needs to develop gradually and immediately apply his knowledge in practice. Information appears in the brain in certain portions, having received which, it can immediately switch to using the knowledge gained. Naturally, these portions cannot describe the whole variety of information, so such training will constantly make random deviations from training based on complete information. But this turns out to be not even bad: if the portion of information (sample) has small deviations, then they will not allow to leave the region of the global minimum of error (corresponding to complete information), but at the same time they will help to leave the region of local minimums.

In the field of artificial neural networks, this approach is called Small Batch Training, and its effectiveness is well described in the article [13]. It also indicates that it is optimal to take a sample with sizes ranging from $m = 2$ to $m = 32$. In other words, even $m = 2$ may turn out to be the most effective option for a specific problem.

If we consider the approach described above from a social point of view, then Small Batch Training can be compared to lessons in schools where students are given information in portions, which allows them to change their unconscious ideas about objects in accordance with new information. But first, the schools teach the most basic, and only then they give more and more complex information, meanwhile the AI should be able to learn on its own—that is, independently go from simple to complex.

In addition, new information or a tactical move made by a person is not always immediately clear, therefore information is transferred from short-term to long-term memory at the end of the day—during sleep (because of which the most ordinary part of life can be well remembered if after it a bright event happened on that day).

These two problems are partially resolved, which we will discuss in Sect. 5.3.

It is worth noting that convolutional neural networks are also quite close to real analysis in the human brain, although the biological implementation of filters that emit abstract objects is rather complicated. It is carried out not only through general training of the entire network (based on internal rewards for good work), but so that similar filters are applied at different places in the layer, self-organization within each layer is applied. If you create an AI model, then you need to develop a similar process of self-organization, so as not to reduce everything to a fixed set of filters. Therefore, as an approximation to real AI, we will use a conventional convolutional neural network, which will immediately train a finite number of filters over the entire area.

Such a process of self-organization is not limited to two-dimensional data. A person can analyze and identify objects of increasing complexity, for example, in a one-dimensional sequence of audio signals. A person also has "layers" responsible for specific images of visual and auditory information. They are contained in the temporal lobe. But for a person's hearing and vision, different architectures are provided. On one hand, hearing is a sequential model of neural networks that leads from sensors to the analysis of complex sound signals and words, on other hand vision is divided into two directions: one leads to an analysis of the image itself, and the other leads to analysis of location where this image is [2].

This is fully consistent with one of the main neural network architectures for image detection [8]. A similar process always occurs in a person, when he, analyzing the image, can focus attention on only one thing. This process is closer to another of the main neural network architectures for pattern detection [19].

In addition, a person has his own maxpool layer, but it works harder than choosing a maximum from several neighbors. There is a choice from an object-oriented structure, due to which an unconscious understanding arises where to choose an object. This corresponds to the self-organization described above, only for significant correlations with more distant neurons. However, if this choice is correctly implemented programmatically, then the programs become more resistant to deformation and changes in the picture that they analyze.

For example, Vicarious, an artificial intelligence company, has developed Recursive Cortical Network (RCN) technology for biologically plausible image analysis. Their algorithm, unlike its predecessors, does not reduce accuracy when the text is deformed, and almost does not decrease when using various effects and styles to complicate the text and overlap it with other objects. This will be discussed in more detail in the Sect. 4.

We believe that transferring the useful properties of natural neural networks to computer models is an optimal resource option, and we also assume that while maintaining the qualitative properties of the neural network, its other more complex properties are also preserved. Researchers have shown that RSN has many more complex properties that they consider to be similar to human ones.

It seems to us that the study of such computer models will help not only to create AI, but to simulate various brain diseases and to study their effect on AI, which may prompt researchers to solve real medical problems with the brain by studying them on the models.

3   Criticism of Detailed Brain Design Projects

When the task is to create an AI similar to human one, there is always a simple desire to simply simulate all the processes of the brain in a computer. However, different people have different views on what the brain consists of—for some it is a set of neurons that interact with each other, for others it is different parts of the brain that act like black boxes and perform some kind of function.

Each of these people, creating their own philosophy, their models—including computer ones—bring to them their understanding of how this happens—electrochemical dynamics on neurons, modeling brain regions and their interactions, expert systems, chat bots, etc.

When comparing the options, it may naturally seem that a model based on electrochemical dynamics should be the most accurate—after all, everything else looks like a description of the higher-level consequences of this dynamics. However, not everything is so simple. There are a huge number of factors that affect this dynamics, and counting all these factors for each neuron and synapse is technically impossible now. It seems natural to simplify the task and reduce the real brain to some model, but at the same time maintaining the basic properties of neural networks. In this subsection, we briefly discuss the case when the brain model is simplified due to the simplifying its general organization, but maintaining intercellular structure.

The Human Brain Project and Blue Brain Project distinguish the neuron column as the basic component of the brain [5]. They analyzed the structural unit of neuron organization in the neocortex and simulated brain electrochemistry, but general architecture of brain [14] was not taken into account in their model. They transferred the connection graph in the part of the brain to a digital model and compared the electrical signals received at the output from the living part of the brain and its model. The columns were fixated, neurons were also fixated. They saved only electrochemical dynamics, from which we can conclude that at least for a short time this model can behave a real fragment of the rat brain.

But during training through the natural selection of neurons and their connections, this model will increasingly diverge from reality. Blue Brain researchers themselves admitted that trying to account for and understand all types of neural cells is a rather difficult task [12]. The process modeling performed by Blue Brain and similar projects very well describes one side—electrochemical dynamics and behavior in a given time frame, but very poorly describes the other—the internal organization of learning processes and neural networks.

Human cells are constantly dying and the human body is constantly being updated at the cellular level. It is appropriate to recall Theseus' paradox—"If all the components of the original object have been replaced, does the object remain the same object?" A human still remains himself, despite the fact that the connections in the human brain are strongly rearranged, and neurons die and new ones appear.

We believe that modeling should preserve qualitatively intelligent processes in the brain, direct transfer at the cell level is incredibly difficult and pointless due to the strong variability at this level, Blue Brain used 8192 processors to calculate a single neural column consisted from a few hundreds of neurons. They still work slower than a real neural column.

If the task is to model a person's thinking at the level of preservation of his intellectual processes, then, in our opinion, this does not have to be done at the cellular level. It is enough to create a system equivalent in processes (in the images that it classifies, memory that it can activate, as well as in the processes that occur during brain function, including emotional).

In the future, we will not talk about the digitization of the mind, since we do not consider such a problem. However, the architecture of AI, which we propose, can inspire someone to create such an AI and, based on analogy, make the transfer of biological processes occurring in the brain to artificial neural networks and various forms of memory, and then perform a comparison with a natural brain (on the basis of which a virtual copy was made, possibly even a human one).

One could set the task of simulating AI so that it develops in the same way as other people, but people have their weaknesses, we will talk about some of them in Sect. 5.2. Modeling AI as a person is also not reasonable, since we want not just the appearance of other essences, but want AI to serve us and solve human problems.

Therefore, the intellectual processes themselves in AI should be equivalent to human ones—while maintaining the general principles of learning and thinking in humans, but in order to ultimately get an assistant, but not a potential enemy or system with unpredictable behavior.

It is worth noting that the achievements of the Blue Brain project are not limited to just a simulation—the transfer of neural connections. In addition, Blue Brain researchers use their models to gain a deeper understanding of the underlying processes in the brain. For example, comparing their model and processes in the real brain, researchers found that artificial neurons appear to have common behavior similar to brain rhythms [18]. Having analyzed the reasons for his appearance in their model, they improved their understanding of real processes.

Those who are interested in the idea of modeling precisely neural networks with similar properties to real ones, as the closest of the classical general models, we suggest exploring the LSTM model and the rhythms in it, especially with regard to long-term information. But if we talk about the closest of the simply classical models, then we propose to consider the rhythms and similar biological characteristics for the BELBIC model. This model we will discuss in Sect. 4.

4 Basic Algorithms for Artificial Intelligence and Their Achievements

Attempts to create AI from scratch with a high probability lead to the creation of what is already there, so it is always useful to examine the idea in parts and see what others have already implemented.

You can break down a task in different ways. You can try to find the programmed analogues of the brain regions and try to connect them, you can try to modify the existing analogues of human thinking that are closest to the model, or you can try to do both.

AlphaZero is a program that, learning from scratch, has won all the people and the rest of the algorithms in chess, segos and Go. Go's game runs on board 19×19 and generates a huge number of options. Conventional algorithms were powerless and reached the maximum level of average amateur in Go. AlphaGo was the first program to defeat the professional, and even the best player in Go. But it was unusual only because of the convolutional neural network as a function of evaluating and determining profitable moves. Its algorithm could be compared with an intuitive assessment of the situation on the board by a person, i.e. with his trained unconscious appreciation. In combination with the Monte Carlo method, this allows successfully simulating the game thinking of a person (we'll talk about this in more detail in Sect. 5.3).

The next version of the AlphaGo program, AlphaGo Zero, learned to train its intuition not on examples of other people's games, but on examples of games with itself. This not only saved it from dependence on external data, but also significantly improved the program. AlphaZero program can in a matter of hours learn to play any game and become the best in it. It itself adapt to the game, which makes it quite flexible and easily customizable to different types of games.

Another example of the Recursive Cortical Network, mentioned earlier, made it possible to correctly guess Captha in 66.6% of cases, and after determining the style and additional training—90%. Moreover, this network uses only 5000 examples of resolved examples and a small number of layers. It simulates the work of the primary visual cortex. A description of the work and experiments with this neural network is given in [7]. What we called the connection with more distant neurons in this article is called the lateral connection.

The emotional component of learning is important for proposed model as it helps to filter and remember important information. Models of emotional brain training are called BEL models (Brain Emotional Learning). Unfortunately, at the moment we do not know any complex models for emotional learning that would adapt to complex mechanisms of thinking.

The most advanced of these models is the BELBIC model [3]. It has practical applications for real-time control systems, since it is computationally efficient and at the same time gives quite acceptable results [15]. It is similar to an RNN network, which predicts the necessary action, which should be chosen, but has one big difference from it—the network learns directly during operation, thus constantly adapting to new conditions. Models of emotional thinking model the functioning of the limbic system and their relationship with other areas during emotional perception and memorization.

Naturally, the readers may have their own views on how to break down the task of constructing AI. In this case we just want to clarify and improve understanding of proposed approach.

5   Proposed Model of Artificial Intelligence

Let us consider the proposed model of artificial Intelligence and its general principles and assumptions. This model can make the complexity of thinking similar in complexity to human and lead to the strong artificial intelligence.

5.1   Fitness Function

As we said in Sect. 3, when trying to create AI, a person follows his own ideas. Part of his ideas is how a person imagines the development of AI and what he wants from it. To successfully characterize the development and the model itself, the researcher must distinguish a set of characteristics. A rather convenient characteristic is the fitness function. Thanks to it, development can be defined as an increase in this function on training data, and the goal can be defined as an increase on its values on testing data.

But the definition of such a function is far from an easy task for a person, because he has a great many goals and desires. The theory of decision-making sees its way out of the situation, considering a whole multitude of factors and introducing a partial order on it. This may help some people make better choices, but does not describe the actual decision making process by a person.

But abstracting from goals and desires, we can still try to define this function. Psychologists, trying to find it, built several theories: the theory of personality of Z.

Freud, in which the objective function is pleasure, the theory of personality of A. Adler, in which the objective function is the pursuit of excellence, etc. But all this does not accurately describe the choice of a person. It is necessary to determine the characteristic that a person is trying to optimize at any given time. In this subsection we shall talk about this.

A person makes decisions for the local maximization of dopamine [2], while the strength of behavior fixing also depends entirely on the amount of dopamine produced. The amount of dopamine is responsible for how much the connection is strengthened, so it is logical to assume that it is him who, as a hormone of desires, tries to maximize. But by default, the cells have fairly stable connections, therefore, for these activated zones, in case of successful completion of the task, serotonin is additionally secreted. It allows increasing synaptic plasticity of the cells and thereby strengthening the connections that led to the improvement of the network. The amount of serotonin can be considered as a parameter of the learning rate of the neural network (similar to the coefficient used for the step during the epoch), which is completely consistent with data from biology and hormonal processes in the brain [11].

It is precisely those neurons and synapses that are trained in neural networks that were involved in the task, as well as those that have not yet been strengthened by their connections (because of this they have high synaptic plasticity). Thus, some basic human skills that are entrenched remain fixed, and new skills of a higher level, using the old ones, can improve. Similarly, if some features emitted by the visual cortex have strengthened, then a new definition of objects in the temporal lobe (the center of work with the semantics of the brain) will already be made on these grounds. This will be discussed in more detail in Sect. 5.4.

5.2 Conscious and Unconscious, or How We Make Decisions Thinking and not Thinking at the Same Time

We will associate the unconscious with various self-organizing neural networks that have an architecture that is adaptable to solving a variety of problems, in particular, image analysis, choice of options in life situations or in various abstract games (intuition).

Consciousness is a mechanism that can direct other parts of the brain to the desired activity. Thinking deeply, a person can model interaction with reality. This process is called "weighing all the pros and cons", "pondering actions", "using common sense", "reasoning" or "planning" (in game with large time scale).

When remembering something, the conscious can reactivate the same neurons as directly observing an object, thereby including unconscious algorithms for analysis. When reproducing in memory of subsequent situations, other neurons are reactivated, due to which a person subjectively senses that he sees other objects or situations, i.e. uses imagination.

Trying to survive, a person must take into account the reaction of the environment, especially that part of it that limits its actions, which he unconsciously defined as optimal.

In humans, consciously accepted behavior forms in the orbitofrontal cortex [22]. She also organizes modeling of scenarios of the future [4]. When we have a specific goal and an important choice, we imagine various options for the development of events. This changes human unconscious assessment of options in the prefrontal cortex. After these mental simulations, the unconscious assessment becomes more thoughtful.

In [4], the choice of a path to achieve a goal is described. This can be perceived as a "game with reality", in which a person is constantly offered to make a difficult and important choice that needs to be considered.

Interaction with society can also be considered as "choosing the right path" of interaction with it. With the help of attention mechanisms, a person can switch to "different games", for example, thinking about choosing the right move in a board game. But it is the hippocampus is responsible for distinguishing possible positions from reality, as well as the rules of the "game" itself [4]. Also, it plays another role. It uses episodic memory, recalling similar situations from the past and thanks to this he immediately gives an idea of which course options will be better. In the future, we will talk about the game with own rules and episodic memory in Sect. 5.3.

To summarize, we note that with conscious behavior, a person using modeling interaction with reality (based on his unconscious assessment and unconscious model of the world), which we will continue to call the "game with reality", can improve his unconscious assessment of choice with maximum dopamine and make more informed choices about the best move. But often a person does not have the time, energy or desire to perform a deep analysis of the situation, therefore he often uses directly unconscious analysis—associations, analogies, patterns, stereotypes, automatic actions developed in society, etc. [6].

AI can think over options instantly, so is it worth giving it the opportunity to choose a less effective—unconscious version of thinking, is a separate issue. Our opinion on this matter: regardless of the task, it is more simple and reliable for people not to let him do this. For this reason, in the future we will consider only a conscious choice of behavior.

5.3 AlphaZero Ideas as a Base for the AI Algorithm

In Sect. 2, many methods were described that allow making AI closer to natural and corresponding to reality. In particular, mini-batch and dropout algorithms were mentioned there. These algorithms have been successfully implemented in the AlphaZero. In addition, AlphaZero, playing with itself, can learn the game and for this it needs enough basic rules. Having gained experience, the program understands how to improve its strategy. "In the evening," it plays with its "morning" version and, if it realizes that after "improvements" it does not play "much better", then these changes are not delayed "during sleep" in long-term memory (although they are all deposited in conscious memory, next "morning" the unconscious memory of AlphaZero will be the same as the previous one).

This can also be compared with how scientists, using new data, try to put forward new hypotheses, and then test them in practice to check whether the theory has become closer to reality, and if not, then completely reject the hypothesis as unnecessary (not using it in the future even to derive other hypotheses).

In the previous subsection, we noted that a person's consciousness works as a simulation of games with reality to better achieve the goal. In Sect. 4 we have already said that the Monte Carlo method with an unconscious assessment resembles a thinking process. In this light, this process is interpreted as consciousness (including conscious thinking) and, in this understanding, AlphaZero has consciousness.

Developing in the game, AlphaZero has its own analogue of game episodic memory—it remembers for each position at which moves it won and lost, so that it immediately has an idea about it without "pondering". As mentioned earlier, AlphaZero defeated all the best programs in the board games of his choice, showing his excellent ability to completely change the type of game he is playing with small adjustments, and to use his own training to achieve better results than the best programs.

But as AlphaZero AI is far from ideal—all reconfigurations to new games are performed by people, not AlphaZero itself, and the most important component of AI for a person is missing—its flexible ability to apply its old experience to new tasks. The changes we propose will be outlined in the next subsection.

5.4 Improvements of the AlphaZero Ideas

As noted in Sect. 5.1, the human brain develops ideas per blocks, leaving the rest fixed. This immediately solves the problem of overfitting, since only a certain unit is trained, and not the entire system.

AlphaZero, performing a simple unconscious assessment, is a common convolutional classification neural network that has two outputs one for assessing the general situation, which for a person can mean a general feeling of the situation on an emotional level (conditionally from very poor to very good), and the second for evaluating options, for which the person is responsible for the prefrontal cortex (see Sect. 5.2).

In addition, the original AlphaZero program analyzes seven previous boards. This can have an optimization effect, thanks to which the network better remembers the tactics that have developed for this particular version of the game.

But, in general, since the games he played were games with full information, it was enough to analyze one board for analysis, although for games with incomplete information, the win in the game may completely depend on important information about past events. As an improvement, we propose the AlphaZero algorithm with a context evaluation over a fairly large time period. In this case, you can break the information into the current situation (board) and the context obtained using the LSTM (long short-term memory) neural network.

As indicated in Sect. 5.2, the hippocampus is responsible for isolating the game from reality and it is he who is responsible for isolating the features of this position, for which he naturally also has to collect information from the context, thereby organizing what is called working memory. As you can see, AlphaZero already analyzes the previous positions, but does it in the form of an analysis of ordered 8 boards, and not as a person who can work with a wider context, and better handle the sequence of position-situations, so we consider this the next step to improve this algorithm and increasing its biomorphism.

With further improvement, we offer the possibility of its "growing up", which is manifested in the gradual training of skills and gaining experience. In the theory of Jean Piaget [16], a person develops in stages and each stage can be considered as a sequence of stages—building up of some skills over others.

For example, a person initially has instincts. A child who controls instincts can use one instinct to use another and thereby get the simplest skills—relatively speaking, "1 level". Having learned to control his body, the child is already learning to manipulate the surrounding reality. To do this, he already needs to manage primitive skills, thereby developing higher-level skills. Subsequently, he already learns to manipulate objects, manipulate some objects with the help of others, and ultimately manipulate representations of objects to successfully manipulate these objects. The next stages and stages become already going on as add-ons over these views.

As an implementation of this principle, the model proposes to conditionally divide neural networks into networks to identify features and networks for the analysis of these features. In the case of convolutional neural networks, we assume that convolutional layers reveal features, and fully connected ones analyze them. Each time, increasing the complexity of identifying features, the neural network increases the level of analysis and actions, and its own fully connected neural network of analysis adapts to each new level, which implements a skill of a certain level. A similar structure can develop hierarchically, and for each specific task there will be its own branch in this hierarchy.

Naturally, there are many cases where the data is not mentally perceived. A person leaves such situations, reducing everything to some kind of images, as well as to formulations that are understandable for his analysis. For example, if AI analysis will be more adapted to mathematics and other exact sciences, then he will need to itself reformulate that inaccurate information that he will receive during development in more convenient terms and images.

For the model, such a learning option is optimal, since not the entire neural network is constantly learning, but only its layer that uses the internal features of already trained neural networks. This not only speeds up learning, but also allows learning from less data and also prevents the neural network from overfitting. In addition, a similar model with small modifications to the network for identifying traits from each neural network makes it possible to better generalize them.

How we can consider the properties of AI from a psychological point of view, we will show in Sect. 5.5.

Based on the growing structure of experience, a logical improvement is the ability to train skills in a close field, similar to quickly learning a person in a similar field with the field in his knowledge. This can be organized as additional training of a layer in a neural network that would translate the features that were identified in this task to be convenient for the already trained layers of the neural network, which are more responsible precisely for analysis and logic.

We propose to implement both a quick option, a conscious and unconscious transition to the existing experience, as well as a more reliable one, the creation of a separate branch in the hierarchy. We believe that such two approaches should simultaneously compensate for each other's shortcomings, attachment to the old analysis and the underdevelopment of the new. This is how people usually learn something new. In addition, if a person begins to take up an area more seriously, then this branch begins to play a larger role than the branches focused on old knowledge, and subsequently becomes completely independent and develops into new visions of this task, and also understanding and analyzing its important cases—new branches in this tree.

As we will show in Sect. 5.5, a separate neural network is responsible for using skills in practice. Thanks to such neural networks, skills can be in a generalized form, which is very important for hierarchy, it can also be said that an approach with more basic and more complex skills fully corresponds to the approach in psychology in the framework of education.

Using the concept of gaining experience, you can optimize learning to understand the context. Having distinguished understanding of the context as growing of experience on the basis of previous experience, we come to the conclusion that first you can learn to squeeze as much information out of what is at the moment, as children do during training, and then, having learned to understand at least what is happening on a given moment, constantly using this understanding, to build a context for the situation, for understanding which you can use, for example, all the same LSTM.

Readers may disagree with proposed visions of a person's skill development. Even agreeing with the modeling of thinking using artificial neural networks, for example, they can represent everything as the development of one complex neural network, or at least a small interconnected number of them, and they have a certain right to do this, considering, for example, the nuclei of structures limbic system with their functions or different areas of the brain (for example, Broadman fields) as separate neural networks.

But in this case, we are talking about a hierarchy of skills, as well as related information. Structural units such as neural columns play an important role in reproducing and memorizing them, we already mentioned them briefly in Sects. 2 and 3 when we talked about the dropout algorithm, and we also mentioned the absence of large co-adaptations in brain neurons.

If we consider everything from the perspective of training these neural columns, then the proposed model can be considered as an effective implementation of this training through a hierarchy of neural layers/ networks. It is also worth adding that in reality parts of several zones are trained at the same time. This ensures the stability of the neural network during dropout of neurons and other processes (which we will implement with the dropout algorithm), due to which, for example, people who have lost part of the brain can restore their functions and memory.

Perhaps the proposed point of view has become more clear after these explanations. But, if readers do not agree anyway, we hope that this work will inspire and help them to create more advanced and versatile AI models.

5.5 Psychology the Artificial Intelligence

To understand the future behavior of AI and how to make it suitable for us, it is advisable to study its behavior with the help of psychology as the behavior of a person, considering the personality not only from the point of view of biological and mental processes, but also from the point of view of its experience and government activities. For this, we propose using Platonov model [17]. In his theory a personality is considered as a dynamic system of interacting structures. The personality in this model is divided into 4 structures: orientation, experience, mental processes and biopsychosocial properties.

The AI model that we described in the previous sections has many uncertain parameters, and as important hyperparameters of proposed model, we suggest initially (before the AI launch) to set the Biopsychic properties, while the remaining properties need to be obtained or supplemented with 4 types of formation associated with the structures: Training, Exercise, Learning, Parenting.

We will not focus on temperament, sexual characteristics and the like, but age-related properties are very important for proposed model. As we wrote in Sect. 5.4, with growing up, more and more highly organized skills appear in order to ensure the active construction of this structure of skills in the process of growing up, so we need a substructure of age-related properties.

We turn to the structure of the features of psychological processes. Attention in a person is provided by the hippocampus, thinking is provided by the whole brain as a system, but if we talk about calculating options and planning, then these processes are organized by the orbitofrontal cortex. Memory, its various types and forms, are provided by the whole brain as a whole. Semantic memory is largely provided by the temporal lobes of the brain, for episodic, in addition to the temporal lobes, the hippocampus is also necessary, etc.

A very important structure is Experience. It is precisely the improvement of the properties of this structure that AI programmers are usually engaged in: skills during human development are built into a hierarchy, skills, i.e. fixed skill management for the optimal solution of a task of a certain type is determined by the neural network responsible for this skill, knowledge, i.e. organization of information in memory develops inextricably with skills, so that they can be well adapted to them. Although AI may not gain its experience with unsuccessful training, but it can gain knowledge that can help it in the future.

The orientation of the personality is determined by its desires, aspirations and midnset. Desires, i.e. specifically formulated needs, when creating an AI, we can direct it in the right direction, and then give it knowledge for the correct (acceptable for us) formulation of these needs. Aspiration, i.e. a persistent desire to do something or to achieve something is an inherently necessary necessity. We think that by making the need "to serve as best as possible" and giving the AI an idea of this, we will open to him levels of service that he has not yet reached, and he will strive for them. A worldview as a system of views and assessments about the world and its role in it can be initially set at the level of knowledge.

Platonov's theory well describes the characteristics of AI, and also gives a clear separation and understanding of how to change them. But for a better understanding of the necessary development of the tree of experience, we rewrite it in the terms of the theory of cognitive development of Piaget [16].

We will divide the development into the following periods and stages:

1. The period of sensory-motor intelligence.

    1.1. The first stage is creating essential reflexes,
         AI needs instructions for minimal work in a field in which it does not understand and is far from it existing skills.
    1.2. The second stage is analysis of its actions and coordination of reflexes,
         For AI, this is the definition of potential action in a new area.
    1.3. The third stage is the analysis of the results and the direction of activity on the target,
         For AI, it's training simple essential skills.
    1.4. The fourth stage is the analysis of the results of the preparation of plans and their implementation,
         For AI these are actions based on a miscalculation of reality one step ahead.
    1.5. The fifth stage is the analysis of consequences and experimentation to obtain more information,
         For AI, this is composing a new game in reality.
    1.6. The sixth stage is internal experimentation, i.e. simulation of interaction with reality,
         For AI, these are internal simulations based on this game, when in the case of a lack of complete information in the game, contextual information is used.

2. The period of preparation and organization of specific operations.
   It is a period of connecting to internal modeling the collection of signs of objects, the possibility of reactivation in the memory of an object according to signs and operating with objects according to their properties, linking to the internal model of reality the properties of various objects.

3. Formal operation period.
   This is the period when it becomes possible to distinguish abstract games in reality using previously selected semantic information.

6   Proposed Architecture

The basic structure of the proposed AI is a hierarchical tree of experience. It should support two functions: execution and development. The execution function provides the implementation of the basic actions of AI, and the development function ensures the construction of new branches when gaining new experience. A new experiance is subsequently gained based on existing experience. This is similar to the concept of mini batch in image recognition, but only on a large scale of different views on information.

6.1   General Principles

According to the Platonov model, the behavior of the proposed system can be described as follows.

1. Define the biological maturation of AI with its characteristics to improve the construction of the tree of experience.
2. Train AI and select information for it with such properties so that it does not have retraining and there is an effective storage of knowledge—memory, as well as other psychological functions.
3. To select AI information on topics that, as in the educational paradigm, will set up skills and knowledge for it—from basic to more complex.
4. To consolidate the internal need for service and to enable it to be properly realized, giving knowledge about how to do it and directing it as a person to service.

In connection with how we characterize the interaction of parts of the brain, we can immediately understand that the first skills that will be trained will immediately translate into attempts to act (Out) on the available input data (Input). In humans, the simplest type of action is reflexes. They are responsible for the survival of newborns and represent their most basic skills. In Sect. 5.5 we also indicated the general patterns of human development of skills, and show their meaning for AI.

The possible tree of experience is presented at Fig. 1, and the proposed architecture of a biomorphic AI is illustrated at Fig. 2.

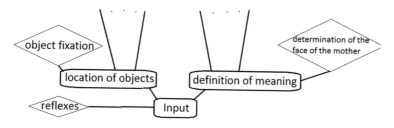

Fig. 1 The architecture of the biomorphic AI (here the diamonds denote experience and the rounded rectangles denote skills)

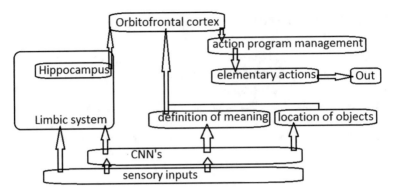

Fig. 2 The architecture of the biomorphic AI

The process of thinking is characterized by the separation from reality of a single "game"—conscious/unconscious focusing only on important objects and their properties and patterns—which in humans is determined by default by the ancient/emotional brain—the limbic system within which the object acts and develops in various ways. AI "thinks" if at each step it consciously models the interaction with reality, calculating the objective function, or acts intuitively (automatically), if conscious "thinking" or modeling is difficult or impossible. AI acts and develops, learning from those options of actions that will ultimately stand out as good or bad within the framework of the "game", and in the learning process using skills developed in this field and skills, destinations for playing this variety.

A similar process of isolating a game and "thinking" simulation of random games with reality is implemented in the improvement of AlphaZero—MuZero program [21]. Her model of reality is embedded in her hidden state. In proposed model, the "digital hippocampus" is responsible for it.

6.2 "Game with Reality"

The main assumption is that the functioning of AI is a "game with reality." He learns the rules of this game and adapts to it in an optimal way, striving for victory. At the same time, it is not limited to one type of game, but transfers the gained experience to the types of games, thereby becoming a universal player. In architecture, the digital hippocampus is responsible for highlighting the game. Already inside the selected game with its rules and general presentation, the training of skills takes place. We'll clarify that, by default, human emotions are responsible for choosing a game, or rather the limbic system, so when a person does not focus on anything, he usually relaxes or does all kinds of entertainment, seeking pleasure, and AI, choosing a game, seeks to maximize cost function.

We emphasize that the choice of a single game at any given time is extremely important, since the choice of several games at once does not allow to achieve perfection and hinders effective learning. In humans, this is manifested in the form of cognitive dissonance.

To illustrate the proposed AI scheme, let us restrict by the visual system, since, as is known, there is the possibility of rearrangement of the auditory cortex to perform visual functions, vice versa [23]. This means that the architecture of the neural networks of the primary analysis, analogues of the auditory and visual cortex, are not configured specifically for their data. They both single out objects of increasing complexity and both have the last "layer" responsible for the "image analysis". In humans, this function is performed by the temporal lobe of the brain responsible for semantics [10].

6.3 Multi-task Learning

The approach described above can find useful applications in multitasking. Tasks are usually related to each other in a hierarchy, so a similar approach can provide significant improvement [25].

Even if a broad topic cannot be divided into subsections, because the tasks within this topic are not related to each other, then even in this case, we can still get a significant improvement in general learning compared to a separate training for each task [20]. The reason for this can be called a more effective implementation of abstract experience to analyze, which is updated differently with different skills. The determination of the structure of subtasks in the case of related tasks is studied in [9].

When a need arises for a new application of a skill (when old applications for the desired activity no longer work effectively), a new skill begins to be determined based on this skill. This can be interpreted as Transfer of Knowledge [1]. If the scope of the new skill is similar to the previous one, then a similar skill is used, but for which an interpretation layer is created, which can be considered as a variant of Transfer of knowledge. And as already mentioned, both of these options complement each other.

6.4 Solving Problems of Strong Artificial Intelligence

There are different definitions of "strong AI," but most researchers agree that AI must have the same general human skills and use them at a certain level. In particular, it is believed that achieving the ideal fulfillment of certain tasks leads to a strong AI. Such tasks are called *AI-complete*. In this section, we describe how to configure AI to solve these problems.

Приведем список задач и их предлагаемое решение с помощью нашей модели.

(1) *Expert assessment of AI* (in a broad interpretation also includes automatic reasoning, automatic proof of theorems, expert system with formalized logic). In the proposed model, automatic proof of theorems can be considered as a game, and instead of training in winning strategies in duels, to teach winning strategies in compiled ratings (with a fixed number of participants for each task) of the options for the proofs. Expert assessment is achieved by mental modeling of reality. Having reproduced the situation mentally, the AI will give an assessment of the situation with its various characteristics and descriptions, as well as optimal solutions in this situation in his opinion.

(2) *Bongard problem.* Some skills of the AI are responsible for knowledge, such as, for example, the ability to determine the convexity of a figure. In an unconscious analysis of objects, we use the correspondingly unconscious selection of signs, but in a conscious analysis we also use various skills for analysis. In the case of the Bongard problem, we use them until we select a sign that separates some objects from others.

(3) *Computer vision* (including recognition of objects). This problem is solved through a parallel definition of semantics and areas during the first analysis, followed by a shift of focus to the necessary information after emotional and semantic processing of information.

(4) *Natural language processing* (including text analysis, machine translation and disambiguation [16]. When reading the text of the AI, it is proposed to simulate the situation as a random game, as AlphaZero does. As we have already said, in the general case, random games when calculating options have their own internal model of the world (in this case, a text model). Thus, the AI, having "sensed" the situation, will analyze this model from different angles (including semantically) and will begin to better understand it and use this understanding taking into account the context: during the conversation—to determine the correct answer (for which he additionally builds random options for the development of the dialogue itself), when studying books and other cognitive literature, uses understanding to identify the most useful information (which is emotionally colored as useful in simulations), etc.

(5) *Reaction to unforeseen situations.* This type of problems involves making decisions in unexpected circumstances when solving any real problem, be it navigation or planning, or even the type of reasoning performed by expert systems. By developing various skills, the AI will be able to deal with various unforeseen circumstances, in addition, each time in an unforeseen situation, an internal miscalculation of the future options (where all the AI capabilities are already involved) is made and the optimal option.

7 Conclusions

In this paper we examined various technologies useful for creating biomorphic AI. We did not want to focus on perfect correspondence with biology, but using the principles of the organization of the human brain and thought processes, we tried to describe the direction for achieving the human level of intelligence. Humanity, striving to create a strong AI, follows a rather complicated and unpredictable path, so it is difficult to say what to expect in the near future from AI.

It is not known to what level AI development will reach, but we believe that all the basic technologies for creating AI, comparable in level of intelligence with humans, already exist. Whether research at this level is achieved or whether a new "Winter of Artificial Intelligence" is coming depends on many factors that go beyond the scope of our discussion.

We do not claim that after this a technological singularity will occur, the system will not be rebuilt immediately, but global processes and patterns will immediately change from that moment, and in the case of a controlled scenario, progress will become even better.

References

1. Argote, L., Ingram, P.: Knowledge transfer: A Basis for Competitive Advantage in Firms. Organ. Behav. Hum. Decis. Process. **82**(1), 150–169 (2000)
2. Baars, B., Gage, N.: Cognition, Brain and Consciousness: An Introduction to Cognitive Neuroscience, 2nd edn. Elsevier/Academic Press, London (2010)
3. Beheshti, Z., et al.: A review of emotional learning and it's utilization in control engineering. Int. J. Advance. Soft. Comput. Appl. **2**(2), 191–208 (2010)
4. Brown, T.I., et al.: Prospective representation of navigational goals in the human hippocampus. Science **352**(6291), 1323–1326 (2016)
5. Buxhoeveden, D., Casanova, M.: The minicolumn hypothesis in neuroscience. Brain **125**(5), 935–951 (2002)
6. Cialdini, R.B.: Influence: Science and Practice, 5th edn. Allyn & Bacon, Boston (2009)
7. George, D., et al.: A generative vision model that trains with high data efficiency and breaks text-based CAPTCHAs. Science **358**(6368), art. no. eaag2612 (2017)
8. Girshick, R.: Fast R-CNN. In: The IEEE International Conference on Computer Vision (ICCV), pp. 1440–1448 (2015)
9. Jawanpuri, P., Saketha, N.: A convex feature learning formulation for latent task structure discovery. In: Proceedings of the 29th International Conference on Machine Learning, Edinburgh, Scotland, UK. http://icml.cc/2012/papers/90.pdf (2012). Accessed 22 February 2020
10. Jung, J., et al.: GABA concentrations in the anterior temporal lobe predict human semantic processing. Scientific Reports, 7, Article number: 15748
11. Kiyohito I et al. (2018) An effect of serotonergic stimulation on learning rates for rewards apparent after long intertrial intervals. Nat. Commun. **9**(2477) (2017)
12. Markram, H., et al.: Reconstruction and simulation of neocortical microcircuitry. Cell **163**(2), 456–492 (2015)
13. Masters, D., Luschi, C.: Revisiting small batch training for deep neural networks. https://arxiv.org/pdf/1804.07612.pdf (2018). Accessed 22 February 2020

14. Mountcastle, V.: The columnar organization of the neocortex. Brain **120**(4), 701–722 (1997)
15. Package with BELBIC controller for Autonomous Navigation of AR. Drone https://github.com/dvalenciar/BELBIC_Controller_ROS. Accessed 22 February 2020
16. Piaget, J.: The Psychology of Intelligence. Routledge, New York (2001)
17. Platonov, K.: A Concise Dictionary of the System of Psychological Concepts. High School, Moscow (2008)
18. Reimann, M., et al.: A biophysically detailed model of neocortical local field potentials predicts the critical role of active membrane currents. Cell **79**(2), 375–390 (2015)
19. Ren, S., He, K., Girshick, R., Sun, J.: Faster R-CNN: towards real-time object detection with region proposal networks. In: Advances in Neural Information Processing Systems, pp. 91–99 (2015)
20. Romera-Paredes, et al.: Exploiting unrelated tasks in multi-task learning. In: Proceedings of the Fifteenth International Conference on Artificial Intelligence and Statistics, PMLR, pp. 22:951–959 (2012)
21. Schrittwieser, J., et al.: Mastering Atari, Go, Chess and Shogi by Planning with a Learned Model https://arxiv.org/pdf/1911.08265.pdf (2020). Accessed 22 February 2020
22. Setogawa, T., et al.: Neurons in the monkey orbitofrontal cortex mediate reward value computation and decision-making. Commun. Biol. **2**(126) (2019)
23. Sharma, J., et al.: Induction of visual orientation modules in auditory cortex. Nature **404**, 841–847 (2000)
24. Srivastava, N., Hinton, G., Krizhevsky, A., et al.: Dropout: a simple way to prevent neural networks from overfitting. J. Mach. Learn. Res. **15**(1), 1929–1958 (2014)
25. Zweig, A., Weinshall, D.: Hierarchical regularization cascade for joint learning. In: Proceedings of the 30th International Conference on Machine Learning, Part 2, pp. 1074–1082. Atlanta, Georgia, USA (2013)

Medical Data Protection Using Blind Watermarking Technique

Abdallah Soualmi, Adel Alti, and Lamri Laouamer

1 Introduction

Nowadays, innovative technologies are well-caught attention of International Health Organizations, where the users could exploit services on demands, and consequently, more time and resources are gained.

The use of services and applications through network could result in some issues that could affect the user's data privacy and security, where unauthorized parties could steal, use, modify or destruct sensitive data such medical information, and this could cause serious problems. To this end, a new mechanism for data protection must be implemented.

In the case of cloud, the data protection approaches could be grouped into three major categories [1, 2], namely: client-side approaches, server-side approaches, and third-side approaches. Roughly speaking, for the first category, the security mechanism is implemented on two client's side (sender, receiver) and the cloud is used just for storing or some processing operation on data. For the second category, the security mechanism is implemented in the cloud server. While for the third category, the system delegated for an external authority to ensure security (e.g. *using a certificate*).

A. Soualmi (✉) · A. Alti
LRSD Laboratory, Department of Computer Science, University of Sétif-1, P.O. Box 19000, Sétif, Algeria
e-mail: sabdallah@univ-setif.dz

A. Alti
e-mail: adel.alti@univ-setif.dz

L. Laouamer
Department of Management Information Systems, Qassim University, P.O. Box 6633, Buraidah 51452, Saudi Arabia
e-mail: laoamr@qu.edu.sa

© The Editor(s) (if applicable) and The Author(s), under exclusive license to Springer Nature Switzerland AG 2021
A.-E. Hassanien et al. (eds.), *Enabling AI Applications in Data Science*, Studies in Computational Intelligence 911,
https://doi.org/10.1007/978-3-030-52067-0_25

In the sequel, most of the telemedicine application use only the encryption techniques to protect data, and even though the encrypted information's are protected during transferring, unfortunately, once it's decrypted the illegal reproduction or the authorship proofing couldn't be ensured. To this end, some application combines cryptography and watermarking techniques for more security and protection.

Digital watermarking consists of encrusting information into cover data (image, video, text file...etc.) [3]. In the last years, lot of image-watermarking methods was presented [4–27]. Where these approaches could be classified as spatial or frequency. In the first-class methods, data are encrusted directly on pixels which make it requires less processing time. However, most of the spatial methods are fragile. In the second method, data are encrusted into the transform coefficients such as DCT (*Discrete Cosine Transform*) [28] or DWT (*Discrete Wavelet Transform*) [29], which make this method more robust but require more processing time [30].

Moreover, the watermarking technique could be robust, fragile, or semi-fragile [31, 32]. For the first class, the encrusted data shouldn't collapse under any attack performed on image [8, 33], for the second class, the encrusted data needs to keep minor manipulation [34, 35]. While, for last, the encrusted data deleted or collapsed after facing any attack [36]. More, The watermarking approach could be grouped as blind, semi-blind, or non-blind techniques [29]. The first type, de-watermarking phase doesn't need the presence of the original image or watermark, so, it will be enough to possess the watermarking key. The second type, the watermark is required and for the last one, it needs the presence of the original image.

Currently, A brief talking about the medical images, and more precisely the DICOM (*Digital Imaging and Communicating in Medicine*), which is a standard developed for data transfer, storage, and designed to cover all aspects of digital medical imaging [37]. The DICOM file is constituted of two principal components (Fig. 1), namely: Header and Body. The header contains the File Meta information [24]. While the body contains the graphical medical data's set (a set of data elements which constitute the medical image), the image pixel data's (Body) are split to regions

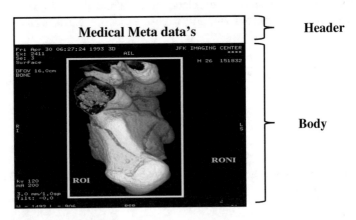

Fig. 1 DICOM file basic structure

of non-interest (RONI) and region of interest (ROI) [20, 36], The ROI area is constituted from the significates pixel (depend on the clinical finding). While the RONI is the insignificant pixel.

This chapter presented a blind watermarking scheme for medical data protection using Diffie-Hellman [38] and Number Theoretic Transform (NTT) [39].

The rest of the chapter is ordered in the following way. Some relevant works are described in Sect. 2. In Sect. 3 presents the basic requirements of the proposed scheme. In Sect. 4, presented the proposed scheme in detail. In Sect. 5, show the experimentation results. Section 6 gives a discussion of the presented method in terms of advantages and disadvantages. And finally, in Sect. 7, the presented conclusion and future.

2 Relevant Schemes

In these last previous years, an important number of new or enhanced watermarking techniques have been proposed. In this section, we review some new relevant available techniques on literature. The methods illustrated are listed according to the robustness offers, into tow major's classes: fragile and robust techniques.

2.1 Fragile Methods

The fragile watermarking techniques are used to authenticate the images or for data integrity [34, 35, 40].

Kannammal et al. [19] presented a blind watermarking scheme for medical images using DWT and QIM (Quantization Index Modulation), the watermark encrusted in *DWT-coefficients* using QIM technique. The proposed method denotes the effectiveness of this scheme in image fidelity and attacks resistance. However, the security and computational complexity are mediocre, also, the authors mention that the technique is used for medical image authentication, but they didn't present how the technique ensures authenticity.

The authors in [20] propose a hybrid method for medical data protection based on watermarking and cryptography techniques, the main idea is to employ Message Digits 5 (MD5) to compute the image *hash value*. After that, ideal groups are obtained from the medical image and then compressed using Huffman technique. Finally, the hash value, the compressed groups, and the patient information are combined and cyphered by Advanced Encryption Standard (AES) algorithm and encrusted in the host image. This technique provides image quality, data protection, and authentication. However, it requires high processing time, and the resistance degree with the attacks was not mentioned (the method was not tested with attacks).

In [9] presented a non-blind fragile method, where the idea is to encrust the watermark pixels into a transform based on Compressive Sensing (CS), NSCT (Non-Subsampled Contourlet Transform), and DWT. This method offers good advantages when it comes to tamper detection, data payload, and authenticity. Except that it requires high execution time.

In [26] presented a blind reversible medical image watermarking method. The method focuses on a set of pixels from the cover image as an emplacement for embedding the watermark using a chaotic key, this method offers a good trade-off between tamper detection and fidelity, but it gives mediocre data payload.

In Soualmi et al. [27] presented a blind fragile method using Schur decomposition, where the main idea consists of encrusting data into the host image after carrying Schur on it, using a new embedding technique which works on embedding the watermark bits on a specific emplacement. Experimental results show their effectiveness when it comes to fidelity and processing time, but, it offers low embedding capacity.

2.2   Robust Methods

Robust watermarking techniques generally propose to use in ownership proving [34, 35, 40].

In [24] present a blind DCT-based scheme, the DCT is performed on some specific location, after that, variation value is employed for watermark encrusting by modifying the DCT coefficients. This scheme gives better fidelity, attack resistance, and execution time. unfortunately, the encrusted data is protected in a mediocre manner, and the data payload is restricted by the embedding blocks number.

Savakar and Ghuli [8] presented a hybrid scheme using SVD and DWT. Their main idea is to embed data into the DWT coefficient of the first image, after that the watermarked image itself encrusted into the DWT-SVD coefficients of the second image. This method offers good fidelity and attack resistance; however, it requires more processing time, also the scheme was tested only against three kinds of attacks: noising, rotation, and JPEG compression.

In Arumugham et al. [23] propose a DWT/SVD semi-blind watermarking technique, the method work on encrusting data on the singular values of DWT coefficients, then encrypt the watermarked image using pseudo-random number sequences. This approach gives good results when it comes to security, attack resistance, and embedding capacity, however, it needs high processing time.

In [11] presented a transform semi-blind zero-watermarking scheme based on tensor mode expansion. First, a zero-watermark image is generated using a watermark logo and the cover image, after that the generated watermark is embedded using DCT, SVD, and tensor mode expansion. Analyzing the experimental results of this method we observe that it's resisted very well against image processing attacks. However, the method needs high processing time, and image quality degree wasn't mentioned.

In [22] presented a blind DWT, Hessenberg decomposition (HD), SVD, and fly optimization-based scheme. The method works on encrust the singular value of data into the DWT-HD coefficients singular values, and then the FOA is applied to get a trade-off between attack resistance and fidelity. This method offers better image quality and attacks resistance, except that, it needs high processing time and gives low security to the watermark embedded.

In [18] proposed a non-blind technique using differential evolution (DE) and DWT, the cover image is decomposed up into blocks using DE, then the watermark bits and then embed DE results into the DWT coefficients of training blocks. This approach gives good robustness and fidelity. However, the scheme wasn't employed any security method to secure the embedded watermark, and consequently, any user could extract the watermark, also, the data payload is depending on the blocks selected by DE.

Ariatmanto and Ernawan [13] proposed a semi-blind transform scheme. First, the cover image is decomposed up into non-overlapping blocks, and the variance pixel value of each pixel is computed to select blocks that will be used for watermark embedding. Finally, the scrambled watermark is encrusted into DCT middle band coefficient using some rules. This method offers high imperceptibility and robustness. However, a bit need 8×8 block to be encrusted, this makes this method gives less embedding capacity.

As mentioned bellows, most of the existing watermarking techniques suffer from the important computational complexity and low imperceptibility, especially when it comes to the medical images. Although one of the basic characteristics of e-health application is the need for low processing time; while respecting the medical image quality.

Through this chapter, our main objective is to develop a secure and imperceptible watermarking technique that requires less processing time. The cause is that the watermarking for medical images have some specific issues due to their high sensitivity caused by the restricted texture number.

In the next part, the basic requirements are explained in detail.

3 Basic Foundations of the Proposed Scheme

To address medical data protection issues in an unsecured channel, a novel blind watermarking technique using NTT is proposed.

3.1 Number Theoretic Transform (NTT)

The NTT [39] is a popular transform decomposition based on DFT. The NTT of a sequence x constituted of S elements presented in the Galois field $G_{F(v)}$ of order v:

$$X_m = \left\langle \sum_{s=0}^{S} x_s \cdot \delta^{sm} \right\rangle_v \tag{1}$$

where $m = 0, 1 \ldots S-1$ and δ is the term generator of S.

In the case of images, the period S is the most important element of NTT, where for at least one $S^{-1} \delta$ exists, to this end we must find modulus v:

$$S^{-1} = v - ent\left(\frac{v}{S}\right) \tag{2}$$

if pgcd $(v, \delta) = 1 \delta$ is guaranteed.

3.2 Diffie-Hellman for Securing Key Exchange

Diffie-Hellman (D-H) [38] is a protocol used to share secrets keys, even though two sides (emitter, receiver) might never have communicated with each other (see Table 1). Diffie-Hellman used basically between two sides to share the symmetric key. When the two sides have the key, they encrypt their shared information between them.

In the proposed technique NTT coefficients are used to embedding area. While Diffie-Hellman algorithm is used to share the key which will be used for encryption/decryption of watermark.

The next section explains the proposed method processes with deep details.

4 Merging NTT and Diffie-Hellman

The combination of NTT and Diffie-Hellman leads to being highly secure in terms of medical data preservation and privacy. In this section, explained in details the

Table 1 D-H technique main idea

Function	Input	Operation	Output
Shared keys	$w, l, C, D, \in \mathbb{P}$	Generate a suitable prime numbers l and C Base w	Public keys: C, D
		Exchange private number b, C Compute $A = w^C \bmod l$ $B = w^D \bmod l$	Secret keys: $K = A^D \bmod l$ or $K = B^C \bmod l$
Encryption	$E \in \mathbb{Z}_n$	Compute C using a public key	$C \in \mathbb{Z}_n$
Decryption	$C \in \mathbb{Z}_n$	Compute E using a secret key	$E \in \mathbb{Z}_n$

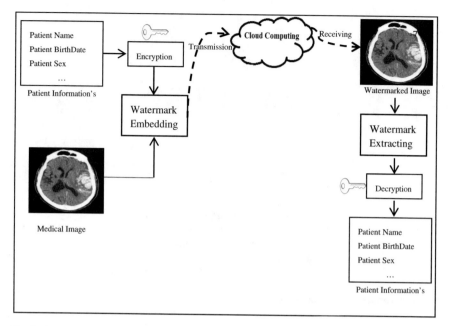

Fig. 2 General model for the proposed solution

proposed technique. In the following the two main processes (Fig. 2) are described: the watermark embedding, and extraction phases. Where the main purpose is to increase data security while keeping image fidelity.

4.1 Embedding Phase

As shown in Fig. 3 the embedding phase is applied in a transform domain by applying NTT on the cover image. Firstly, encrypt the watermark using Diffie-Hellman protocol, and then each block of the cover image is used to embed the encrypted bits. In the following the embedding phase steps:

1. Decompose the cover image into Non-overlapping 16×16 blocks B_i;
2. Perform the NTT on the cover image.
3. Encrypt the watermark using Diffie-Hellman algorithm. The encrypted data are obtained after performing *XOR* operation between the key and data to be encrusted.
4. Select the coefficients of B_i, then select every four coefficients as a group G_j.
5. Perform the NTT on the cover image.
6. Encrypt the watermark using Diffie-Hellman algorithm. The encrypted watermark is obtained after performing *XOR between* key and encrusted bits.
7. Select the coefficients of B_i, then select every four coefficients as a group G_j.

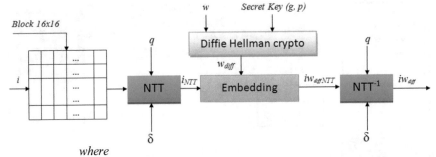

where

  i: the cover image.
  w: the used watermark.
  δ: the NTT generator term.
  $iw_{diffNTT}$: the NTT watermarked image.
  i_{NNT}: the NTT image.
  w_{diff}: the encrypted watermark.
  q:parameter of NTT.
  $iw_{diffNTT}$:the encrypted watermarked image.

Fig. 3 Watermark embedding framework

8. Compute the Weight of G_j as follow:

$$WE(G_j) = \sum_{k=1}^{4} (C_{kj} \times K) \tag{3}$$

where C_{kj} is the coefficient of group j at indice k.

9. A watermark Bit W_r is embedded in G_j as the following case:

If mod $(WE(G_j), 2) \sim = W_r$, then:

$$MAX(C_{1j}, C_{3j}) = MAX(C_{1j}, C_{3j}) - 1 \tag{4}$$

where C_{1j} and C_{3j} are the coefficients of the group G_j in indices 1 and 3 respectively (Fig. 4).

10. Repeat with $r + 1, j + 1$, and $i + 1$ until encrusting all watermark bits. Apply NTT^{-1} to get the watermarked image (Fig. 5).

Figure 6 illustrates the watermark embedding phase.

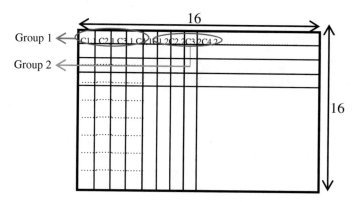

Fig. 4 Illustration of block decomposition into set of coefficients groups

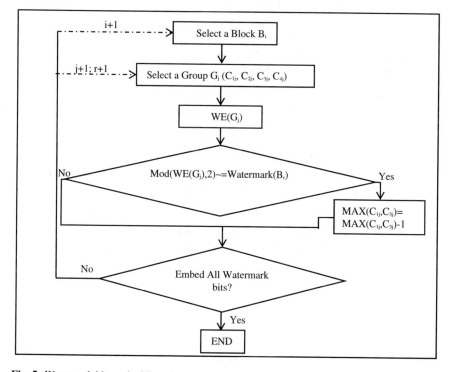

Fig. 5 Watermark bits embedding phase

4.2 Extraction Phase

A shared image could be altered during its transmission, the receiver of the transferred image must be identified. The extraction phase (Fig. 5) is described as follow:

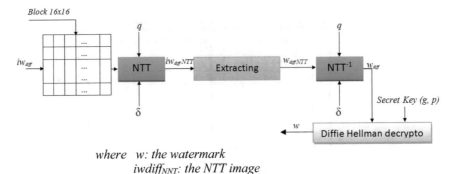

where w: the watermark
iwdiff_{NNT}: the NTT image

Fig. 6 Watermark extracting framework

1. Divide the watermarked image into blocks B_i of 16×16.
2. Perform NTT on the watermarked image.
3. Select the coefficients of B_i, then select each four coefficients as a group G_j.
4. Compute the Weight of G_j WE(G_j) using (Eq. 3).
5. A watermark B_r is extracted from G_j:

$$B_r = \mathrm{mod}\left(WE(G_j), 2\right) \tag{5}$$

6. Repeat with $r + 1, j + 1$, and $i + 1$ until all bits extracted.
7. Apply NTT^{-1}.
8. The encrusted watermark is deciphered using the encryption key shared with Diffie-Hellman algorithm; the decrypted bites are obtained using *XOR between* encrusted bits and key bits.

   Figure 7 illustrates the watermark bits extracting process.
   The next section shows in detail the effectiveness measuring of the proposed scheme.

5   Performances Measurement

The performances and strong points of the proposed scheme are well clarified in this section.

5.1   Performance Results

The proposed approach is implemented using the programming languages JAVA and MATLAB. All tests are performed on medical image in DICOM format of size

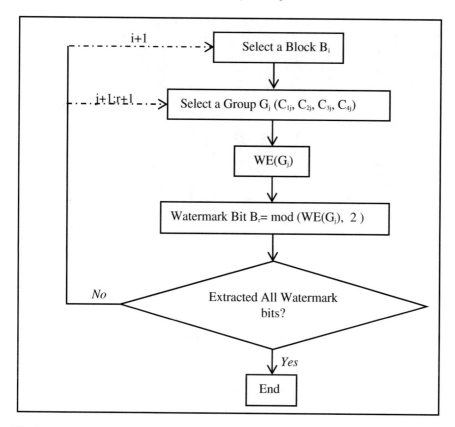

Fig. 7 Watermark Bits extracting phase

256×256 downloaded from [41, 42, 43], while the watermark data are string type of size 32 bytes, cover images sample, and watermark are shown in Fig. 8, Table 2 illustrates the execution steps. Figure 9 displays the patient's information, the cyphered watermark, and the amount of data needed for embedding. Figure 10 displays the received patient's information by an expert after extracting and decrypting the watermark and the processing time required for extracting.

5.1.1 Measure the Imperceptibility

Imperceptibility is how much the watermarked images are similar to the original one in terms of dB [29]. The PSNR is used to measure the similarity degree between the cover and the watermarked images. It is defined via the Mean Square Error (MSE), The relative metrics are shown in Eqs. (6) and (7).

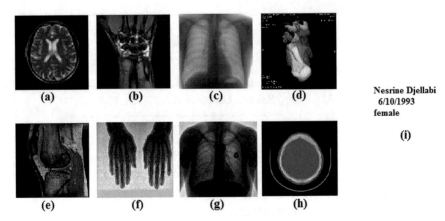

Nesrine Djellabi
6/10/1993
female

(i)

Fig. 8 a–h Medical image **i** patient information (watermark)

$$MSE = \frac{1}{H \times W} \sum_{l=0}^{H-1} \sum_{m=0}^{W-1} (OI(l,m) - WI(l,m)) \tag{6}$$

where H and W both are the image size, OI and WI are the original and watermarked images.

$$PSNR(dB) = 10 \log_{10} \left(\frac{255^2}{MSE} \right) \tag{7}$$

From the experimental values of PSNR, as shown in Table 3, we observe that the proposed keep the image fidelity.

5.1.2 Robustness Measurement

The robustness means how much the watermark could resist after any kind of alteration [29]. The proposed method is tested against most popular serious attacks such as rotation, adding noise, median filtering, and average filtering attacks. The first attack affects the watermark by rotating the image from 0° to 360° [44]. While the main purpose of the second attack is to increase the difficulty of the watermark extraction process, by adding noise to the watermarked image. The noise value is varied from 0 to 1 [24, 35]. The third attack changes the values of center pixels with the median values of the sorted pixels [44]. The last attack (e.g. average filtering attack), replaces each pixel of the image with the neighbor pixels Average Value [44].

In our experimentations, we use Bit Error Rate (BER) measure to evaluate the resistance to attacks. It gives the probability of watermark binary data that are incorrectly received [15]. The BER could be bounded value between 0 and 1, or in terms of percentage (0–100), and the BER low value means a good resistance against attacks.

Table 2 The important execution steps of the proposed method

Cover image	Apply NTT	Watermark data
		Nesrine Djellabi 6/10/1993 Female

Encrypted watermark data	NTT⁻¹	Extracted watermark data	Decrypted extracted watermark
34DE8D0DFCEF50E198675E2EE3147E34		34DE8D0DFCEF50E198675E2EE3147E34	Nesrine Djellabi 6 /10/1993 Female

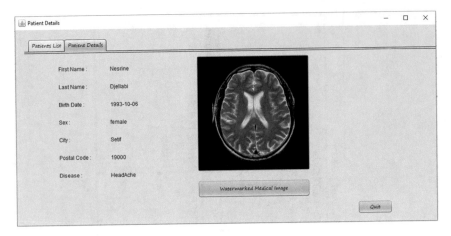

Fig. 9 Watermark embedding application illustartion

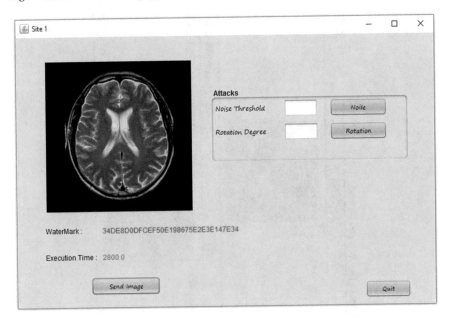

Fig. 10 Watermark embedding illustration 2

Table 3 Fidelity measuring using PSNR

Images	(a)	(b)	(c)	(d)	(e)	(f)	(g)	(h)
PSNR(dB)	43.32	44.18	43.93	43.58	43.64	42.46	42.84	43.41

The BER value is computed as follow:

$$BER = 100 \times \frac{Cr}{ABr} \qquad (8)$$

where Cr is the number bit corrupted and ABr is the watermark bits number.

The processing time, the BER for both retrieved watermarks, and the watermarked images are shown in Table 4, which clarifies that the proposed approach resists against some dangerous attacks such noise, rotation, median filtering, and average filtering. This is proved through the low BER of the retrieved watermarks. The attacks are performed on the watermarked image using Stirmark benchmark software [14].

5.1.3 Measure the Processing Time Required

The processing time means the time required for encrusting and de-watermarking phases. Figure 11 demonstrates the execution time (ms) needed to watermark embedding/extracting in/from different images (a..h). The processing time directly depends on image's characteristics (*textures, smoothness...etc.*).

5.2 Comparative Study

The performance of the proposed scheme is evaluated by the effect of constructing an evaluation study of the methods presented in [13, 22] in terms of imperceptibility explained in Sect. 5.1.1. Figure 12 gives the comparison results. Our technique preserves the image's quality comparing to other works [13, 22]. The reason is that the proposed technique uses an embedding technique which modifies slightly a coefficient's value (it's enough to subtract one coefficient by 1 for embedding a watermark bit).

Table 5 shows the comparison results with works presented in [19, 24, 25] in terms of major's drawbacks and advantages, where the security, blindness, computational complexity, and robustness are the main criteria of evaluations.

6 Discussion

In this work, a blind watermarking technique is proposed for medical data protection to support telemedicine. The proposed method leak from the good performances of NTT. Hence, it offers medical staffs with many advantages: good fidelity, high Security, and low execution time. However, it still suffers from the limited robustness degree with many geometric and signal processing attacks such: JPEG compression

Table 4 Obtained results after noise, rotation, median filtering, and average filtering

Original image	Watermarked image	Attacked Image
	Embedding time: 2800ms	Noise 0.01, Extracting time: 1560 ms, BER=0.197.
	Embedding time: 2800ms	Noise 0.01, Extracting time: 1570 ms, BER=0.197.
	Embedding time: 2750ms	Noise 0.02, Extracting time: 1710 ms, BER=0.249.
	Embedding time: 2910ms	Rotation 1°, Extracting time: 1723 ms, BER=0.199.

(continued)

Table 4 (continued)

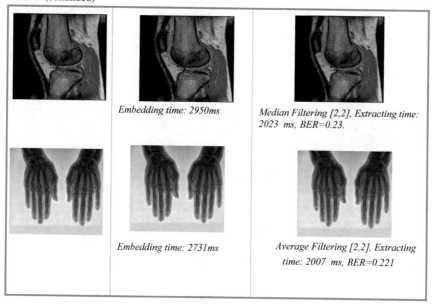

| | Embedding time: 2950ms | Median Filtering [2,2], Extracting time: 2023 ms, BER=0.23. |

| | Embedding time: 2731ms | Average Filtering [2,2], Extracting time: 2007 ms, BER=0.221 |

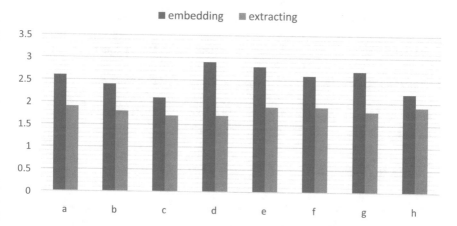

Fig. 11 Execution time (ms) measure for embedding/extracting operations

attack, cropping attack, translation attack …etc. This necessitates an improvement in future works.

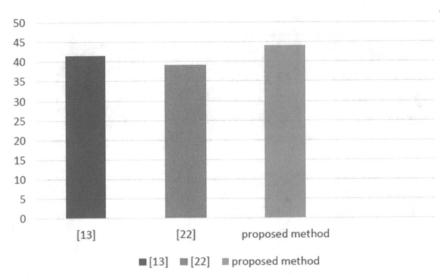

Fig. 12 Imperceptibility comparison with the works presented in [13, 22]

Table 5 Comparison with works presented in [19, 24, 25]

Method	Advantages	Drawbacks
Singh, D., Singh [24]	Good imperceptibility and security	Require high processing time, semi-blind
Sadeghi et al. [25]	Blind, good attacks resistance and fidelity	Mediocre protection
Kannammal and Rani, [19]	Good fidelity, good attack resistance	Non-blind, *Security Mediocre*
Proposed method	Blind, good imperceptibility, and computational complexity	Low robustness

7 Conclusion

In this work, our purpose was to propose an effective watermarking method that could ensure medical data security. The proposed method is employed to encrust the data into the transform domain-based NTT. This watermarking approach allows fast embedding and extracting of watermark in/from the watermarked image, the watermark is cyphered using the secret key shared with Diffie-Hellman then embedded in the NTT coefficients. The proposed technique is blind, this means that the data embedded could be extracted only with the key used in embedding, without the presence of the watermark or host image.

Future works will focus to enhance the robustness against geometric and signal processing attacks The future works will focus also on validating the watermarking technique for real-time telemedicine applications This presents open issues that hinder the progress of telemedicine systems.

References

1. Agarwal, N., Singh, A. K., Singh, P.K.: Survey of robust and imperceptible watermarking. Multimedia Tools Appl. 1–31 (2019)
2. Liu, Y., Zhang, Y., Ling, J., Liu, Z.: Secure and fine-grained access control on e-healthcare records in mobile cloud computing. Fut. Gene. Comput. Syst. (2017)
3. Byun, S.W., Son, H.S., Lee, S.P.: Fast and robust watermarking method based on DCT specific location. IEEE Access 1–17 (2019)
4. Singh, S.P., Bhatnagar, G.: A robust blind watermarking framework based on Dn structure. J. Amb. Intell. Human. Comput. 1–19 (2019)
5. Kumar, S., Jha, R.K.: FD-based detector for medical image watermarking. IET Image Process. 13(10), 1773–1782 (2019)
6. Phadikar, A., Jana, P., Mandal, H.: Reversible data hiding for DICOM image using lifting and companding. Cryptography 3(21), 1–19 (2019)
7. Gao, T., Jiang, F., Li, D.: A robust zero-watermarking algorithm for color image based on tensor mode expansion. Multimedia Tools Appl. 1–16 (2020)
8. Savakar, D.G., Ghuli, A.: Robust invisible digital image watermarking using hybrid scheme. Arab. J. Sci. Eng. 1–14 (2019)
9. Thanki, R., Borra, S.: Fragile watermarking for copyright authentication and tamper detection of medical images using compressive sensing (CS) based encryption and contourlet domain processing. Multimedia Tools Appl. 1–20 (2018)
10. Hsu, L.Y., Hu, H.T.: Blind image watermarking via exploitation of inter-block prediction and visibility threshold in DCT domain. J. Vis. Commun. Image Represent. 1–20 (2015)
11. Jiang, F., Gao, T., Li, D.: A robust zero-watermarking algorithm for color image based on tensor mode expansion, Multimedia Tools Appl. 1–16 (2020)
12. Vaishnavia, D., Subashini, T.S.: Robust and invisible image watermarking in RGB color space using SVD. Procedia Comput. Sci. 46, 1770–1777 (2015)
13. Ariatmanto, D., Ernawan, E.: An improved robust image watermarking by using different embedding strengths. Multimedia Tools Appl. 1–27 (2020)
14. Su, Q., Niu, Y., Wang, Q., Sheng, G.: A blind color image watermarking based on DC component in the spatial domain. Optik Int. J. Light Electron Opt. 124(23), 6255–6260(2013).
15. Radharani, S., Valarmathi, D.: A Study on watermarking schemes for image authentication. Int. J. Comput. Appl. 2(4), 24–32 (2010)
16. Gomathikrishnan, M., Tyagi, A.: HORNS-A homomorphic encryption scheme for cloud computing using residue number system. IEEE Trans. Parallel Distrib. Syst. 23(6), 995–1003 (2011)
17. Mohanta, B.K., Gountia, D.: Fully homomorphic encryption equating to cloud security: an approach. IOSR J. Comput. Eng. (IOSRJCE) 9, 46–50 (2013)
18. Salimi, L., Haghighi, A., Fathi, A.: A novel watermarking method based on differential evolutionary algorithm and wavelet transform. Multimedia Tools Appl. 1–18 (2020)
19. Kannammal, A., Rani, S.S.: Authentication of DICOM medical images using multiple fragile watermarking techniques in wavelet transform domain. Int. J. Comput. Sci. Iss. (IJCSI) 8(6)(1), 181–189 (2011)
20. Abdeldayem, M.M.: A proposed security technique based on watermarking and encryption for digital imaging and communicating in medicine. Egypt. Inform. J. (EIJ) 1–13 (2012)
21. Ernawan, F., Kabir, M.N.: An improved watermarking technique for copyright protection based on Tchebichef moments. IEEE Access 1–20 (2019)

22. Liu, J., Huang, J., Luo, Y., Cao, L., Yang, S., Wei, D., Zhou, R.: An optimized image watermarking method based on HD and SVD in DWT domain. IEEE Access 80849–80860 (2019)
23. Arumugham, S., Rajagopalan, S., Rayappan, J.B.B., Amirtharajan, R.: Tamper-resistant secure medical image carrier: an IWT–SVD–Chaos–FPGA combination. Arab. J. Sci. Eng. **44**, 9561–9580 (2019)
24. Singh, D., Singh, D.: DWT-SVD and DCT based robust and blind watermarking scheme for copyright protection. Multimedia Tools Appl. (2016)
25. Sadeghi, M., Toosi, R., Akhaee, M.A.: Blind gain invariant image watermarking using random projection approach. Sig. Process. **163**, 213–224 (2019)
26. Rahman, A., Sultan, K., Aldhafferi, N., Alqahtani, A., Mahmud, M.: Reversible and fragile watermarking for medical images. Comput. Math. Methods Med. **2018**, 1–7 (2018)
27. Soualmi, A., Alti, A., Laouamer, L., Benyoucef, M.: A blind fragile based medical image authentication using Schur decomposition. In: International Conference on Advanced Machine Learning Technologies and Applications, pp. 623–632. Springer (2019)
28. Rakhmawati, L., Wirawan, W., Suwadi, S.: A recent survey of self-embedding fragile watermarking scheme for image authentication with recovery capability. EURASIP J. Image Video Process. **61**, 1–22 (2019)
29. Tao, H., Chongmin, L., Zain, J.M., Abdalla, A.N.: Robust image watermarking theories and techniques: a review. J. Appl. Res. Technol. **12**(1), 122–138 (2014)
30. Guikema, S.D., Aven, T.: Assessing risk from intelligent attacks: a perspective on approaches. Reliab. Eng. Syst. Saf. **95**(5), 478–483 (2010)
31. Kamran, A.K., Malik, S.: A high capacity reversible watermarking approach for authenticating images: exploiting down-sampling, histogram processing, and block selection. Inform. Sci. **256**, 162–183 (2014)
32. Lee, C.F., Shen, J.L., Chen, Z.R., Agrawal, S.: Self-embedding authentication watermarking with effective tampered location detection and high-quality image recovery. Sensors 1–18 (2019)
33. Wang, C., Wang, X., Zhang, C., Xia, Z.: Geometric correction based color image watermarking using fuzzy least squares support vector machine and Bessel K form distribution. Sig. Process. **134**, 197–208 (2017)
34. Agarwal, N., Singh, A.K., Singh, P.K.: Survey of robust and imperceptible watermarking. Multimedia Tools Appl. 1–31 (2019)
35. Ortiz, A.M., Uribe, C.F., Beltran, R.H., Hernandez, J.J.G.: A survey on reversible watermarking for multimedia content: a robustness overview. IEEE Access 1–21 (2019)
36. Mousavi, S., Naghsh, A., Abu-Bakar, S.: Watermarking techniques used in medical images: a survey. J. Digit. Imag. **27**(6), 714–729 (2014)
37. http://dicom.nema.org/dicom/2013/output/chtml/part03/PS3.3.html. Accessed 15 Oct 2017
38. Kallam, S.: Diffie-Hellman: Key exchange and public key Cryptosystems. Master degree of Science, Math and Computer Science, Department of India State University, USA, pp. 5–6 (2015)
39. Laouamer, L.: Toward a robust and fully reversible image watermarking framework based on number theoretic transform. Signal Imag. Syst. Eng. Indersci. **10**(4), 169–177 (2017)
40. Roy, S., Pal, A.: A blind DCT based color watermarking algorithm for embedding multiple watermarks. AEU Int. J. Electr. Commun. **72**, 149–161 (2017)
41. http://www.barre.nom.fr/medical/samples
42. http://deanvaughan.org/wordpress/2013/07/dicom-sample-images/
43. http://www.barre.nom.fr/medical/samples/
44. Kandi, H., Mishra, D., Gorthi, S.: Exploring the learning capabilities of convolutional neural networks for robust image watermarking. Comput. Secur. **65**, 247–268 (2017)
45. http://w.petitcolas.net/fabien/watermarking/stirmark/

An Artificial Intelligence Authentication Framework to Secure Internet of Educational Things

Ahmed A. Mawgoud⬩, Mohamed Hamed N. Taha⬩, and Nour Eldeen M. Khalifa⬩

1 Introduction

The rapid growth technology in IoT industry led to the increment in the computing capacity of devices. Many generations of IoT devices where defined based on many factors and the ability of computing at the edge is providing huge independency without the need of having a central management [1]. The first time the definition of IoT identified was by Kevin Ashton in 1999 [2]. Simultaneously, devices are being developed to adapt the new cloud-scale developments in many fields of IoT. The expansion of using Internet of Educational Things systems–and their rapid growth development devices with various application types–represents the next level of a strong connected community [3].

However, many security risks and vulnerabilities have appeared due to:

(1) The intricacy level of the IoET systems.
(2) The wide-usage of intelligent machines in industry procedures.

Because of the smart applications fast development and the automation integration in different industry fields nowadays, IoT and Artificial Intelligence have contributed in security part for the development form both academically and practically [4]. Figure 1 shows the initial three generations of IoT starting from the classic router

A. A. Mawgoud (✉) · M. H. N. Taha · N. E. M. Khalifa
Faculty of Computers and Artificial Intelligence, Information Technology Department,
Cairo University, Giza, Egypt
e-mail: aabdelmawgoud@pg.cu.edu.eg

M. H. N. Taha
e-mail: mnasrtaha@cu.edu.eg

N. E. M. Khalifa
e-mail: nourmahmoud@cu.edu.eg

A.-E. Hassanien et al. (eds.), *Enabling AI Applications in Data Science*,
Studies in Computational Intelligence 911,
https://doi.org/10.1007/978-3-030-52067-0_26

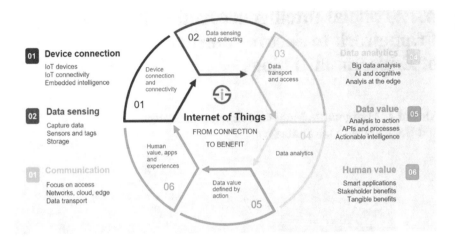

Fig. 1 IoT development steps from connected devices to human value

at home reaching to sophisticated small sensors. The result of merging those three definitions (network of objects, computing at the edge and cloud-scale) is to enable the transformation to IoT in each industry [5].

Internet of Educational Things (IoET) is a definition that describes a huge range of addressable unique smart devices in the education field. The devices can communicate with each other in one or many nodes through the internet [6]. The connected devices or things represent a new concept level of the internet usage. A node of devices that can be connected anytime, anywhere with the internet can provide a smart environment. The connected devices can communicate with each other in education environment (Labs-Classrooms-Research Centers-Lecture Halls). Also, it can share critical data that are involved in the layer of decision making. The whole previous processes describes the definition of IoE. The concept of (Internet of Things) at its beginning generally was defined in many previous researches as: Internet of processes, Internet of Data, Internet of People or Internet of Signs [7], as [8] stated in their research, that the initial concept of IoT depends on the connectivity of many objects through the internet based on a pre-defined architecture, those objects can be physical or virtual. There is an enormous increment of the connected devices number, as a result, there was a need from the scientific research field to come out with new solutions; to face the new rised challenges with this regards. As shown in Fig. 2 below, (Gartner's forecast) estimated that the number of connected devices wordwide will increase from 6 billion devices (2015) to about 27 billion smart device by 2025. Most of those connected devices will be connected through Wireless Personal Area Networks (WPAN). It is estimated that by 2025:

(1) The revenue in the IoT industry will be 3 trillion US$.
(2) Over two zetta bytes of data will be generated from IoT devices [10].

The IoT challenges are mostly related to security and privacy, reliability, scalability and management [11]. Those challenges were studied and analyzed in many

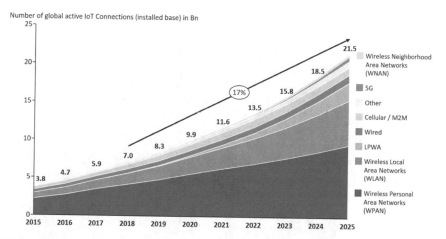

Fig. 2 Statistics shows the expected number of active IoT devices between 2015 till 2025 [9]

previous surveys such as [12–14]. IoT systems generally are being considered an essential infrastructure for future smart cities, smart transportation, heavy industries, smart hospital, higher education and military operations [15]. The main discussion in this paper is the role of IoT in the higher education field and how to provide a secure authentication scheme to suit the characteristics of this environment. The concept of IoET was studied previously in may different researches, the discussions took the idea of how to integrate the IoT system with the education facilities. IoET is considered the bridge that connects the physical side with the cyber side in education industry; However, IoET provides a vision of an educational network of connected smart devices. Figure 3 shows various types of end points (i.e. actuators and sensors), places or even the environment conditions for providing on-demand services [16].

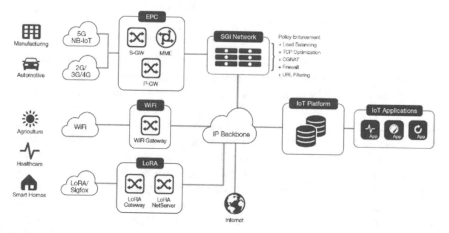

Fig. 3 An example of a general IoT system architecture

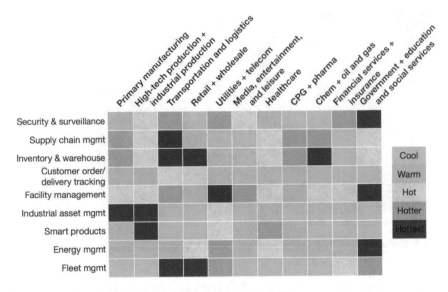

Fig. 4 IoT opportunities heat map classified by industries [18]

It is estimated that the market of IoT size will grow up to reach to 3.8B$ by 2020. This growth is mostly achieved through the sellings of IoT vendors in softwares, hardwares and IoT solutions [17]. It will take no time till each known industry is touched by the technology of IoT. As shown in Fig. 4, Forrester's heat map illustrated the IoT opportunities in many different industries and applications. Although there is a noticable growth for IoT opportunities in many fields, the map shows different sectors to use IoT in. The highest opportunities for IoT are as follow:

1 Security and surveillance in both government and education sectors.
2 Supply chain and fleet management in transportation.
3 Inventory and warehouse in retail.
4 Industrial assest management in high technology production.

Lufthansa Airlines, is an example of smart airline environment; they use real-time aircraft and weather sensor data to enhance performance and optimize operations. However, there were a lot of warnings from buying huge numbers of IoT devices in such an early stage of its development; to avoid the negative consequences that the scientific field did not spot the light on [19]. No doubt that the IoT market is growing fast in many fields that provided new business opportunities, but the predictions about the IoT growth in industries are still un realistic; as the fear many companies may take the risk of running out of time and mone before gaining profits from their investments. In this paper, we focus on the security part of IoT in general and the educational environment in specific. A multi factor authentication is proposed to secure the connected devices in IoET systems, we have provided a general introduction about the IoT technology to highlight the importance of security and privacy in this field, this paper is organised as following: Sect. 2 discusses previous related work and overview of IoT in educational environment. Section 3 discusses the IoT-based smart environment and its components. Section 4 discusses some of the challenges related to the IoT integration. Section 5 is the research conclusion summarizing the whole paper idea.

2 Literature Review

Although conventional internet connects humans to a network, IoT has a exclusive method in which it offers machine-to-machine (M2M) and Human-to-machine (H2M) connectivity, for heterogeneous sorts of machines to be able to assist variety of applications (e.g., wi-fi, locating, monitoring, tracking, and controlling) [20]. Connecting a big variety of heterogeneous machines results in a huge traffic [21], therefore the need to address the storage of massive records [22, 23]. consequently, the TCP/IP structure, which has been used for a long period for network connectivity, does no longer suit wireless the desires of IoT regarding various aspects which includes privacy and protection (e.g., data privacy, machine's protection, information confidentiality, data encryption, and network security) [24], even though numerous architectures have been proposed for IoT, there is still a need for a reference structure [25, 26]. The primary structure model proposed in the literature is a three-layer structure, they are as following:

- **Perception Layer**: It is the physical layer that senses the environment to understand the physical specifics (e.g., temperature, humidity, velocity, vicinity, etc.) the use of end-nodes, via the use of different sensing technology (e.g., RFID, GPS, NFC, etc.).
- **Network Layer**: This layer is responsible for the information from the perception layer and transmitting it to the application layer via diverse network technologies (e.g. 4G, 5G, Bluetooth, Zig-Bee, etc.). As well as it is responsible for information management from storing to processing with the assist of middle-wares including cloud computing.
- **Application Layer**: It is the layer that is responsible for delivering application-specific service to the person. The main target of this layer is to cover numerous markets (e.g., smart cities, smart apartments, health care, infrastructure automation, etc.) [27]. The five layers are from highest to lowest: business, application, processing, transport, and perception layers.

As shown in Fig. 5, it represents an architecture for an smart educational environment with all the IoT objects including in it such as: Lighting, Elevator, BLE Marker and Human Monitor Sensor. There are possible security challenges in such an environment that can cause both a disastrous management concerns and deep damages to the infrastructure itself, this usually occurs due to precarious parts the IoT presents in providing support for different applications through the connection of immense heterogeneous devices [28], in addition to the sequential response from the huge symmetric correlation in IoT [29].

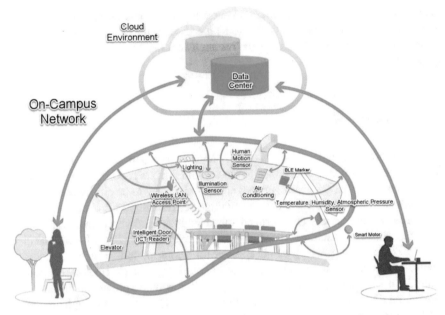

Fig. 5 A simple structure of integrating an IoT system with an educational environment

- **Smart Classroom Management**

The concept of smart classrooms is defined as a smart education room equipped with advanced learning IoT devices. Those smart devices can be sensors, smart boards, cameras and 3d visual effects, those objects are used to obtain statistics about the learning metrices. Integrating the IoT in educational environment became a trend; as it provides innovative methods for both learning and teaching management. Some of those smart objects were already used in smart schools such as (eBooks—Temperature Sensors—Eyes Biometric System—3D Printers and Attendance Tracking Objects). Smart classrooms provides the professors with the needed statistics to manage the learning attitude and provide early solutions for any detected problem.

- **Smart Attendance System**

It takes time and effort to have the students' attendance each course, using IoT can provide the consumption of both time and efforts. Previous studies have focused on efficient smart classrooms using IoT systems to record the attendance for both students and teachers after every lecture accurately, for each student's ID there is an attached RFID tag; to let the installed (Student Record System) in every classroom and lecture hall to count and identify each student attached with their timing. As a result, statistics can be extracted periodly to show the attendance ratio for each student [30].

- **IoT-based Smart Lab**

The essential target from smart labs is teaching needed learning skills for students in a PiP environment. The lab is equipped with a node of IoT devices to provide the needed data and statistics. Researches such as [31] have showed the result of testing the smart labs with IoT. It has provided the suitability for students from different age and backgrounds to learn faster.

- **IoET Authentication Problem**

In education environment, security and privacy are important to secure critical data of both students and professors, these data can be: Modules lectures, Grading History, Student Record Systems and Attendance Applications. Giving the needed authorization to a certain group members through traditional authentication approaches have showed some security breaches.

Researches such as [32, 33] have studied the authentication security problems in IoT environment in general. However, there was a huge space for improvement for the proposed approaches; as it can provide more effectiveness in enhancing the security in IoT systems. Particularly, in dynamic IoET networks, there was an authentication scheme, this was proposed in [34] by using machine learning algorithm based on the physical layer, to be as a defensive solution against spoofing attacks through tracking the events in physical layer and application layer in dynamic environments. Additionally, [35] has provided an authentication provision in physical layer based on extreme machine learning for improving the accuracy of the attack detection process. Hasan et al. [36] designed watermarking methods for detecting attacks based on deep learning method for dynamic authentication that enables the IoT objects to get a set of stochastic qualities from the generated signals and watermark those qualities into a signal. Suzuki and Koyama [37] have studied a Physical non-cloneable Function (PUF) in their research, this was based on existing circuits array, they use an extreme learning approach during the authentication phase [38]. While in [39] a developed deep learning method was proposed to recogne the behavior of denial of service and malicious attacks based on previously calculated data.

In his research, [40] has developed a learning-based model for centralized servers in order to extract the mobility features and differentiate between the (Sybil Attacks) and (Benign Vehicles) through analyzing their behaviors. Although, the previously mentioned approaches contribute to security enhancement through proposing, exploring and developing new machine learning methods, they still insufficient in facing the security challenges in the IoT systems. The taken time by machine learning methods in training, detecting and preventing attacks can lead to an increment of long time-latency, this is due to the overload on communication and communication [41].

Yu et al. [42] have proposed an authentication approach using ECC-based IoT scheme, they have created their own scheme based on a secure ID-verifier protocol, the main focus in their research was on the security on the server and its relation with the Radio Frequency Identification (RFID) tag [42].

Chen [43] in their study have provided a framework using a user authentication for IoT systems and key agreement approach through a smart card, after the registration process in the proposed scheme, the authorized user will obtain a (Smart Card) to help him/her into server access [43].

Cui et al. [44] have proposed ML-attack predictive model with PUFs. This was done by investigating the Ring Oscillator PUF architectures strength in normal PUF softwares. The proposed attack model is based on different machine learning methods. The numerical simulations based on PUFs digital approaches shows the CRPs results. The silicon CRPs results were nearly to the one obtained from the simulated CRPs, the finding of this work led to the need to have new additional requirement to secure the electrical PUFs. Thus, the results of this research will help both PUF attackers and designer at their future work [45].

Most of the proposed machine learning solutions are based on a binary technique which means in the case of authentication success, the user will gain access to 100% of the resources. While in the case of authentication failure the user will be incapable to communicate with any of the system components overall [46]. Consequently, the previous researches were far away from providing a speed and lightweight authentication. Therefore, the focus in this paper is to (1) Envision an machine learning-secure methodology to overcome the security threats, (2) Achieve fast and progressive authentication in IoET networks.

3 The Challenges in IoET Environment

The unique features of IoET systems in education systems brings various vulnerabilities in security provisioning. Apparently, IoT devices that suffer from resource restrictions would not provide the required security mechanisms for the marvelous devices included in the IoET systems demand low-delay transmissions to assure the performance of their communication [47]. Therefore, to provide protection against those security challenges, this paper focuses on examining the challenges of the traditional authentication schemes in IoET systems and providing an method for enriching security in IoET systems. Furthermore, the usage of (Resource-Constrained Devices) can be risky and represent a huge threat to the whole IoET network by forging, tampering, data injection and spoofing attacks [48]. With the consecutive effect, those risks can lead to IoET network failure. Specifically, for the applications that depend in cooperation with different entities [49]. Authentication has been defined as a security key mechanism for IoET architecture design; as the hackers need to have access to the IoET system to start their attack [50]. This method secures the communication within the whole IoT network through approving the identities and their right for accessing to the authorized resources in the network [51].

3.1 Traditional Authentication

The traditional authentication methodology including physical layer key generation mechanisms and cryptography mechanisms, can suffer from both high latency and complexity and can fail in adapting in a complex dynamic environment especially in IoET systems [52]. Significantly, leakage may occur to the generated keys during the security management operations [53]. The generated keys challenges in the IoET systems can be summarized as follow:

- **Security Complexity**

The traditional cryptography methods need a protracted process and improved over-head; in order to increase the security level. Therefore, it will lead to high overhead in both computation and communication [54].

- **Communication Latency**

There are unendurable for IoET systems that has an expressively high number of smart devices and resource-constrained machines with synchronized communications [5]. Furthermore, the traditional statistical methods for authentication need sufficient time and high resources for attaining the statistical resources. Consequently, it will lead to limited proficiency in detecting attacks instantly [55]. Therefore, there is a necessity to have an effective multi-factor authentication method to be provided for the applications in the IoET systems.

- **Adapting Failures**

Concisely, failing to adapt any dynamic environment can happen due to lack of security; so enriching the security is necessary in IoET networks, mostly when it comes to the data that uses Machine Learning [56].

3.2 Machine Learning as an Approach to Authenticate Devices

Machine learning is used as an effective approach to overcome the security threats by developing new methods in (Internet of Educational Things) environment. This can be achieved through thinking, learning and self-adjustment. İt can be browbeaten at gateways and the plentiful data contained in IoET environment can be used for learning [57]. As a result, it will be a valid method for achieving fast and secure channel for packets transmissions in the whole network. Furthermore, Machine Learning methods are using self-correction to the exported dataset by the passing time, then the progressive security can be achieved to adapt to the dynamic IoET environment [58]. Specifically, machine learning can achieve security enrichment in IoET systems because of the following reasons.

- **Improve security management based on multi-domain information**

The routers and gateways can use machine learning management in IoET systems, such as information gathering, training and maintenance. Hence, the processes overload in the IoT environment can be reduced in low-power devices. Machine learning methods can use data history to simplify security management [59]. Therefore, the IoT devices authentication processes can be enhanced as well as security management. Hence, machine learning can achieve continues acceleration of identity authentication in dynamic IoET networks relying on multi-dom Machine Learning information [60].

- **Provide real-time learning under limited statistical data**

Practically, enhancing security became very complicated with a modified statistical model; because of the global and un-predictable nature of the dynamic IoT systems in addition to the resource and time limitations that can affect the detection accuracy of real-time attacks, providing continuous protection and gMachine Learningning a specific statistical data [4]. Although, many previous researches discussed statistical hypothesis testing; the machine learning methods provide prediction results accuracy for the software applications based on trMachine Learningning method of the collected data without a previous algorithm [61].

4 Problem Solution

Generally, security enrichment in IoT systems represents a big challenge for both researchers and developers, the authentication method should be secured from leaking any confidential data [62]. With the machine learning methods, the provision of intelligent security can be designed by using channel reciprocity as well as the precise connections, devices and biometric properties deprived of the confidential data transmission [63]. Machine Learning methodologies can help in strengthening the channel reciprocity, with the purpose of acquiring the exceedingly private information on both the transmitter and the receiver. Furthermore, the machine learning method can track certain authentication behavior features without the transmission of any confidential data [64].

5 Experiment

In the proposed approach, in which the routing protocol is mainly relying on the Leach Protocol [65], there is a new proposed protocol for the existing sensor networks, a set of networks are formed through the usage of power data for better efficiency. There are many characteristics for the IoT environment, each sensor network in a certain pre-defined zone can send the receiver sensor with one move.

Concerning the battery life, every sensor node has the ability to choose a middle node (MN) and a network from this connection. In the time the sensor node makes this connection with the chosen node, it starts collecting the needed information to be transmitted for the middle node. From its side, the middle node confirms the needed information and send it back. With the passing time, MN uses comparatively higher power with a comparison to the other nodes, it can obstinately affect the whole network in case some nodes became overloaded. Therefore, after a pre-defined certain time, a specific MN is chosen for distributing its usage of power by the surrounding nodes. In order to measure the power consumption for the sensor nodes, each node that exist in every set will start transmitting their power usage data, after that the MN will calculate the overall used power and send back the collected data to the sensor node. The sensor node from its side will choose about 40% of the nodes which is not representable in the MN calculations in a probability algorithm, as shown in the equation below.

$$Pi(x) = \frac{Kopt}{N - Kopt\left(r \bmod \frac{N}{Kopt}\right)} Ci(t) = 0 \, or \, 1$$

- i represents each node, it includes a value of the overall sensor nodes N that exists in the system.
- t represents the time.
- $Zi\,(t)$ is the event in which the selected i as MN at a certain time t;
- r represents the overall sensor network;
- K_{opt} represents the overall rounds number which reach its end at Round Set;
- $C_i\,(t)$ is a binary value, it is either or 1, depending on whether it is chosen as MN;
- The round is defined as a period between the cluster creation to its termination.
- Round Bunch is defined as the state in which all the nodes are qualified to be cluster heads.

As the round advances, since the sensor's number engaging in the middle node choice decreases, there is a probability of getting to be a middle node MN increments. At this moment, the possibility to become a middle node is K_{opt}.

- The sensor which make the $Z_i\,(t)$ calculation will choose a z a number value between 0 and 1;
- In case $z < Z_i\,(t)$, it chooses itself as MN.
- In case $z > Zi(t)$, the node is utilized to shape a cluster.
- In case the existing nodes got to become a middle node (MN) at Round Group, $Z_i\,(t) = 0$, so it is inconceivable to turn into a middle node (MN) before having another round.
- In the last round, $Z_i(t) = 1$, as a result the nodes that have never became middle node will be chosen as middle node.

Table 1 Proposed authentication model (Constraints Abbreviation)

Parameter	Meaning
MN	Middle Node
CA	Certificate Authority
ID	ID of the normal node
C_{id}	ID of the middle node (MN)
R_i^1, R_i^2, R_i^3	3n Bit Divided Value
Sk	Session Key
Nk	f () Function Shared Key
N_i	After distance bounding remaining bits
C_i	Random bit
$g(x)$	Polynomial Key
$e(x,y)$	Encryption Polynomial Key

The main constraints that are used in our proposed protocol is used as following (Table 1):

Stage 1. The nodes which have received a spam message creates M_0 and H_0 as a verification step followed by sending the spams to the gateway.

Stage 2. The gateway which has gained both M_0 and H_0, has to create M_1 and H_1 and transmits both of them to the Certificate Authority origin.

Stage 3. The Certificate Authority which has gained data, this data was sent from the gateway and it has two irregular numbers using deciphering and approves H_0 and H_1. It creates M_2 and M_3 and then it transmits them to the gateway.

Stage 4. The gateway which has already verified both M_1 and M_2 gets R_1 from deciphering M_3 which is more checks to whatever slip in. Those accepted esteem through a hash capacity. It conserves $3 \times$ n-bit data produced by two irregular numbers for confirmation after that transmits M_2 to the smart node.

Stage 5. The node which obtained M_2 gets R_2 and approves the despatched value via a hash feature. If the approval is completed, the node saves $3 \times$ n-bit data.

Stage 6. The smart node generates random variety C_1 to carry out the verification step and sends it through bit. In this part, the time test for stopping relay attacks occurs.

Stage 7. The gateway that acquired bits from the smart node sends the ith little bit of R_0 if $C_1 = 0$ in acknowledgment, and sends the ith little bit of R_1 to the smart node if the acknowledgment is 1.

Stage 8. The smart node generates R_i^{cn} depends on the c despatched to the gateway and compares it to R_i^{cn}, that is the collected value of the cluster head's response to confirm whether the data was sent from the right node. After "time off," it guesses the gap via time measurement and prevents communication if that is extra than a specific time.

Stage 9. The smart node that validated the gateway sends the obtained Ri^{cn} values the use of the $f()$ feature to the gateway.

Stage 10. The smart node that received a certification cost from the gateway generates a certification cost in same manner and compares it to the value obtained from the cluster head to validate the smart node.

Stage 11. The both node and the gateway generate a session key through the left n bit and a random range from the $3 \times n$-bits and ends the verification as shown in Fig. 6

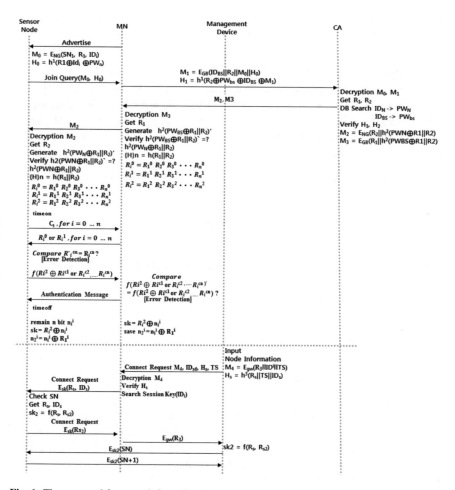

Fig. 6 The proposed framework for authentication process and device registration

- **Smart Device Key Updating Method**

Stage 1. After finishing certification sensor set formation and session key distribution, the sensor node includes out the set key distribution and renewal technique as follows. the use of the polynomial distribution of PCGR, it makes MN do various of the calculation, distributes portions of the particular value via the $f(y)$ function, and verifies whether or not the node is infected or should to be withdrawn.

Stage 2. The middle node defines and generates polynomials $g(d)$ and $f(z)$ for the set distribution and node verification. After that, it generates verification values S and D_n subsequent, it generates P_1 for transmitting to nodes and deletes $g(d)$ and $e(d)$ to stop them from being uncovered through an attacker.

Stage 3. The sensor node that obtained P_1 decodes this, and obtains the group key and mystery piece. It then informs middle node that the set key changed into effectively acquired.

Step 4. After a specific time period, middle node transmits a message to update the set key to the nodes inside the set. Nodes that acquire the key update message send middle node an encrypted mystery piece with the session key in acknowledgment, the middle node tests the validity of the obtained value through a lagrangian polynomial and sends a verification message. If the value is different, it will inform certain nodes of this as an intrusion.

Step 5. Nodes that obtained the message from middle node transmit nodes around P_n.

$$L\,(c) = L'\,(c) + e_u(c, N_{id})$$

Step 6. After the key is updated, it finishes the verification procedure with nodes via the set key.

6 Discussion and Analysis

6.1 Overhead Analysis

Blundo's protocol that is considered as the sensing security approach in sensor networks, changes depend on the number of the network nodes; if the number of nodes is O, then the polynomial expression with $(O - 1)$ variables of degree p are calculated. Every sensor node can save the polynomial expression of length O $(p + 1)$ L that exists in the network server, with a need of L messages. That is, Blundo's Protocol could be secured except t members were attacked, but it can load the network, as every node achieves important calculation for an IoET environment that frequent sensors exist. Blundo's protocol delivers high security in a time the damaged nodes number is lower than λ; if the damaged nodes number is higher than λ, most of the secure data are in-danger. In the case of the PCGR approach, data can be saved and the needed calculations can differ based on the neighboring nodes created the group key. Presumptuous that the neighboring nodes number is O and the possibility that the neighboring nodes cause the group key is $Pro(d)$, it provides a polynomial expression of the encrypted group key with $(p + 1)$ L-bit length and bit data of n $(p + 1)$ L, that is fractional data of the neighboring nodes.

6.2 Power Efficiency

The following experiment that was made through using the Matlab application to measure the time efficiency for every IoET device for proposed solution and the earlier encryption techniques. The initial values were set to perform the simulation and the simulation time, the amount of node power were put into consideration. For simulation, the efficiency of the time was measured based on the IoET devices increment by setting a total of 20 to 300 nodes, and the experiment was compiled by setting an independent environment for every encryption technique and the proposed approach. The distribution of the sensor nodes was created randomly in a 50 m × 50 m area in spite of the nodes number. The gate was allocated in a certain position with the consideration of the placement area.

Based on the placement, the space between the gateway and the sensor nodes continued between 50 and 20 m. Every round was configured to stay for 40 s. Figures 7 and 8 illustrates the simulation result after the network set was set. The results of the simulation have proved that in the situation of the proposed approach, the operation

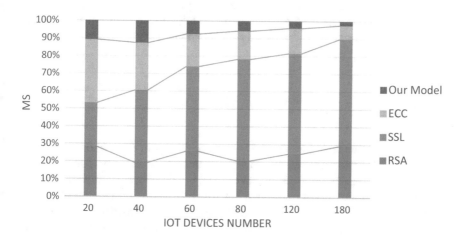

Fig. 7 Client performance between authentication time rates and IoT devices number

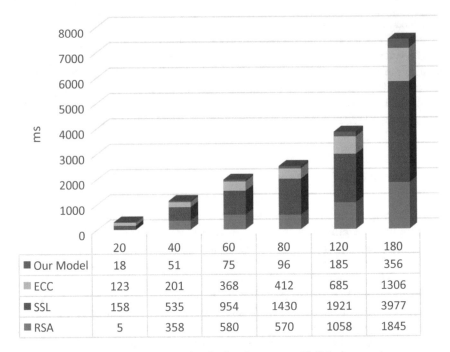

	20	40	60	80	120	180
■ Our Model	18	51	75	96	185	356
■ ECC	123	201	368	412	685	1306
■ SSL	158	535	954	1430	1921	3977
■ RSA	5	358	580	570	1058	1845

Fig. 8 Server performance between authentication time rates and IoT devices number

being completed in the device was depending on a primitive hash process and arithmetic process with less operation usage time in comparison of other authentication approaches.

Finally, during the disturbance of the group key and renewal operation, all polynomial process was made in the server, thus the ability to reduce the device load and as long as an inconsequential characteristic in comparison to other authentication approaches.

7 Conclusion

In this paper, both the definitions and the characteristics of Internet of Educational Things (IoET) were introduced with illustrations of the security challenges and threats in the IoET environment. However, there is a possibility for false authentications to occur for devices in distributed IoT networks during the compromisation of the trusted nodes. İt may lead to potential leakage and high-security threat. Machine learning can represent a significant contribution for designing a proper multiple authentication schemes for IoET systems. In this paper, a proposed model solution through using easy hash calculations to authenticate different types of IoT devices in educational environment with hardware limitations. Additionally, the routing protocol is considered plan in current sensor system environment for the protocols that was designed to simply accept IoT devices that cannot provide encrypted modules in a heterogeneous IoET system environment, which is a problem of existing IoT environment in general.

References

1. McKnight, M.: IOT, Industry 4.0, Industrial IOT... Why connected devices are the future of design, vol. 2, pp. 197. KnE Engineering (2017). https://doi.org/10.18502/keg.v2i2.615
2. Gabbai, A.: Kevin Ashton Describes "The Internet of Things". In: Smithsonian Magazine (2020). https://www.smithsonianmag.com/innovation/kevin-ashton-describes-the-internet-of-things-180953749/. Accessed 26 Jan 2020
3. Singh, A.: Implementation of the IoT and cloud technologies in education system. SSRN Electron. J. (2019). https://doi.org/10.2139/ssrn.3382475
4. Miyajima, H., Shiratori, N., Miyajima, H.: Proposal of security preserving machine learning of IoT. Artif. Intell. Res. 7, 26 (2018). https://doi.org/10.5430/air.v7n2p26
5. Hwang, S., Seo, J., Park, S., Park, S.: A survey of the self-adaptive IoT systems and a compare and analyze of IoT using self-adaptive concept. KIPS Trans. Comput. Commun. Syst. 5, 17–26 (2016). https://doi.org/10.3745/ktccs.2016.5.1.17
6. Mawgoud, A.A., Taha, M.H.N., Khlifa, N.E.M.: Security Threats of Social Internet of Things in the Higher Education Environment, pp. 151–171 (2020)
7. Abbasy, M., Quesada, E.: Predictable influence of IoT (Internet of Things) in the higher education. Int. J. Inf. Educ. Technol. 7, 914–920 (2017). https://doi.org/10.18178/ijiet.2017.7.12.995
8. Gronau, N., Ullrich, A., Teichmann, M.: Development of the industrial IoT competences in the areas of organization, process, and interaction based on the learning factory concept. Proc. Manuf. 9, 254–261 (2017). https://doi.org/10.1016/j.promfg.2017.04.029
9. Bring the "Smart" into Smart Things at CES 2020- MicroEJ-Market-Leading Solutions for Embedded and IoT Devices. (2020). Retrieved 15 July 2020, from https://www.microej.com/news/microej-brings-the-smart-into-smart-things-at-ces-2020/
10. Ismail, N.: Gartner's 2017 forecasts-information age. In: Information Age (2020). https://www.information-age.com/gartners-2017-forecasts-123463932/. Accessed 26 Jan 2020
11. Pinka, K., Kampars, J., Minkevičs, V.: Case study: IoT data integration for higher education institution. Inf. Technol. Manag. Sci. (2016). https://doi.org/10.1515/itms-2016-0014
12. Ayare, M.: A survey on IoT: architecture, applications and future of IoT. Int. J. Res. Appl. Sci. Eng. Technol. 7, 1235–1239 (2019). https://doi.org/10.22214/ijraset.2019.4221
13. Moinuddin, K., Srikantha, N., Lokesh, K.S., Narayana, A.: A survey on secure communication protocols for IoT systems. Int. J. Eng. Comput. Sci. (2017). https://doi.org/10.18535/ijecs/v6i6.41
14. Mewada, D., Dave, N., Prajapati, P.: A survey: prospects of Internet of Things (IoT) using cryptography based on its subsequent challenges. Aust. J. Wirel. Technol. Mobil. Secur. 1, 1–5 (2019). https://doi.org/10.21276/ausjournal.2019.1.3
15. Condry, M., Nelson, C.: Using smart edge IoT devices for safer, rapid response with industry IoT control operations. Proc. IEEE 104, 938–946 (2016). https://doi.org/10.1109/jproc.2015.2513672
16. Lamonaca, F., Scuro, C., Grimaldi, D., et al.: A layered IoT-based architecture for a distributed structural health monitoring system. ACTA IMEKO 8, 45 (2019). https://doi.org/10.21014/acta_imeko.v8i2.640
17. Top 10 Insights of 2018. (2020). Retrieved 19 November 2019, from https://www.mckinsey.com/about-us/new-at-mckinsey-blog/top-10-insights-of-2018

18. Forrester Research Reports 2017 Fourth-Quarter And Full-Year Financial Results. Forrester. (2020). Retrieved 3 September 2019, from https://go.forrester.com/pressnewsroom/forrester-research-reports-2017-fourth-quarter-and-full-year-financial-results/

19. Hussein, D., Hamed, M., Eldeen, N.: A blockchain technology evolution between business process management (BPM) and Internet-of-Things (IoT). Int. J. Adv. Comput. Sci. Appl. (2018). https://doi.org/10.14569/ijacsa.2018.090856

20. El Karadawy, A.I., Mawgoud, A.A., Rady, H.M.: An empirical analysis on load balancing and service broker techniques using cloud analyst simulator. International Conference on Innovative Trends in Communication and Computer Engineering (ITCE), pp. 27–32. Aswan, Egypt, IEEE (2020)

21. Lee, C.: An Adaptive traffic interference control system for wireless home IoT services. J. Digit. Converg. 15, 259–266 (2017). https://doi.org/10.14400/jdc.2017.15.4.259

22. Yaswanth Sai, P.: Illustration of IOT with big data analytics. Int. J. Comput. Sci. Eng. (2017). https://doi.org/10.26438/ijcse/v5i9.221223

23. Revathy, R., Aroul Canessane, R.: IoT based decision making system to improve veracity of big data. Int. J. Eng. Technol. 7, 63 (2018). https://doi.org/10.14419/ijet.v7i3.1.16799

24. Dyagilev, V.: Target Attacks on IoT and network security vulnerabilities increase, pp. 72–73. LastMile (2018). https://doi.org/10.22184/2070-8963.2018.75.6.72.73

25. Capella, J., Campelo, J., Bonastre, A., Ors, R.: A reference model for monitoring IoT WSN-based applications. Sensors 16, 1816 (2016). https://doi.org/10.3390/s16111816

26. Vukovic, M.: Internet programmable IoT: on the role of APIs in IoT. Ubiquity 2015, 1–10 (2015). https://doi.org/10.1145/2822873

27. Sun, X., Ansari, N.: Traffic load balancing among brokers at the IoT application layer. IEEE Trans. Netw. Serv. Manage. 15, 489–502 (2018). https://doi.org/10.1109/tnsm.2017.2787859

28. Majchrowicz, M., Hufnagiel, M.: Management of IOT devices in a smart home through the application of an interactive mirror. Image Process. Commun. 22, 43–50 (2017). https://doi.org/10.1515/ipc-2017-0020

29. Guo, L., Zhou, B., Sun, Z., Liu, X.: Correlation of symmetric raised cosine keying signals. J. Electron. Inf. Technol. 34, 1793–1799 (2013). https://doi.org/10.3724/sp.j.1146.2011.01361

30. Dedy Irawan, J., Adriantantri, E., Farid, A.: RFID and IOT for attendance monitoring system. In: MATEC Web of Conferences, vol. 164, pp. 01020 (2018). https://doi.org/10.1051/matecconf/201816401020

31. Maria de Fuentes, J., Gonzalez-Manzano, L., Solanas, A., Veseli, F.: Attribute-based credentials for privacy-aware smart health services in IoT-based smart cities. Computer 51, 44–53 (2018). https://doi.org/10.1109/mc.2018.3011042

32. Khoureich Ka, A.: RMAC-a lightweight authentication protocol for highly constrained IoT devices. Int. J. Cryptogr. Inf. Secur. 8, 01–14 (2018). https://doi.org/10.5121/ijcis.2018.8301

33. Braeken, A.: PUF based authentication protocol for IoT. Symmetry 10, 352 (2018). https://doi.org/10.3390/sym10080352

34. Wu, X.: Embedded physical-layer authentication in cognitive radio requires efficient low-rate channel coding schemes. IET Commun. 11, 400–404 (2017). https://doi.org/10.1049/iet-com.2016.0812

35. Chavan, A., Nighot, M.: Secure and cost-effective application layer protocol with authentication interoperability for IOT. Proc. Comput. Sci. 78, 646–651 (2016). https://doi.org/10.1016/j.procs.2016.02.112

36. Hasan, M., Islam, M., Zarif, M., Hashem, M.: Attack and anomaly detection in IoT sensors in IoT sites using machine learning approaches. Internet of Things 7, 100059 (2019). https://doi.org/10.1016/j.iot.2019.100059

37. Suzuki, H., Koyama, A.: An implementation and evaluation of IoT application development method based on real object-oriented model. Int. J. Space-Based Situat. Comput. 8, 151 (2018). https://doi.org/10.1504/ijssc.2018.10018388

38. Huang, Z., Wang, Q.: A PUF-based unified identity verification framework for secure IoT hardware via device authentication (2019). https://doi.org/10.1007/s11280-019-00677-x

39. Praseetha, V., Bayezeed, S., Vadivel, S.: Secure fingerprint authentication using deep learning and minutiae verification. J. Intell. Syst. **29**, 1379–1387 (2019). https://doi.org/10.1515/jisys-2018-0289

40. Gokula Krishnan, C., Suphalakshmi, D.: An improved MAC address based intrusion detection and prevention system in MANET sybil attacks. Bonfring Int. J. Res. Commun. Eng. **7**, 01–05 (2017). https://doi.org/10.9756/bijrce.8315

41. Mawgoud, A.A., Taha, M.H.N., Khlifa, N.E.M.: Cyber security risks in MENA region: threats, challenges and countermeasures. In: International Conference on Advanced Intelligent Systems and Informatics, pp. 912–921. Springer, Cham (2020)

42. Yu, S., Park, K., Park, Y.: A secure lightweight three-factor authentication scheme for IoT in cloud computing environment. Sensors **19**, 3598 (2019). https://doi.org/10.3390/s19163598

43. Chen, J.: Hybrid blockchain and pseudonymous authentication for secure and trusted IoT networks. ACM SIGBED Rev. **15**, 22–28 (2018). https://doi.org/10.1145/3292384.3292388

44. Cui, Y., Gu, C., Ma, Q., Fang, Y., Wang, C., O'Neill, M., & Liu, W.: Lightweight modeling attack-resistant multiplexer-based multi-PUF (MMPUF) Design on FPGA. Electron, **9**(5), 815 (2020)

45. Mawgoud, A.A., Ali, I.: Statistical insights and fraud techniques for telecommunications sector in Egypt. In: International Conference on Innovative Trends in Communication and Computer Engineering (ITCE). IEEE, pp. 143–150 (2020) (Amar, S., Deep, P.: A review on various IOT analytics techniques for bridge failure detection in fog computing. Int. J. Comput. Appl. **169**, 38–43 (2017). https://doi.org/10.5120/ijca2017914586)

46. Elazhary, H.: Internet of Things (IoT), mobile cloud, cloudlet, mobile IoT, IoT cloud, fog, mobile edge, and edge emerging computing paradigms: disambiguation and research directions. J. Netw. Comput. Appl. **128**, 105–140 (2019). https://doi.org/10.1016/j.jnca.2018.10.021

47. Anuradha, K., Nirmala Sugirtha Rajini, S.: Analysis of machine learning algorithm in IOT security issues and challenges. J. Adv. Res. Dyn. Control Syst. **11**, 1030–1034 (2019). https://doi.org/10.5373/jardcs/v11/20192668

48. Brous, P., Janssen, M., Herder, P.: The dual effects of the Internet of Things (IoT): a systematic review of the benefits and risks of IoT adoption by organizations. Int. J. Inf. Manag. 101952 (2019). https://doi.org/10.1016/j.ijinfomgt.2019.05.008

49. Ferrag, M., Maglaras, L., Derhab, A.: Authentication and authorization for mobile IoT devices using biofeatures: recent advances and future trends. Secur. Commun. Netw. **2019**, 1–20 (2019). https://doi.org/10.1155/2019/5452870

50. Ullah, I., Tila, F., Kim, D.: Access rights management based on user profile ontology for IoT resources authorization in smart home. Int. J. Control Autom. **11**, 1–12 (2018). https://doi.org/10.14257/ijca.2018.11.3.01

51. Xue, H., Wang, Q.: Authentication of the traditional Tibetan medicinal plant Lygodium japonicum using MALDI-TOF spectrometry. Planta Med. (2010). https://doi.org/10.1055/s-0030-1264303

52. Yang, J., Yang, P., Wang, Z., Li, J.: Enhanced secure low-level reader protocol based on session key update mechanism for RFID in IoT. Int. J. Web Grid Serv. **13**, 207 (2017). https://doi.org/10.1504/ijwgs.2017.083386

53. Parashar, N.: Design development and performance evaluation of low complexity cryptographic algorithm for security in IOT. Int. J. Res. Appl. Sci. Eng. Technol. **7**, 2481–2485 (2019). https://doi.org/10.22214/ijraset.2019.3454

54. Nkenyereye, L., Jang, J.: Design of IoT gateway based event-driven approach for IoT related applications. J. Korea Inst. Inf. Commun. Eng. **20**, 2119–2124 (2016). https://doi.org/10.6109/jkiice.2016.20.11.2119

55. Kannan, G., Manoharan, N.: Force multiplier effect of futuristic battlefield preparedness by adapting the Internet of Things (IoT) concept. Indones. J. Electric. Eng. Comput. Sci. **9**, 316 (2018). https://doi.org/10.11591/ijeecs.v9.i2.pp316-321

56. Kaur, G., Dutta, R.: Sentiment mining based on products reviews using machine learning. J. Adv. Sch. Res. Allied Educ. **15**, 185–191 (2018). https://doi.org/10.29070/15/57679

57. Rekleitis, E., Rizomiliotis, P., Gritzalis, S.: How to protect security and privacy in the IoT: a policy-based RFID tag management protocol. Secur. Commun. Netw. **7**, 2669–2683 (2011). https://doi.org/10.1002/sec.400

58. Sung, M., Shin, K.: An efficient hardware implementation of lightweight block cipher LEA-128/192/256 for IoT security applications. J. Korea Inst. Inf. Commun. Eng. **19**, 1608–1616 (2015). https://doi.org/10.6109/jkiice.2015.19.7.1608

59. Jogdand, G., Kadam, S., Patil, K., Mate, G.: Iot transaction security. J. Adv. Sch. Res. Allied Educ. **15**, 711–716 (2018). https://doi.org/10.29070/15/57056

60. Yasuda, S., Miyazaki, S.: Fatigue crack detection system based on IoT and statistical analysis. Proc. CIRP **61**, 785–789 (2017). https://doi.org/10.1016/j.procir.2016.11.260

61. Lin, H., Bergmann, N.: IoT privacy and security challenges for smart home environments. Information **7**, 44 (2016). https://doi.org/10.3390/info7030044

62. Punithavathi, P., Geetha, S.: Partial DCT-based cancelable biometric authentication with security and privacy preservation for IoT applications. Multimed. Tools Appl. **78**, 25487–25514 (2019). https://doi.org/10.1007/s11042-019-7617-1

63. He, T., Jiao, L., Yu, M., et al.: DNA barcoding authentication for the wood of eight endangered Dalbergia timber species using machine learning approaches. Holzforschung **73**, 277–285 (2019). https://doi.org/10.1515/hf-2018-0076

64. Behera, T., Samal, U., Mohapatra, S.: Energy-efficient modified LEACH protocol for IoT application. IET Wirel. Sens. Syst. **8**, 223–228 (2018). https://doi.org/10.1049/iet-wss.2017.0099

65. Alshowkan, M., Elleithy, K., & AlHassan, H.: LS-LEACH: a new secure and energy efficient routing protocol for wireless sensor networks. In: 2013 IEEE/ACM 17th International Symposium on Distributed Simulation and Real Time Applications (pp. 215-220). IEEE, (2013)

A Malware Obfuscation AI Technique to Evade Antivirus Detection in Counter Forensic Domain

Ahmed A. Mawgoud, Hussein M. Rady⊙, and Benbella S. Tawfik

1 Introduction

Counter Forensic is a domain where any developed technique, device or software can obstruct an investigation in computer science field, there are multiple ways which can be used to hide specific data through fooling the computer and mostly this can be done by changing the file header. Encryption is another way of hiding data through using complex algorithm to make data unreadable, to read the data it is must to use the encryption. Other ways to manipulate data is by using tools that have the ability to alter the files' metadata, the malware success level can be measured by passing many different antiviruses' detection layers [1]. In this paper, an experiment was implemented to illustrate how the malware can evade every detection layer the antivirus uses, the first protection layer called static signature, it is an algorithm that keeps looping on the disk file contents; in order to find a pre-defined sequence of hexadecimal values [2]. There were many previous trials to diverse motivations to exploit IoT systems. Recently, IoT malwares that were designed for IoT systems have grown with over thousands of malware amendment. Although the most popular

A. A. Mawgoud (✉)
Faculty of Computers and Artificial Intelligence, Information Technology Department, Cairo University, Giza, Egypt
e-mail: aabdelmawgoud@pg.cu.edu.eg

H. M. Rady
Department of Information System, National Telecommunications Regulatory Authority, Smart Village, Egypt
e-mail: hrady@tra.gov.eg

B. S. Tawfik
Computer Networks Department, Faculty of Computers and Information, Suez University, Suez, Egypt
e-mail: benbella@gmail.com

A.-E. Hassanien et al. (eds.), *Enabling AI Applications in Data Science*, Studies in Computational Intelligence 911, https://doi.org/10.1007/978-3-030-52067-0_27

motivation was to design botnets that can be used to ease DDoS attacks, some of those malwares were having a high detection rate and could be easily detected by anti-malwares due to their low evasion rate [3]. The signature length usually equals bytes. Another signature type has the ability for calculating the entire file hashes, it uses MD5, SHA-1 and similar hashing function for calculating the file control sum [4].

The technique which sequels the signature-based scan contains the API analysis functions which is imported or exported by the software. Mostly, APIs are used to escape from being detected (i.e. "Sleep Windows API", "IsDebuggerPresent Windows API", "CreateRemoteThread Windows API" and "Portable Executable"), the static analysis also looks at the header analysis into the code enumeration section and ask for non- standard names [5].

Khaja et al. [6] have investigated a case study using BIM model through parametric design tools, they studied more in-depth the metadata manipulation mechanism in the 'Portable Executable—PE' header that has the ability to trick anti-malware tools. In general, it shows the effectiveness of bytes changing which identify the file type and trick the tools of ant malware to use a wrong tests set [6].

Tahir [7] has analyzed different evasion rates through using different payloads, those payloads were created with metasploit/msfvenom and Veil, through his experiment he changed the shell connection destination port and try different encoding techniques in order to obfuscate the payload [7]. The weakness of this technique is that by using both products of metasploit/msfvenom and Veil, it will be easily detected by most of the antiviruses; as anti-malware industry is monitoring them and developing techniques to detect malware binary, which is being generated by them [8].

Li et al. [9] have combined both antivirus' frameworks msfvenom and Veil-Evasion to create a payload that has the ability to bypass the control of any antivirus, using these tools have a huge advantage, they are having the ability to convert an existing executable file into obfuscated one [10]. Obfuscated executable has a higher chance of bypassing the antivirus detection [11]. However, those tools are considered as off-the-shelf products by information security vendors.

This paper is organized as following: Sect. 2, a study on previous researches for different detection approaches on malware evasion. Section 3, it states the common problems of previous researches related to obfuscation methods. Section 4, it represents the proposed methodology through four main stages for malware obfuscation. Section 5, the experiment requirements to apply the proposed method by using three different malware scanning engines, followed by two tables that explain the characteristics of each sample. Section 6: A conclusion about the general idea of the paper, the results of the proposed methodology and a comparison with other similar methods.

2   Literature Review

Malware developers faced a difficulty in stating a mechanism or criteria for evading static analysis. Specifically, the signature recognition [12]. As a result, there is a need to come out with an idea for developing an algorithm that has the ability of changing its signature with every new execution. Which later led to the appearance of new techniques (e.g. Code Obfuscation, Metamorphism, and Polymorphism) using compression [13]. The information security industry took some countermeasures to face the possible risks of these kind of techniques [14].

Kong et al. [15] have developed a semantic-aware analysis, the semantic aware inspection is having the ability to detect code manipulation such as renaming the processor registers or instruction reordering–all common obfuscation techniques. However, the disadvantage is that this tool can only identify a limited set of obfuscation tricks (i.e. the tool will not detect the manipulation in mathematical operations as (y = y * 2) will not be recognized in "y = y << 1") [16].

Goh and Kim [17] have proposed a unique malware approach insight that is commonly used, it can identify and load, the API's based on hashes instead of using the LoadLibrary/GetProcAddress standard [17], the hashing technique is an approach that hides specific API functions names, this approach facilitates the algorithm obfuscation once it compiles into the OllyDbg or Ida Pro debuggers [18]. None of the debuggers will have the ability to solve or show the API symbolic names called in the Import Address Table—IAT.

Luckett et al. [19] have proposed in their study the emulator checks environment attributes—API discrepancy, time difference and Inconsistencies—in CPU instructions execution mechanism—this was done using an emulator—those attributes may deduce the information about an antivirus [19].

Kumar [20] has studied a practical coding snippets implementation, it does not fingerprint antivirus. However, it attacks with an implemented pre-defined window APIs in the emulator that will return a result inconsequent compared to the executed results outside the emulator [20]. For example, usually opening an unreal URL will return 'true', but it would return an error in multi-processor function, the main purpose behind this work is to highlight the gap between the emulator of the sand box and the APIs implementation of a fully operating system and use these vulnerabilities to evade the detection [21].

Pektaş and Acarman [22] have proposed new techniques in their research—that this paper is inspired from-, they have created a simulation of an environment where the malware is executed when the user interacts with keyboard and mouse [22].

Joo et al. [23] have presented in their research a malware mechanism for virtual machine/sandbox detection, some certain details illustrate the different attributes that is found on the registry keys, system devices or commands output that fingerprints the exact VM/Sandbox [23].

Maestre Vidal et al. [24] their research is similar to the proposed study in [22] research, they have proposed an enhanced payload analyzer for malware detection robust against adversarial threats [24].

Although the proposed techniques in the related work are based on dynamic analysis, they still suffer from high detection rate from the new version of anti malware/antiviruses. Such obfuscation approaches can represent a huge threat to cyber security; because of the simplicity level in applying those mechanisms on existing malicious codes and the lack of effective detection methods. Thus, there is a critical need for an effective method that can provide the ability to achieve a high level of both performance and accuracy in malware detection.

3 Problem Statement

From all the previous studies, the software used in the studied researches do not have the ability to modify the payload once it is decrypted in the memory; because the same malware set is being used in their experiments.

Nevertheless, Iwamoto et al. [25] used in their study a mechanism that obscure the shellcode, anti-dynamic analysis techniques such as mathematical functions for total compilation time increment. Captivatingly, the 64-bit payloads appeared to be having a low detection rate [25]. Although, the findings of this study specifically confirms the outputs of the previous studies, they share the common weakness that is mentioned before; the off-shelf products create some limited malicious code and evasion techniques. There is no doubt that the dynamic analysis has advantages over the static one, but it still has disadvantages [26], the emulators of the sandbox are not the perfect choice for the operating system; they do not achieve many features as the operating system does, as Kruegel [27] stated that the majority of sandboxes made their detection based on system calls. However, it is not sufficient; as these tools mainly miss a huge amount of possibly relevant behaviors.

The other dynamic analysis challenge is the suspicious behavior that can be detected by a certified software. There are some restricted detection policies for the emulator that leads to false positives [28]. On the contrary, other emulators have resilient rules that leads to false negatives. Machine assistedanalysis is one of the new trends in the security industry for malware recognition [29].

Mahawer and Nagaraju [1] have proposed a framework to identify the malware novel classes, this framework uses:

(1) Clustering Techniques.
(2) Automatic Classification.

Their study uses the behavior resemblance for a certain malware [1], but their method has a weakness related to the used assumption for detecting the malware execution using the CWSandbox. Although, many different techniques are available for circumventing the sandbox environment which is deployed by the modern malware.

Both papers [30, 26] had similar contribution idea to the one presented in [31], the main idea behind their studies is to use sandboxes in order to decrease the false positives.

Zatloukal and Znoj [32] have studied the file format of windows PE attributes, this research analyses a machine learning detecting mechanism for evaluating the malware evasion success rate [32]. However, their study only targeted the attributes of PE, the conclusion of their study is only limited to the attributes of static PE header.

Tomasi et al. [33] have studied both correlation optimized warping and dynamic time, they have used the Dynamic Type Warping—DTW algorithm to detect the system call injection attacks [33]. The technique of System Call Injection is used by malware for confusing the antiviruses through injecting irrelevant system calls [34].

Modern malware has the ability to apply multi-techniques for anti-dynamic analysis detection techniques [35]. The main idea of our contribution is to propose an effective malware obfuscation methodology; to provide a high evasion rate towards anti-malware engines. However, the proposed study has some weakness; as it still is not sufficient to challenge some of nowadays detection technique standards.

4 Proposed Solution

The research has three main targets:

Firstly: The purpose is to build a malware that has the ability to evade every detection layer of the antivirus. Each layer is static signature and dynamic detection. The malware is a sample of a reverse TCP shell that is a code function, which can establish a connection to the control server and extracts the privilege escalations payload.

Secondly: The mutation engine algorithm that has the ability for changing the attributes of the malware PE 'Portable Executable'.

Thirdly: To profile the antiviruses' engines through monitoring the malware sample logs and defining the ones that is missed.

The design phase of the proposed malware is working around the TCP connection, which uses port number 443 to imitate encrypted web traffic. The type of traffics the reverse shell is using is not a web traffic but a control connection that is used for downloading the second stage payload. This payload will allow privileges escalation for the attacker, the implementation of our proposed solution consists of four main stages. Each stage targets for evading a certain technique set which is used by the antivirus for the malware detection.

Figure 1 illustrates the high-level implementation for code development mechanism as well as the used methodologies and tools used in each of stage the four stages. In the first and second step, the plain shell is made from both msfvenom and modified syntax in binary code (Online and offline binary + Customized API hashing algorithm). The obfuscated shell in third step is also modified in binary code (Customized obfuscation and de-obfuscation routines). However, the obfuscated shell in the fourth step is modified in addition to Existing anti dynamic techniques.

4.1 Stage 1

The msfvenom generates a reverse TCP shellcode without obfuscating the source code. The payload generation command is

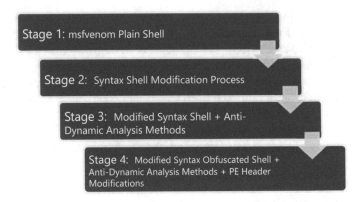

Fig. 1 A representation of the proposed obfuscation four stages methodology

*msfvenom –p windows/meterpreter/reverse_tcp LHOST 81.21.106.117 LPORT 443
–f c > evilexp.c*

4.2   Stage 2

In this stage the shell code already existing from phase one will be modified with two different methods. The shell's binary will be obfuscated while keeping its functionalities, the main source code objective is to preserving all the current.

In order to achieve those objectives, the source code is compiled into binary for showing the byte-level performance, the binary byte-level shows the assembler symbolic instruction output into hexadecimal [36]. As an example, *global _start* will be represented into hexadecimal in one byte. In addition to syntax modifications, the code will make some changes to the used algorithms for calculating the Windows API functions hashes. The algorithms will load every character in the *DLL library name* module. Then, the module name will be normalized through a lower cases conversion into upper cases, after that the character bits will be rotated and print the calculated value summation for every subsequent character. This will be an iterative process until reaching the end of the module name.

The API hash can be calculated through a similar algorithm. This original shell-code uses the hash to get the needed API in the execution phase. Hashing is considered as a different strategy that uses *GetProcessAddress API's* to get the function's pointer. After the checksums is being calculated by the original shell code, the shell code includes them in its body. Then the algorithm checks if the API hash are found by the search loop, the search loop will keep looping through the modules and the functions sets iteratively. After that, it measures the hashes for each function, the original algorithm alterations not just only change the shell code syntactic fingerprint, but also re-compile the original hashes.

```
7e: ac          loda   al,BYTE PTR ds:[esi]
7f: c1 c7 10    rol    edi,0x10         ◄── Ror changed with rol 16 bits
82: 01 c7       add    edi,eax               rotation instead of 19
84: 38 e0       cmp    al,ah
86: 75 f6       jne    0x7e
```

Fig. 2 API hashing algorithm change effect on the calculated checksum values

In this paper, a new obfuscation methodology was introduced through multiple techniques to calculate the new checksum, the checksums overwrite the values of the original ones. The following methods are considered for the syntax modification strategy:

a. NOP sequences insertion.
b. Instruction sequences making Push/Pop.
c. Instruction sequences for making no register changes (with the exception of instruction pointer).
d. Un-conditional different checks insertion.

Figures 2 and 3 shows the modifications effect on the API pre-calculated hashes values (*LoadLibrary, connect, VirtualAllocation*.... etc.).

The new API hash value is calculated using a supplementary designed code in addition to the essential code, there are three main steps as illustrated in the Fig. 2:

a. **Hashing Algorithm**: it creates the modified hashing function.
b. **Hash Values**: the new API hash values are calculated by supplementary C code.
c. **Inserted New Hash**: The new hash values are inserted into the binary code/byte representation.

Many different ways exist for making a binary code changes. As an example, to increment the register value "*ecx*" by one, it can be done by *inc ecx* or add *ecx*, 1 or sub *ecx*, −1. The de-obfuscation technique compiles, loads and de-obfuscates the obfuscated shell code into the memory, a separate supplementary obfuscation code implements this conversion at the same time the de-obfuscation code is the part of the malware. The next Pseudo-Code includes both obfuscation and de-obfuscation for the symbolic version.

1. *Deobfuscation => (bytes_obfuscated_shell [y] > 5) + string_obfuscate [4]) (xor string_obfuscate [3]) − string_obfuscate [2])*
2. *Obfuscation => (byte_plain_shell [y] + string_obfuscate [2]) (string_xor obfuscate [3]) − obfuscate_string [4]) < 3*

4.3 Stage 3

There are the algorithms implementations for bypassing AV engines the dynamic analysis portion. The dynamic analysis detection became a complicated problem; as the detection is not accurate for its ability to get a high level of predictable output

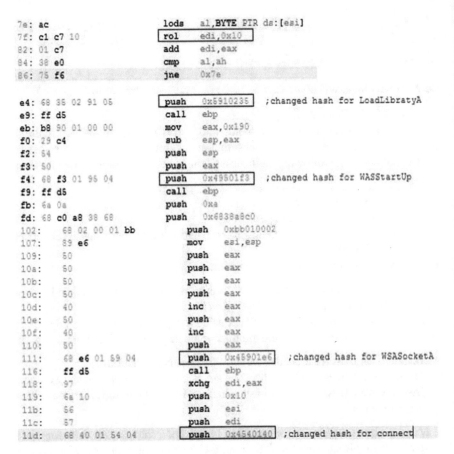

Fig. 3 Changing the API hash algorithm calculation causes incorrect original hard-coded values

using the identical rules set. There is no existence for a malware that has a capability of avoiding the entire dynamic analysis technique; the same applies for the antivirus capability of detecting all types of malware. In addition to the anti-dynamic analysis technique that is illustrated through pseudo-code, this paper provides another two detection techniques to detect the sandboxed environment. The first technique uses the audio drivers' detection that is setup on the victim machine. In addition to the check for any previous audio driver setup. If there is existence for a (Windows Primary Sound Driver) then the detection algorithm will conclude that there is no existence for a sandbox environment.

The second technique verifies an existence for a USB connected devices. If the number of USB connected devices is less than or equal one, then the detection algorithm will conclude that there is no existence for a sandbox environment.

```
C:\masters>sound.exe
Device description = Headphones (High Definition Audio Device)
Driver name = (0.0.0.000000000).(5e486e3e-b9c5-4706-80e3-5dde3001f2e4)

Device description = Speakers (High Definition Audio Device)
Driver name = (0.0.0.000000000).(5461cfaf-adee-4deb-826e-26cfe53b12de)
```

Fig. 4 A CMD screenshot shows an example of the process of Audio devices enumeration

4.3.1 Audio Devices Detection

This method assumes that there is no implementation for devices enumeration APIs by any anti-malware tool. If the APIs are implemented, the sandboxes will not detect any audio device except the windows default one that is usually known as (Primary Sound Driver). As a result, the sandboxes may delete them in the implementation stage, the developed method uses DirectSoundEnumerate API that calls the windows callback function, this callback retrieves the device (description—name—version). However, the audio detection algorithm keeps searching for any additional audio driver. If the search result finds another audio driver in addition to the default one, the audio check algorithm then keeps iterating to proceed with anti-dynamic checks.

Figure 4 illustrates a developed code for testing the enumeration process of audio device (the output is from Windows 7 Desktop). The changed algorithm of the *reverse_tcp* ignores the Windows default audio driver API. Most of recent windows' versions have a default audio driver installed on them.

4.3.2 Existing USB Devices Detection

Many studies such as [37–39] did not use the method of existing USB devices enumeration. The main purpose behind this check is as same as searching for audio drivers. The main hypothesis is that there will be no developed APIs for the USB device in the sandbox. In a case the APIs are developed, there will be no single entry returned. The USB authenticating method in the main algorithm enumerates all the existing USB devices. If the mapped devices number bigger than one, then it assumes automatically that it is a normal windows desktop Installation and the anti-dynamic analysis keeps executing for further checks.

4.4 Stage 4

This stage is inspired by both researches [39, 40]. Both of them have provided a modified static PE header as an effective method for antivirus evasion. To achieve this evasion, a small algorithm was developed to import the PE file input then mutates its attributes (PE fields name, version and data stamp). These values are changeable

Fig. 5 Modification process of the portable execution (PE) header

with every execution. The mutation engine changes the input file static signature; the change happens when the identical PE file is detected again.

Figure 5 illustrates PE header manipulation method process. The mutation engine of PE attribute is having a mapping file method from the disk to the memory. This process is done through using *CreateFileMapping* and sometimes using *MapViewOfFile* APIs.

A Portable Execution (PE) is a format of a file that is considered as an identifier for files commonly in windows operating systems such as (exe, png, dll, dbo, pdf...etc.). The definition of "portable" indicates the formats' inconstancy in many different operating systems' environments. The format of PE in mainly a data structure contains the important information for the operating system loader for managing the executablecode. The PE is having a dynamic library references—DLR for linking API importing and exporting tables and resource attributes. The PE file consists of many different section, those sections indicate the dynamic library reference to the file mapping into the memory [41]. The executable figure consists of many sections; each section needs a specific type of protection. Import Address Table (IAT), is used to search for a different module attribute in a table; as the application compilation cannot recognize the libraries location in the memory and a jump is needed in an indirect way in case pf API call [40].

The UnmapViewOfFile writes these values to a file. As a result, the engines of the antivirus that depends on the file hashes calculations will not be able to identify any of the.exe file posterior mutation. The pseudo-code below clarifies the method that is used for compiling the three steps above.

1. *Begin*
2. *{*
3. *IS_Debuger_exist:*
4. *If yes > End Compilation*
5. *If no > Continue next check*
6. *IS_big_memory_block_success*
7. *If no > End Compilation*
8. *If yes > Continue next check*
9. *If_Audio_Driver_Check*
10. *If yes > Continue next check*
11. *If no > End Compilation*
12. *IS_USB_Listing_Check*
13. *If yes > Continue next check*

14. *If no > End Compilation*
15. *Compile_Idle_Loop_for_2_mins*
16. *IS_mutex_with_name_exists*
17. *{*
18. *{*
19. *Load obfuscated_mutex_string;*
20. *De-ofuscate_mutex_string;*
21. *Close_Handle_with_invalid_ID;*
22. *}*
23. *If yes > End Compilation*
24. *If no > Continue next check*
25. *}*
26. *Shell_De-obfuscate;*
27. *Execute_Shell;*
28. *End.*

Figure 6 shows the fields' random compilation names for both PE Header and data stamp, the bytes that are shown after the execution of fileattrib2.exe randomize the section names ($0 \times 5a$, 0×06, 0×03 and $0 \times 3f$).

When the file memory mapping time is complete, there will be a change to the intended PE attributes through the mutation in memory, those changes are done through a PE values random routine upon each compilation. The files PE attributes—with different extensions—can be altered using the tool fileattrib2.exe. The design of transition engine is responsible for:

- Generate byte's value randomly from a predefined set.
- Copy the generated bytes values and store them into the file's structured memory that is created by CopyMemory API.
- Write the altered PE attributes to the disk using UnmapViewOfFile.

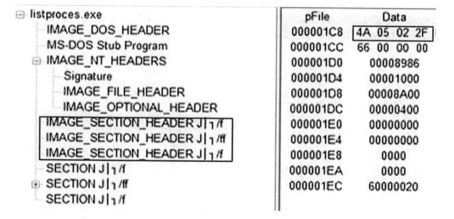

Fig. 6 A random compilation names for both PE header and data stamp

5   Experiment

Static analysis observes the malware without executing it, while the dynamic analysis actually executes the malware in well monitored environment to monitor its behavior. The experiment process of the malware consists of two phases.

First Phase: is to submit the created samples of the TCP shell to a Kaspersky 2018 build 18.0.0.405. Kaspersky was installed on ten virtual machines (VirtualBox Platform) with (Windows as an operating system. Each antivirus' application version is compatible with the neoteric malware signatures (July 2018).

Second Phase: submitting the samples of TCP shell into two online scan engines, those online scan engines are:

a. **Virustotal.com**: The website uses 68 online anti-malware engines.
b. **Virusscan.com**: The website uses 42 online anti-malware engines.

The main reason for using two different online scan engines is to compare between the detection rate and the correlate the contradictions in the results, while the reason for using Kaspersky as a local antivirus on the virtual machines is to identify the differences of evasion rate between the local antivirus and the online antivirus engines for each sample.

The samples consist of different phases from the code development process; those samples are described with their characteristics and generation methods in both Tables 1 and 2.

Testing dynamic and static analysis together provides a feedback analysis to identify the malware capabilities through analyzing a sequence of technical indicators that cannot be achieved through the simple static analysis alone, as the focus in this paper to make the detection rate of the dynamic analysis harder we will perform some tests through various antivirus' engines to measure the performance of our proposed methodology. The evasion rate presents the AV's ratio that does not have the ability to detect each sample from both Tables 1 and 2 divided by AV's overall number. Figure 7 shows the evasion rate results from every sample that being scanned by the antivirus. Three anti-malware scanners categories have tested each sample from Table 1:

Table 1 Modified samples of scanned samples by antivirus engines

Sample #	Shellcode type	Generation method	Characteristics
X2.exe	Modified	Assembler level	• API hashing algorithm modification
X4.exe	Modified	Assembler level	• Custom-coded obfuscator
			• Anti-dynamic analysis behavior
X6.exe	Modified	Assembler level	• API hashing algorithm modification
			• NOP sleds
			• Anti-dynamic analysis behavior
X8.exe	Modified	Assembler level	• Custom-coded obfuscation
			• Anti-dynamic analysis behavior

Table 2 Plain samples of scanned samples by antivirus engines

Sample #	Shellcode type	Generation method	Characteristics
X1.exe	Plain	Msfvenom	• No modifications to the shell
			• No modifications to obfuscation
X3.exe	Plain	Msfvenom	• No modifications to the shell
			• Special-coded obfuscator
X5.exe	Plain	Msfvenom	• No modifications to the shell
			• Anti-dynamic analysis behavior
X7.exe	Plain	Msfvenom	• Anti-dynamic analysis behavior
			• No modifications to the Shell
X9.exe	Plain	Msfvenom	• No modifications to the shell
			• Anti-dynamic analysis behavior
X10.exe	Plain	Msfvenom	• No obfuscation
			• No modifications to the shell
			• Anti-dynamic analysis behavior
			• Realized via nested for-loop
X11.exe	Plain	Msfvenom	• No obfuscation
			• No modifications to the shell
			• Anti-dynamic analysis behavior
			• Creation child process
X12.exe	Plain	Msfvenom	• No obfuscation
			• No modifications to the shell
			• Anti-dynamic analysis behavior
			• Check the success of memory
X13.exe	Plain	Msfvenom	• No obfuscation
			• No modifications to the shell
			• Anti-dynamic analysis behavior
			• USB device enumeration
X14.exe	Plain	Msfvenom	• No obfuscation
			• No modifications to the shell
			• Anti-dynamic behavior
			• Sound device enumeration
X15.exe	Plain	Msfvenom	• Modified using a coded PE metadata engine

(1) **Kaspersky**: The first category represents the installed local antivirus which is Kaspersky 2018 (build 18.0.0.405) on ten virtual machines.

(2) **Virustotal.com**: The second category represents the samples' scan result through antivirus online engine.

(3) **Viruscan.com**: The third category represents the samples' scan result through antivirus online engine.

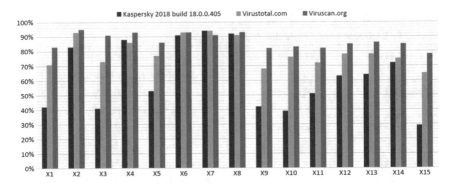

Fig. 7 The evasion rate level results for samples from Tables 1 to 2 using three scanners engines

The experiment was made through virtual machines in private cloud network in four labs at Suez University:

Lab 1A: 45 PCs
Lab 2A: 40 PCs
Lab 5B: 24 PCs
Lab 6B: 24 PCs.

It is important to clarify that:

- The local antivirus (Kaspersky) has been tested in an isolated environment.
- The experiment using the local antivirus engines lasted for 480 min (8 h).
- Each sample was tested in a separate device (all devices with are having the same system requirements), the evasion rate is measured in percentage.

6   Discussion and Analysis

From Fig. 7, it shows the samples' evasion rate with anti-dynamic techniques through being scanned by three different scanners engines, there is a symmetry in the evasion ratio results in the three scanners engines; as the samples the had a low evasion rate result in the local antivirus also had a proportionately evasion rate result in the online scanning engines and vice versa with the samples that had a high evasion rate result. The reason behind that is the techniques which are used by both local antiviruses and online engines.

This study mainly focuses on the malware evasion rate development and propose a new methodology. However, it does not intent to highlight the differences between the local antiviruses and the online engines. Kaspersky shows a lower evasion rate of the samples comparing to the online scanners' engines. Hence, the accuracy level of the locally installed antivirus is having more detection capabilities.

Hidost is categorized as SVM a classification model. It is the best margin classifier which attempts to search for a small data points number, which divides the whole data points of two sets with a hyperplane of a dimensional space. With kernel tricks, this can be extended to suit complex classifications problems using nonlinear classifier. Radial basis function (RBF) is used in Hidost for mapping data points into endless dimensional space. In experiment, the data point distance (positive or negative) to the hyper plane is a result of a prediction output, the positive distance is classified as a malicious while the negative one is classified as a benign. Precisely, the most smart evasion technique was successful according to Hidost—for only 0.024% of malicious samples tested against nonlinear SVM classifier with radial basis function kernel through the binary embedding.

Our experiment uses Hidost for testing the malware evasion rate, the experiment took 74 h (around 3 days) to execute. Even though Hidost was mainly designed for resisting evasion attempts, our method has achieved a rate that reached more than 75% for a lot of samples. We have tested 1749 generated samples (from the samples in both Tables 1 and 2) in total for 400 seeds (4.4 evasive samples for each seed).

Trace Analysis: Each mutation analysis trace has been analyzed in the same way for Virustotal. Figure 8 shows both length and success rate for every mutation trace.

Generally, it needs shorter mutation traces to be able to achieve about more that 75% evasion rate in attacking Hidost than it has achieved for Virustotal. We detected two main differences compared to Virustotal.

Firstly, there is no any increment in trace length for the new mutation traces, unlike Virustotal where the trace length directly proportional with the trace ID.

Secondly, there is a relation between the trace length and the success rate. The longer the traces become, the more success in generating evasive variants.

Figure 9 shows the success rate results of the scanned samples from three different antiviruses' engines, the success rate in our experiment is measured by the average rate of the three scanners' engines results from Fig. 7. It is initially expected that the modified samples could achieve a noticeable success evasion rate than the pure shell. However, the success rate of the modified shellcode is higher by levels than the plain shellcode that is created by msfvenom.

Additionally, the anti-dynamic techniques proved low relevant to the modified assembler code (i.e. Samples with low evasion rate 'X2 and 4' versus samples with

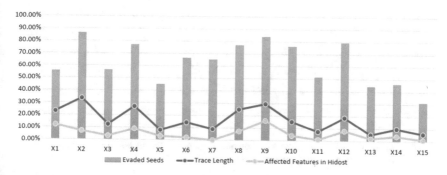

Fig. 8 The mutation traces length and success rate for evading Hidost

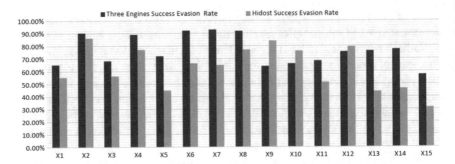

Fig. 9 The average evasion success rates for each sample using three samples against Hidost rates in our experiment

high evasion rate 'X9 and X14'). Unexpectedly, the modified syntactic samples that achieved a high evasion rate such as X2, X6 and X8 proved that the most of the antiviruses' engines struggle in recognizing the alteration in code syntax. There is an equalization with the distributed samples, starting from sample X9 to sample X14.

Those samples are using anti-dynamic technique and show an equalized evasion rate. Meanwhile, there is no noticeable evasion success in the enumeration of sound devices, there is no any type of anti-dynamic techniques shows considerable failure or success over one to another.

Xu et al. [42] in their study, they first classify the detected JavaScript obfuscation methods. Then they created a statistic analysis on the usage of different obfuscation methods classifications in actual malicious JavaScript samples. Based on the results, the have studied the differences between benign obfuscation and malicious obfuscation as well as explaining in-detail the reasons behind the obfuscation methods choice When we compare the implemented fifteen samples using the proposed methodology with fifteen implemented data obfuscated samples from [41] as shown in Fig. 10.

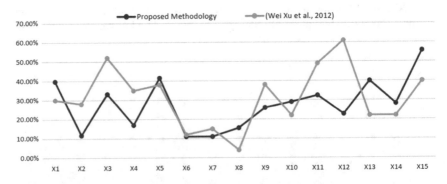

Fig. 10 A comparison of the average detection ratio results from the proposed methodology samples to data obfuscated samples from Xu et al. [42]

There is a weakness from our proposed methodology; as some samples achieved high detection rate compared to Xu et al. [42] such as (X3, X8, X10 and X15) while the rest samples have succeeded in achieving lower detection rate than [42].

However, there are noticeable rate differences between samples (X4, X11 and X12). The reason behind comparing the results of our proposed methodology with the results in [42] is to measure the effectiveness of our proposed methodology with the actual obfuscation techniques that already exist and have been studied in their paper, our research differs from their one in the main contribution purpose; as our main aim is to propose a new malware obfuscation methodology while their target was to study and analyze different types of malware obfuscation techniques. Although there were noticeable developments that were made on anti-malware industry since their experiment, it proves that our proposed methodology has achieved positive results overall, but it still is not sufficient to challenge or evade nowadays detection technique standards.

Our future work will be focused on developing each stage from the proposed four stages; to provide more flexibility with different anti-malware engines' and provide higher success rate.

7   Conclusion

The main purpose of this paper is to propose a malware obfuscation methodology with a high evasion ratio, antidynamic techniques have the ability for evasion increment rate with some limitations; as audio and USB enumeration have the ability of fulfilling the same evasion level effect of any antidynamic method. However, the changes in the algorithm on the assembler level proved to be the most vigorous technique.

In this paper, we have introduced methodology that aids a malware to avoid anti malware tools, these techniques were mainly developed to avoid detection of a malware via static analysis technique. On the other hand, the proposed intuitive anti-virtualization techniques that will avoid analysis under a sandbox. The evasion increment and decrement level depends on the code obfuscation. It still not known from our evaluation the reason of this behavior. Even though many developed malware methods were studied in previous researches, the wellknown antivirus engines keep showing interests of such techniques to test their engines' detection power level. The ineffectual method based on dynamics and static analysis is the common deployed method.

Finally, the provided technique that is used for developing the malware samples—armlessly modified programs—proved its efficiency after scanning them on multiple antivirus engines and the results showed minimum proportion of false positives.

References

1. Mahawer, D., Nagaraju, A.: Metamorphic malware detection using base malware identification method. Secur. Commun. Netw. **7**(11), 1719–1733 (2013)
2. Nai Fovino, I., Carcano, A., Masera, M., Trombetta, A.: An experimental investigation of malware attacks on SCADA systems. Int. J. Crit. Infrastruct. Prot. **2**(4), 139–145 (2009)
3. Mawgoud, A.A., Taha, M.H.N., Khalifa, N.E..M.: Security Threats of Social Internet of Things in the Higher Education Environment, pp. 151–171. Springer, Cham (2019)
4. Xu, D., Yu, C.: Automatic discovery of malware signature for anti-virus cloud computing. Adv. Mater. Res. **846–847**, 1640–1643 (2013)
5. Kumar, A., Goyal, S.: Advance dynamic malware analysis using Api hooking. Int. J. Eng. Comput. Sci. **5**(3) (2016)
6. Khaja, M., Seo, J., McArthur, J.: Optimizing BIM Metadata Manipulation Using Parametric Tools. Procedia Engineering **145**, 259–266 (2016)
7. Tahir, R.: A study on malware and malware detection techniques. Int. J. Educ. Manag. Eng. **8**(2), 20–30 (2018)
8. El Karadawy, A.I., Mawgoud, A.A., Rady, H.M.: An empirical analysis on load balancing and service broker techniques using cloud analyst simulator. In: International Conference on Innovative Trends in Communication and Computer Engineering (ITCE), pp. 27–32. IEEE, Aswan, Egypt (2020)
9. Li, J., Sun, L., Yan, Q., Li, Z., Srisa-An, W., Ye, H.: Significant permission identification for machine-learning-based android malware detection. IEEE Trans. Industr. Inf. **14**(7), 3216–3225 (2018)
10. Li, Q., Larsen, C., van der Horst, T.: IPv6: a catalyst and evasion tool for botnets and malware delivery networks. Computer **46**(5), 76–82 (2013)
11. Suk, J., Kim, S., Lee, D.: Analysis of virtualization obfuscated executable files and implementation of automatic analysis tool. J. Korea Inst. Inform. Secur. Cryptol. **23**(4), 709–720 (2013)
12. MaHussein, D.M.E.D.M., Taha, M.H., Khalifa, N.E.M.: A blockchain technology evolution between business process management (BPM) and Internet-of-Things (IoT). Int. J. Advanc. Comput. Sci. Appl. **9**, 442–450 (2018)
13. Malhotra, A., Bajaj, K.: A survey on various malware detection techniques on mobile platform. Int. J. Comput. Appl. **139**(5), 15–20 (2016)
14. Kritzinger, E., Smith, E.: Information security management: an information security retrieval and awareness model for industry. Comput. Secur. **27**(5–6), 224–231 (2008)
15. Kong, D., Tian, D., Pan, Q., Liu, P., Wu, D.: Semantic aware attribution analysis of remote exploits. Secur. Commun. Netw. **6**(7), 818–832 (2013)
16. Khalifa N.M., Taha M.H.N., Saroit, I.A.: A secure energy efficient schema for wireless multimedia sensor networks. CiiT Int. J. Wirel. Commun. **5**(6) (2013)
17. Goh, D., Kim, H.: A study on malware clustering technique using API call sequence and locality sensitive hashing. J. Korea Inst. Inform. Secur. Cryptol. **27**(1), 91–101 (2017)
18. Pandey, S., Agarwal, A.K.: Remainder quotient double hashing technique in closed hashing search process. In: Proceedings of 2nd International Conference on Advanced Computing and Software Engineering (ICACSE) (2019, March)
19. Luckett, P., McDonald, J., Glisson, W., Benton, R., Dawson, J., Doyle, B.: Identifying stealth malware using CPU power consumption and learning algorithms. J. Comput. Secur. **26**(5), 589–613 (2018)
20. Kumar, P.: Computer virus prevention & anti-virus strategy. Sahara Arts & Management Academy Series (2008)
21. Yoshioka, K., Inoue, D., Eto, M., Hoshizawa, Y., Nogawa, H., Nakao, K.: Malware sandbox analysis for secure observation of vulnerability exploitation. IEICE Trans. Inform. Syst. **E92-D**(5), 955–966 (2009)
22. Pektaş, A., Acarman, T.: A dynamic malware analyzer against virtual machine aware malicious software. Secur. Commun. Netw. **7**(12), 2245–2257 (2013)

23. Joo, J., Shin, I., Kim, M.: Efficient methods to trigger adversarial behaviors from malware during virtual execution in sandbox. Int. J. Secur. Appl. **9**(1), 369–376 (2015)
24. Maestre Vidal, J., Sotelo Monge, M., Monterrubio, S.: EsPADA: enhanced payload analyzer for malware detection robust against adversarial threats. Fut. Gene. Comput. Syst. **104**, 159–173 (2019)
25. Iwamoto, K., Isaki, K.: A method for shellcode extraction from malicious document files using entropy and emulation. Int. J.Eng. Technol. **8**(2), 101–106 (2016)
26. Zakeri, M., Faraji Daneshgar, F., Abbaspour, M.: A static heuristic method to detecting malware targets. Secur. Commun. Netw. **8**(17), 3015–3027 (2015)
27. Kruegel, C.: Full system emulation: achieving successful automated dynamic analysis of evasive malware. In: Proc. BlackHat USA Security Conference, 1–7 August 2014
28. Zhong, M., Tang, Z., Li, H., Zhang, J.: Detection of suspicious communication behavior of one program based on method of difference contrast. J. Comput. Appl. **30**(1), 210–212 (2010)
29. Mawgoud, A.A.: A survey on ad-hoc cloud computing challenges. In: International Conference on Innovative Trends in Communication and Computer Engineering (ITCE), pp. 14–19. IEEE (2020)
30. Eskandari, M., Raesi, H.: Frequent sub-graph mining for intelligent malware detection. Secur. Commun. Netw. **7**(11), 1872–1886 (2014)
31. Barabas, M., Homoliak, I., Drozd, M., Hanacek, P.: Automated malware detection based on novel network behavioral signatures. Int. J. Eng. Technol. 249–253 (2013)
32. Zatloukal, F., Znoj, J.: Malware detection based on multiple PE headers identification and optimization for specific types of files. J. Adv. Eng. Comput. **1**(2), 153 (2017)
33. Tomasi, G., van den Berg, F., Andersson, C.: Correlation optimized warping and dynamic time warping as preprocessing methods for chromatographic data. J. Chemom. **18**(5), 231–241 (2004)
34. Vinod, P., Viswalakshmi, P.: Empirical evaluation of a system call-based android malware detector. Arab. J. Sci. Eng. **43**(12), 6751–6770 (2017)
35. Saeed, I.A., Selamat, A., Abuagoub, A.M.A.: A survey on malware and malware detection systems. Int. J. Comput. Appl. **67**(16), 25–31 (2013)
36. Mawgoud, A.A., Ali, I.: Statistical insights and fraud techniques for telecommunications sector in Egypt. In: International Conference on Innovative Trends in Communication and Computer Engineering (ITCE), pp. 143–150. IEEE (2020)
37. Mawgoud A.A., Taha, M.H.N., El Deen, N., Khalifa N.E.M.: Cyber security risks in MENA region: threats, challenges and countermeasures. In: International Conference on Advanced Intelligent Systems and Informatics, pp. 912–921. Springer, Cham (2020)
38. Ismail, I., Marsono, M., Khammas, B., Nor, S.: Incorporating known malware signatures to classify new malware variants in network traffic. Int. J. Netw. Manag. **25**(6), 471–489 (2015)
39. Toddcullumresearch.com: Portable executable file corruption preventing malware from running—Todd Cullum Research (2019). https://toddcullumresearch.com/2017/07/16/portableexecutable-file-corruption/ Accessed 10 June 2019
40. Blackhat.com (2019). https://www.blackhat.com/docs/us-17/thursday/us-17-Anderson-Bot-VsBot-Evading-Machine-Learning-Malware-Detection-wp.pdf. Accessed 9 Apr 2019
41. Ye, Y., Wang, D., Li, T., Ye, D., Jiang, Q.: An intelligent PE malware detection system based on association mining. J. Comput. Virol. **4**(4), 323–334 (2008)
42. Xu, W., Zhang, F., Zhu, S.: The power of obfuscation techniques in malicious JavaScript code: a measurement study, 9–16 (2012). https://doi.org/10.1109/malware.2012.6461

An Artificial Intelligence Resiliency System (ARS)

Ali Hussein, Ali Chehab, Ayman Kayssi, and Imad H. Elhajj

1 Introduction

Future network advancements are mainly focusing on discovering solutions to the constantly expanding challenges due to the exponential increase in network traffic and devices. The lack of programmability, remote management, and scalability led to the development of new networking technologies such as SDN. SDN Provides flexible control over network devices through orchestrating switches in the data-plane under a centralized controller. Nevertheless, a reliable consistent network security solution is still out of reach. Late advances in artificial intelligence and deep learning are facing the concerns associated towards the network security and their correct behavior.

Before, and if, a full SDN transition is possible; SDN is anticipated to co-exist with traditional networks for a certain period. Accordingly, our proposed work discusses the integration of AI and SDN to provide, on one hand, security, and extend the work targeting, on the other hand, the consistency and correctness of hybrid networks. The proposed solution consists of a multilevel distributed security architecture with a Central Artificial Intelligence (CAI) based controller. This is possible through

A. Hussein (✉) · A. Chehab · A. Kayssi · I. H. Elhajj
Electrical and Computer Engineering Department, American University of Beirut, Beirut 1107 2020, Lebanon
e-mail: aih10@aub.edu.lb

A. Chehab
e-mail: chehab@aub.edu.lb

A. Kayssi
e-mail: ayman@aub.edu.lb

I. H. Elhajj
e-mail: ie05@aub.edu.lb

A.-E. Hassanien et al. (eds.), *Enabling AI Applications in Data Science*, Studies in Computational Intelligence 911, https://doi.org/10.1007/978-3-030-52067-0_28

featuring multiple levels of security measures throughout different nodes over the network.

Network administrators handle various applications on the control side to perform different management tasks such as firewall, monitoring, routing, and others. Most of these applications have complex interactions among each other, creating difficult challenge when reasoning about their behaviors. We argue that there is no security without consistency, such that enforcing security policies to a network without the ability to verify any misbehavior is a challenge by itself.

Our fundamental motivation was the fascinating tactics of the human immunity system, that is based on a double line of defense. A first line responsible for the deflection of unfamiliar attackers at various exterior contact points in the body without the necessity to identify their type. A second line responsible for the detection of the attacks that passed through the first line, by distributing watchers throughout the blood stream, keeping an eye on any abnormal. This system earned its efficiency from one layer using minimum energy, and a second layer requiring distributed techniques and more processing.

Another inspiration was our brain functioning as a single controlling unit that, aside from its millions of functions, controls the consistency of our body through keeping in touch with all our organs to make sure that they are functioning correctly. In simple words, our brain handles this task by knowing ahead the correct function of each organ, then alerts our awareness if any part of the body is malfunctioning after comparing the intended function and the actual one.

This paper is an extension of our series that started with [1, 2], proposing a general solution to maintain network resiliency in terms of security and consistency, and correctness of hybrid networks, while minimizing the processing overhead, in a software-based network with remote management environment. Our proposal is a multi-layer AI-based resiliency enhancement system that builds its strength from both a consistency establishment system, and a hybrid (centralized and distributed) security technique in terms of accuracy, efficiency, scalability, robustness, and others.

The rest of this paper is organized as follows: Related AI-based SDN solutions are reviewed in the next section. Section 3 introduces our architecture design and explains how it would achieve a balance between security, consistency and performance in an SDN network. Finally, Sect. 4 presents the system analysis and simulation followed by system testing and results. Finally, we conclude the paper in Sect. 5.

2   Related Work

Intelligent networking in modern systems started to rely on both AI and soft computing. The integration of the abstract concepts in SDN and AI techniques can evolve to a more adaptive behavior of network elements dealing with both SDN related issues and old traditional ones [3].

Fig. 1 Security research
distribution between
different SDN layers

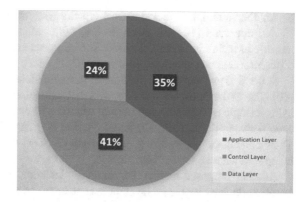

The controller is a major player in soft networking, therefore it is always targeted by programming vulnerabilities, DDoS attacks, and error configurations on the secure channel [4].

The security related SDN research community distributed their work between solutions related to the different layers of SDN such as control, data, and application layers as shown in Fig. 1. Only limited number of researchers took their work a step forward and proposed techniques that acted on securing multiple layers simultaneously through distributing the load among them.

Data mining techniques have previously been used for routing problems and performance optimization. Alongside AI, and after the introduction of the programmable network capabilities, data mining have been recently used to play an important role in SDN networks [5–7].

2.1 Traditional AI Techniques

The authors of [8] worked on combining a fuzzy inference system and both of Rate Limiting and TRW-CB to create an SDN based information security management system.

A survey presenting a classification of intrusion detection systems (IDS) was discussed in [9]. The authors reviewed anomaly detection techniques and the performance of machine learning in such domain. They also included a study included a on the false and true positive alarm rates. In addition to discussing feature selection and its importance in the classification and training phase of machine learning IDS.

The authors of [3] proposed a DDoS attack detection technique through traffic flow tracing. The authors relied on different ML algorithms including: K-Nearest neighbor, Naive Bayes, K-medoids and K-means to classify traffic as normal and abnormal. The authors discussed how DDoS attacks can be detected using these techniques to classify incoming requests.

Another attempt to tackle DDoS attacks was using the Self-Organized Maps (SOM) approach [10–12]. when dealing with unlabeled input vectors, SOM can perform as a classification technique. The proposed approach was compared against different methods on the well-known KDD-99 data set.

SOM also participated in the field of intrusion detection as shown in the work of [10]. The author's experiments showed that anomalous behavior could be detected using a single SOM trained on normal behavior. They argued that the ratio of difference between normal and abnormal packets is greater by an order of magnitude. Their conclusion showed that the strength of SOM based IDS comes from it not being told what abnormal behaviors is.

Another proposed method for anomaly detection is MLP with a single hidden layer neural network [13]. The authors tested their work on the DARPA 1998 dataset and achieved a DR of 77%. The authors of [14] used selected generic keywords to detect attack preparations, achieving a DR of 80% using the same dataset.

Another IDS work was the proposal of [15], who presented a 57 gene chromosome, with each gene representing a single connection feature such as: destination or source IP address. The authors of [16] achieved a 97.04% on the DARPA dataset for their IDS work on linear genetic programming (LGP). However, the main disadvantage of this kind of IDS, as concluded by the authors of [17], is the high resource consumption involved.

The authors of [18] also presented an IDS based on fuzzy c-means and rough sets theory, achieving an average of 93.45% accuracy on the KDD'99 dataset.

The authors of [19] proposed an IDS based on KNN with similarity as a quantitative measure for distance, achieving a DR of 90.28% on the KDD'99 dataset. Similarly, the authors of [20] achieved a DR of 91.70% on the DARPA dataset while testing their implementation of KNN based IDS. The authors of [21] proposed a Naïve Bayesian based IDS and achieved a 91.52% on the KDD'99 dataset. Similarly, the authors of [22] achieved a 94.90% on the same dataset.

The authors of [23] discussed the effect of principal component analysis (PCA) on the work of Decision Tree (DT). The processing time was reduced by a factor around thirty on the KDD'99 dataset, with a slight decrease in accuracy from 92.60% till 92.05%.

The authors of [24] proposed dimension reduction through preprocessing using the rough set theory (RST) before applying SVM for intrusion detection. They achieved an accuracy of 86.79% on the KDD'99. The authors of [25] concluded that the combination of SVM and DT techniques results in higher results compared with each individual technique.

The authors of [26] discussed their Random Forest (RF) based security model, achieving high DR and concluded with a stable set of important features. The authors of [27] reached an average DR of 92.58 and 99.86%, on the KDD'99 the balanced dataset, respectively. The balanced dataset was tested to increase the DR of minority intrusions through over-sampling the minority classes and down sampling the majority.

2.2 Deep Learning AI Techniques

The authors of [28] proposed a deep learning approach based on Self-taught Learning (STL) to implement a Network IDS. They tested their work on the NSL-KDD dataset. The authors of [29] proposed a new method that illustrates network traffic as images using convolutional neural network for malware traffic classification. The authors argued that applying CNN on images rather than raw data results in higher classification accuracy.

Deep learning techniques showed much improvements in the DDoS detection domain. The authors of [30] proposed a deep learning multi-vector-based DDoS detection system in an SDN environment. They discussed feature reduction derived from network traffic headers. Their results show an accuracy of 99.75% between normal and attack classes.

A list of ML-based proposed techniques is presented in Table 1. As shown, AI-based techniques are being used, offering high results. Thus, the problem remains in systemizing these techniques, making such solutions inflexible, and vulnerable. The main issue is that these solutions are focusing on increasing accuracy regardless of

Table 1 Previous IDS projects in terms of accuracy and Data set

References	ML technique	Attack type	Data set	Accu. (%)
Gupta [31]	K-means + KNN + DT	R2L, U2R, DoS and probe	KDDCup99	96.5
Jongsuebsuk et al. [32]	FL + GA	Dos and Probe	Real life	97
Senthilnayaki et al. [33]	GA + SVM	Dos, Probe U2R R2L	KDDCup99	99.1 99.1 97 96.5
Masduki et al. [34]	SVM	R2L	KDDCup99	96.1
Enache and Sgârciu [35]	SVM + BAT Algorithm	Malicious	NL-KDD	99.4
Akbar et al. [36]	GA	DoS, R2L, U2R and Probe	KDDCup99	92.6
Aziz et al. [37]	DT	Dos Probe	NSL-KDD	82 64
Lin et al. [38]	Cluster centre + K-NN	Probe Dos	KDDCup99 (6 dimension)	99.9 99.9
Aburomman and Reaz [39]	SVM + K-NN	DoS, R2L, U2R and Probe	KDDCup99	87.4–91.7
Horng et al. [40]	SVM + hierarchical clustering algorithm	DoS, R2L, U2R and Probe	KDDCup99	95.7
Hodo et al. [41]	ANN	DDoS/Dos	Real Life	99.4

the other feature of a network that may be affected. First, through redirecting traffic to a fixed point for processing and analysis, such as the Controller, introducing high traffic overhead and security risks. Second, through redirecting only packet headers, which decrease the overhead, but still inherits similar weaknesses.

2.3   Consistency Establishment Solutions

As previously mentioned, ensuring consistency in a network is essential for a security solution. To the best of our knowledge, this is the first work of integration between consistency establishment and security solutions towards a more resilient network. Different consistency-check techniques have been proposed in the literature for various purposes as described below.

The consistency related work in the literature focused on several main aspects. The authors of [42] worked on state synchronization to ensure that redundant controllers have the same sate information. While, the authors of [43] focused on inter-flow consistency by proposing a scheduling process for multiple network updates to be checked for overlaps or contradictions that may cause network failure or even security threats.

The work of [44] proposed a third-party consistency verification check module that includes route correctness and network security isolations in a single domain. Other researches, such as [45], handled tunable consistency models to enable controllers to tune their own configurations to enhance the performance of applications running on top of them.

Several network verification tools, such as [46–48], have been developed to check for network inconsistencies such as: loops, dead ends and network unreachability alerts. The main drawback of most of these techniques lie in the security concerns of the solutions themselves, and their security side effects on the network. In addition to the extra network traffic being exchanged and redirected from one party to another.

In another work, we focused on what consistency preservation can offer to network security, and how can a consistent network create a solid foundation for a security system to perform with better efficiency.

3   ARS Architectures and Designs

Our vision to enhance network **security** and **resilience**, in addition to the influence of AI-based systems over the network domains, drove us to the combination of AI and programmable networking such as SDN. Hence, we began researching the diverse AI techniques alongside an efficient data management system, so as to supply our system with the required information while maintaining the processing load and overhead as low as could be expected under the circumstances. Our proposals and contribution are summarized as follow:

Fig. 2 Architectural distribution

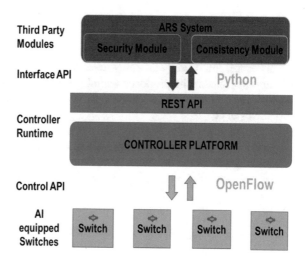

1. A general system architecture for Network resiliency in a Software based network.
2. An AI based multi-layer security solution adopting new techniques with higher efficiency.
3. An effective technique aiming to provide our systems with the required data while maintaining a decent processing overhead.
4. An integrated consistency solution as a pre-security measure, allowing a more resilient system.
5. An AI optimization system for better AI decision confidence and unknown attack identification.
6. The proposed designs were implemented and tested as later discussed, with the results presented at the end.

3.1 The General Architectural Design

Our proposed system relies on a hybrid architecture that includes both centralized processing at the controller level and distributed processing at the nodes level over the data plane. Figure 2 illustrates a general overview of our system.

3.2 Research Investigation Phases

1. **Edge feature extraction** to abstain from diverting traffic to the controller to limit the security risks. We are considering an *edge node* being any relevant node in the network, specified by to network environment and differs from one scenario to another. The Feature extraction process includes only the relevant features,

which are then forwarded in a vector to the following destination, every time
cycle, *tc*.

2. **The multi-layer detection technique** provides consistency verification as a first
   measure, followed by anomaly detection and finishing with an attack specification approach. The proposed technique allows us to distinguish anomalies and
   unknown attacks at a faster rate and lower processing overhead, while preserving
   consistency.
3. **Feature marking** at this stage the extracted features undergo a marking process
   at the ingress points. Including the mark as a feature in the training and decision
   making allows the detection of slight changes in the global flow or randomness
   of the network through triggering a change in the output of the AI system.

3.3   Consistency Establishment as a Precaution Measure

Inspired by the brain-awareness human consistency techniques, that is also part of the
total human security system, we are proposing a network consistency establishment
system that relies on the controller's complete view of the network to check for
any misbehavior in the network nodes. Such feature would increase our confidence
regarding the correctness of the network nodes, before starting an external network
security analysis.

For the controller to check for inconsistencies every time cycle, *tc'*, it should
be able to know what each node should be doing at every time *t*. The consistency
module at the controller would enable the controller to save an updated image of
the matching (flow) table of each node. Before any consistency check, the module
should make sure that no updates are being injected at this time. At *tc'* the module
would probe the security agent at the node level for a hash of the local matching
(flow) table. This process ensures the security and privacy of this sensitive data not
to be exchanged through the network. At the same *tc'*, the module would calculate
its own local hash outputs. Since no updates are done at this time, the output of each
network node (actual behavior) should match the output of its corresponding image
at the controller (intended behavior).

Deploying this technique would allow us to identify any compromised or misbe-
having nodes, therefore leading to inefficient and unreliable security analysis at that
point. In case of such events, a specialized module should be alerted in order to
identify the cause of such inconsistencies and take necessary actions. Such solution
is kept for future work in the following issues of this series of articles.

3.4   ANN for Anomaly Detection

Diverse solutions have been explored preceding work with a fundamental obstacle
regarding the traffic and processing overhead, as a result of port mirroring, packet

cloning, even header extraction. Such techniques have opened a backdoor for vulnerabilities and security risks, while transferring duplicate data to the controller.

Hence, to avoid duplicated and extra traffic we proposed feature extraction at the data plane level. At this point, participating input nodes will handle the task of extracting the required features from incoming packets, forming the feature vector, and forwarding it, each time cycle *tc*, to the following destination. In this paper we are proposing two techniques:

1. Enabling the participating nodes to upload one vector, containing the necessary extracted features, to the controller each *tc*.
2. Designing an NN based overlay network over the control plane, where every node handles a sample processing portion (e.g. the processing load of a single or multiple neurons). Instead of the vector being uploaded to the controller, it would be sent to the next hidden layer, and so on, reaching the output layer for decision making.

The extraction stage and feature marking guarantee little overhead and real-time processing while examining the traffic flow and analyzing its randomness. Such a technique process equips the security module with the ability to detect distributed attacks including DDoS and multi-sequence attacks.

Regarding the first technique the AI processing is handled at the controller side after receiving the required input vectors. Here we would gain both faster processing and quick mitigation after the detection stage.

Regarding the second technique, an independent overlay solution, managed by the controller, acts autonomously starting from the feature extraction reaching the detection stage. The processing load assigned to each node is minimal, with no duplicated traffic.

3.5 A Two-Stage Solution for Anomaly Detection and Attack Identification

Another unique technique regarding our work is the two-step process security solution design. We intend to distinguish irregularities inside the system traffic, which might be activated by a security risk, a system architectural change, or system administrative change. It has been proven in the literature that anomaly detection needs fewer features and less processing for it to work. Thus, it is more effective and efficient to depend on anomaly detection to monitor the traffic flow and its randomness. It is for these reasons that we are proposing to apply anomaly detection as a first line of defense. In the case of an anomaly the second line is activated, applying an attack specification process to the flagged features.

Detecting unknown attack is an important feature of such an architecture. The attack identification model was trained to classify according to both the general attack class followed by its specific type. Hence, if an anomaly is detected and the second model was unable to identify the type of this attack, or even if it identified its general class and was unable to specify the type, then we can assume that we are dealing with an unknown attack. At this stage, new techniques should be used to label these features and teach the system of the new attack. Another feature of such an architecture is shown in the case of new attacks, where there is no need to train the whole system again. In this case, only the second system is retrained offline, while the first stage remains online.

3.6   General ARS Architecture Overview

The two proposed architectures are based on either:

1. **D**istributed Extraction, **C**entralized Processing, and **C**entralized Management (D~C^2).
2. **D**istributed Extraction, **D**istributed Processing, and **C**entralized Management (D^2~C).

The general architectural design of D~C^2 is presented in Fig. 3. The Artificial Intelligence Resiliency System (ARS) agent resides on the edge nodes for feature extraction and marking. The ARS security and consistency solution resides on the remote management level, where the monitoring and management processing is handled.

The general architectural design of D~C^2 is presented in Fig. 4. The ARS AI agent resides on the edge nodes for feature extraction, marking, and handling the processing of part of the AI model. In the D~C^2 architecture the monitoring and management processes are handled by the controller. Another task includes the optimization and distribution of the processing tasks in the overlay network (Fig. 5).

D~C^2 architecture

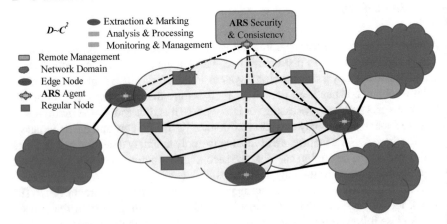

Fig. 3 D~C^2 architecture

D²~C architecture

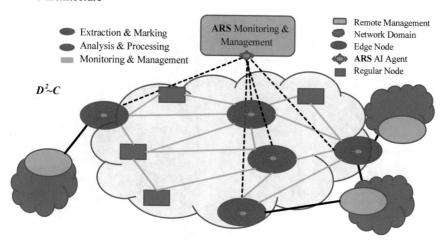

Fig. 4 D²~C architecture

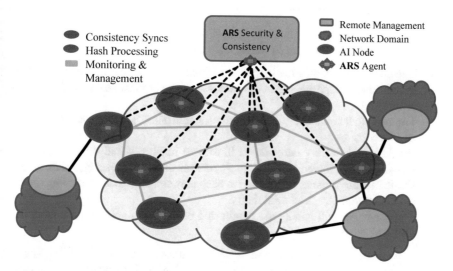

Fig. 5 System consistency architecture

The D~C² architecture can be described as follows:

(a) The controller handles the monitoring and management of the network including specifying the edge nodes.
(b) The edge nodes handle the distributed extraction phase including vector marking and forwarding.
(c) The controller performs the centralized processing and analysis part.

System overlay architecture

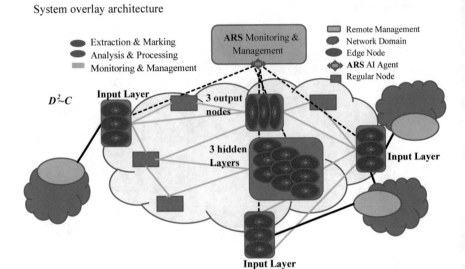

Fig. 6 System overlay architecture

The D^2~C architecture can be described as follows:

(a) The controller handles the management and control of the network including specifying the edge nodes and the ARS AI agents responsible for the distributed processing.

(b) The edge nodes handle the distributed extraction phase including vector marking and forwarding.

(c) An AI based overlay network, as illustrated in Figs. 4 and 6, is designed to handle the distributed processing. A portion of the processing, relative to a single neuron or more, is assigned to each participating node depending on the network parameters and availability.

(d) The controller will handle the monitoring of the entire system throughout the different stages.

An illustration of a simple neural network architecture consisting of 3 inputs, 1 hidden layer (3 neurons), and a single output layer. The idea is to overlay this scenario over our D^2~C architecture (Figs. 4 and 6) relying on 5 nodes to participate in this test. Therefore, a single ARS AI agent will handle the processing of a hidden node and the output node at the same time. The advantage of this technique relies in its scalability, that is we can test any large Neural Network with multiple nodes and layers over a small network. This can be effective regardless of the number of node willing to participate.

Figure 5 shows the ARS agents distributed over all network nodes in order to take part in the consistency establishment phase. The ARS agents are equipped with a communication channel with the ARS security and consistency system that includes our consistency module. Figure 6 also illustrates how both the security and consistency solution work in parallel without interfering with one another to achieve a more resilient network.

Figure 6 presents an illustration on our distributed NN overlay network that integrates 3 overlay systems each defending an ingress point of the network. Each overlay consists of 3 layers (input, 1 hidden, output), with each layer distributed over a single node. The distribution width depends on the network parameters and availability, which ensures the scalability of such an approach. In the presented example we chose a single node to handle the processing of each hidden layer of each overlay.

3.7 System Optimization

When dealing with AI based security, decision uncertainty is an important condition that needs to be given the proper consideration. Deep learning models generally provide point estimate predictions. These predictions lack any confident measures regarding each decision. Thus, a deep learning system will output a class decision according to the highest probability, even if another class produced a less but comparable probability. In such a case, we should assume that the module is giving a prediction with low confidence. Therefore, a probabilistic view of AI offers confidence boundaries.

In order to calculate the model's confidence, we reengineer our models to return a set of predictions each time we perform inference (repetitions **r** on the same input, while enabling dropout mode). After that we use the distribution of these predictions to calculate the model's confidence intervals. As mentioned, we implemented dropouts: a popular method of regularization of neural networks. Such that, every unit of our neural network is given the dropout rate **p** of being temporarily ignored or "Turned off" in each iteration. Dropout was initially used in training to help prevent overfitting through reducing co-adaptation between units and force them to generalize well to unseen data. From a single neuron perspective: since in each iteration, any input value can be randomly eliminated, the neuron tries to balance the risk and not to favor any of the features.

For our system, we implemented dropouts in testing mode to calculate, offline, a confidence interval for every prediction. This allows us to further calibrate our AI system in case of low confidence in some situations. Also, allows us to detect unknown attack that could hold similar signatures to known attacks, tricking an unoptimized system to falsely classify the attack class or even classify it as a normal behavior.

In order to optimize the processing and traffic overload, we enabled the optimization module to handle the network node selection in two modes: fixed and dynamic. The second job of this module is to regularly probe the network for specific network parameters such as available bandwidth between nodes, processing load on each node, update frequency on each node, the priority of each node on the general functionality of the network, the distribution of the node in the network, the status of each node and others. At this stage, regarding the security system on one hand, the module should select a set of nodes to act as the edge nodes, and should these network features change, the module should activate a new set of nodes, if possible. Another technique is to schedule multiple sets, each working for a specific period of time. These techniques would allow to minimize the traffic and processing load, compared with activating a large set of nodes working at the same time, all the time. The system could always fall back to the full analysis technique when needed or in case of an attack. On the other hand, regarding the consistency system, the module would follow a similar technique were the checks are divided into multiple sets of nodes from all over the network and activate a different set during each time cycle. The system can also fall back to full system check if needed or if triggered by the security system.

4 System Discussion

Considering AI-based techniques being part of our solution; the first step involves finding different sets of efficient ensembles of AI models that would achieve high accuracy rates under the proposed conditions. Afterwards integrating these techniques to become part of our proposed system.

Next would be to analyze and implement the consistency establishment module and providing the ARS agents with ability to securely communicate and cooperate with the consistency module, in addition to handling the necessary processing

4.1 System Datasets

At this point of our work we are considering two datasets. The first being the benchmark NSL-KDD dataset, a modified real dataset proposed to solve a number of the existing problems of the older KDD'99 data set mentioned in [49]. The second being the Intrusion Detection Evaluation Dataset (CICIDS2017) [50], a more recent dataset published by the Canadian Institute for Cybersecurity (CIC), providing a more reliable dataset that covers the variety of recent attacks.

The NSL-KDD dataset consists of around 150,000 records including both normal and anomaly traffic categorized into 4 attack classes:

1. Denial of Service (Dos).
2. User to Root (U2R): unauthorized access to root privileges.
3. Remote to Local (R2L): unauthorized access using a remote site.
4. Probing (Probe): traffic monitoring, surveillance and other probing such as port scanning.
5. Normal.

The different attack classes are presented in Table 2, including a set of specific attack types that fall under those class.

Table 2 Attack classes their types

Attack class	Attack name
DoS	Smurf, Land, Pod, Teardrop, Neptune, Back
U2R	Perl, buffer_overflow, Rootket, Loadmodule
R2L	Ftp_write, Gess_pass, Imap, Multihope, phf, spy
Probe	Ipsweep, nmap, portsweep

The training and testing simulation took part on different sections of the dataset as follows:

1. The training subset: 60%.
2. The testing and evaluation subsets: 40%.

Our current tests are also performed on the 41 features included in the dataset and in the literature for network security.

The CICIDS2017 dataset contains 350,000 records including normal traffic as well as anomaly traffic classified as attack types rather that attack classes, which leads to a more accurate result regarding our second security layer that aims towards identifying specific attack types. The attack types include Brute Force FTP, Brute Force SSH, DoS, Heartbleed, Web Attack, SQL Injection, Port scanning, Infiltration, Botnet and DDoS. In addition, the data set offers 80 network features, and traffic is categorized according to time stamps, which allows us to extract more statistical features that would enhance the reliability of the system.

Table 3 presents a sample of attack types and the extracted set of features for each attack detection process.

For the purpose of our consistency system, we have generated our own real traffic to be able to extract the necessary data from our simulated SDN network represented in the following figure. The network consists of 9 nodes with 3 border nodes, 3 core nodes, and 3 edge nodes.

After the first traffic flow have ended and the network have converged, the data extraction stage starts. The flow table of each node contained around 100 flows. At the same time t, the controller was keeping a database of each update being inserted on each node for future checks.

At time tc, the first extraction took place from each node. This process was repeated for 100 traffic flows with each traffic flow = tc giving a total of 100tc data collection duration. Hence, the total data collected consists of 100 instances of each node with each instance containing around 100 flows.

During each traffic flow, we performed multiple consistency-based attacks on multiple flows on each node in the network. Our data was classified into 6 classes as shown in Table 4.

Table 3 Attack types and their features

Attack type	Features
Benign	B.Packet Len Min
	Subflow F.Bytes
	Total Len F.Packets
	F.Packet Len Mean
DoS GoldenEye	B.Packet Len Std
	Flow IAT Min
	Fwd IAT Min
	Flow IAT Mean
Heartbleed	B.Packet Len Std
	Subflow F.Bytes
	Flow Duration
	Total Len F.Packets
DoS Hulk	B.Packet Len Std
	B.Packet Len Std
	Flow Duration
	Flow IAT Std
DoS Slowhttp	Flow Duration
	Active Min
	Active Mean
	Flow IAT Std
DoS slowloris	Flow Duration
	F.IAT Min
	B.IAT Mean
	F.IAT Mean
SSH-Patator	Init Win F.Bytes
	Subflow F.Bytes
	Total Len F.Packets
	ACK Flag Count
FTP-Patator	Init Win F.Bytes
	F.PSH Flags
	SYN Flag Count
	F.Packets/s
Web attack	Init Win F.Bytes
	Subflow F.Bytes
	Init Win B.Bytes
	Total Len F.Packets

(continued)

Table 3 (continued)

Attack type	Features
Infiltration	Subflow F.Bytes
	Total Len F.Packets
	Flow Duration
	Active Mean
PortScan	Init Win F.Bytes
	B.Packets/s
	PSH Flag Count
DDoS	B.Packet Len Std
	Avg Packet Size
	Flow Duration
	Flow IAT Std

Table 4 Consistency module classes

Class	Type	Description
1	Consistent	Flow is consistent
2	Traffic redirect	Traffic changed its original path (e.g. Action = New destination IP)
3	Deadlock	Traffic is deviated to a blocked destination (e.g. Action = Virtual Port)
4	Loop insertion	Traffic have entered a network loo (e.g. Action = InPort)
5	DoU	Denial of update (e.g. an overlap flow is added by an attacker with higher priority but different action, each time a specific update is inserted)

4.2 System Training

In general, the training algorithm for a BP neural network consists of two part. First the outputs of each layer are calculated successively by the end of the forwarding phase. Second weights for all connections are recalculated in order minimize the backward error generated from the test using the training subset.

For our BP network training we considered the resilient backpropagation approach (RPROP) [51]. The RPROP converges faster than the gradient-descent [52] algorithm. For our scheme, the objective function for learning mode is:

$$E = \frac{1}{2} \sum_p \sum_q \left(y_{p,q} - d_{p,q} \right)^2$$

- p represents the pth output.
- q represents the qth training sample.
- $y_{p,q}$ represents the real output.
- $d_{p,q}$ represents the excepted output.

The gradient-descent approach depends directly on the learning rate η and the size of partial derivative dE/dW to determine the size of the weight update. While the RPROB approach depends only on the derivative to determine the direction of weight update, Δω. This creates an update-value $Δ_{ij}$, directed by the following rule:

$$
Δ_{ij}(t) = \begin{cases} \min\left(n^+ \cdot Δ_{ij}, Δ_{max}\right), & \text{if } \frac{\partial E(t)}{\partial \omega_{ij}(t)} \cdot \frac{\partial E(t-1)}{\partial \omega_{ij}(t-1)} > 0 \\ \max\left(n^- \cdot Δ_{ij}, Δ_{min}\right), & \text{if } \frac{\partial E(t)}{\partial \omega_{ij}(t)} \cdot \frac{\partial E(t-1)}{\partial \omega_{ij}(t-1)} < 0 \\ Δ_{ij}(t-1), & \text{else} \end{cases}
$$

where $0 < η^- < 1 < η^+$.

If the sign of partial derivative $\partial E/\partial \omega_{ij}$ stays the same, an increased by $η^+$ can be done to speed up the convergence. While if $\partial E/\partial \omega_{ij}$ flips its sign, then the previous update of ω_{ij} is too large and it jumped over a minimal error value. Thus, the update-value $Δ_{ij}$ of the weight ω_{ij} in hand should be decreased by $η^-$.

As a next step comes weight-update correction. When a constant sign is preserved, the update of Δω follows the following rule:

$$
Δ\omega_{ij}(t) = \begin{cases} -Δ_{ij}(t), & if \; \frac{\partial E(t)}{\partial \omega_{ij}(t)} > 0 \\ +Δ_{ij}(t), & if \; \frac{\partial E(t)}{\partial \omega_{ij}(t)} < 0 \\ 0, & else \end{cases}
$$

While if $\partial E/\partial \omega_{ij}$ undergoes a sign change, the backtracking [51] method would be activated. Then the update of Δω follows the rule:

$$
Δ\omega_{ij}(t) = Δ\omega_{ij}(t-1)
$$

Also, set $\partial E(t)/\partial \omega ij(t) = 0$, meaning that if the algorithm took a large missing a minimal error value, it will be able to return to the previous location.

Finally, the weights are corrected under the following rule:

$$
\omega_{ij}(t+1) = \omega_{ij}(t) + Δ\omega_{ij}(t)
$$

- $\omega(t+1)$ represents the updated weight of $(t+1)$th iteration.
- $\omega(t)$ represents the previous weight.

4.3 Consistency System Analysis

Our research towards an efficient consistency establishment system focused on overcoming the main security issues in the techniques that were proposed in the literature. These issues mainly concern privacy due do sensitive data being exchanged for processing in a remote location, or the lack of authentication of the output of such checks. In addition to processing overhead and some latency introduced by these solutions.

We have relied on the existing structure of our security system, in particular our ARS agent, and equipped it with the ability to perform an authenticated hash function of the local flow table. In order to ensure both security, authentication, and also preserve privacy, we have chosen the HMAC-SHA2 to perform this task. Such a choice provides a good balance between a secure hash and the required processing overhead. The SHA-2 (SHA-256) function accepts variable size inputs and produces a fixed 256 bits (32 bytes) size output, which can be carried out in our security vector to the controller, in order to minimize the required traffic overhead. Another import feature of HMAC is providing authentication of the delivered hash, therefore protecting against reply and man in the middle attacks. This feature was possible through generating two keypads from the original key for both the authentication and hashing processes. Finally, to secure the node from physical attacks and to prevent an attacker from generating her own fake hashes, we propose to pre-install each key in a tamper proof device on each node. This device will be accessed explicitly by our ARS agent and a set of these keys would be also stored on the controller to complete the consistency checks.

At this point, the hash function is applied to the three feature classes. The features pertaining to each class are concatenated in one string and thus, the hash is applied to a string resulting in a 32-byte hexadecimal output.

After hashing the node and the controller data, the hexadecimal output obtained for each features class is transformed to bits. Consequently, after concatenating the binary outputs of the three feature classes, we obtain a binary vector containing $16 * 16 * 3$ elements for each flow entry. An XOR operation is performed between two corresponding vectors of each flow entry. The first vector is the one obtained based on the controller data and the second vector is the one obtained based on the node's flow table. As a result, we obtain a $16 * 16 * 3$ vector.

At this stage, the ARS agent is able to feed the HMAC-SHA2 function with a binary input of the flow table at each predefined time cycle, *tc'*, and extracting a 32-byte output and forwarding it in our security vector to the controller. At the same time, the controller would calculate the same hash output using each key and the flow table image of each node. After this stage, the consistency module would compare each hash with its corresponding actual hash from the network nodes in order to check for any inconsistencies between them. The security module will be informed of the output of each check to act accordingly.

In addition to flagging inconsistent flows, we have enabled our first consistency verification layer to visually represent the output data for more clarity. In this context, the visualization of the obtained XOR-ed data, as illustrated in the following figure, shows clearly how our technique, with low processing of bits detection, helps in the differentiation between consistent and non-consistent flows. Thus, to visualize these vectors, we transform them to RGB images of size $16 * 16$, with samples of the resulting images shown in the figure. It can be noticed that the inconsistent flows present different RGB colors while the consistent flows present black images. An important parameter extracted from these images, aside from being colored or not, is the color itself. A RED image implies that the altered flow belongs to the first class of features (constant class), while the GREEN color implies that the altered flow belongs to the second class (time class), final the BLUE color represents the third class (statistical class).

Once an inconsistency has been detected, a second layer in the system will be triggered. This layer will take as input the vector of XOR-ed data that was responsible for triggering the inconsistency in order to specify the class of attack that may have caused this issue. Furthermore, knowing the attack class and the color of the image of this specific flow allows us to identify the features that were modified within that corresponding flow. The consistency system is presented in Fig. 7.

Our human brain has a distinguished technique of learning new attacks (viruses and others) through learning their symptoms. These attacks, that could affect our body on their first attempt, would be recognized by our immunity system in later attacks. Thus, it could be treated or mitigated according to similar symptoms in the future. Inspired for this technique and benefiting from what AI has to offer in the area on known and unknown attacks, we have modified our system to detect unknown consistency attacks. A third layer was implemented with the purpose of handling attacks that were flagged by the first layer but failed to be recognized by the second. The third layer would detect such an attempt an unknown new attack and integrate it with the set of altered featured to be taught, offline, to our AI system. The new weights are integrated and updated to the current system at a specific time chosen by the administrator as a network idle time (Fig. 8).

5 System Simulation

The setup consisted of a Linux Intel Core i7 CPU (8 cores) (Ubuntu 16.04) server with Python3 installed. All tests were performed on both balanced and unbalanced data, in order to undergo the train process over a more realistic traffic environment. Hence, we would be able to study the response of each technique under different network conditions. Principle Component Analysis (PCA) was performed for feature reduction in order to reduce the entire set of features into 4 input features for the different techniques.

The next step on our list was the edge node feature extraction. We assigned an agent residing next to each edge node. The agent was programmed to analyze specific headers of each incoming packet and extract certain features. The extraction was limited to only the participating interfaces. The agents are also responsible for relating the extracted features to an individual connection in a statistical manner. The future goal is to reach an efficient statistically based set of features that is resilient against attacker's interventions and manipulations. This technique helps our system to focus on the general functionality and randomness of the network rather than on a specific entry point. Overall, allowing better detection of unknown distributed behavior.

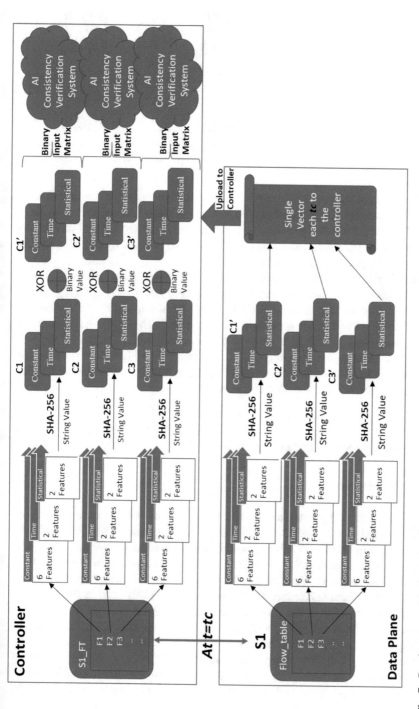

Fig. 7 Consistency module data extraction and preprocessing

Fig. 8 Consistency visual
data representation

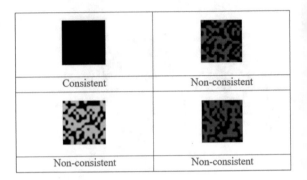

| Consistent | Non-consistent |
| Non-consistent | Non-consistent |

At the end, all agents would form a vector of extracted features and send it securely to the ARS module at the controller. The feature marking techniques allows a single server to handle and track the work of multiple clients. If any further processing is required on the uploaded set of features, it would be handled by the ARS module before preparing the features for the AI model.

Another phase, working in parallel, is the consistency check, where we implemented a second client for each agent with the main purpose of waiting for a command from the controller that includes the time cycle *tc' + x with (x < tc')*; *x* being a period that varies according to the network condition, such that for the controller to ensure that no updates will be injected at this moment, which helps to sync the time between the ARS agent and controller. The controller could send a new set of *tc'* and *x* according to the update frequency of the network. Next, the client will handle the output checks to be forwarded to the controller.

5.1   Simulation Tests and Results

5.1.1   Security System Tests

The following section presents a comparison between the different techniques tested for both stages on both datasets, using balanced and unbalanced data, BD and UBD, respectively. The Results in Tables 5 and 6 show that random forest achieves better results as compared to the other techniques tested at this stage, followed by the neural network-based technique. Figures 9 and 10 represent the random forest test confusion matrix (CM) for both unbalanced (UBD) and balanced (BD) data, respectively. We can notice an increase in the precision of the DNN from the first dataset to the second. This reflects the strength of deep AI techniques when handling different attack types and as the dataset increases in size. Note that although the Random forest technique showed the higher precision, yet the DNN showed faster processing with competitive precision. This sheds light on part of our future work were our system should be equipped with an algorithm aimed for chosen the best ensemble

Table 5 NSL-KDD based test results

Tech.	Anomaly detection (BD) (%)	Anomaly detection (UBD) (%)	Attack identification (BD) (%)	Attack identification (UBD) (%)
Random Forest	99.2	99.3	98.3	99.61
SVM	96.74	96.93	90.64	93.6
KNN	94.5	96.4	90.19	90.5
DT	95.76	96.4	89.5	90.89
MLP	97.5	97.0	94.3	92.73
BPNN	98.7	98.6	77.2	75.8
DNN	96.11	96.13	95.03	93.67

Table 6 CICIDS2017 based test results

Tech.	Anomaly detection (BD) (%)	Anomaly detection (UBD) (%)	Attack identification (BD) (%)	Attack identification (UBD) (%)
Random Forest	98.1	98.0	97.3	97.6
SVM	95.4	95.5	89.1	90.1
KNN	94.0	94.4	88.2	89.5
DT	93.6	94.0	87.2	88.0
MLP	96.5	96.0	93.7	92.1
BPNN	97.9	97.6	77.9	75.9
DNN	97.8	97.35	97.11	96.70

Fig. 9 RF CM for UBD

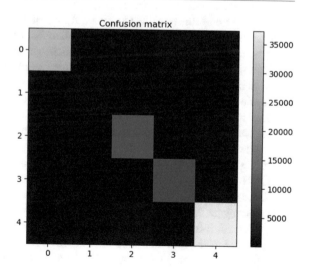

Fig. 10 RF CM for BD

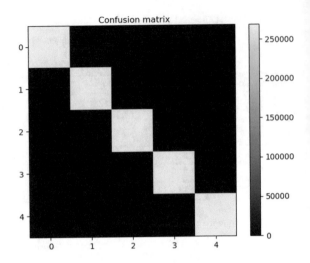

of AI techniques for the first and second security layers for a prestored database according to different network parameters and conditions.

Another test scenario that was done on the full system till this point aimed to simulate both the security and consistency systems working together. The simulation was based on Mininet [53], an SDN network emulator, and mini-edit [54], an extension to Mininet for graphical network topologies, were we constructed a fully connected tree network topology that consists of 20 OVS switches and 14 virtual hosts. We manipulated the OVS code in order to connect two physical servers on two different OVS switches as shown in Fig. 11.

The purpose of this test was to lunch different attacks at the same time from each external physical server on the SDN controller. For this task we assigned the edge nodes being the directly connected OVS switches as shown in the figure.

The tests were based on the $D{\sim}C^2$ architecture, were the edge nodes are responsible for extracting the specified features and uploading them to the controller during each time cycle. Since the network converged and no frequent updates were needed, we activated the consistency client within the same time cycle as the security client such that the consistency client would include the hash output in the same vector sent each time cycle to the controller.

The tests were done on both the RF technique for the anomaly detection layer, and DNN for the attack identification process. Both systems were trained by the CICIDS2017 dataset. After time t_0 we initiated both a DoS attack from server-1 and port scanning on the controller from server-2. It took around 6 s from t_0 for the first layer to detect a change in the network traffic; at this point the second layer was activated due to the first alert, and the abnormal traffic were injected in the second system. It took around 3 s from $t_0 + 6$ for the second layer to identify both attacks. Even though the second layer requires more processing time, but as the results show, it took less time. This is due to the double layered technique since we are injecting

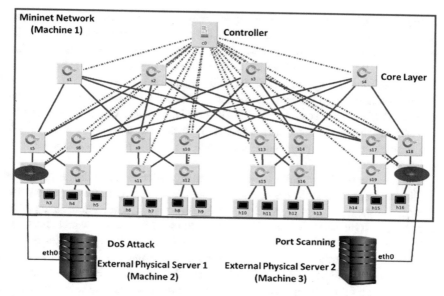

Fig. 11 Security test scenario

the second system with a subset of the traffic that were flagged by the first system. Therefore, resulting in a faster and more efficient detection process.

We added an extra security feature to our system: during an attack, the security module is able to block the attack source from the directly connected edge node. This process was possible by injecting a new rule on the edge nodes forcing them to block all traffic matching the attacker's source-IP. Through monitoring the attacker's traffic on the egress ports of the edge node, we can see the specific traffic being blocked as illustrated in Fig. 12.

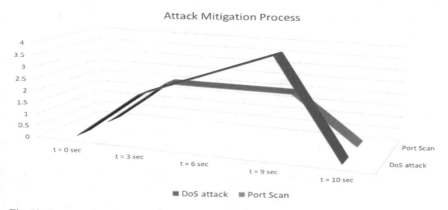

Fig. 12 System mitigation process

Figure 11 shows how the DoS attack intensity (packet numbers) increases exponentially throughout the attack period while the port scanning traffic is constant. The attack started at $t = 0$ s and at $t = 6$ s, the anomaly detection flagged the incoming traffic before the attack identification system detected the attack type at $t = 9$ s and injected the attack mitigation rule to block the attack before $t = 10$ s.

After we have tested the centralized mode of our work, we started our tests on the distributed NN overlay security layer. We selected the Neural Network AI technique for this stage due to its high accuracy results in the anomaly detection tests discussed earlier. Also due to its architecture that made it possible to consider such a new distribution technique.

The security module in this case with run an algorithm to select the best suitable candidates to participate in the distribution process. The algorithm can query the network for the number of physical connections n of each node, and distance from edge node d. If the network is already running, the security module would query the ARS agent for the CPU **load** percentage (e.g. iostat -c). The other parameter should be inserted by the administrator, which are the processing capability $PC = $ CPU Speed (Mhz) + Number of CPUs + Ram (Ghz). Finally, $PC = (PC \times (1 - load))/n$. The nodes with the highest parameters would be chosen as candidates. Furthermore, we would refer to d, for nodes with equal PC, such that node with shorter d will have a higher probability to be chosen. The edge nodes are excluded from this procedure since they are included by default. The number of participating nodes from the chosen candidates depends on the number of NN to be overlaid, which also depends on the number of input points on the network. Another feature was implemented was a database for manual selection by the administrator, which would override this stage.

After the participating nodes are set, the ARS agent is enabled to play its new role. The controller would send a packet to each ARS agent to let it know which part of its NN script to enable (hidden or output), along with other parameters (number of layers, number of neurons in each layer, and the number of NN network to participate with). After this stage, each edge agent with be informed of the ID of the participating internal agents in its own NN. Each edge agent with then send a packet to the corresponding internal agents to be linked as their next hop in the AI processing phase. At this point all the NN networks are set, each protecting a specific entry point of the network.

We tested the distributed system for abnormal traffic using the same test done on the first layer of the centralized security mode. The test was done on the network of Fig. 8.1, with 2 edge nodes. To protect the 2 ingress points we create 2 NN overlays. The 2 edge nodes played the role of the input layers and (s5, s8) along with (s18, s19) played the 2 hidden layers. while s6 and s17 where the output layers. The same NN architecture was tested for the centralized test. After the overlay was set, the weights were injected in the network. The weights are calculated after training the same NN network offline using the 60% of the CICIDS2017 dataset. We then injected the remaining 40% of the dataset as traffic in the network from both external servers. The overall results of the 2 NNs was around 94.8% accuracy. The same latency test was done as the centralized mode, were a DDoS attack was launched from one server into s7 and a port scan attack was launched into s20. The attack started at t = 0 s and at t = 4 s the alarm was triggered detecting an abnormal traffic. The detection latency will vary between the centralized and distributed security layer depending on the network itself and the node capabilities. The higher the capabilities the more a node can handle hidden layer, thus less processing time. On the other hand, the network congestion and the distance between the controller and the edge nodes play a role in the latency of the centralized mode. Another test was to estimate the CPU processing overhead introduced by the distributed NN on each participating node. The node working on a hidden layer showed an **11%** increase in CPU load, each time a processing phase took place; while each output layer nodes showed a **3%** increase.

5.1.2 Consistency System Tests

This section discusses all the necessary simulations and results that provided our consistency system with the proof of concept and effectiveness. The following sections includes our data collection phase, which presents the flow-based attack classes and data structure. Followed by the data preprocessing phase, which also discusses the first consistency verification layer till the point where our system is able to flag any inconsistent flow.

Next, we present the classification results, which include the second layer of the system where the AI module is able to identify the class for each inconsistency flagged by the first layer. Also, a third consistency layer is discussed, which is responsible for detecting unknown consistency attacks in case an attack is triggered and not classified. The first simulations where done on our dataset, followed by real time consistency check test scenarios presented in the following sections.

After the data preprocessing stage, we obtain our labeled data of the XOR-ed output between the nodes and controller. The obtained 16 * 16 * 3 vectors are passed to the classifier.

In our test, we compared Different deep learning architectures: LeNet5, AlexNet, ConvNet, GoogleNet, ResNet, RNN, and DNN. Cross-validation was applied with 4 folds. At each fold, the different architectures were trained with 50 epochs and 50 as batch size. Moreover, the data was split into 60% for training, 20% for validation, and 20% for testing. Moreover, at each fold, the following performance measures were recorded and at the end, the average over the 4 folds was computed.

The comparison results, presented in Table 6, show that the ConvNet architecture gives the best results with 99.39% as accuracy, precision, recall and f1-score. In fact, ConvNet was able to differentiate between different attack classes even though the images resulting for the same attack, may be caused by altering features from different classes and thus, resulting in images of different colors. As such, ConvNet has shown its ability to recognize the patterns contained in the XOR-ed data of each type of attack (Table 7).

Table 7 Deep learning consistency results

	Accuracy (%)	Precision (%)	Recall (%)	F1-score (%)
AlexNet	83.02	85.35	82.93	82.79
ConvNet	99.39	99.39	99.39	99.39
LeNet5	72.91	77.09	72.90	69.72
GoogleNet	69.65	58.48	69.43	62.45
ResNet	95.96	96.27	95.98	95.97
RNN	83.56	89.1	83.53	83.62
DNN	83.78	84.27	83.69	83.66

Another real-time consistency test was done on our system. The test network environment was the same of Fig. 10. The purpose of this test was to lunch different configuration-based attacks from the two external servers on different switches in the network. At this point, our AI system is fully trained offline as discussed earlier. After we attacked the flow table of 4 switches (s5, s12, s15, s17) in the network (12 flows in each switch), we wait for tc = 10 s, as programmed, for the consistency module to start the next check. The altered 12 flows in each switch are based on multiples flow-based attacks belonging to different attack classes. We attacked each switch with two types of classes as shown in Table 8.

The test was extended to include an unknown attack through altering a random set of features. This attack was done on S4. This set was not taught or included in any of the previously mentioned attacks. Such a test would show us the precious and effectiveness of the third consistency check layer.

We have chosen the best three deep learning AI techniques to be tested for this second scenario. The techniques are CNN, DNN, RNN.

The test started with feature extraction, followed by the first layer of consistency verification after the first vector of hashed features was upload to the controller at tc = 10 s. After the first layer, the system was able to detect the inconsistencies found in the targeted 4 switches while all other 16 switches returned a consistent result.

Table 8 Switch/attack classes scenario

Switch	Attack class 1	Attack class 2
S5	Traffic redirect (redirecting traffic for sniffing purposes)	Traffic Deadlock (blocking all Controller packets)
S12	Traffic loop insertion (inserting loops through matching egress to ingress port)	DoU of security updates (denial of update of any new flow in the security flow)
S15	Traffic redirect (redirecting traffic towards controller)	Traffic loop insertion (inserting loops through matching egress to ingress port)
S17	Traffic Deadlock (blocking all incoming packets exiting a specific port)	Traffic redirect (redirecting traffic randomly to disrupt the functionality on the network)

Fig. 13 S1–S5 visual
comparison

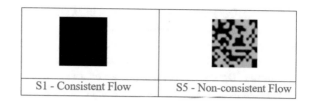

| S1 - Consistent Flow | S5 - Non-consistent Flow |

A sample of the graphical representation of inconsistencies, shown in the following Fig. 13, show a comparison between flows of switch S1 and S5.

As the work of the first layer is finished and an inconsistency is flagged, the second layer starts taking as input the same binary matrices of the 4 switches that flagged the inconsistency in the first layer. The results of the 3 chosen deep learning techniques are shown in the following Table 9.

At the end of the test we checked the database of the third layer that included the new attack with its related set of altered features. Also, we checked the resulting new weights that resulted for the new training process. In order to verify these results, we rerun the same test, but with including the new attack in S17 instead of the deadlock attack. The new attack was given the name "unknown1" with id $= 7$. The results were the same as the previous one shown in the previous table, and the updated system was able to classify the new attack as a known rather than unknown attack.

The final results show that CNN have also given the highest accuracy and precision values and was able to classify the attack class even in real-time scenarios where only few flows were modified (6 flows per attack class).

Other test results are the processing times of the two layers. The attack started at $t = 0$ s and was finished at $t = 7.5$ s, followed by feature extraction, hash and vector composition, which was finished at tc $= 10$ s. At the controller side, the Processing time of the first verification layer for all switches, including the graphical representations, was 3 s. Regarding the second verification layer, the processing time of the CNN system was 2.5 s to verify a single switch in the network (4 switches $= 10$ s). Note that only the flagged switches are transferred to the next layer, and hence, minimizing the processing load and time. Another technique is the ability to perform multiple CNN modules in parallel for each switch to further minimize the processing time.

Another assessment that was done to evaluate our system, was a traffic overhead review. Regarding our test that included two distributed NN overlays working alongside our consistency module, we calculated the following:

Table 9 Consistency results

	Accuracy (%)	Precision (%)	Recall (%)	F1-score
ConvNet	96	96	93	90
RNN	96	83.4	78.3	78.5
DNN	95.5	93	81.6	77.5

- In centralized mode: we are introducing a single extra packet every time **tc**, shared by for security and consistency related data.
- In distributed mode: we are introducing an extra 26 packets every tc, to provide our network with both security and consistency. The packets are distributed as follow: 20 packets (a single packet every tc from each switch to the controller for consistency measures), and 6 packets every tc forwarder from each layer to the next, in our two distributed (3 layered) NN overlays.

5.1.3 Optimization Test Results

We used our test network environment to further optimize both the DDN layer for attack specification, and CNN layer for consistency class specification. The systems were previous trained using the CICIDS2017 and the consistency dataset respectively. In order to optimize the AI system, we needed to optimize the dropout technique to best fit our data first. This is done through test repetitions to choose the best dropout probability **p**. Another parameter was the number of repetitions **r** that should be done on each input to set a good balance between the processing overhead and confidence interval convergence (the difference between the probabilities of the top classes). For our data and network, the tests showed an optimal **p** $= 0.375$. In order to test the efficiency of the optimization phase we created a new security attack with similar features to both a DoS attack and port scanning attack. Also, we created a new consistency attack with similar features to both a deadlock and DoU attacks.

After testing these attacks while enabling the optimization system, we noticed a security confidence interval of $\{53, 0, 1, 46, 0\}$ and a consistency interval of $\{0, 1, 52, 0, 47\}$. We need to mention that the AI system classified the attacks as DoS and Deadlocks attack. This test can prove that AI system would predict and output even with low confidence, such as this case, which will provide an incorrect result. Such false classification would occur regardless of the very high accuracy result that was obtained during the training phase.

Regarding the probabilities in the confidence interval of the security test, they varied between 15% and converged at 7%, after 13 repetitions. While during the consistency test the probabilities varied between 12 and 5%, after 16 repetitions. These percentages control the threshold decision that will be set to check each confidence interval for decision confidence. Such that, if the probabilities of two classes are less than a threshold then a warning will be set. In the cases of the normal class being one of these classes then an alert will be triggered.

This module enhanced our systems abilities to detect unknown attacks though retraining the system with any features that results in low confidence measure. These new features that may belong to a new attack or a new legal behavior in the network, would be given a new ID and require administrative intervention in the future, for a more accurate class name.

6 Conclusion

This paper we presented a full system analysis including the general architectural design. We discussed modifying and managing AI techniques to design a general system solution to protect our network against different types of attacks based on a combination of both centralized and distributed capabilities with the least possible overhead. Our goal is to deploy a state-of-the-art adaptive security system over a consistency verified network for better network resiliency.

We proposed and implemented a virtual ANN network overlay as a first layer of security. It is from the simple and parallel computational capabilities of neurons in an ANN, that it is possible to distribute the processing of a traditional ANN over the network. The ARS agents played role of different layers in the neural network. Every node/agent contributes a free part of its storage and computational capacities to virtualize one or more NN layer. These agents will connect to each other using logical links to exchange data and results; hence, leading to an ANN-based overlay network. Such a design minimized traffic overhead through enabling independent processing, achieving real-time detection. The second security layer consists of an ensemble of AI techniques centralized at the controller with higher processing power for attack specification and mitigation.

We presented a new AI-based consistency verification system integrated with our general ARS architectural design. We discussed how adopting distributing data extraction techniques can provide the necessary information to any AI system while keeping a low traffic overhead and preserving privacy.

The consistency solution provides a double layer consistency verification system inspired by the human brain-immunity cooperation system. The first layer works on comparing a hashed version of specific extracted features from the flow table of each network node with its corresponding image at the controller. The comparison is based on a simple XOR between the two hash vectors. We equipped our system with a graphical representation of the consistency results.

In case of any inconsistency in any node, a second AI-based layer is triggered to identify the attack class that triggered these inconsistencies on all the flagged switches. Our results show that by adopting a double layer technique, we can perform faster checks, and in-depth classification only when necessary, minimizing processing time and overhead.

At the end of our research we have proof of concept of our work after presenting our ARS system through both implementation and test results. We have proven that with both multiple layers of security and consistency constructed in efficient techniques, we can reach a well resilient network. Our work has provided network consistency with real-time protection, while preserving privacy and minimizing traffic overhead.

Acknowledgements This research was funded by TELUS Corp., Canada, AUB University Research Board and Lebanese National Council for Scientific Research, Lebanon.

References

1. Hussein, A., et al.: Machine learning for network resilience: the start of a journey. In: IEEE, 2018 Fifth International Conference on Software Defined Systems (SDS), Barcelona, Spain (2018)
2. Hussein, A. et al.: SDN security plane: an architecture for resilient security services. In: 2016 IEEE International Conference on Cloud Engineering Workshop (IC2EW), Berlin, Germany (2016)
3. Barki, L., et al.: Detection of distributed denial of service attacks in software defined networks. In: 2016 International Conference on Advances in Computing, Communications and Informatics (ICACCI), pp. 2576–2581 (2016)
4. Akhunzada, A., et al.: Securing software defined networks: taxonomy, requirements, and open issues. IEEE Commun. Mag. **53**(4), 36–44 (2015)
5. Bai, H.: A Survey on Artificial Intelligence for Network Routing Problems. University of New Mexico, Albuquerque, NM (2007)
6. Mustafa, U., et al.: Firewall performance optimization using data mining techniques. In: Wireless Communications and Mobile Computing Conference (IWCMC), 2013 9th International, pp. 934–940 (2013)
7. Mukherjee, D., Acharyya, S.: Ant colony optimization technique applied in network routing problem. Int. J. Comput. Appl. **1**(15), 66–73 (2010)
8. Dotcenko, S., Vladyko, A., Letenko, I.: A fuzzy logic-based information security management for software-defined networks. In: 2014 16th International Conference on Advanced Communication Technology (ICACT), pp. 167–171 (2014)
9. Hodo, E., et al.: Shallow and deep networks intrusion detection system: a taxonomy and survey. arXiv Preprint arXiv:1701.02145 (2017)
10. Rhodes, B.C., Mahaffey, J.A., Cannady, J.D.: Multiple self-organizing maps for intrusion detection. In: Proceedings of the 23rd National Information Systems Security Conference, pp. 16–19 (2000)
11. Braga, R., Mota, E., Passito, A.: Lightweight DDoS flooding attack detection using NOX/OpenFlow. In 2010 IEEE 35th Conference on Local Computer Networks (LCN), pp. 408–415 (2010)
12. Yan, Q., et al.: Software-defined networking (SDN) and distributed denial of service (DDoS) attacks in cloud computing environments: a survey, some research issues, and challenges. IEEE Commun. Surv. Tutor. **18**(1), 602–622 (2016)
13. Ghosh, A.K., Schwartzbard, A.: A study in using neural networks for anomaly and misuse detection. In: USENIX Security Symposium, p. 12 (1999)
14. Lippmann, R.P., Cunningham, R.K.: Improving intrusion detection performance using keyword selection and neural networks. Comput. Netw. **34**(4), 597–603 (2000)
15. Li, W.: Using genetic algorithm for network intrusion detection. In: Proceedings of the United States Department of Energy Cyber Security Group, vol. 1, pp. 1–8 (2004)
16. Mukkamala, S., Sung, A.H., Abraham, A.: Modeling intrusion detection systems using linear genetic programming approach. In: International Conference on Industrial, Engineering and Other Applications of Applied Intelligent Systems, pp. 633–642 (2004)
17. Garcia-Teodoro, P., et al.: Anomaly-based network intrusion detection: techniques, systems and challenges. Comput. Secur. **28**(1), 18–28 (2009)
18. Chimphlee, W., et al.: Anomaly-based intrusion detection using fuzzy rough clustering. In: International Conference on Hybrid Information Technology, 2006. ICHIT'06, pp. 329–334 (2006)
19. Ma, Z., Kaban, A.: K-nearest-neighbours with a novel similarity measure for intrusion detection. In: 2013 13th UK Workshop on Computational Intelligence (UKCI), pp. 266–271 (2013)
20. Liao, Y., Vemuri, V.R.: Use of k-nearest neighbor classifier for intrusion detection. Comput. Secur. **21**(5), 439–448 (2002)

21. Amor, N.B., Benferhat, S., Elouedi, Z.: Naive bayesian networks in intrusion detection systems. In: Proceedings Workshop on Probabilistic Graphical Models for Classification, 14th European Conference on Machine Learning (ECML) and the 7th European Conference on Principles and Practice of Knowledge Discovery in Databases (PKDD), in Cavtat–Dubrovnik, Croatia, 23rd September, p. 11 (2003)
22. Panda, M., Patra, M.R.: Network intrusion detection using naive bayes. Int. J. Comput. Sci. Netw. Secur. 7(12), 258–263 (2007)
23. Bouzida, Y., et al.: Efficient intrusion detection using principal component analysis. In: 3éme Conférence Sur La Sécurité Et Architectures Réseaux (SAR), La Londe, France, pp. 381–395 (2004)
24. Chen, R., et al.: Using rough set and support vector machine for network intrusion detection system. In: First Asian Conference on Intelligent Information and Database Systems. ACIIDS 2009, pp. 465–470 (2009)
25. Mulay, S.A., Devale, P., Garje, G.: Intrusion detection system using support vector machine and decision tree. Int. J. Comput. Appl. 3(3), 40–43 (2010)
26. Kim, D.S., Lee, S.M., Park, J.S.: Building lightweight intrusion detection system based on Random Forest. In: International Symposium on Neural Networks, pp. 224–230 (2006)
27. Zhang, J., Zulkernine, M.: Network intrusion detection using random forests. Pst (2005)
28. Javaid, A., et al.: A deep learning approach for network intrusion detection system. In: Proceedings of the 9th EAI International Conference on Bio-Inspired Information and Communications Technologies (Formerly BIONETICS), pp. 21–26 (2016)
29. Aminantoa, M.E., Kimb, K.: Deep learning in intrusion detection system: an overview
30. Niyaz, Q., Sun, W., Javaid, A.Y.: A deep learning based DDOS detection system in software-defined networking (SDN). arXiv Preprint arXiv:1611.07400 (2016)
31. Gupta, S.: An effective model for anomaly IDS to improve the efficiency. In: 2015 International Conference on Green Computing and Internet of Things (ICGCIoT), pp. 190–194 (2015)
32. Jongsuebsuk, P., Wattanapongsakorn, N., Charnsripinyo, C.: Network intrusion detection with fuzzy genetic algorithm for unknown attacks. In: 2013 International Conference on Information Networking (ICOIN), pp. 1–5 (2013)
33. Senthilnayaki, B., Venkatalakshmi, K., Kannan, A.: Intrusion detection using optimal genetic feature selection and SVM based classifier. In: 2015 3rd International Conference on Signal Processing, Communication and Networking (ICSCN), pp. 1–4 (2015)
34. Masduki, B.W., et al.: Study on implementation of machine learning methods combination for improving attacks detection accuracy on intrusion detection system (IDS). In: 2015 International Conference on Quality in Research (QiR), pp. 56–64 (2015)
35. Enache, A., Sgârciu, V.: Anomaly intrusions detection based on support vector machines with an improved bat algorithm. In: 2015 20th International Conference on Control Systems and Computer Science (CSCS), pp. 317–321 (2015)
36. Akbar, S., et al.: Improving network security using machine learning techniques. In: 2012 IEEE International Conference on Computational Intelligence & Computing Research (ICCIC), pp. 1–5 (2012)
37. Aziz, A.S.A, et al.: Multi-layer hybrid machine learning techniques for anomalies detection and classification approach. In: 2013 13th International Conference on Hybrid Intelligent Systems (HIS), pp. 215–220 (2013)
38. Lin, W., Ke, S., Tsai, C.: CANN: an intrusion detection system based on combining cluster centers and nearest neighbors. Knowl. Based Syst. 78, 13–21 (2015)
39. Aburomman, A.A., Reaz, M.B.I.: A novel SVM-kNN-PSO ensemble method for intrusion detection system. Appl. Soft Comput. 38, 360–372 (2016)
40. Horng, S., et al.: A novel intrusion detection system based on hierarchical clustering and support vector machines. Expert Syst. Appl. 38(1), 306–313 (2011)
41. Hodo, E., et al.: Threat analysis of IOT networks using artificial neural network intrusion detection system. In: 2016 International Symposium on Networks, Computers and Communications (ISNCC), pp. 1–6 (2016)

42. Sakic, E., et al.: Towards adaptive state consistency in distributed SDN control plane. In 2017 IEEE International Conference on Communications (ICC), Paris, France (2017)
43. Liu, W., et al.: Inter-flow consistency: a novel SDN update abstraction for supporting inter-flow constraints. In: 2015 IEEE Conference on Communications and Network Security (CNS), Florence, Italy (2015)
44. Hussein, A., et al.: SDN verification plane for consistency establishment. In: The Twenty-First IEEE Symposium on Computers and Communications, Messina, Italy (2016)
45. Aslan, M., Matrawy, A.: Adaptive consistency for distributed SDN controllers. In: IEEE, 2016 17th International Telecommunications Network Strategy and Planning Symposium (Networks), Montreal, QC, Canada (2016)
46. Khurshid, A., et al.: Veriflow: Verifying network-wide invariants in real time. ACM SIGCOMM Comput. Commun. Rev. **42**(4), 467–472 (2012)
47. Ball, T., et al.: Vericon: towards verifying controller programs in software-defined networks. In ACM SIGPLAN Notices, pp. 282–293 (2014)
48. Skowyra, R., et al.: A verification platform for sdn-enabled applications. In: 2014 IEEE International Conference on Cloud Engineering (IC2E), pp. 337–342 (2014)
49. Tavallaee, M., et al.: A detailed analysis of the KDD CUP 99 data set. In: IEEE Symposium on Computational Intelligence for Security and Defense Applications. CISDA 2009, pp. 1–6 (2009)
50. Sharafaldin, I., Lashkari, A.H., Ghorbani, A.A.: Toward generating a new intrusion detection dataset and intrusion traffic characterization. In: Th International Conference on Information Systems Security and Privacy (ICISSP), Purtogal (2018)
51. Riedmiller, M., Braun, H.: A direct adaptive method for faster backpropagation learning: the RPROP algorithm. In: IEEE International Conference on Neural Networks, pp. 586–591 (1993)
52. Xu, D., et al.: Convergence of gradient descent algorithm for a recurrent neuron. In: Advances in Neural Networks–ISNN 2007, pp. 117–122 (2007)
53. Mininet: Available: http://mininet.org/
54. MiniEdit. Available: https://github.com/mininet/mininet/wiki/. Mininet-Apps